A Rebel in Defense
of Tradition

A REBEL IN DEFENSE OF TRADITION

The Life and Politics of Dwight Macdonald

MICHAEL WRESZIN

BasicBooks
A Division of HarperCollins*Publishers*

Frontispiece: Dwight Macdonald, 1968. Photo by Mariette Pathy Allen, New York City. Used by permission.

Copyright © 1994 by Michael Wreszin.
Published by BasicBooks, A Division of HarperCollins Publishers, Inc.

Designed by Ellen Levine

Library of Congress Cataloging-in-Publication Data
Wreszin, Michael.
A rebel in defense of tradition : the life and politics of Dwight
Macdonald / by Michael Wreszin.
    p. cm.
Includes bibliographical references and index.
ISBN 0–465–01739–8 (cloth)
ISBN 0–465–06857–X (paper)
1. Macdonald, Dwight. 2. Intellectuals—United States—Biography.
I. Title.
E748.M147W74   1994
973.91'092—dc20
                                                                    93–33286
                                                                        CIP

95 96 97 98 PS/RRD 9 8 7 6 5 4 3 2 1

For Carol and Frank and in memory of
Bill McLoughlin

Contents

III: CRITICISM AS A SUBSTITUTE FOR POLITICS

IV: A RETURN TO POLITICS

Acknowledgments

THIS PROJECT BEGAN TEN YEARS AGO AND IN THAT DECADE I ran up many debts. Relatives and friends had to listen to my one-track mind, to my calling them "Dwight" and calling on them for all kinds of help and encouragement. Frank Warren is one of the people to whom the book is dedicated. A close friend for over forty years, he has been involved with this project from the beginning. He read several drafts, helped cut what Dwight would have described as a "lumbering dinosaur," and gave advice, suggestions, and warnings throughout. He is not to blame for the final product. But he may be censured for continuing to encourage me, and I am forever indebted to him. Carol did her share of editing, listened to several sections, and invariably offered penetrating insights and suggestions. My daughter and son, Sarah and Danny, have suffered my obsessive preoccupation with amiable tolerance. As for friends, there are many who have given me help and support; I cannot name them all here. Chief among them are John and Alberta Head, who read sections of the manuscript and offered invaluable criticism; Miriam Braverman and Albert Fried, who listened to long sections over lunch, never complaining, always encouraging; B. J. Widick, an old friend of Dwight's who had much to do with getting me started; and Raymond Franklin, Paul Avrich, Leo Hershkowitz, Harry Levine, and many of the members of the Monday Group at Queens College, all of whom showed great interest and infinite patience. Time and time again, Donald Fleming, Alan Davis, and Jack Thomas were ready with a recommendation for a grant or fellowship proposal. I am much indebted to Dwight's former wife, Nancy Macdonald, who was always ready to help. Their two sons, Michael and Nicholas, and daughter-in-law Elspeth Macdonald, in addition to granting extensive interviews, went out of their way to provide me with photographs and other information in their possession. Michael gave me a careful chronology of Dwight's life and long and extensive written memories of his father. The late Gloria Macdonald was keenly interested in the project and spent many hours talking to me. Her two daughters, Mary Day and Sabina, also were ready and willing to grant interviews and provide me with information and insights.

Librarians are indispensable to the researcher's work. I am particularly indebted to Judith Schiff, chief research archivist at the Sterling Library, Yale University, and Ruth Ann Carr, chief of the U.S. history and genealogy section at the New York Public Library. I am also indebted to the staffs of the Library of Congress, and the libraries at Harvard, Brown, Columbia, New York University, and Vassar College. I am also grateful to the nearly hundred people I interviewed, most in person, some by phone; they are listed following these acknowledgments. The National Endowment for the Humanities and the Woodrow Wilson International Center for Scholars awarded me fellowships that relieved me of my teaching obligations for extended periods. My year at the Wilson Center in Washington was one of the most rewarding of my academic career. I also profited from the Biography Seminars at New York University. The City University Research Foundation gave me several travel and research grants over the course of my work, and Queens College honored me with its President's Award, which allowed me to take a semester off to pursue my research and writing. Steve Fraser, my editor, did the very best he could to make the book manageable, and I have tried to meet his exacting criticism. I am indebted to Anne Montague, the copyeditor who worked beyond the call of duty. Many people and institutions have shown an interest in this project, and I am indebted to them for their help and encouragement. Finally, without Carol's infinite patience and the inspiration and example of my great friend, the late Bill McLoughlin, there is no way I could have completed this job.

INTERVIEWS

I am grateful to the following people for in person or telephone interviews, and in several cases, correspondence. Angus Cameron, Sigmund Diamond, Penn Kemble, Carey McWilliams and Arthur Schlesinger, Jr., were interviewed during the initial work on the "non-Communist left," which led to this biography.

Daniel Aaron, Joel Agee, Anastasia Anrep, Boris Anrep, Annabel Anrep, James Atlas, Stefan Bauer-Mengelberg, Saul Bellow, Alfred Bingham, Katherine Stryker Bingham, Bowden Broadwater, Shirley Broughton, William Buckley, Gertrude Buckman, Angus Cameron, Joel Carmichael, Virginia Chamberlain, Lewis Coser, Tom Dardis, Sigmund Diamond, Barbara Dupee, Ernest Eber, Jason Epstein, Leslie Fiedler, Bea Friedland, Theodore Gill, Todd Gitlin, Albert Glotzer, Paul Grabbe, Beatrice Grabbe, Clement Greenberg, Toni Greenberg, Myrna Greenfield, Janet Groth, Philip Hamburger, Michael Harrington, Nika Hazleton, Daphne Hellman, Robert Hivnor, Gertrude Hooker, Irving Howe, Joshua Kahn, Alfred Kazin, Murray Kempton, Irving Kristol, Madi Lanier, Robin Lanier, Sabina Lanier, Raymond Lindquist, Louis Lombardi, John Lukacs, Conrad Lynn, Mary McCarthy, Elspeth Macdonald, Gloria Macdonald, Michael Macdonald, Nancy Macdonald, Nicholas Macdonald, John McDonald, David McReynolds, Jean-Noel Mahoney, Norman Mailer, Anna Matson, Felix Morrow, Malcolm Muggeridge, Daniel Okrent, Jack Phillips, Norman Podhoretz, Elizabeth Pollet, Dachine Rainer, Helen Rattray, Alistair Reid, Selden Rodman, Adolph Rosenblatt, Daniel Rosenblatt, Suzanne Rosenblatt, Bob Ross, Vera Russell, Nora Sayre, William Shawn, Mark Shechner, James Shenton, Michael Silverblatt, John Simon, Eileen Simpson, Alan Spiegel, Peter Steinfels, Thomas Tansele, Yvonne Thomas, Diana Trilling, Niccolo Tucci, Arturo Vivante, Manuel Villarrubia, Adelaide Walker, Margot Walmsley, Erik Wensberg, B. J. Widick, Brenda Wineapple, Mary Day Lanier Wollheim.

Preface

Early in 1984 I first made my way to the Sterling Library at Yale. I was researching the "non-Communist left," having grown impatient with those students who thought that anyone opposed to Stalinist Communism was a right-wing McCarthyite. I had already written a polemical critique of a liberal anti-Communist. I wanted to study and inform my students about radicals who opposed Stalinist Communism but who were also critics of the liberal cold warriors and of much of American foreign policy, as well as capitalist consumer culture.

I went to the Sterling to peruse several collections, but I was chiefly anticipating a look at the Dwight Macdonald Papers. The Dwight Macdonald I knew was a radical intellectual journalist with a splendid prose style and an acerbic, penetrating wit. I had read selections from his *Memoirs of a Revolutionist* and used them in my classes. I looked forward to wading into this motherlode of material over seventy-three linear feet long.

It was an exciting experience: the collection was so rich and varied, touching on nearly every major intellectual, political, and cultural controversy of my lifetime. The correspondence was breathtaking. Macdonald was on a first-name basis with T. S. ("Tom") Eliot and Abbie Hoffman. He corresponded with Henry Regnery, the conservative publisher, and Noam Chomsky, the outspoken anti–Vietnam War critic. Dwight had been chided by Leon Trotsky and replied in kind. He was a close friend of Hannah Arendt, Nicola Chiaromonte, a character out of *Man's Hope*, and Mary McCarthy—the diversity of the correspondence was astonishing. For a person interested in politics and culture in modern America, this collection was pure nirvana, and I was engaged in a labor of love. I was winding down an academic career with no publish-or-perish sword hanging over me. I could engage in the research and writing just for the pleasure of it. I felt familiar with the terrain. I had written workmanlike intellectual monographs on Oswald Garrison Villard, a *Nation* editor, and Albert Jay Nock, another intellectual journalist and also a fine prose stylist. Writing about such good writers is intimidating, but the work is such a joy. It quickly occurred to me that I might do a full-scale biography of Dwight Macdonald. I could say all I wanted about the "non-Communist left" and I would have a more easily managed focus.

I checked around and contacted a group of young Macdonaldites, who were writing, or had written on, various aspects of his career or his circle. It is remarkable that since the sixties a number of scholars have become admirers of Dwight Macdonald and see him as a pivotal figure in twentieth-century American intellectual history. They all urged me to undertake a full-scale biography. Soon after my introduction to the Macdonald Papers, I saw Diana Trilling, who had interviewed Dwight not very long before his death in 1982. When I told her I was considering a biography of him, she also gave me encouragement, assuring me that Dwight was a significant participant in the internecine warfare on the left before, during, and after World War II and that a book on him would be important.

It may have been a labor of love when I began the work, but I had no idea what a monumental task a full-scale biography was, compared to the monographic studies I had done. Nor was I aware how living with a subject night and day would invade my life and often make me impatient, carping, unfair, and judgmental, not only toward my subject but toward anything or anyone that interfered with the work. Nor did I know at the start that writing biography has much to do with studying oneself and that the self-knowledge can be unnerving. But I had further incentive. I identified with Dwight. Even when I strongly disagreed with him, I identified with his rebellious, anti-authoritarian, anti-administrative mentality, and his persona. One of my best moments was when his younger son Nicholas approached me after a talk I had given about Dwight and said that at times I reminded him of his father. I have called Dwight by his first name throughout this book because of that personal identification and also because Michael Harrington told me I should title the book *DWIGHT!*, since everyone addressed him by his first name. I gave it serious thought, imagining the hundreds of thousands who might purchase the book believing it to be about Eisenhower or Gooden. Dwight was always disappointed that his books didn't sell. I'd had the same regrets.

Early on, I thought I knew Dwight, at least his politics and his personality. I had encountered him once or twice at anti–Vietnam War meetings. I had an impression of the man. But I soon realized that he was impossible to pigeonhole, difficult to categorize, wildly unpredictable. The major consistency in the man and his life was paradox and contradiction. On the one hand, he was a great success. He had established an enduring place in the volatile pantheon of the New York intellectuals. He was admired and respected for having edited and published *Politics*, one of the most imaginative and provocative dissenting journals in our intellectual history. It had honestly confronted the vital issues of the times, and raised the right questions long before others tackled them. As a cultural critic he had few peers, and if some of his views seem dated in this postmodernist age, the opponents of the notion of a high culture still feel obliged to take Dwight on. He, as much as anyone else, shaped the dimensions of the debate over mass culture and was acutely aware of its relationship to totalitarianism. And yet, despite these recognized achievements, Dwight Macdonald was a diffident man. As his historian friend John Lukacs has perceptively written, Dwight's "self-distrust was deeply hidden, at profound variance with his strong confident sure-footed opinions." Dwight often thought that his life had been a series of failures and he kept a careful record of his disappointments. These self-destructive feelings often led to debilitating bouts of depression and writer's block.

There are more paradoxes. For many people who remember Dwight in the thirties and forties and then in the sixties, he remains an American radical. He flirted with the Communist Party in the thirties, probably voted in 1936 for the Communist candidate, Earl Browder. He briefly became a vociferous Trotskyist, insisting that World War II was a capitalist, imperialist adventure that would only further brutalize the world; he then became an anarcho-pacifist opponent of the United States' role in the conflict. He was a strong civil rights advocate, gave a platform in *Politics* to gay and feminist writers long before it was fashionable, supported the New Left opposition to the war in Vietnam, and expressed sympathy for and identification with the counterculture. The last section of his essay "The Root Is Man" became a seminal document of "the movement." Dwight not only wrote against the war, he marched, carried picket signs, advocated, plotted, and practiced civil disobedience. He never accepted the idea that one could simply profess one's beliefs. Conduct was the measure.

Despite this political activism, he always had a strong conservative streak, from his earliest thoughts on politics and social and cultural issues. His anarchical anti-authoritarianism was not compatible with disciplined collectivism. He called himself a "conservative anarchist," and many radicals recognize the phrase as redundant. He argued that he was a democrat politically and a conservative culturally. No one had more contempt for the popular culture. He stood for standards as determined by members of an educated elite like himself. He longed for the order and civilized limits of an elite community with respect for the traditions of the past. Paradoxically, the author of some of the most trenchant and acerbic criticism of the popular culture had a strong interest in it, and the bulk of his criticism was devoted to analyzing mass and "midcult" trash.

Besides being a cultural conservative and self-proclaimed elitist, Dwight was a practicing egalitarian. If he championed high culture and looked to the creation of a protected elite of appreciative citizens, his doors were open to all who showed an interest by making the effort to study and enjoy the culture he promoted. Dwight treated people as equals, regardless of background or education. In the course of his work he naturally got to know students, secretaries, elevator operators, tradespeople. He was interested in their ideas and problems and was always ready to discuss politics, philosophy, and current affairs with them; he was never patronizing. He did not suffer fools gladly, but he was willing to entertain the notion that no one was a fool or uninteresting until given a chance to prove it.

All these things mattered to Dwight. He took ideas seriously. He was an enthusiast; he lived for intellectual conversation and thought it was an art worth cultivating. But there was no pomposity in him. He was not a name dropper; he didn't flaunt his learning and education. Nor did he take himself so seriously as to fail to see the humor in the human predicament. He appreciated the comic in life's tribulations. Malcolm Muggeridge liked Dwight for the way he pursued his cultural essays with the commitment of a scholar, "but a comic one, which is the only honorable kind." For Dwight there was always something humorous in nearly any situation, and he had no trouble finding the joke, even if it was at his own expense.

One could go on with the apparent paradoxes, but my book is also concerned with what was consistent about Dwight. He was a rebel in the tradition of Randolph Bourne, the early cultural critic and eloquent opponent of the State, and more recently of Albert Camus and Vaclav Havel. It is the tradition of the rebel who resists and says no to the intolerable absurdities of life, and by doing so makes an affirmative statement. Naysaying was Dwight's consistent style and he was certain that practicing it was what made a person fully human. In his very rebellion and resistance to war (any war), racism, injustice, and inequality was an implicit assertion of values. Dwight Macdonald knew of no transcendent authority to point the way to an ethical life. For him, as for Bourne, Camus, and Havel, values and worth lay within oneself. Although Dwight was a devoted cosmopolitan who had no truck with any form of nationalism, the Polish poet Czeslaw Milosz, who understands the menace of human inertia and accommodation, sees him as "a totally American phenomenon" out of the tradition of "the completely free man, capable of making decisions at all times and about all things strictly according to his personal moral judgment."[1]

Dwight invariably turned against the orthodoxies of his own causes, usually shortly after enlisting in them. The focus of much of his rebellion was the dominant liberal culture, for its hypocrisy, sanguine delusions, and pragmatic accommodation to the most appalling events. The "lib-labs" of the New Deal, the Wallaceites, the popular front, the cold war liberals all became his targets. This explains why he sometimes seemed to be mouthing conservative arguments, but his instinctive contrary stance allowed him to make the most penetrating and necessary critiques of the status quo. Even his criticism of World War II was important and exposed the tawdry compromises and sorry expediency behind many wartime policies. Dwight was one of the few voices to speak out unequivocally at the time against the atomic bombing of Japanese cities. And Dwight was one of the first to try to grasp the meaning and implications of the Holocaust and the menace of the totalitarian state.

Anti-statism was another recurring target of his rebellious politics. Dwight consistently attacked the mass, managerial, bureaucratized society. From the early thirties, with his critique of American capitalism, his study of the rise of dictators in the totalitarianism of Nazi Germany and

Stalinist Russia, to the fifties, with his analysis of the Eichmann controversy and his views of mass culture, Dwight tried to find the way of
achieving human, individual survival through the creation of what Bourne
called the beloved community. It was this strain of thought that led him
away from Marxism to "The Root Is Man," in which he juxtaposed a common humanity against the overwhelmingly anti-human forces of the modern world, with its brutal and irresponsible nationalisms and routinized
horror. This perspective led to his most persuasive insights and significant writing. Dwight asked the important questions in a clear, precise
prose: Just how do people lose their humanness? How are they made into
things? What is the process? How may we tell good from evil? In asking
these questions, in seeking the answers, Dwight Macdonald, as Daniel
Bell noted, "called attention to changes that were taking place in the
moral temper the depths of which we still incompletely realize." That was
written in the 1950s, but Dwight's "remarkable eye for the significant
detail" and the record of observation he compiled remains a seminal
source for those who would try to grasp our history and what has made us
what we are.

Of course a person so passionately engaged made mistakes, shot from
the hip, acted on hasty and poor judgment. I do not neglect these matters.
On occasion Dwight played the holy fool, appearing at best unreasonably
naive and at worst ridiculous. Dwight was the first to recognize his mistakes, his absurd judgments, and he took pleasure in exposing them in
carefully documented footnotes. But despite the bad calls, Dwight Macdonald took risks, was invariably honest, embarrassingly candid, more so
than many of his contemporaries. He did not calculate his career opportunities or cut corners to "make it." He did not sell out, and he needed the
dough.

Dwight Macdonald's story is one of an upper-middle-class white male,
schooled in the elite institutions of the Wasp establishment, who started
out with the prejudices and provincialism of his class—anti-Semitism,
racial stereotyping, sexist patronizing—and through the force of his
inquiring mind and his wife Nancy's humanist spirit, jettisoned a footlocker full of dandyish pretensions and became one of the most penetrating critics of politics, society, and culture in twentieth-century intellectual history. He possessed a pluralism of spirit which enabled him to keep
an independent critical intellect alive in a time of strident political and
cultural orthodoxy.

It is a small hope that after reading this book, some may see how much we need the Dwight Macdonalds. There were others who were more learned and profound, more scholarly and reflective, who had more "original ideas," but few, if any, who spoke more eloquently against the mechanized terror of the modern world, about the separation of means from ends, or who tried harder to make their deeds compatible with their creed.

New York City, April 1993

I

THE EDUCATION OF A REBEL

1

Schoolboy Aesthete

WHEN DWIGHT MACDONALD WAS IN HIS SIXTIES, HE READ through his lengthy correspondence with his closest friend from school and college days, Dinsmore Wheeler, and observed:

> What a friendship D and I had, how generous and loving and concerned I was and what a high regard I had for him. . . . And that pic of [James] Agee and Via [Agee's wife] at table with me in Brookfield [Connecticut], so smiling and friendly and eager and so long gone, long long off and died. Of What? How did we get into this? How did this happen to us all? What fault or weakness trapped us, destroyed us. . . . Age of innocence, friendship, hope, Oh God. . . .

Perhaps these despairing thoughts are no more than the anxious reflections of an aging man looking back on what might have been. But Dwight's youth, a time when, bursting with energy and ambition, he was convinced he controlled his future and could make a difference in the world, now looked like a wonderland from the other side of an abyss.[1]

Dwight Macdonald was born March 24, 1906, on the Upper West Side of Manhattan, at Riverside Drive and 122nd Street. He grew up in the neighborhood. His family did not move to the East Side until some time

in the 1920s, when Dwight was at Yale. Dwight's father, Theodore Dwight
McDonald, came from what Dwight described as "a lower-middle-class,
shabby-genteel Binghamton, ny, family," which may account for Dwight's
lifelong suspicion of wealth. His father's family was, however, able to
send Theodore to Exeter and Yale. After graduating, he attended New
York University Law School at night and ran a "'coachin' school' to get
rich dullards into college." As a lawyer he had a strong interest in the
movie industry, served on the board of several motion picture firms, and
at one point in the 1920s was chief legal counsel to the Triangle Film Co.
He also lectured at Yale in 1919 on the history and development of the
film business.

Dwight always saw his father as lovably ineffectual, too open, too generous to make much of a mark in the business world. The two had much
in common and were remarkably close. Dwight loved his father because
of his "open, imaginative, emotional, impractical nature, strong on sympathy and weak on guidance." He felt that his father had an instinctive
resistance to the "grown-up, conventional, upper-class social forms" that
Dwight saw himself resisting in the 1920s.

Dwight's mother, Alice Hedges, an attractive woman who resembled
the Gibson Girl, was the daughter of a wealthy Brooklyn merchant. She
was much more conventional than her husband, status-conscious, deeply
concerned with social connections and appearances. Her family was
hardly enthusiastic about her choice of a mate. Dwight remembered stories of her many suitors who had much better prospects than those of the
provincial tutor-lawyer from Binghamton. Dwight considered her falling
for his father the single act of rebellion against her "haute-bourgeois
social climbing family." Dwight admired his mother's high-spirited
strength and determination and even her practicality, but he never felt the
same warmth and love for her that he had for his softer and more compassionate father.[2]

On his father's side, the family traced its lineage to Scotland. His
father's grandmother was Bessie Dwight of the New England Dwights;
they were descendants of John Dwight of Fullam, England, who had settled in Dedham, Massachusetts, in the 1630s. It was the same family that
had produced two presidents of Yale College. If Dwight Macdonald's
father did not have the financial success to impress Alice Hedges's parents, she was taken with the prestige of his family. She decided he should
change the spelling of his name from McDonald to Macdonald; the Scottish spelling had more class than the Irish.[3]

Dwight was a precocious child, and his parents were intensely interested in his education and that of his younger brother by two years, Hedges. Dwight went to Mrs. DeLancey's private kindergarten at the age of five and then on to the Collegiate School at 77th and West End Avenue. Two years later, because his parents thought it would be pleasant for him to get out of the city, he was enrolled at the Barnard School at Fieldstone and West 242nd Street, on a hill overlooking Van Cortlandt Park in the Bronx. Dwight commuted to the school by subway. Summers were spent in Sea Girt, on the Jersey shore. In his teen years Dwight attended camps in the Poconos and the Adirondacks. Judging from the sparse records of these early years, he was a contented child and adolescent, who enjoyed his life and his schoolwork.

He remained at the Barnard School for five years, where he was secretary of his class, was an enthusiastic Boy Scout, played Rosenkrantz in *Hamlet*, and was nicknamed "the human sky-scraper," as he was already well on the way to reaching his adult height of six feet two. Rather than continue on to the Upper School, at Barnard, he was sent off to Exeter in 1920 at the age of fourteen.[4]

Dwight thoroughly enjoyed himself at the Phillips Exeter Academy. After a very brief period of homesickness, he got into the spirit of the New Hampshire prep school and found many congenial friends. Dwight was a good student from the start, but he didn't worry about grades; they ranged from Ds to Bs. Although he always did his best and had high personal standards, he really studied for the information, for the excitement of what interested him. At fourteen he was already a serious reader committed to educating himself. He wrote weekly reports to his parents; sometimes they were twenty pages long: "I have decided I should get a good foundation in the Greek and Latin classics before tackling modern literature, so I have given up Swift and am now engaged in laboriously translating Livy and Thucydides." He did master Latin and Greek as well as French while continuing to work his way through the nineteenth- and twentieth-century classics of English prose and poetry.

He always kept lists—the 100 best books on China, for example. Endless schedules— "I am going to read all about the writers of the first half of the 18th century and then I will read their works." A weekly log of time spent—"Made a list of the chief works of Defoe, Addison and Steele, read in W. Merrill's book, read the next chapter of Froude's Caesar and topped off the evening (a Saturday night no less) with an hour and a half session with the immortal Thackeray."[5] In his choice of friends, a love of books was paramount. "Met Dinsmore Wheeler," he reported to his parents. "He's a rather literary bird. . . . I lent him my Christopher Morley essays."

Dwight always recalled his days at Exeter with enthusiasm. Unlike so many who appreciated the rigorous training but never forgot the cold loneliness, the harsh intellectual discipline, the fear of not measuring up, Dwight remembered only a place where his voracious desire to read, read, read and write, write, write was amply satisfied. At one point Dwight's mother expressed concern that Hedges, two years behind Dwight, was floundering at Exeter. Dwight wrote reassuringly that there was no cause for worry. As soon as Hedges learned that courses and grades were a necessary evil, he could then concentrate on "educating himself" and everything would be fine.[6]

At Exeter Dwight found a dedicated faculty who admitted him into their company and friendship, inviting him to dinner, lending him books, discussing literature, encouraging his enthusiasms. Dwight was soon writing for all the school publications. He was especially interested in the *Phillips Exeter Monthly*, a serious literary magazine. From his freshman year, he published stories, poems, sonnets, essays, and literary criticism, much of it derivative and imitative of the classics he was reading. He helped put out several issues of *Masquerade*, an independent magazine of parody and satire. In his senior year he walked off with three literary prizes and was elected class poet.

Although Dwight deliberately concentrated on gaining a solid foundation in the classics, he maintained a lively interest in contemporary writers and reviewed new acquisitions for the library. At one point he developed a heart murmur and was released from compulsory gym classes, for which he had little enthusiasm anyway, except for tennis. He used the time to read the entire works of Edgar Allan Poe, who remained one of his favorite American authors and with whom he always felt an identification. Dwight later published an anthology of Poe's writings and planned a critical study.

Dwight became part of a circle of bookish young men who saw themselves as set apart from the majority of the students. These "literary boys," a tiny enclave of teenage intellectuals, were not hazed or bullied, nor were they made to feel like "so many Ishmaels cast off from the tribe." On the contrary, they were amicably tolerated. Dwight recalled that the "big men on campus" called him "Mac" when they encountered him. "We were licensed lunatics."[7]

Within that group, Dwight, Justin O'Brien, and Dinsmore Wheeler founded their own inner circle, the Hedonists. Mencken, Baudelaire, George Jean Nathan, and Oscar Wilde were their patron saints. Their motto: *"Pour épater les bourgeois."* Their notepaper bore the slogan "Cynicism, Estheticism, Criticism, Pessimism." They sported monocles and batik ties and carried canes on their walks in the surrounding countryside. They were, Dwight later admitted, "frightful snobs."[8]

One is hard put to find any trace of social conscience in Dwight during his school years. He was fascinated by the image and lifestyle of the upper-class English gentleman. He read with enthusiasm John St. Loe Strachey's *Adventure of Living* for its portrait of aristocratic public school types. He was concerned about his dress and bought his fashionable suits and jackets from de Pinna. Dinsmore Wheeler thought Dwight dressed like an English schoolboy, "a college sophomore, and a literary freak." Dwight's letters to his parents frequently described his new friends as boys of background, and he assured his mother that he would "cultivate" the wealthier students of good breeding.[9]

Dwight sometimes expressed a patronizing, even contemptuous, attitude toward women. "My only pronouncement on the feminist question is: WOMEN ARE ON THE WHOLE INFERIOR TO MEN." He complained of young female correspondents who could not write interesting letters and had nothing to say. The short stories he wrote at the time often featured the image of a young, beautiful, intelligent girl. But he saw this as a fantasy that one could seldom expect to encounter. "Middle aged women—with exceptions of course—are loathsome to me," he confided in his journal. "Especially if they talk too much, and most of 'em do." They were constantly preoccupied with running the house and taking care of children; it "atrophied their brains." To his credit, Dwight did have more than an inkling of the culture that created such circumstances. He spoke of loving wild young girls who were like high-flying birds. "But life in the form of a husband and a family comes along and clips her wings . . . the virgin is stamped by the ugliness of life . . . and those once pretty, intriguing, graceful girls" were turned into "forbidding matrons." It was an unusually reflective perception for a young man of nineteen in the 1920s. Dwight understood how the girl became the woman in a man's world—and the cruelty of it all.[10]

These years spent at boys' schools, Barnard, Exeter, and Yale, offered Dwight very little actual experience with young women. This lack of experience produced in him and countless others of his generation sexual anxieties that persisted into adulthood. There were chaperoned theater and dinner parties in Boston with girls who were friends of the family. There were tea dances during New York vacations. But he complained of his failure to find interesting girls.

In his senior year at Exeter, Dwight met a lowerclassman about whom he wrote to his parents, "Vernon is the only person that really attracts me."[11] Vernon and his roommate, Dwight reported, were both charming chiefly because they both seemed to have so much of their mothers in them. "They aren't especially effeminate, but they are responsive and delightfully young."[12]

Dwight became Vernon's literary mentor. They took long walks in the countryside, discussing books and their own literary aspirations. One event, riding the rapids of the Exeter River, Dwight recorded as an exciting sensual experience. He "stripped to the buff," hauling the boat up the river against the roaring current, the water rushing by as high as his knees, "the sun beaming on my bare body!" The ride down was "like being borne on the wind." The trip was "good—very good. Partly to be alone with each other, partly on account of the river and woody banks, partly because of the adventure of it, and the sunny day."[13]

Dwight's intense relationship with Vernon was the subject of "The Wall," a story Dwight wrote during his senior year that was published in the *Phillips Exeter Monthly*. It was a well-crafted account of a warm, trusting, loving friendship between an older boy and younger one, and the loneliness and despair its breakup caused. James Agee, who started Exeter a year after Dwight graduated, wrote Dwight that the story had affected him profoundly, so closely did it mirror his own experience. Agee declared that he had "been the one who suffers" and had drowned himself "in psychological meandering" even to "the verge of suicide." The story was "more to me than Gospel." Agee assumed that Dwight had written from personal experience. Dwight did write in the margin of the manuscript that the story was based on his own experience of a romantic friendship.[14]

Although the crush and its pain passed, during the following year, as a freshman at Yale, Dwight wrote Dinsmore Wheeler, whom he often addressed as "Dearest Dinsie":

I'm in love and not with a girl either. . . . It's largely a matter of looks as it was with [Vernon] . . . He is the most enchanting boy I have ever known . . . except Vernon . . . the added attraction is he has intelligence, culture and breeding. . . . If you think I am a homosexualist, please disburden yourself. My amatory feelings are not physical and I would also hate to have little boys like you at Oxford Street [Cambridge] think about such things. It's just not manly that's all.[15]

His friend's quick response was in keeping with his membership in the Hedonist Club: "Your love for those two boys is undoubtedly homosexual. But why worry about it? It isn't a sin. It's part of your make up: Therefore don't repress or deny it." When Dwight reread this correspondence twenty years later, he questioned his denial of a physical attraction to the young men. In a marginal note he added that the attraction "may have been in more than their looks (maybe also social position. Both much richer than me)!"[16]

When James Agee discussed his own attraction to young men with Clement Greenberg, much the way Dwight had with Dinsmore Wheeler, Greenberg dismissed it as a kind of affectation. In an age obsessed with Freud, "it was the fashion to suspect one's self of homosexuality." Given the circumstances of all-male schools, it was perhaps commonplace to cultivate the notion and to stress a heightened sensitivity. Dwight's exchange with Dinsmore Wheeler and his journal entries indicate an openness to sensual experience and exploration, a new and free approach to relationships. There was also an almost classical reverence for the Greek notion of friendship among kindred spirits. Both Dwight and Dinsmore were forever examining their closeness to each other and treating their friendship as if it were a rare and fragile flower. Dwight saw himself and his friends as superior to the majority of students in matters of culture, sensitivity, breeding. For members of this elite, the rules of conventionality did not apply. At the same time, Dwight had a fear and unease concerning homosexuality and was not averse to referring to gay men in the vernacular of the day as "pansies" and "fairies."[17]

During his school days Dwight frequently expressed the prevailing ethnic and racial slurs of his class and background. He could use the word *nigger* with no hesitation, as when he described to Dinsmore hearing the Fletcher Henderson band in Harlem: "Those niggers played like men possessed." He could be patronizing about blacks' demeanor, as when he told Dinsmore that the butler at the Elizabethan Club at Yale was not like other "niggers . . . doesn't brag." When he encountered a bright and literary Irish student, he took pains to point out to his parents that Lenihan was indeed Irish, "but of the better class."[18]

Dwight also expressed a Waspish disdain for Jews at this time. However, he developed a close and enduring friendship at Yale with Geoffrey Hellman, who was Jewish. Dwight's later membership in the largely Jewish community of New York intellectuals suggests that he jettisoned a good part of his anti-Semitic background and training. It was his mother's social pretensions that had encouraged his youthful prejudices. She admonished him to cultivate friends of good background. Dwight came to resent this snobbery, but as a young student he took pains to reassure his parents of the acceptability of his friends and associates. "I am fascinated by a fellow named Kepple—don't think for a moment he is a Jew, because he is not." Reporting on a party in Cambridge: "I was the only Christian present; however, it wasn't bad fun and I met a rather intelligent fellow."[19]

Dwight's early anti-Semitic attitudes came to the fore when Dinsmore Wheeler became seriously involved with a young woman at Radcliffe, Miriam Isaacs, the daughter of the editor of *Theatre Arts*. Dwight informed his parents that Dinsmore was madly in love with a Jewess. It was not long before she had become "the terrible Jewess." In the summer of 1925 Dinsmore invited Miriam Isaacs to the family farm in Ohio, where Dwight was also vacationing at the time. Apparently Miriam's stay caused a strain and unease which finally erupted when Mrs. Wheeler "upbraided" the young woman and made what Dwight described as an "ungracious and nasty attack on her morals, race etc." Reduced to tears, Miriam quickly left, much to the relief of everyone, for "she was out of place at the farm."

When Miriam wrote Dwight asking him to explain his own hostility, he responded in a six-page letter that was candid to the point of cruelty. While acknowledging the obviously tense situation, Mrs. Wheeler's resentment at her presence, and Miriam's understandable discomfort, Dwight scorned what he described as her fawning efforts to ingratiate herself. He criticized her for allowing Mrs. Wheeler to humiliate her and for not responding more aggressively. "And then there is the fact you are a Jewess, and are rather obviously one, to make me react unfavorably. For I dislike rather violently the Jews as a race, and although you are in my opinion superior to your race, there are certain Jewish characteristics that are present in your make up." Patronizingly arrogant, Dwight explained her superiority by expressing appreciation for her tendency to wear her heart on her sleeve. He also conceded that despite her failings he did not think her "malicious in intent." The fact that he had taken the trouble to write six pages of criticism, he felt, was proof of his respect for her; his letter might seem offensive, but he stated that it was his nature to make his feelings clear. Dwight was so proud of his letter that he wrote out an exact copy in his notebook.[20]

The impact of the incident lingered on. Dwight, like Mrs. Wheeler, felt threatened by Dinsmore's affection for Miriam and appealed to his friend not to let his family's opposition throw him into Miriam's arms. When Dinsmore asked Dwight how Miriam had responded to his letter, Dwight wrote that she had taken his strictures self-consciously and had not bene-fited from them. Dwight did concede that Miriam was "of greater intelli-gence than most of her sex." Dinsmore's interest in Miriam persisted, and Dwight came to admire her for her wit, free spirit, and unconventionality. When his parents encountered Dinsmore and Miriam in the city and wrote Dwight about it, Dwight responded: "I bet you thought her a com-mon little Jewess. I thought so for the first five times I saw her. But she isn't entirely—or even mostly what she looks."[21]

Later in life, Dwight was accused on several occasions of being anti-Semitic during boozy encounters with Jewish colleagues and friends, arguing far into the night about Israel, Palestine, Jewish nationalism, and the meaning and political implications of the Holocaust. Jewish friends and associates when questioned have been circumspect. No, Dwight was not anti-Semitic, but he could be terribly insensitive and express himself in very harsh and direct tones that invariably prompted such charges.[22]

While Dwight and his Hedonist literary circle set themselves apart from the philistines, they were not immune to the deep cultural orthodox-ies of their society and class. Eventually Dwight's unusual intelligence, perceptions, and decency did allow him to break out of that cultural cage far more successfully and completely than such distinguished contempo-raries as Edmund Wilson, F. Scott Fitzgerald, and Ernest Hemingway. If Dwight came to school with a footlocker full of the prejudices and stereo-types of his class and background, he certainly discarded those preju-dices more readily than most of his gentile associates did.

Dwight's parents, particularly his father, were thrilled with his achieve-ments at Exeter and with his decision to go to Yale. Dwight had some reservations, since several of his closest friends had chosen Harvard and Princeton; he was really acquiescing to his father's request to attend his alma mater. Dinsmore Wheeler had gone up to Cambridge with Justin O'Brien, and Dwight later regretted not joining them. He calmed his anxi-ety by spending an idyllic summer at the Wheeler family farm in Ohio.

Dwight was quick to make his mark at Yale. Before the end of the first semester he had poems and a story accepted by the *Yale Literary Magazine* (known as the *Lit*). By his sophomore year he was a member of the Lantern and Elizabethan clubs, both prestigious literary societies. He eventually became an editor of the *Lit*, a columnist for the *Yale Daily News*, and a chief editor of the *Record*, Yale's humor magazine.

Dwight found Yale a breeze compared with the far more rigorous training of Exeter. He easily maintained honor roll grades until a period of rebellion during the spring of his junior year. He thought little of the faculty and not much of the student body, complaining to Dinsmore:

> This is a hell of a college. . . . If I don't meet someone soon with brains, individuality, distinction, something, I shall go mad. . . . All the intellectuals here are either "nice boys" with a gentlemanly inclination toward the better known works of Keats and Arnold, or embryo pedants and bores. To be intellectual, say intelligent, reserved, individual, and sensitive, is to be solitary. About 4/5ths of the undergraduates aspire to be Yale Men. The remaining are either the Peoria High School mental incompetent types, or "cranks." I am of necessity driven back on myself—from which I escape by reading and writing. . . . thus I get a lot more reading done than otherwise.[23]

It was not long before Dwight became a part of a small literary coterie with serious writing ambitions. There was Wilder Hobson, a trombone-playing comic who ultimately went to *Time* magazine and then the Associated Press; G. L. K. (George) Morris, who subsequently became a fairly well known abstract expressionist and the angel of the *Partisan Review*; and Fred Dupee, who became a professor of English at Columbia and remained a lifelong friend.

While these friendships helped make life at Yale more tolerable, Dwight's real refuge remained his literary efforts and, perhaps even more important, the long letters he wrote once or twice a week to Dinsmore Wheeler. These letters were serious attempts at literary expression, analysis of himself and his friends, and general philosophical ruminations on life, its meaning, and his own aspirations. Dwight expressed a buoyant zest for life, often prompted by the self-consciously morbid, world-weary letters of Dinsmore. Dwight complained that Dinsmore's letters were "one long melodious whine (though perhaps the word melodious is misplaced)." They repeatedly assured each other that they were superior to those around them, but for Dinsmore this meant loneliness and pain; for Dwight it meant exploiting his isolation. Dwight claimed that the diversity of his interests, in Browning, the *Lit*, China, friends, the attractive wife of a prof, his Victrola, "and my blooming self, have kept me preoccupied. I refuse to become a lifeless lump of clay, dull, uninterested, stupid, sketchily read and impervious to 1001 of the finer sensations and experiences. To be open to all thoughts and experiences, loves . . . is my one aim."[24]

Despite Dwight's catholicity of mind and spirit and a penchant for eclecticism, he read in a disciplined way, keeping extensive bibliographies and cataloging his thoughts. "I am going to keep a notebook . . . all great writers do . . . I have a prose mind. . . . I want to write serious criticism." Although interested in literary criticism, he disapproved of its current practice. He was ecstatic over Henry James's treatment of Hawthorne: "If you want to read a piece of criticism, what is criticism and not the useless slop that is ladled out by [Stuart] Sherman, [Henry Seidel] Canby, and [Mark] Van Doren & Co. you should read James." Dwight was convinced that "in criticism the primary requisite is not sympathy, cleverness, or Menckeness, but INTELLECT, vigorous, clear, and above all, subtle."[25] James appealed to Dwight because he never tried to convey the impression that "art is holy." Literature should be measured by common standards, Dwight believed.

As for his own opinion of Hawthorne, he complained to Dinsmore: "[Hawthorne] is one of the most irritating writers I know—his possibilities were so great, and he remained to the last, except for the *Scarlet Letter*, a purveyor of quaint, gloomy, beautiful allegories and allegory is undoubtedly an inferior branch of art." But more to the point for Dwight, Hawthorne was "so damn lacking in fire, in vitality . . . so provincial, so complacent, so fatalistic and so goddam upright."[26]

Dinsmore received these literary pronouncements with his usual weary cynicism, but he became irritated by what he thought were often arrogant, abrupt, and ill-informed opinions—a kind of shooting from the hip before Dwight had a firm grasp of the subject. Dwight conceded the justice of Dinsmore's complaint, but Dwight's instant contempt or enthusiasm followed by backtracking remained a problem throughout his life.

In college his readiness to rush to judgment was prompted by his disdain for the blandness and lack of critical judgment on the part of his professors and the mainstream critics of the time. Henry Seidel Canby, who was a visiting professor at Yale and had encouraged Dwight's literary ambitions, nevertheless served as a model of what Dwight felt was wrong with the guild of critics, who, in his view, should be "guardians of the culture." This was a role he came to assume for himself. In his view, Canby took his cue from others and as a result praised only those writers who had already arrived, such dodos as Galsworthy, Christopher Morley, and A. A. Milne, while his pedantic common sense prevented him from recognizing deep or original writers. If, however, an original writer was touted by a Mencken or a Nathan, then Canby dutifully jumped on the bandwagon. "If the future of our letters rests in such hands there is simply no hope," Dwight wrote in his private notebook. But then he observed: "I often feel sickened and depressed by thoughts of what a superficial, ridiculous, valueless little twerp I am. Then I think of English literature and French language and the other fields of knowledge in which I am interested . . . and I am consoled because the unworthiness of the instrument is sunk in the glory of the possible achievement." Dwight compared himself with the early church fathers, Augustine, Ambrose, Francis, who found transcendence of themselves in their religion. "Literature and knowledge, wisdom and understanding, intellect, call it what you will, it is my religion."[27]

There was in his early love for literature and for the craft of writing a strong notion that literary art was valuable in itself and not for its political or social content. Standards of technical and aesthetic judgment existed. "A writer's moral and philosophical views are not an intrinsic part of his work—they have nothing to do with how good it is. They form it morally and philosophically, but not aesthetically."[28]

Since Dwight's denunciations and enthusiasms were invariably extreme, they brought him quickly to a confrontation with the Yale authorities. In his sophomore year he wrote President James Rowland Angell regarding compulsory chapel: "To be forced to listen to such puerile, stupid twaddle is an insult to any intelligent person." Resorting to a Sheridan quip, Dwight asserted that the sermonizer in question, a Yale faculty member, had "said much that was original and much that was sound—unfortunately the original parts were not sound and the sound parts were not original." He was called before the dean and reprimanded, mostly for the sloppy, careless appearance of his letter, with its strikeovers and cross-outs. This was a practice he continued all his life.[29]

From his first days at Yale, Dwight repeatedly expressed contempt and even hatred for the "whole damned business of officialdom and authority." His letters to Dinsmore railed against the administrators for their lack of imagination and intellectuality. As for the faculty, with the exception of the great classics scholar Michael Rostovtzeff and the seventeenth-century expert A. M. Witherspoon, who loved and made Dwight love seventeenth-century poetry, Dwight viewed them as mediocrities. He wrote several well-researched pieces for the *Yale Daily News* comparing Yale's curriculum unfavorably with that of Harvard and most of the other Ivy League colleges. He singled out Yale's English department as a haven for inept, simpleminded entertainers and popularizers. Dwight's bête noire was William Lyon Phelps, a master of ceremonies, after-dinner speaker, and all-round booster of Yale civilization. Dwight kept a careful record of Phelpsian banalities, pasting in his notebook an especially absurd piece of Phelps profundity that the good professor had recently shared with the local press: "If I should die now and go to heaven as I hope to and I should be asked who I should like to room with, either Napoleon, Lincoln, Washington or any other great man, I would pick Jesus at once. He had the most attractive personality of anyone that ever lived." Dwight circled the clipping and wrote: "Oh God, sacrilege, bad taste, silliness, sloppy mush, filthy, beyond comment. That's our Billy. Say Christ, old fella, who'er ya roomin with? Billy can never live this down."[30]

When Dwight learned that Phelps was scheduled to teach English 31, Shakespeare, he wrote a protest for his regular column in the *Daily News*. It was red-penciled by the editor as too insulting. Dwight then printed a handbill which was to be inserted in the *Yale Record*. The printer was so shocked at this insolence that he rushed off to the dean's office. Dwight had written that Phelps and the world knew he was not competent to teach the course. Dwight denied the charge of impertinence, stating that he felt deeply about the matter and the reputation of the college.

Dwight was again summoned to the dean's office. When Dwight pointedly asked whether the Yale administration meant to suppress his right to free speech, the dean replied that he certainly could publish the piece and leave the college immediately or not publish it and remain. He chose the latter, but remembered the incident for years as a moment of bad faith when he had compromised his principles.[31]

Dwight's mother grew increasingly alarmed at his lack of respect for the Yale authorities and the faculty. She was shocked when he wrote a devastating review of a sentimental evocation of England published by the Yale University Press, *The Book of Crowns and Cottages* by Robert Peter Tristram Coffin, a Yale faculty member. Dwight wrote disdainfully that the book would appeal chiefly to "underpaid English teachers" who

yearned to visit the promised land. It "reeks or stinks . . . of Merrie England." Early on, Dwight had developed a nose for trash and a fear that toleration of it threatened the entire culture. When his mother admonished
him for not being more politic and diplomatic, Dwight responded that he
agreed one should not injure another's feelings wantonly:

> But if you mean that one should keep silent in the face of cant and lies
> and evil and silliness and stupidity and injustice in order to save one's
> own hide, I do not. I believe that no one of any intelligence or justice is
> going to think worse of one for speaking out his mind. . . . And as for oth-
> ers I don't much care what they think. . . . If one's fellow creatures are
> asses and swine—I am going to take the hide off such silly charlatans.
> This book made me sick to my stomach.[32]

In his junior year Dwight aroused the ire of the New Haven police, who
removed one of his editions of the *Record* from the stands on the grounds
of obscenity. Dwight's crime was that he had published a parody of a contemporary movie magazine, *Film Fun*, illustrated with prints of Titian,
Correggio, Goya, and classical nude statuary. Dwight assumed the classics would provide protective covering for the double entendres of the
captions. This was too much for the new dean, Clarence Mendell, who
wrote Dwight's mother (his father had died the previous year of a heart
attack at the age of fifty-one), informing her that if Dwight continued with
these antics, he would not be permitted to return. It was also true that
Dwight's grades had fallen from their previous high level. Mrs. Macdonald saw her son's conduct as disastrous and in a hysterically angry letter
she charged that he had destroyed his "splendid chance to make a good
record at Yale." All the sacrifices that "Father and I have made for years
have been ruthlessly disregarded by your ridiculous pose of free-thinking
radicalism and general agin the government" attitude.[33]

The dean charged that Dwight was deliberately building a poor reputation, despite his real quality. Dwight had written a series of articles
attacking the entire success ethic promoted by Yale and constantly
ridiculed its anti-intellectualism. This was indirectly a denunciation of
his mother's values. She warned him that he could not "remain a poser, a
gutter type of man, unconventional and disheveled mentally." Dwight's
rebelliousness may well have been unconsciously aimed at his mother.
Surely he did want to succeed as a campus intellectual, and he even
expected to be chosen by one of the prestigious clubs, which would have

verified his success and delighted his parents. But at the same time, he was compelled to repeatedly express his contempt for the snobbish social system. As Robert Cummings has suggested, Dwight may well have wanted to defy his mother and identify with the "failures" of his father, with whom he had felt such a close bond.[34]

Dwight did identify with his father's impracticality, his softness, in contrast to his mother and his brother, Hedges, who thought Theodore Macdonald lacked judgment. Dwight agreed, but "so what of it. I loved him all the more, as we cherish those we love for their frailties as well as their good qualities. My father was the most human person I ever knew and . . . I didn't care about anything else." During this troubled period, when discussion of his leaving Yale arose, Dwight sorely missed his father. He wrote Dinsmore Wheeler at the time that a gulf had always existed between him and his mother: "For instance if I had a joke to tell I would always tell it to father even though I seemed to be telling it to both. [His] laughter always seemed more genuine than Mother's. I always suspected Mother of laughing just to be polite."[35]

Dwight had always gone to his father when he wanted appreciation for things he had written. "I knew that mother would applaud . . . just because she loved me, not because she saw what I wanted her to. . . . Therefore I was always much closer to my father than to her. In fact I loved my father more than anyone else I ever knew." Dwight recalled how much they had enjoyed each other's company, appreciated each other's humor. He had admired his father's "active . . . inquiring mind . . . I could always get him to listen to a new idea or a new fact . . . he was always interested in what I was saying, just as I was interested in what he thought of things. I loved him for his humanity, he loved me because I was his son." Dwight felt a terrible loneliness after his father's death, and years later he recalled the desolation of arising in the New York apartment after making only two cups of coffee, for his mother and himself.[36]

After the death of Dwight's father, Alice Macdonald's brother, a wealthy businessman living in the posh suburb of Westbury, Long Island, took on the schooling expenses of Dwight and Hedges. Uncle George was appalled by Dwight's disrespectful conduct at Yale and immediately threatened to cut off funds. He insisted that Dwight work in a bank during the summer of 1927 to prove that he could be practical and obedient. Dwight did agree to take a job in the Guarantee Trust, but he managed to get out of it early and spent most of the summer at the Wheeler family farm. He had passed his final exams at the end of his junior year, so he was permitted by the dean's office to return for his senior year in exchange for his promise that he would mend his ways.

When he returned, he was strapped for funds and worked briefly in the checkroom of the library. Then he took a job as a waiter in a student club and received scholarship money from Yale. Dwight hated the waiter job. "The greasy hot dirty kitchen, the negro cooks who were very arrogant and unpleasant, so that to get one's order one had to push and shove and yell them out until one of the cooks heard . . . [and] grudgingly filled the order." But what really humiliated Dwight was having to "wait on my classmates for the first time." Swallowing his distaste, he managed to finance his last year at college.[37]

Despite his mother's warnings to be more accommodating, Dwight continued to express contempt for the Yale authorities, the regulations of the college, and most of the faculty. There was very little political content to his rebellion; however, in 1924, while a freshman, he cast a vote for William Foster, the Communist candidate for U.S. president, in a campus straw poll "just to see what would happen." There were two other votes for Foster cast by "practical jokers." Dwight did attend a big political rally that year organized by the Liberal Club. He assured his parents that he thought it "a bit absurd to take politics so seriously in college," but that he did expect to be amused.[38]

Dwight read the little magazines, the *Dial*, the *American Mercury*, *Hound and Horn*. He saw himself, as he had at Exeter, as part of a small band of unconventional literary intellectuals. He relished that feeling, and during these years he developed a fantasy in which he and a select group of kindred spirits would retire to a rural commune to "live together a life such as few could live in this rotten age." Dwight, Dinsmore, and Dwight's new friend Fred Dupee would repeatedly refer to this pastoral dream. Less a political conception than a vision of a utopian retreat for aesthetes, it may have reflected a fear that the day of reckoning when they would be forced to leave their sheltered niche on the campus for the real world was not far off. But it was an ideal that Dwight would hold on to throughout his life and repeatedly tried to realize.[39]

During the spring vacation of his senior year, Dwight consulted career counselors and wrote to advertising agencies and business friends of his Uncle George, finally landing a position in the executive training program at Macy's department store. His enthusiasm, nearly without qualification, leaps from the pages of his correspondence and shows a shocking departure in attitude. Previously contemptuous of the "business mentality," he suddenly appreciated its power. His interviews in New York corporate offices appear to have had an enormous effect on him. "All my life here at college seems a pale dream." He wrote Dinsmore that his whole aim in life had changed. "Literature was my be all and end all and my greatest ambition was to some day create it. Now I honestly don't care if I ever set pen to paper again." He was "really interested in business and had a real ambition to succeed in the business world."

His initial glimpse of "the business machine had terrified and depressed" him, but it had also fascinated him. The executives were doers, men of achievement. They were so cold, so keen, so confident, so certain of their values that Dwight felt obliged to question his own. He confessed to Dinsmore that the power these businessmen had was the dominant power in American society and he wanted it for himself. "I tell you, Dinsmore, that American art, letters, music, culture is done."

Dwight's argument was based on professionalism, on his own deep sense of craft. In his judgment, thousands of businessmen were "better in their line than the best poet or painter we have today." He had come to see "just how feeble, shallow, infinitely weak American art and letters are besides American business." His teachers, even at Exeter, not to mention Yale, made a very poor showing beside these men, who "were keener, more efficient, more sure of their power than any college prof I ever knew."

In capital letters Dwight conceded that his comments were "ALL TOO EXAGGERATED," that he hadn't given up entirely the idea of a retreat to a Brook Farm kind of community, nor had he totally abandoned his interest in literature, but "I'm just right now very eager to start in business and see what it is all about." Dwight's notebook reads like an update of Elbert Hubbard, replete with pictures of the titans of finance and industry, the captions touting their power, moral probity, and achievements.[40]

Dwight's plan was to gravitate from Macy's into a high-powered advertising agency, where he could combine his newfound business knowledge with his way with words. He pleaded with the Oblomovesque Dinsmore, who viewed this new enthusiasm with a jaundiced eye, to come along with him to New York and Macy's—"it was the most up and coming store in the city . . . a great organization." It would be fascinating for them to study it at close range. "And after we have had enough of it we could get into some agency and stand the advertizing [sic] industry on end."[41]

Dwight was graduated from Yale with a B.A. in History in June 1928

somewhat disappointed by his time there. Despite winning several literary prizes, he felt that his confrontations with the deans, his sharp criticism of Billy Phelps and the faculty, and his unruly editorship of the
Record, particularly the *Film Fun* prank, had kept him from achieving
major class distinctions and membership in Skull and Bones. For years
the only good thing Dwight could say about the college was that the
authorities did not prevent him from using the library. He claimed that at
Yale he educated himself, with the exception of one or two good professors. During the period of graduation festivities, Dwight concentrated on
writing a long critical study of Sherwood Anderson, which appeared in
the *Lit* the month after graduation. Almost as a parting shot, he published
a piece in the June issue of the *Lit* once again characterizing the teaching
of English at Yale as devoted to entertainment and showmanship rather
than serious scholarship. Dwight made a mark at Yale and years later was
named one of the most influential student intellectuals by the college's
official historian, but he clearly felt that his contribution had not been
appreciated.[42]

Dwight spent the better part of that summer at the Wheeler farm. Dinsmore had become seriously ill with a blood disease and was still in the
hospital in Cambridge. Although Dwight missed his friend, he made up
for the loss by enjoying a summer romance with Dinsmore's fourteen-
year-old sister, Dot. Dwight's involvement was so intense, although nearly
platonic, that he put off his start at Macy's until September. His first note
to Dinsmore expressed confidence that he would "like selling a lot . . . if I
ever learn how to make out a sales check."[43]

2

From Luce to Lenin

"SEE 'THE ROMANCE OF RAYON,'" ANNOUNCED A slender, awkward young man with just a trace of a stutter. It was Dwight Macdonald, poet-essayist turned apprentice retailer, guiding customers to daily showings of a film touting the new miracle fabric. By October 1928, after little more than a month on the job, Dwight was sending reports to the bedridden Dinsmore Wheeler laced with contempt for the customers as saps and anger about the "blank" days spent trying to sell things he neither knew nor cared about. Dwight gave it his best, spending his evenings reading the likes of Paul Mazur's *Principles of Organization Applied to Modern Retailing*. He found it heavy going, since the author was "unequal to the task of sustaining his paragraphs beyond three sentences." Dwight wondered why a man who could organize a department store couldn't organize a mere book. He was irritated by the constant boosterism of Macy's, with their pretensions of great service when they were in fact devoted to tricking the public into buying whatever it was they were promoting on any given day.[1]

It took only three months for total disillusionment to set in. Dwight realized he was not cut out for a career as a retailing tycoon. He was not practical and he was incapable of command; the details bored and irritated him and he took no pleasure in performing the work. Spengler's distinction between Men of Truths and Men of Facts explained his situation. Dwight now realized he was in the first category and that it was "insanity to have attempted to function in the world of action." He felt he could work for "an abstraction, a theology, or an art, but never for a business concern, which does not seem to me at all of first-rate importance." Business, like Dickens's Gradgrind, cared only for the facts. Business cared nothing for abstractions, for morals, for principles or "truths"; it was simply a field for "adroit opportunists." He could not master the facts, he explained, because he was "inhibited by a thousand doubts, self-conscious hesitations, theories, principles," which constantly retarded him.

His situation was very dark and he never experienced so many weeks of uninterrupted depression: He rose in the morning "with loathing for the day ahead." The view from Macy's enhanced his memory of Yale; even an academic career seemed tolerable. "At college I would have laughed if someone had told me that in less than a year I would look back to the academic barnyard as a delightful mode of existence. I will probably end up by going back to the universities as the dog returns to his vomit and having despised the race of college professors above all others, end up by becoming a second-rate one myself."[2]

By mid-January 1929, only four months after he had reported for work at Macy's, his prospects had gone "from bad to worse." The chief of personnel expressed doubt that Dwight would ever become a Macy's executive. The advertising director did not find the following copy persuasive: "If an interesting room means as much to you as an interesting book, then Macy's home furnishings department is the place for you." Dwight left at the end of the month, declaring his departure "a great relief on both sides."[3]

Dwight's friend and Yale classmate Wilder Hobson soon landed him a job with Henry Luce's Time Inc. He was assigned to the finance and business department of the editorial staff. Dwight thought ruefully that it was "curious how business seems to stalk my footsteps." He was also to begin planning and research for Luce's new venture, *Fortune* magazine, a "lushly illustrated dollar-a-copy monthly dreamed up by Luce to celebrate the 'saga' of American business." It was only a few weeks before Dwight was complaining about the atmosphere at *Time*. They cared only for flashy stuff, he said. They wanted a personal tone to the copy. It was like the cigarette ads with the testimonials by sea captains to "kid the strap-hanger—the homo boobiaensis"—that when they picked up an Old

Gold they were entering into some kind of personal relationship with the movers and shakers of the world. Dwight was also engaged in working up portraits of corporate leaders for *Fortune*: Robert Patterson Lamont, Secretary of Commerce and former president of American Foundries; Lenore Loree of the Delaware Hudson Railroad; Claude Porter, author of the Interstate Commerce Commission's forthcoming railroad consolidation plan. All pretty dull fellows, he reported.[4]

Although his letters to Dinsmore Wheeler were full of long descriptions of research trips to Washington, interviews with captains of industry, and editorial luncheons, Dwight wrote that he was not "doing a man's work in the world," nor did he feel on "the broad highway to success." His job wasn't particularly unpleasant, it simply wasn't satisfying. It didn't allow him to do "what God intended [him] to do—read, talk, think, write." He was still doing all those things, of course, but only in a "hasty, hurried way."[5]

Dwight sent Sherwood Anderson the long essay about him he had published in the *Yale Literary Magazine*. Anderson was impressed by the flattering appreciation—at one point Dwight declared that Anderson would be genuinely remembered longer than Joyce or Proust. Anderson was in fact so grateful for the piece that he called on Dwight during a trip to New York. This visit prompted Dwight to think again of the Wheeler family farm in Ohio as the seed ground for a "new intellectual and literary movement" from which might develop an American Renaissance. "By God, Dins, we have a group as talented, as full of intellectual and artistic juice, say, as any in these United States." He felt certain that he, Dinsmore, and Fred Dupee could start this movement by themselves. The need was great for a place "where people of wit, intelligence, understanding [could] live together."[6]

In place of this elitist retreat, Dwight, G. L. K. Morris, and Dupee started a little magazine, the *Miscellany*. Dwight wrote essays, film and theater reviews, and a long, ambitious analysis of Robinson Jeffers. At the time, Dwight considered the extended essay on Jeffers his last piece of genuine literary criticism. Again he was not cautious in his judgment, arguing that Jeffers was a far more accomplished poet than Eliot, who in comparison to Jeffers was like "a sophomore writing clever verses for the 'Lit.'" The Jeffers piece was a continuation of Dwight's infatuation with Anderson and with his utopian notion of escaping the false intellectuality of the city to find a purity in the earthier life of rural America.

The *Miscellany* marked the beginning of Dwight's formal interest in popular culture. He was at this time fascinated by the movies, he was an enthusiast of the sports writing of Ring Lardner and Westbrook Pegler, and he was considering a dictionary of American slang. Thinking that his boss, Henry Luce, would be "pleased and interested by this evidence of cultural enterprise," he mistakenly sent him a copy of the *Miscellany*. Luce would have none of it. Working for Time Inc. was a twenty-four-hour-a-day profession demanding all of a man's effort. Why, the very name *Fortune* was thought up by a dedicated staffer in the middle of the night. "You can't waste time on some damn little magazine," Luce decreed.[7]

In addition to putting out the *Miscellany* (a bimonthly that lasted for nearly two years) and working overtime for Luce, Dwight was doing the city: theater, museums, concerts, nightclubs, and parties, parties, and more parties. Remarkably, he showed no concern about the crash of the stock market and the Depression. He had taken little interest in politics except at election time. He did attend a Democratic rally in the city and described the bowler-hatted ward heelers with benign amusement in a long letter to Dinsmore. Dwight cast his first vote for Al Smith and the straight Democratic ticket in 1928 on the grounds that the Democrats were the party of "progress and individuality and Al Smith stood for the human quality in government which the Republicans have deliberately destroyed in the interest of efficiency and economy."[8]

In 1931 Dwight, who was making the then princely sum of $5,000 a year, was far more absorbed in his own suffering at *Fortune*, his compensatory interests in books and films, and his ceaseless effort to find an attractive, companionable woman to share these interests. Dwight still lived with his mother in a nice apartment on East 68th Street. In the winter of 1930 he had shared a bachelor flat on Lexington Avenue with Fred Dupee and Geoffrey Hellman "for the purpose of giving parties and seducing girls." The first object was accomplished but not the second, much to Dwight's chagrin.[9]

"I'm lost without a woman—or some little animal about," Dwight complained to Dinsmore. From the time he was at Exeter through his years at Yale, Dwight's relationships with girls and then young women were almost always unsatisfactory. When he was younger he was awkward and stiff with them. At Exeter his emotional ties had largely been with other young men; he looked back on those friendships as platonically homosexual. At Yale he dated rather infrequently, usually finding the women uninteresting intellectually; he had no success sexually. In his junior year Dwight fell "madly in love" with the daughter of an English professor. She had graduated from Vassar two years before, had had a failed affair with a "big man" on the Yale campus and was still harboring an unrequited love. When she told Dwight that she could never love him and wanted only a friendship, he was deeply hurt. "I don't give a damn for all the friendship in the world from *her*," he wailed to Dinsmore.

After the postgraduation summer at the farm, his relationship with Dinsmore's sister cooled down and their correspondence was infrequent. Back in the city, he was constantly on the lookout for women, always searching for his ideal mate, an intellectual equal and a sexual partner. Things never seemed to work out and the failures took their toll on his confidence. Of course, he had his own inner barriers. He met a "young Jewish girl from Chicago who was poised, witty, had charm and was good looking. But she was very Jewish." Dwight wondered how Dinsmore could have loved Miriam, because "hitherto Jewish beauty hasn't appealed to me." He added that he was "insinooatin nothin."[10]

Dwight did see a series of women, including the von Hoershelman sisters, Lola and Natasha, and a "pretty provincial girl" from Baton Rouge whom he tried unsuccessfully to maneuver into bed. Another, who had a lot of "Jewish blood in her," enticed him so aggressively on the dance floor of a New Canaan country club that he felt she might be able to release the "sensuous, physical side of myself. . . . My mind has grown fat but my body is hungry." In most cases nothing came of the relationships except damage to his ego.

In 1932 he did have an affair of sorts with a young woman he found his intellectual equal. She had a vivacious, imaginative mind; she was up to the semi-sophisticated cultural level of those he went around with; she was voluptuous and "very 'sexy' in appearance and manner." Dwight was taken with her "carefree, arrogant way of living" and by the fact that she had always been rich, which seemed glamorous to him. There was no reason it shouldn't have worked out. But it didn't. She proved to be an unstable personality, given to drunken escapades and bizarre episodes. After one boozy night on the town, Dwight ended up sleeping with her alongside her dead-drunk former boyfriend.[11]

Dwight later analyzed his failures with young women during this period of his life, comparing himself to other men he saw as far more successful. They made "an emotional drama out of life." They were better able to make "life interesting emotionally for their girls." Women, Dwight concluded, wanted to have a "RELATION with a man above all. . . . not just a sexual relation." Dwight felt he had shied away from emotional attachments. He wanted to move quickly from the "impersonal-intellectual plane" into bed, but was often rebuffed for these hasty demands and then felt a loss of self-confidence in his sexual attractiveness.[12]

After a series of these social mishaps, he lamented in a letter to Dinsmore: "What the hell is wrong with me? I can understand why people go mad, commit suicide. One thing wrong: I fall in love with witches. . . . God I hate them. The hatred of rejection. And I hate *Fortune*—the hatred of the underdog. And I hate the world. The hatred of the outcast." The stresses of the job and city life in general, added to his feeling of failure with women, encouraged him to drink a good deal. "It's a great feeling you get from whisky," he confided to Dinsmore. "All your cold inhibitions and awkwardnesses vanish and you . . . become another person. I feel brilliant, warmhearted, free—and I become these things to some extent." An after-work routine was established. The martinis before dinner, the wine during the meal, the liqueurs after—all eased the tension. "Only sleep and liquor can really unwind me," he said. He began to experience fits of depression, complaining to Dinsmore that he was "passing through a mental and spiritual hell. The work on *Fortune* is more unsatisfying every minute. And yet I can find nothing to cling to in its place. . . . The only times I feel alive now are when I'm either drunk or God-awfully miserable." In the spring of 1932 he drank so much that he developed a high fever and a kidney ailment and had to lay off drinking, but that did not ease his tension and dissatisfaction at work.[13]

A six-week vacation to Paris and London that spring of 1932 did seem to rejuvenate him. During this trip he and his friend George L. K. Morris, via an introduction from Sylvia Beach, visited James Joyce at his Paris apartment. It was one of those classic awkward encounters, with Joyce mostly silent, offering one-word responses to forced questions.[14]

On his return from Europe, Dwight plunged back into his work at *Fortune* with a bit more enthusiasm. First there was a study of trade associations, which he found boring. But then came an assignment on the American Federation of Labor, which ultimately turned into an article on the Communist Party. What could Dwight bring to a study of the labor unions? Shortly before receiving this assignment, he had been enthusiastically grateful to "Massa Luce" for allowing him to do a piece on ancient Athens—"entirely without reference to economics or the present." Since leaving Yale, Dwight had expressed practically no interest in politics except for his vote for Al Smith in 1928. Nor had he shown any interest in the collapse of the economy. Perhaps that was because he was making over $5,000 a year. Eunice Clark, a young Vassar graduate, remembers that Dwight was looked upon in astonishment at parties as the young man in their circle who earned $10,000 a year! Dwight did report every raise to Dinsmore and listed his investments in American Can, Texas Gulf, Standard Brands, Safeway Stores—"speculation pure and simple."[15]

Dinsmore Wheeler had relatives in Chicago whom he visited often, and his letters frequently brought up the desperate condition of the poor in "stinking wretched industrial sectors" and ruefully added that their exploited condition made possible the palaces that lined the lake in Chicago and its wealthy suburban enclaves like Lake Forest. Dwight was not prompted to respond, except to describe his whirlwind social life, attending dinner dances on Long Island and parties in the city.[16]

Only upon leaving Macy's had Dwight reflected on the nature of the capitalist economy and its implications for society. In a letter to Dinsmore he repudiated an Exeter friend for being "full of the class struggle and the rights of labor against capital . . . he all but sang the International. This sort of radicalism seems to be a dead issue in America today." Dwight rejected any class-conflict analysis. He did feel that the individual was becoming increasingly powerless to affect his own destiny, and ruminated at some length on the deplorable materialistic culture. He was certain that things were "too complex and over-developed to be resolved into any capital-labor conflict." It would imply a willful human power "directing our destinies."

But that was just the problem. Never before, he argued, had the human will played so small a part in the course of events as in this century. In a tone reminiscent of Randolph Bourne, the pre-war critic of Deweyan instrumentalism, Dwight accepted the "marvelous efficiency" of businessmen but was appalled by their lack of ideas concerning the ends of all their energy. Herbert Hoover and the rest were silent "as to any abstract ideal end of liberty, truth, or love to be attained." Adopting a view frequently attributed to the literary lost generation, Dwight lamented the lack of values: they had been corrupted and debased during the wartime hysteria. He saw only forces fighting battles "as meaningless as those of animals or savage tribes. Liberty, Justice, Morality, Love, Faith, Truth—these have gone hollow." This philosophical outburst reflected a literary anarchism which he would pursue much more seriously in the next decade. He ended his rumination with a Nietzschean observation: "The full circle is rounding itself out and we are once more as 'beyond good and evil' as the barbarian from whom we have developed."[17]

Shortly after the Wall Street crash, Dwight had recommended a piece by John Dos Passos in the *New Republic* to Dinsmore. The article attacked the limited vision of American entrepreneurs as epitomized by the mythical heroes Ford and Edison. Dinsmore dismissed Dos Passos as a "parlor pink" who regarded everything from a "half-baked Greenwich Village Communist point of view." Dwight, who had once compiled a scrapbook of entrepreneurial heroes, now came to the defense of Dos Passos's portrait of the mechanical geniuses who had no imagination. "They are practical men, so damned practical they are insane."

Dwight saw no conspiracy, nor did he see them as representatives of their class; these two, Ford and Edison, had changed the face of America, but without an inkling of what they had done. "If you took Ford out of his factory and Edison out of his laboratory you would have two individuals indistinguishable in spirit and taste, in intellectual scope, from millions of Americans. . . . My God! If these men are the Lincolns and Napoleons of today, the human race has gone to hell." These were not the lamentations of a radical appalled by the unequal distribution of power and wealth so much as they were the concerns of a guardian of the culture, whose chief worry was over the decline of the educated elite's role in society.[18]

The Wall Street crash had done much to vindicate all the critics of the "business mentality." Edmund Wilson recalled that the writers of his generation had always resented the barbarism of the big-business era; they had been exhilarated at the "sudden collapse of that stupid, gigantic fraud."[19] But Dwight had mixed feelings. After spending a few weekends at George Morris's father's estate in Lenox, Massachusetts, he concluded that "the poor and oppressed were more apt to be comparatively rough, unkind, unfeeling, brutal in their relationships with human beings." Dwight found among "the well-born a consideration for others, a courteous repression of the ego—however superficial—you don't find in the bourgeois or peasantry."[20]

Thus armed with a growing skepticism of capitalists (if not for the rich and the well-born), Dwight approached an investigation of the labor unions with curiosity and interest. He had had quite enough of the business leaders.

It was at this point, in the early spring of 1934, that Dwight met Nancy Rodman, a 1932 Vassar graduate, Phi Beta Kappa, an art history major with a minor in math. She was the sister of Selden Rodman, who along with Alfred Bingham edited the independent, quasi-socialist magazine *Common Sense*. They met at the Rodmans' apartment, where a group of *Common Sense* contributors gathered to talk politics and socialize. Unlike her brother, who was egotistical and affected, Nancy was self-effacing, soft-spoken and not entirely at ease among these aggressive intellectual hotshots. Dwight, still somewhat shy and insecure with young women, was taken with Nancy. She was attractive, intelligent—and well-bred! "She's a sweet girl," Dwight reported to Dinsmore, "even though she does let me in for drearily long-winded left wing political meetings." This was apparently a reference to the frequent *Common Sense* discussion meetings where he sought Nancy out and was soon asking her for dates. Actually, Nancy was not "a political person," in the sense of taking much of an interest in mainstream politics or political theory. As a privileged young graduate of the best schools (Brearley and Vassar), who had traveled frequently in Europe and had come out as a New York debutante in 1928, she was disturbed by the egocentric attitudes of much of her class and genuinely appalled by the plight of the poor in the midst of the Depression.[21]

Nancy's deep social conscience, which she attributed to her mother, underlay her outward appearance of passivity and diffidence. At the time, she had a special interest in Asian art, had worked as a volunteer at the Persian Institute, and shared Dwight's literary tastes. But forces of circumstance had turned her attention to the flood of radical literature that attempted to explain the collapse of the economies of the United States and Western Europe. She read Trotsky's autobiography before Dwight did and praised it to him. More attuned and receptive to these radical analyses than Dwight, she soon suggested books on radical politics to him, introducing him to the *Communist Manifesto* and the American radical tradition to be found in Lillian Symes and Travers Clement's *Rebel America: The Story of Social Revolt in the United States*.

There was some resistance on Dwight's part to the overbearing seriousness and turgid prose of the ideologues. He complained that *Rebel America* was written "in a wordy academic lifeless style. I wonder why people think that a stiff style must always go with an impartial point of view."

Despite the fact that Dwight was seeing two other women, and was still very interested in Dot Wheeler (who was in school in Chicago), after three or four dates with Nancy he asked her to marry him. She declined on the grounds that they really did not know each other. But since she'd always felt somewhat in the shadow of her flamboyant brother, she did think that this proposal, coming as it did from one of the intellectuals, would give her status. In any event, she put him off, but they continued to see each other, and she continued to have a strong influence on his growing interest in politics and social issues.[22]

In the summer of 1934, while Nancy was vacationing with her mother in New Hampshire, Dwight was researching the Communist Party for a *Fortune* article. His initial interest in the unions had led him to the Party, which he found much more interesting. The Communists were "a maddening race" for an outsider because of their "great air of mystery and conspiracy." They would never answer a question directly and invariably found a pretext for "preaching a Communist sermon instead. . . . And they are shocked if one is blasphemous enough to question any one of the innumerable dogmas they rigidly hold." The trouble with them, he felt, was that they "are always in church. . . . They'd all make good Jesuits." An entire day was spent interviewing the sectarian opposition, the Trotskyites, Gitlowites, and Lovestoneites, winding up with the Weisbordites "(which I am told consist of Weisbord and his wife.)" This trip to the radical opposition had been undertaken with Selden Rodman, Nancy's brother, whom Dwight described to Nancy as "an extremely self-centered young man with little desire or capacity to take in phenomena in the outside world." Dwight, on the other hand, wanted to show Nancy that he was a good reporter and could get inside his subject.[23]

On a hot Saturday in July, Dwight, the photographer Walker Evans, and Geoffrey Hellman, who was a staff writer for the *New Yorker,* drove up to Westchester County to visit Nitgedaiget "(Yiddish for 'not to worry')," a Communist Party training and recreation center. Dwight reported to Nancy, still vacationing in Dublin, New Hampshire, that visiting the camp almost "made a fascist" out of him. He had discovered in himself an aristocratic revulsion—"as to bathing in a slightly dirty pool with the other comrades and eating off slightly soiled plates with the other comrades and applauding mass-level violin solos with the other comrades"— that surprised him. He added by way of justification that he had a "fundamental dislike of living as one of a herd." The comrades were "99 44/100% pure Yiddish and they had that peculiar Yiddish love of living in each other's laps that you can observe any day at Coney Island."

Dwight must have been slightly defensive about these class and ethnic slurs, for he quickly added that he felt "disgusted by humanity, whether Yiddish or Racquet Clubbish, when it presented itself as a squirming mass." In any event, he and his cohorts could not stand the camp beyond Sunday noon, brunch time, so they beat a hasty retreat to the Westchester Embassy Club, where Geoffrey Hellman was a member, and they "bathed in a clean capitalist pool and drank a couple of Tom Collinses in capitalistic solitude."[24]

Nancy did not seem to take offense at this glib, Waspish mockery, responding that she could well understand how he "might have turned Fascist" by attending the camp. "My stomach has often turned at continual radical meetings and dirty radicals and self-interested radicals." But she was quick to add, "Still I believe in communism as the only way out of the political disruption that is going on nowadays." Nancy had just read Ella Winter's *Red Virtue: Human Relationships in the New Russia*, which she thought very good. Winter's Russia sounded pretty idyllic despite the hardships. "There at least everyone has some definite goal. An attempt is made to let everyone do work for which they are fitted and to adjust those that are not fitted. And stupid conventions are forgotten and everyone has a personal freedom that we never dreamed of."[25]

Despite Dwight's close call with the "squirming Yiddish masses," he was quick to agree with Nancy that "all else aside, Communism is the only way out of the mess our society is in. Whether it's a blind alley is another question, but it's the only alley that has any chance of not turning out to be blind." He was incensed over the way such prominent New Dealers as Hugh Johnson, head of the National Recovery Administration; Senator Wagner of New York; and Secretary of Labor Frances Perkins, whom Dwight thought a "liberal and intelligent dame," had "applauded the brutal and illegal persecution of the 'reds' as a result of the San Francisco [longshoreman's] strike." He promised Nancy he would "flay them" in his *Fortune* article. It was not long before he was confronting Luce, charging that his magazines lacked social conscience, and predicting his own dismissal as a red.[26]

Dwight's article on the Communist Party, accompanied by Walker Evans's photos from the Westchester training camp, appeared in the September 1934 issue of *Fortune*. It maintained the same sardonic tone as his description to Nancy of life at the camp. He adopted the well-worn ploy of the disinterested observer who separates himself from the buffoonery of the State as well as from the comic antics of the revolutionary

zealots. In his view, the American Legion, the Fish Committee (a forerun-
ner of the House Un-American Activities Committee), and all the other
guardians of the American way of life were as foolish and ineffective as
the targets of their accusations—the various Communist sects and parties—were. He did not neglect to point up the violent consequences of
Communist militancy: "broken heads, mass picketing, baseball bats and
bricks." But he also scorned the liberalism of the New Dealers who tried
to play down the effects of Communist organizing and fomenting of
strikes.

He denounced the persecution of the Communists, insisting that such
tactics were futile as long as the "capitalist system continued to permit
gross social injustice." He wrote that the Party did genuinely work and
fight for the interests of "the down and out." The Communists' leadership
of the oppressed and their articulation of grievances forced "the capitalist
system to adjust its more glaring inequalities. Whatever else it does or
will do, the Communist Party in the U.S. provided a vigilant and persistent Opposition."

This was a remarkably sympathetic statement for the pages of a Luce
publication. Dwight's repeated assertion that the capitalist system's gross
injustice to some social groups gave Communism its appeal lent the Party
political legitimacy. One suspects it was the Party's militancy that was the
drawing card for him. It made the establishment, the "whole gamut of
officialdom," nervous and angry. A party that did that couldn't be all bad,
despite its self-righteousness and pretentious revolutionary posture.[27]

Dwight felt that his Communist Party piece might be his valedictory,
but it was accepted without much emasculation. (Generally he found
dealing with his "simple-minded" editor, Ralph McAllister Ingersoll,
dispiriting: his "brain [was] so convoluted that it was impossible to discuss anything with the man.") Even Dwight's demand that Luce make his
publications take a stronger stand on social issues did not prompt his
walking papers, although it is entirely possible, given Dwight's increasing
dissatisfaction, that he was hoping to provoke a dismissal.[28]

Discussion of politics and particularly radical politics was intensifying
among Dwight, Nancy, and their friends Selden Rodman, Dinsmore
Wheeler, and Fred Dupee, who had recently returned from an extended
stay in Mexico and within a year was to join the Communist Party. While
Nancy and Dwight seemed to be fellow-traveling with the Party, still as
skeptical liberals, Dinsmore, who was living on his relatives' farm in
Ohio, was the first to express concern about the Soviet Union under

Stalin. He told Dwight that the "Nazi blood purge showed how hideously blind and barbaric fascism could be," but then he added that "the recent executions and banishments by Stalin and Co." were being shamefully explained away "as the brilliant successful effort of the Great Russian Experiment [a mocking reference to the current liberal cliché] to clean itself of festering elements that would impede progress, the onward movements of Communism." Dwight agreed with Dinsmore about the partisan hypocrisy of American fellow-travelers, acknowledging a vast difference between the "theory and ideal of Communism and the practice of Stalin. I can imagine dying for the former and at the hands of the latter (neat, what?)."[29]

However, Dwight and Dinsmore's early recognition of the direction of Stalinism was overshadowed by the imminent threat of fascism. By 1935, Dwight was calling *Time*'s politics "mildly fascist" or "proto-fascist" and predicting that it would not be long before Roosevelt would be wearing a Sam Browne belt and "selling black shirts to all Democrats." When Dinsmore continued to condemn the partisan apologies for Stalin, Dwight insisted that for all the Communists' failings, their "point of view is more often right than any other when it came to a realistic interpretation of current political and economic trends." Dwight applauded their criticism of Roosevelt and the New Dealers. They had seen right away that Roosevelt "wasn't strong enough to redress the balance between capital and labor," whereas it had taken liberals like Dinsmore and him two years to wake up to reality.[30]

Dwight's arguments, despite the obvious contradictions, were essentially anti-statist. Distrusting federal power as a dangerous weapon that would be used by the strongest side in any conflict of class interests, he felt it was obviously being used most effectively by corporate capitalists rather than weak-kneed liberal reformers. "Roosevelt had gotten into good fighting trim only to be taken over by the capitalists." Dwight was moving rapidly to the left. In his last college notebook he had placed photographs of Leon Trotsky and Andrew Mellon side by side. Between the two portraits was the caption "An Interesting Contrast in Faces" with a comment below: "Trotsky—keen, alert, vital, electric. Mellon—the wornout man of breeding and banker. Which of these men possess the future? Can there be any question?"[31]

Dwight's journey from Luce to Lenin was briefly interrupted by his marriage to Nancy Rodman in November 1934. Through the summer he had courted her with a barrage of letters. He had spent a week with her and her mother in New Hampshire and Nantucket. At the same time, he was still trying to find out whether there was any future in his relationship with Dot Wheeler; late in the summer he made a trip to the farm to see her. Dwight later recorded that he had gone through some sort of "a weighing and judging process between Dot and Nancy with scales evenly balanced up to the last minute." After he had decided to marry Nancy, he wrote Dot to say that his marriage would not make any difference "except, of course, outwardly. The living part of my feeling for you and interest in you goes on living."

By the fall, Nancy had decided that she would marry Dwight, and when he returned to the city she agreed to spend the weekend with him at a cottage he had bought with a *Time* writer, Ralph (Del) Paine, in Brookfield Center, Connecticut. Dwight recalled later that Nancy had responded to him "sexually and emotionally more than any other woman except . . . Dot" and Nancy was more "stable, faithful, dependable." For her part, Nancy was taken with Dwight's energetic intellectuality.

The weekend alone together at the cottage obviously cemented the relationship, and shortly afterward, on October 16, 1934, Nancy's mother announced the engagement in the *New York Times*. They were married a month later in the apartment of Nancy's aunt and uncle by a senior leader of the Ethical Culture Society. Dwight's mother was delighted that he was marrying such a well-bred young woman. Just before the wedding, however, she warned Nancy of the risk she was taking because her son "was a drinker."[32]

Dwight and Nancy set off on their honeymoon in the most conventional way, with a shower of rice in his old Ford jalopy, with a Just Married sign hanging off the skewed rear bumper. Their wedding night was spent in an elegant hotel in Philadelphia. The room had louvered doors, which proved to be inhibiting to the newlyweds. Fortunately they were equipped with the most widely read marriage manual, *Ideal Marriage: Its Physiology and Technique*, by the Dutch physician Th. H. Van de Velde, a wedding gift from Dinsmore. Dwight wrote to thank him and said they had both found it enthralling throughout their wedding trip.[33] Given Dwight's fear of failure, the Dutch doctor's insistence on the virtuosity of the male to bring satisfaction to his passive mate may well have been the wrong medicine.

There is much to suggest that Dwight and Nancy, though they had much in common intellectually and politically, were never passionate lovers. Nancy recalls that very shortly after their wedding, Dwight insisted on going off to a party and leaving her at home sick in bed with a bad cold. When she complained of being abandoned, he responded that they should not be dependent on each other and should go their own ways when they felt like it. It was tantamount to suggesting an open marriage, whereby they might maintain relationships with other partners, and it brought Nancy to tears. She was astonished that he could consider such a possibility so soon after their wedding. Dwight had great respect and admiration for Nancy, but after the early years of the marriage he found it difficult to give her the warmth and affection she expected and needed. She wondered whether she had married an "idea," recalling that logic ruled and that "lovemaking was an intellectual rather than emotional process."[34]

They took a short trip by car through Pennsylvania and Virginia, sightseeing at Monticello, the University of Virginia, and Williamsburg (which they found disappointingly artificial). After spending their last honeymoon night in an old-fashioned inn in Harpers Ferry, West Virginia, the couple returned to the city, where they lived very briefly with Dwight's mother before finding their own place in the new, fashionable, art deco Beaux Arts Apartment Hotel on East 44th Street. Dwight threw himself with renewed vigor into his work at *Fortune*; Nancy returned to work at the *Common Sense* offices, and also made an unenthusiastic effort to learn how to cook.

But Dwight remained unsatisfied with most of his work at *Fortune* and requested a six-month leave of absence. He actually hoped he might work out an arrangement to work for six months and take six months off every year. The idea was to enable him to pursue his own writing interests instead of being tied down to the limiting requirements of the magazine. For his first sabbatical, Dwight planned to tour Italy and then to spend three months in Majorca. It was perhaps to be a second honeymoon, as their first had been brief and hectic because of the pressure of *Fortune* deadlines. If this was the plan, it was going to be a keenly cerebral one. They packed a steamer trunk with books (Max Weber and Lenin, Freud and Bertrand Russell), which they shipped ahead to Trieste, where they would pick it up after travels in the Azores, Lisbon, and Italy and before heading to Spain. They also toted a traveling valise with Nancy's art histories and Dwight's eclectic library, which included Burkhardt, Machiavelli, Auden, Spender, *The Oxford Book of Verse*, Proust, Stendhal, Joyce, James (*The Art of the Novel*), Jane Austen

(*Emma*), and Rainer Maria Rilke (*The Journal of My Other Self*).[35]

The initial plan was to do as little traveling as possible, stopping in quiet, peaceful places for extended days in order to get some real work done on their literary projects. Dwight hoped to return with a briefcase full of manuscripts, articles, and outlines for future writings while Nancy looked forward to pursuing her interest in Oriental art. From the start there was always a tension between being bourgeois tourists on a holiday versus serious, questing scholars. Both Dwight and Nancy felt a compulsion to work and achieve and a sense of guilt if they could not point to some scholarly production. Dwight complained of being in a state of flux. Comparing himself to Heraclitus, who had lamented that on "all sides [his] deficiencies appear," Dwight thought he wasn't measuring up. He constantly questioned his knowledge, sensibility, egotism, taste, and lack of direction. He had come to Europe intending to do little if any sightseeing, but with his trusty Baedeker in hand he couldn't pass up a church or monument without close inspection. It would add to his store of references to illustrate his work. He was right—he did draw on these sources for the rest of his life—but he took little satisfaction in the sightseeing at the time because it conflicted with his puritanical drive to be doing something immediately productive, creative, imaginative.[36]

There is a poignant photograph of Nancy taken in their Villa Soleil, just a fifteen-minute walk out of Palma, Majorca. She looks happy and contented as she draws a bucket of water up to her second-floor window. But Nancy also rejected the possibility of contentment in the simple life of this Spanish village. She too was driven to work diligently, taking constant notes, as Dwight reported proudly to Dinsmore. She too hoped to write, and complained of a "frightful style . . . one cliché after another." She accused herself of not doing enough work and lacking discipline.[37]

It is revealing that these two well-to-do Americans had so much difficulty giving themselves over to the pleasures of travel. Perhaps the contrast between their own situation with that of the rest of the world, which was suffering through a deep depression, made them uncomfortable. Soon Dwight would be accusing some of his friends of smug and lazy indifference and a failure to feel for the powerless victims of the economic collapse. Both Nancy and Dwight had an ascetic, puritanical streak, which charged them a high price in guilt for every moment of pleasure. The last thing they wanted was to look like a wealthy young couple lolling away their time at a Mediterranean resort. Thus, absorbing "culcha," as

Dwight's friend Wilder Hobson put it, and even producing it were serious obligations. If Dwight could not focus on a topic and sit still at a desk, he would insist that even sightseeing had to be done seriously. It was all part of his fashioning an image of himself, defining himself as an "intellectual." He had to do what an intellectual does.

It was Nancy who offered Dwight a subject to focus on. She was taking extensive notes for him, and at her suggestion he decided to concentrate on his fascination with the exercise of power, choosing as his topic modern dictatorships, which he found to be "extraordinarily suited to [his] talents." Dwight was off and running, quickly producing an eight-page outline of the subject. He wanted to deal with the concrete (what life was like in a dictatorship) and the theoretical (how dictators came to and kept power). What were the overriding economic, political, cultural forces that gave rise to them? How did dictators differ from tyrants and despots of the past? Did dictators simply reflect the economic, social, and political conditions of modern industrial society as an evolutionary development? Or were they a drastic departure, a unique phenomenon? Dwight read in the classic texts and the most recent literature. With Nancy's help, he took detailed notes and transcribed passages from Hitler's *Mein Kampf*, William Henry Chamberlin's *Russia's Iron Age* (1934), Trotsky's *My Life* (1930) (initially suggested to him by Nancy), Ortega y Gasset's *Revolt of the Masses* (1932), Nietzsche's *Beyond Good and Evil* (1886), Konrad Heiden's *History of National Socialism* (1934), John Dewey's *Liberalism and Social Action* (1935), Berle and Means's *Modern Corporation and Private Property* (1932), and a book that had a profound effect on him, *Middletown: A Study in American Culture* by Robert S. Lynd and Helen Merrell Lynd. Dwight was also dipping into Marx but did not attempt to read him seriously until the summer of 1936.[38]

Dwight proceeded on a premise that he would hold throughout the World War II years: Liberal democracy in the Western world was finished. It had died with World War I, though atavistic liberals refused to read the writing on the wall. That conflict had made it impossible for a reasonable person to believe in democracy's effectiveness. The decision for war had not reflected the will of the people. The die had been cast by a ruling clique in each nation. The chancellories had declared the war, but the people had fought it. They'd had nothing to say, even remotely, about the decision. They were bamboozled by propaganda, which had become the major tool of governing elites—in democracies as well as in the dictatorships. Dwight concluded that if the democracies failed to protect their citizens from a war which decimated them, what good were they? Why not try dictatorship? He noted that modern dictatorships had begun in the defeated nations, Germany, Italy, and starving Russia, because they were necessary to provide order where economic and political chaos reigned. The victorious nations were economically viable and could maintain the mask of democracy, but in fact they were increasingly led by a dictatorship of the corporate elite. Dwight's fascination with corporate leaders and their similarities to and differences from modern dictators remained a constant theme throughout his notes on dictators.[39]

During this period Dwight saw himself as a mild supporter of Communism, at least in the West, but his notes show little grounding or interest in economic analysis. While he dipped into Marx, his comments were invariably critical, because Marxism, like liberalism, was rooted in what he felt was an outmoded nineteenth-century rationalism and belief in progress. It failed to appreciate the role of individuals and of accident in shaping the modern world:

> The Marxists are bound by their concept of history as determined by economics (a product of 19th century thought by the way); to look on (non-Marxist) dictators as capitalists cats' paws. That ideas should triumph over interests, emotion over economics, idealism, however perverted, over the profit motive—this wasn't allowed for in Marxist speculations.

In Dwight's view, capitalism and socialism were systems based on rational *interests*. The creative and imaginative dictator's strength and power rested on ideas. Racism, for example, tapped the wellspring of resentment and vengeance lying under the surface and blossomed forth in incredible power.

At this point Dwight even challenged the notion of class struggle. It too was dismissed as a pre-war concept, based on nineteenth-century rationalism. It paid too much attention to materialistic factors and ignored the "deeper psychic forces which cause men to act without reference to their interests."

Dwight had an absorbing fascination with the dictator as a creative genius, a strongman who sweeps the masses off their feet like an impatient lover. This became a guiding metaphor, hardly flattering to the masses, or women, for that matter. Dwight saw democratic liberalism as weak, ineffective, and no competition to the virile dictators:

Liberal Democracy displays an anxious, idealistic respect for the feeling and opinion of the masses similar to the reverence of a romantic young man [for] the woman he loves. But the masses, like the woman, prefer a more brutal, overbearing lover, who would tell them what to think and feel. To most people, and to almost all women, such a choice is a burden. They have no equipment with which to make it, and no energy to carry it off. Dictators have discovered the basic fact about the masses. Their love-making, imperious, impetuous, brutal, has easily won the masses away from their tediously high-minded liberal democratic lovers. The nineteenth century was a long honeymoon in which the masses became increasingly bored (and, after the war, skeptical of their powers) with their democratic bridegroom. The dictator is the man of the world who broke up the happy but dull home of the romantic young couple.

It would appear as though Van de Velde's marriage manual had an impact beyond its ostensible purpose, so passive did Dwight construe both women and the masses to be. In his notes on *Mein Kampf* Dwight observed that it was difficult to tell which Hitler despised most, the masses or women. Dwight wondered whether Hitler's own sexual nature might not be involved in this imagery.

Dwight continually expressed an ambivalent attitude toward the dictators. He ridiculed their bombast and posturing. But a sneaking admiration for them often emerges, especially in contrast to his portraits of the well-meaning but illusionary politics of Keynes, Dewey, and even Ortega y Gasset. Mussolini comes off pretty well, for Dwight identifies him as an intellectual. Even Hitler's abilities are not dismissed, particularly his capacity for exploiting ideas, perverted as they may be, to serve his interests.

But if Dwight was ambivalent about the dictators, his notes are marked by a consistent contempt for the masses. In a tone reminiscent of Mencken, he insists that it is simply impossible to underestimate their ability, taste, character, and potential. Emotional and unthinking, they are putty in the hands of powerful leaders.

It was the concept of democracy, he argued, that had elevated the masses to a role beyond their capacity and thus created the material for dictatorial success and totalitarian control and exploitation:

The growth of popular democratic government is at the root. Up to Rousseau the world was always divided into small bands of rulers and huge unconscious bodies of the ruled. Then Rousseau shattered the walls of class divisions, followed the French Revolution, the first popular revolt. The 19th century gave more and more power to the people, until by 1914 power was more widely diffused, more people took part in the government than ever before. Everyone read the newspapers, actively supported politicians, considered himself as much at the controlling center of the nation as any other citizen. No space here for the failure of democracy, only to note it was based primarily on the diffusion of control over a vast number of individuals who (1) weren't intelligent enough or (2) hadn't time enough or (3) hadn't any opportunity to wisely direct the course of government. Democracy breeds demagogues and demagogues easily become dictators. In the old 18th century aristocratic government, the rulers were a body small enough to know each other, to combine against common enemies, to penetrate frauds. But the 1935 PEOPLE is so vast, so stupid as to be helpless, a dinosaur which drags its tons of flesh wherever a small directing brain tells it to go.

The deplorable state of education contributed to the rise of a controlled and banal mediocrity under the guise of democracy. Once again the liberals were the target of Dwight's abuse, for their naive faith that popular education led to progress. In the past, "democracy" meant a ruling class with slaves; under it all was well. Government was by the upper classes; that is, the experts. Once popular education roused people to demand votes, then government was by amateurs and the system began to collapse. Repeating a line he must have read in the Stanley Report on industrial concentration, he practically quoted the head of U.S. Steel in the 1920s by arguing that the "most urgent problem of contemporary democracies is to repair the ravages of popular elections." For Dwight, widespread literacy in a society with such a low level of education was a dangerous thing, because it simply made the mind of the mob more accessible to the corporate manipulators and the dictators.

This simplistic historicism and schoolboy cynicism is surprising, given Dwight's future desire to make a connection with the common people. It reflects the elitism of the Yale aesthete when confronting the mob. Dictators knew how to pander to the "stupidity and baseness of the mob. That is their genius." On this score Dwight took both Nietzsche and Ortega to task. He read them carefully, as he did Hitler, and it is the latter who seemed more astute to Dwight. Dwight appreciated Hitler's bilious contempt for the mass man and his idolization of supermen. Nietzsche's failure lay in his nostalgia for a return to the past and his blindness to the fact that the masses were an essential element of modern industrial soci-

ety that could neither be ignored nor throttled. Dwight admired Nietzsche's prophetic understanding of what was actually shaping world
events, but his barbaric, warlike ethos of power had been taken up by the
very people Nietzsche most despised, the demagogues, and used by them
to sway those masses he so profoundly condemned. "A stroke of irony we
won't see again."

If Dwight faulted Nietzsche for failing to see the permanence of the
masses and the need for the modern man of action to exploit them, he was
equally critical of Ortega y Gasset, whom he considered to be another
spokesman for the obsolete nineteenth-century liberal mind. Dwight was
drawn to Ortega as he was to Nietzsche, for he confirmed Dwight's own
fear of the masses as a threat to culture and civilization. Ortega diagnosed
the modern sickness but "failed to recognize that the remedy lay not in
liberalism but in some authoritative system, be it communism or fascism." Ortega couldn't see that liberalism had produced the evils of the
age and couldn't cure them.

Dwight raked through the writings of John Dewey, William James, and
other contemporary liberals, criticizing their refusal to acknowledge that
it would take power and the methods of authoritarianism to achieve some
kind of equitably ordered society. Liberals were simply trying to ignore
the revolutionary realities recognized by both Communists and fascists.
Dwight attacked the likes of Selden Rodman and Alfred Bingham for
their rational independent socialism and utter naïveté. "They seem to
think that they can change the government into a socialist state by sweet
reason alone without causing anyone any suffering. They're much too nice
little boys to get anywhere with the strong capitalist gang."

Dwight's notebooks, journals, correspondence indicate the degree to
which he had joined with so many other disillusioned intellectuals in
their condemnation of the liberal democracies for lacking the will or
capacity to confront the economic and political crisis of the age. By the
mid-1930s, Dwight looked on the New Deal as the politics of illusion.
This encouraged his ambivalence toward dictators, not to mention his
veiled admiration for men of power who could get the job done. He
echoed the tone of Lincoln Steffens, John Dos Passos, Henry Miller, and
countless others who derided liberal ineffectiveness and embraced both
Stalin and Mussolini as forceful men who did not shrink from the exercise
of ruthless power to achieve their ends. Some did so with little enthusiasm; it was simply an exercise in realism. A letter received by Granville
Hicks during the period is illustrative:

It is a bad world in which we live, and so even the revolutionary movement is anything but what (poetically and philosophically speaking) it
ought to be: God knows I realize this as you do and God knows it makes
my heart sick at times ... it seems nothing but crime and stink and
sweat and obscene noises and the language of beasts. But surely this is
what *history* is. It is not made by gentlemen and scholars, and "made"
only in the bad sense by the Norman Thomases, Devere Allens and John
Deweys.[40]

Many of Dwight's ideas as they applied to the United States came
together in his close reading of Robert and Helen Lynd's *Middletown: A
Study in American Culture* (1929). *Middletown* allowed Dwight to contrast
the corporate-controlled mass society in America with the new totalitarian states in Europe. It was an early premonition of the idea of "friendly
fascism": "Europe has its Hitlers but we have our Rotarians." Dwight
made extensive comparisons between the forced consensus achieved
through ceaseless advertising in America and the propaganda machines
of dictatorial states. It was his purpose to contrast the naked brute force
of the totalitarian leader with the more benign but just as efficient "fascism" of corporate leaders and their bought press.

Middletown, based on a study of Muncie, Indiana, was a graphic
"record of the results of 150 years of democracy." The subject of the
Lynds' study was a bland, thoughtless community mired in an unimaginative conformity. Its ideas and ideals were shaped and controlled by the
Rotary Club. The main purpose of such control was to maintain the status
quo, which served the interests of the dominant business class. Democracy had brought Middletown to a two-class society, the businessmen and
the workers. The former ran the society, drumming up enthusiasm for
their ideals and cracking down on troublemakers and dissenters. The
workers lived in a world growing ever darker, deprived of control over
their lives, "robbed of all possible interest in their work by modern
machine methods."

Dwight made much of the Lynds' analysis of the rise of the Ku Klux
Klan in Middletown. Brought to town initially by some of the town's leading business men as a vigilance committee to police the Democratic
administration and to "clean up the town," it was used to rally the working class and deflect their resentments away from their exploiters and
toward religious and racial scapegoats. If the business elite could use
such methods, Dwight wondered why radicals failed to adopt similar tac-

tics. Dwight was critical of the "liberal" Lynds for failing to see the lesson for radicals in such a demagogic ruse. He detested the bland "rabbit warren" atmosphere of the typical middle class community, but saw that when a dynamic rallying cry, like the Klan's in Middletown, exploded in such an unbearably dull community, all hell broke loose.[41]

Dwight was arguing that the techniques of the reactionary demagogue, the dictator, could be profitably employed by the revolutionist who meant to change society and make it more equitable and just. He was dealing with the crucial issue of means and ends, which was to become a primary philosophical preoccupation, and agreeing with the radicals that the ends justify the means. He was trying to pinpoint just what it was that allowed energetic, imaginative, and manipulative leaders to change society.

Throughout his speculations his approach to the subject was clinical and detached. In his notes on *Mein Kampf* Dwight noted Hitler's imaginative use of propaganda, his wild and amoral use of scapegoats to mobilize masses, his adroit collaboration with German capitalists only to put them under his iron will later. There was something to be learned from a recognition of Hitler's "brilliant exploitation of ideas [such as racism], as perverted as they might be."

Sometimes Dwight's attitude toward Hitler was similar to his feeling for men of artistic genius who were willing to take risks, go out on a limb, seize the day. He cited Machiavelli's dictum that "audacity succeeds better in the world of action than caution":

The Alexanders and Napoleons make war precisely as they like to and the greater a general, or any other creator, is the more he imposes his will on his materials. The banality "One can't change human nature" is used by lethargic leaders to justify inaction and to dispose of political or social innovations, Communism for example. It is precisely the talent of a great leader that he can change human nature. What else is Mussolini doing? And how did Lenin gain power? Men will always have the same qualities of mind and spirit, perhaps, but these can be directed to widely different ends if their nature is understood by would-be directors. And today of all eras, when propaganda has reached a development hitherto unheard of, is the change of human nature being applied everywhere.

A love-hate relationship with the subject emerges in these notes. While Dwight emphasized the power and verve of audacity, will, and risk-taking, he complained of the hype and posturing of the dictators: "These Hitlers and Mussolinis bellowing and posturing all over the place, harrying the elephantine masses along like yipping fox terriers, crushing whole people into forms to satisfy their barbaric love of violence." Despite ridiculing the liberal faith in progress and the power of rational persuasion, Dwight sanguinely looked forward "to the day twenty or thirty years from now when they [the dictators] will be quietly dead." He anticipated the arrival of a "deep wave of relaxation" which would sweep over the land, when people would be "steeped in contentment" for awhile.

But what Dwight feared most in the coming age of mass society was the boredom of inertia, the "life of quiet desperation." The men of caution never responded to a challenge, never felt capable of transforming the world into a just and humane place. Dwight expressed a contempt for the "dull utopias," rational schemes for the ordering of societies which, in addition to ignoring the irrational element in mankind and life itself, also ignored the evil in human nature. Sounding much like Reinhold Niebuhr or Joseph Wood Krutch, both of whom he had read, Dwight wrote of the "dark irrational passions and prides which [were to history] as the urge to create something beautiful is to architecture. The difficulty in demanding that mankind forsake its present evil, lustful, unjust worship of force and walk in the ways of justice and reason is precisely that such ways are so damnably dull."

In an imaginative comment on literary models, Dwight provocatively argued that Milton's Satan was much more interesting than God. In an accompanying examination of *Gulliver's Travels*, he dismissed Swift's Houyhnhnms as "monsters of reason." Their society was a reductio ad absurdum of rational utopias. In a self-revealing letter to Dinsmore Wheeler in 1929, he had offered a scathing criticism of William James:

What revolts me about him is the vulgar show he makes of being open-minded, liberal etc. Oh, my, yes, we are just too sensible and free from prejudice. Bring us all your problems, children, and we will put on our nice clean thinking cap and think them through for you. Bertrand Russell, H. G. Wells and in some of his stuff, Havelock Ellis, these belong to the same priggish school. The reason they are so dangerous is that they make such a pretense of thinking things out when really they are quite without imagination or insight. They remind me of the Jeremy Bentham school of philosophy—weren't they the famous "utilitarians?"—and of that prize hygienic-scientific-level-headed and open-minded prig, Herbert Spencer. Such men are essentially vulgar, materialistic (i.e.) James and his pragmatism, and Wells and his hygienic-sinless-incubative utopias. They are so damn sensible about things that they see no good in anything unless it makes for the Betterment of the Race. No thinker satisfies me who does not meet the tragedy of life with something better than an expression of regret and dismay.[42]

Dwight's "dictator notes" constitute well over 200 handwritten pages, with several potential outlines for a book. The tone throughout is, if not that of an aggressive graduate student, a cocksure confidence in wide-ranging generalizations and speculations, most of which reflect the common currency of liberal disillusionment with the effectiveness of democracy. Despite his simultaneous pilgrimage to the left during these very same years, his notes contain a consistent repudiation of Marxism as a nineteenth-century hangover, rooted in a delusionary optimistic faith in the rational order of events. As one might expect in the formative years of a young intellectual, contradictions appeared constantly. At one point Dwight equated businessmen in America with dictators in Europe, and at another point he denounced businessmen as unimaginative and timid as compared to the dictatorial leaders.

Dwight's dictator notes, specifically in their application to the United States, contained nearly all the themes that were to characterize his political and cultural criticism for the next three decades. The Tocquevillian tone, the traditionalist stance on high culture, the contempt for the business elite, the fear of the enervating boredom of a mass society, and the relationship of such mass organization to totalitarianism—all can be found in these notebooks.

Dwight had hoped he might return to New York with a bundle of manuscripts under his arm. He had produced voluminous notes, but nothing for publication. He was not discouraged, however. The time had been well spent and deeply satisfying. He had passed his days reading and writing in the sun, "free of the tensions of dynamic American and European 1935 civilization." His normally diffuse energies had been concentrated. He wrote Dinsmore that he had "never felt so perceptive, so teeming with ideas, so able to look down on the world from a lofty point of view, the attitude one of intellect rather than morality." He and Nancy had "never felt so mentally alive—which is our peculiar way of being alive—ever before in our lives."

Despite this enthusiasm for being abroad, they went home earlier than planned, not because Dwight had been summoned by his editors at *Fortune*, as he and Nancy explained it, but because he had tired of the relatively isolated reflective life. He wrote Esther Dette, a good friend of theirs, that in addition to needing to consult books not available in Spain, he and Nancy decided to return "chiefly because we want to see the U.S. and our friends." Dwight complained that six months abroad was "longer than it sounds when one is in an alien land, friendless and hence, except what can be contrived tete-a-tete, conversationless. I am a great deal more of a social animal than I realized." They left in the middle of August rather than waiting until September as they had originally intended.[43]

Dwight planned to take up his work on the dictatorship project in the three weeks left to him before his scheduled return to *Fortune*. He did take sporadic notes, made trips to the 42nd Street library, and continued reading. The temptations of the city were too much for him, however, and he was quickly caught up in seeing friends, taking weekend trips to the Hamptons and Martha's Vineyard. There was little time for reading and writing.

Back in New York, he found his leftward pilgrimage, despite his traditionalist philosophical speculations, picking up considerable speed in the heady atmosphere of the city's intellectual life. Lionel Abel has written that in these years New York "became the most interesting part of the Soviet Union. For it became the one part of that country in which the struggle between Stalin and Trotsky could be openly expressed, and was! And how!"

When Dwight returned to Selden Rodman's salon for the well-heeled "radical gentry" at his summer place on the Vineyard, he found that one was expected to "express deep pink political opinions or suffer social ostracism" and one "must not object to nude bathing." Dwight accepted the first requirement with ease. As for the nude bathing, although he was later to become an advocate and practitioner of it on Cape Cod, at this point he had some questions about it as a violation of privacy.[44]

During the fall of 1935 Dwight and Nancy found an apartment on the East Side in the shadow of the Queensboro Bridge. From their window they could see the lower half of the pedestrians walking to and from Queens. Their apartment became known as the site of frequent, almost weekly cocktail parties. Mary McCarthy recalled that at these gatherings Dwight would invariably get into a row with Geoffrey Hellman, whom he viewed as an extreme Tory and member of the "NYorker/genteel/social" set. So angry would the two become that Dwight would throw Hellman out. There would be a great ringing for the elevator and Hellman would be escorted to the street. They just as inevitably made up during the week and the whole thing would be repeated at the next party.[45]

When Dwight returned to *Fortune*, he began the research for a serious analysis of the steel industry, beginning with Republic Steel and then taking on "The Corporation," U.S. Steel. If Dwight had consciously or unconsciously hoped before his trip to Europe that his article on the Communist Party might provoke his dismissal, there is little doubt that this series of articles was designed to do the trick.

Dwight left for the steel mills with his attractive researcher, Natasha von Hoershelman, a young woman he had been keen on just prior to his marriage. In October they visited Pittsburgh, Youngstown, Gary, Chicago, and then down to Birmingham and into West Virginia. It was to be a trip that Dwight never forgot, and he credited it as having much to do with his subsequent radicalization. What he saw in the mills and among the workers as well as management confirmed his arguments concerning the frailty of liberalism before the power of the modern corporation. It made him much more receptive to the notion of class struggle, if not entirely in a Marxist sense. He was convinced that the corporate leaders would not relinquish their inordinate power without a struggle. His contempt for the business mentality was fueled by what he saw as management's ineptitude and lack of imagination. These men, with some exceptions, were not the movers and shakers he had waxed so enthusiastic about while preparing for Macy's and in his initial articles for *Fortune*.

Although Dwight was aware of "the contradictions of capitalism," he was still ambivalent about the potential role of individual leadership. In his notes and in these articles he compared the power of the steel leadership with that wielded by the current political dictators. At the same time, he saw these powerful men as inept, uncreative, unresponsive, and unimaginative, whereas he has seen the dictators as just the opposite. It was the labor organizers and the leaders in the mills and mines who were indeed men of skill, competence, and genuine vision. Very early in his trip he wrote Nancy: "I am learning how capitalism works out in practice. . . . Between your Marxian dialectic and my [journalistic] experience we should beat John Strachey all hollow." Strachey was the author of *The Coming Struggle for Power*, an influential Marxist tract of the time that

analyzed the contradictions and inevitable decline of capitalism and the
need for a socialist challenge to fascism. Nancy had encouraged Dwight
to read the book prior to their European trip and it had made a strong
impression on him. That and his reading of Lawrence Dennis's *The Coming American Fascism* in 1936 helped to shape his growing apocalyptic
view that fascism was indeed a real possibility in America. He was sure
that most intelligent Americans would be quick to agree with Dennis's
argument that fascism was "superior to liberal capitalism as a technique
for running modern society."[46]

Dwight was astonished at the efforts managers made to wine and dine
him during his travels, hoping to influence his interpretation. A group of
them held a luncheon reception for him at the Duquesne Club in Pittsburgh. "Imagine me sitting at the head of a long line of beefy, predatory
steel magnates," he wrote to Nancy. The next day he had dinner in the
"humble cottage" of Clarence Irwin, a left-wing leader of the steelwork-
ers' union. The following day he was picked up by a company limousine
and was escorted around the Braddock, Duquesne, and Charleston works
by some "blockhead" management flunky, then made a quick escape
back to meet with the radical union organizers. Dwight was annoyed that
his researcher got more chances to visit with the workers and organizers,
while he was forced, as *Fortune*'s representative, to spend time with the
dim-witted bosses. The squalid company towns, the ramshackle houses
with peeling paint, no plumbing, and leaking roofs shocked him. It was
his first view of this world of suffering during the Depression, and it had
an enormous influence on him.[47]

Dwight seized every opportunity to talk to the radicals. He had meetings with Harvey O'Connor, the Marxist activist scholar, who gave him
"the lowdown" on U.S. Steel and labor. He spent a morning with the Communist organizer Bill Mosely and a group of radicals. He took special
delight in having a company limousine drive him to the mill towns to
meet with the likes of Irwin, O'Connor, and Mosely.[48]

As Dwight later saw it, when he returned from his trip he "was looking
for trouble and [he] got it." In his first article on Republic Steel, largely
a portrait of chief executive officer Tom Girdler, Dwight was forced to
edit out harsh criticism of Republic's brutal labor policies while playing
up Girdler's efficiency and the competence of his management team.
When Dwight's copy was leaked to the corporation's executives, there
were "violent objections." Dwight fought tooth and nail with *Fortune*'s
managing editor, Eric Hodgins, who wanted to "cut most of the guts out."

After a direct appeal to Luce, a compromise was engineered by Archibald MacLeish, then a senior contributor and good friend of Luce. Dwight, however, felt that the "sting" had been taken out of the piece and that it ended up reading like a "panegyric to Girdler and his merry men."[49]

Even when Girdler saw the revised, toned-down manuscript, he banged his fist on the desk and said, "Why, this is socialism." At that Dwight exclaimed, "God knows what he would have done if he ever came across a real socialist." Dwight did have a grudging admiration for Girdler and his associates as efficient engineers of a money-making machine. But he noted that "they are as efficient as steel operators as they are reckless as stockholders' trustees . . . it sounds too much like puffing to dwell on their managerial ability."[50]

Simultaneous with his work on the next company in the series, U.S. Steel, Dwight, with the support of Archibald MacLeish and Wilder Hobson, sent a detailed analysis of *Time*'s biased reporting with specific examples to Luce. Dwight said he had been reading the *Nation*, the *New Republic*, and the *New Masses* and found their coverage of domestic and foreign affairs more complete and revealing than *Time*'s. He accused *Time* writers of "implying things by shrewdly chosen adjectives and neatly turned phrases" in order to put a "prejudiced twist" on a story without adequate documentation. He had done it himself when he wrote for *Time* in 1929, so he knew what he was talking about. He was intent on challenging Time Inc. on every front. He obviously didn't think he could transform *Fortune* into a journal critical of capitalism, but he could not pass up the chance to goad them at every opportunity.

Dinsmore Wheeler was amused by Dwight's "boring from within" and wrote him commenting on the phenomenon of radicals and quasi-radicals like Dwight, MacLeish, John Chamberlain, and Robert Cantwell tying up with that "gaudy laurel wreath on the bulging brow of finance capitalism." He conceded it might be all right if Dwight were willing to admit with Lincoln Steffens that he was a "solid bourgeois who wanted the same things the average American wants." Wheeler noted that Dwight was now making $10,000 a year. Dwight was quick to take offense and protested that he was not bourgeois and did not want *things* like Steffens: "I don't want comfort or new cars or a radio or a frigidaire or a Tudor house with a two-car Georgian garage. In fact I loathe such things—and the people who like them." As for radicals working on *Fortune*, they weren't doing it to acquire things. Dwight insisted that it was a way to learn something about American capitalism. He noted that Selden Rodman's wife Eunice was taking a job at *Fortune* rather than serve as a secretary for "some society for improving relations with the Soviet Union," because she was more interested in American capitalism than Russian Communism; Dwight said he could see her point but that he would take the USSR job if it meant a trip to Russia.[51]

Dinsmore's charge that Dwight was bourgeois, pulling down a big salary as a capitalist tool for *Fortune* in the midst of the Depression, weighed heavily on Dwight and Nancy's consciences. They were to make a fetish of eschewing the comforts of wealth. Nancy was invariably to be found in the most casual, even shabby, attire, and Dwight's carefully cultivated image was increasingly one of bohemian eccentricity. Sporting a Leninist goatee and also looking a little like Trotsky, he had taken to wearing blue chambray work shirts to his office in the Chrysler Building. Harry Roskolenko, a Village character and Trotskyist, recalls that "nature and man had conspired to give Macdonald the voice of a North American Screech Owl, the beard of a Russian Revolutionist and the iconoclastic mind of a *Fortune* magazine writer."[52]

Dwight reveled in the constant contentiousness. Sniping at Harry Luce, whom he genuinely liked, challenging the magazine, provoking his editors was his shtick. It harked back to his attacks on his professors at Yale and his secret ridicule of the executives at Macy's. From his youth to his dying days, Dwight took pleasure in challenging the nearest authority. The issues might be, were in fact to him, extremely important, but it was a purely personal satisfaction, what Jane Addams once described as the "subjective necessity," that sparked his dissent and provocation.

When Dwight's first draft on U.S. Steel hit his editor's desk, the battle was on. The first two parts of the article were pieces of investigative reporting and by and large acceptable with some minor changes. The third part, on the corporation's labor policy, had been written by Robert Cantwell at Dwight's request. Cantwell had been a member of the Communist Party and had written a classic proletarian novel, and Dwight just assumed he would do a good job on labor's behalf. But Cantwell had moved substantially to the right, and he bent over backward to defend the the corporation's policies. Dwight felt betrayed and embarrassed, and he told Dinsmore Wheeler that his own fourth installment, divided into two parts, one on the corporation and the other a scathing portrait of Myron Taylor, its head, would be "a grand finale full of high philosophical speculations."

During this period Dwight was signing petitions for Angelo Herndon, an imprisoned black Communist in the South, holding a cocktail party to raise funds for the Share Croppers Union, and supporting a strike by elevator operators against the landlords in New York apartment buildings. Because the landlords were acting with such arrogance and stubbornness, Nancy and Dwight wrote and circulated a petition in their own building and managed to persuade twenty-two out of sixty apartments to support the operators. They withheld their rent and even picketed. He reported to Dinsmore Wheeler that he and Nancy were on the "friendliest possible terms with the elevator people." It was "swell to feel such a warm sympathy flowing back and forth between us. The class struggle makes you love your fellow man, if you're on the right side, that is the left side."[53]

Dwight's militancy and newfound activism led to a serious falling out with Geoffrey Hellman and George Morris. On two successive nights Dwight became so outraged with their refusal to support the elevator operators that the acrimonious arguments strained these old friendships. Hellman's anti-union stand, his pleasure at crossing picket lines, inflamed Dwight. George Morris, the detached abstract artist, seemed to "think one could simply ignore conflict and it would go away or work itself out." Dwight regretted his argument with Morris, because he was intelligent and open-minded, but Hellman could go to hell as far as he was concerned. Dwight told Dinsmore Wheeler that Hellman was both stupid and ignorant of the issues and had become soft and lazy. Dwight noted that the more he saw "those who stand—or rather squat—in the way of radical progress, the more [he learned] about the conservative businessmen who run this country and the more [he saw] of the injustices done people under this horrible capitalist system," the more intolerant he became and the more difficult he found associating with those of his class who failed to move left with him. He hoped Wheeler was keeping him company on his pilgrimage to the left. After his rows with Hellman and Morris, it would be just too much if his oldest friend "appeared among the enemy."[54]

In the spring of 1936 Dwight's mounting anger and impatience reached a climax during his confrontation with his *Fortune* editors over the last installment of the U.S. Steel articles. His chief antagonist was Ralph McAllister Ingersoll, who challenged both his theory and criticism of the corporation at every point. In this final article, which Dwight saw as the pièce de résistance of the series, he leveled a double-barreled assault on the ineptitude of the company's managers and offered a quasi-Marxist analysis of the contradictions of the capitalist system. The article reflected his belief, by this point, that the system was doomed and that the captains of industry were hastening its demise with their refusal to deal with its injustices and paradoxes. Dwight actually started the article with a quotation from Lenin's *Imperialism:* "Free competition is the fundamental property of capitalism. . . . Monopoly is the direct opposite of free competition. . . . Monopoly is the transition from capitalism to a higher order."

It is doubtful that Dwight ever believed his editors would accept this revolutionary indictment of monopoly as the last stage of capitalism, leading inevitably to socialism. But they went much further than eliminating these incendiary sentiments, cutting every note of criticism, every dart of sarcasm, every suggestion that the system had deep structural failings. At one point in this war of editorial notes, Dwight wrote Ingersoll, whom he detested: "God damn it I'm getting sore. What earthly purpose is served by extracting all the points out of every sentence? If stories are to be edited like this why hire writers at all? There is a limit to this petty interference." In an imaginative metaphor, Dwight had compared U.S. Steel to a "mighty hulk becalmed in the Sargasso sea . . . bereft of both the social intelligence of Communism and the dynamic individualistic drive of capitalism." The line was completely eliminated. Pages of theory about monopoly were slashed. Drawing on his study of dictators, Dwight described Myron Taylor as exercising a power equal to that of Mussolini or Stalin: The stockholders had no voice; Taylor commanded private armies to bully the workers into submission.

None of this was acceptable to Luce and his obedient editors, who dutifully tore Dwight's manuscript apart and left it "a shambles, bleeding internally at a dozen points where vital organs had been excised by the fawning editorial scalpel of Luce's rewrite men." As for the second part of the article, the sardonic attack on Myron Taylor, the editors threw it out entirely and replaced it with a "a full-throated burst of lyrical eulogy, to one of the nation's most far-seeing industrial statesmen."[55]

So Dwight quit. He walked out of a $10,000-a-year job in the midst of the Depression, much to the mortification of his mother. But Nancy had no qualms about his leaving. She was the beneficiary of a fairly sizable trust fund. Dwight had long been miserable at *Fortune.* He wrote Dinsmore Wheeler that the bitter taste of the Steel sell-out was reducing me to a state of psychic impotence. . . . A weight has been lifted off my shoulders. . . . I was just dragging out a lifeless dried-up existence. . . . I now feel free to see what life is all about . . . to really read scores of books I've long wanted to. Especially to orient myself politically. I know that I don't believe in capitalism. But I am still hazy as to what course to take from here.[56]

3

A Distinguished Goy
Among the Partisanskies

I T WAS A GREAT RELIEF FOR DWIGHT TO BE FREE OF *FOR-tune*. He and Nancy felt that they could live on the $2,000-a-year income from her inheritance, supplemented by savings and dividends from investments he'd made while at *Fortune*. Both he and Nancy were inordinately careful about spending money. They had been embarrassed by the appearance of wealth when they were well-off, so becoming genuinely impoverished intellectual bohemians fit their persona.

Although the plan was for both of them to concentrate on serious writing, neither did. Dwight was still gathering material for the dictators project, but his enthusiasm had waned. He thought he should read American history in a systematic fashion in order to prepare himself for writing a more technical history of the steel industry. But he wasn't certain. Nancy no longer spoke of her work on Asian art. She was prepared to serve as her husband's researcher. But things were always moving in fits and starts. Dwight didn't really know just what it was he wanted to do with his life. It was not clear to him whether he had the desire or the discipline to pursue scholarly work. He had a wide-ranging intellect, a great deal of energy, and an increasingly engaging prose style. But his interests were eclectic, if not dilettantish.

Dwight at thirty felt he was at that point in his life when he ought to establish an identity, decide what he wanted to do, and get on with it.

There was an impatience, almost a panic about it all. "It's a critical period in my life," he wrote Dinsmore. Two weeks after leaving *Fortune* he complained of "a new kind of weight" that had descended on him, "the responsibility of doing things on one's own." He felt like a man just discharged from the Army: "How difficult to decide, to act, instead of doing what one is told. . . . Ordering one's life so as to get something accomplished isn't easy."[1]

Dwight was also drinking too much in the evening and waking with severe hangovers. It was nothing to be alarmed about; everyone he knew drank, at a constant round of cocktail parties, fund-raisers, and impromptu get-togethers. It was of some concern to Nancy, who sometimes complained of the boozy parties at which the loudness of the conversation far exceeded the quality.[2] Contributing to the decibel level was Dwight's political enthusiasm. It had reached the point where his friends had to pass a political means test. He found it difficult to spend time with those who did not share his contempt for the inequities of the capitalist system.

Earlier, Geoffrey Hellman and Dwight had faced off symbolically on *New Yorker* versus *Fortune* letterhead. Hellman was himself a moderate who conceded that the country should become "more socialist than it is now." Presumably he meant that a more equitable distribution of wealth would help. But he interpreted Dwight's militancy as favoring the "abolition of private property" and his support of the union shop as an acceptance of the "sacrifice of liberty and individualism etc. for the sake of the least fit." He accused Dwight of spending his time discussing nothing but politics and "ramming your opinions down the throats of everyone, whether they're much interested of not." As an added insult, he charged that Dwight's conduct was worse than annoying, it was boring. Dwight's "sneering nasty streak" made him unable to be with people who didn't agree with him without resorting to an "infantile sarcasm."[3]

Dwight was particularly piqued by Hellman's assertion that he was interested only in politics and had lost his appreciation for culture, literature, and the arts. He denied this, pointing to his current review of a biography of Raleigh in the *Nation*. What Dwight demanded, which is typical of the new convert, was a sense of urgency. How, in the face of economic oppression and imminent disaster, could one not be more involved? Silence was complicity. He asked Hellman: "Isn't the real truth that you aren't interested much in anything, that you've let yourself get shockingly slack and smugly comfortable since you left college?"

This accusation was simple projection, for it was Dwight's greatest fear during his years at *Fortune*. His assigned articles in the magazine, which ranged from prep schools to wild horses to lighthouses, not to mention the puff pieces on noted businessmen, became an embarrassment; they seemed so fatuous during a devastating Depression. Both he and Nancy had a deep, puritanical conviction that one should do serious work of consequence. Dwight had had little social consciousness, but he had always been driven by the notion of a cultural obligation, to stand up for things of value. As he became more politically aware, politics became a form of culture. Politics, literature, and the arts were all part of the same thing, the quality of life in society.

Early on, Dwight had felt an almost Henry Adams despair over the low intellectual and moral level of politics. Dwight longed for intellectuals to make a difference, to make history. Expediency, pragmatic compromise irritated him. He was attracted not so much to abstract ideological frameworks as to commitment, enthusiasm, dedication to one's beliefs. In his quarrel with Hellman, he was sensitive to the charge that he had abandoned a belief in individualism and had embraced a thought-controlling fanaticism. He insisted that he was as individualistic and nonconformist as Hellman: "I hate submitting to authority." He declared that he cherished individual liberty, his own and others', but "in this country today we live under the dictatorship of a small class of capitalists—mostly those who control the great corporations—and the only way we can get free is to submit to discipline."

Dwight conceded that he was hard-boiled and fanatical, but that growing right-wing movements like the Liberty League made it necessary to be so. He accused Hellman of not being personally affected by "the economic autocracy which weighs so heavily on most people in this country. Nor am I." But Dwight insisted that he could feel for those whom Hellman had categorized as the "unfit," and that only by meeting "the conservatism of the ruling class with an equally fanatic and intense radicalism will the masses ever get control in their own hands. You can't fight dictatorship with anything but dictatorship." Dwight closed the discussion by arguing that Hellman's position as a "detached lover of fair play" no longer had any meaning. "Whichever side wins, the capitalist (i.e., Fascist) or the democratic (i.e., Communist), you will get it in the neck."[4]

This was the high point of Dwight's brief fellow-traveling period. He apparently accepted the current ingredients of the Party line: class warfare, dictatorship of the proletariat, the necessity of Party discipline, and even apologies for Stalin himself. Charles Biederman, a painter Dwight admired, recalls that when Dwight visited his studio he dismissed Biederman's rejection of the Stalinist dictatorship by "vociferously defending Stalin with rationalizations." So intense were their differences on this issue that Dwight stopped seeing Biederman.[5]

By the late summer and fall of 1936, Dwight and Nancy had stepped up their political activism. He joined the consumers' union, had lunch with the editor of *Fight*, the house organ of the popular front League Against War and Fascism, and helped organize a unit of the Newspaper Guild at Time Inc. At the same time, he finally settled on a writing project. It would be the steel industry in the United States. He and Nancy would go to Cambridge to use the Widener Library at Harvard. After preparation there, they planned an extended trip to Washington to work in the National Archives and the Library of Congress. Nancy was now totally absorbed into Dwight's project and would accompany him to aid in the research.

It was during this period that Dwight attempted to read Marx and Marxist literature in a systematic way, though it is doubtful whether he ever achieved his goal of mastering the texts. He reported to Dinsmore Wheeler that he and Nancy, living in a little $7-a-week furnished room in Cambridge, had just finished the first 200 pages of volume one of *Capital*: "They are tedious, repetitious, hard to follow and yet somehow they produce a feeling of awe. Marx goes so directly to the heart of the problem and his argument, as much as I can follow it, seems irrefutable." What attracted Dwight was "the awesome learning of the man." Reading Marx returned him to his schoolboy practice of making lists of books that had to be read.[6]

Dwight had sent Felix Frankfurter, the Harvard Law professor soon to be appointed to the Supreme Court, a copy of his *Fortune* steel articles. Frankfurter took the Macdonalds to lunch. Dwight found him stimulating, dynamic, and full of intelligent questions. Still, to Dwight, Frankfurter was no radical, just an old anti-monopoly liberal type of the Borah–Brandeis school. Dwight recalled that Frankfurter showed little enthusiasm when told that Dwight planned to write his book from a Marxist perspective. Frankfurter warned him not to get mired in abstract theory and to really describe the actual workings of the industry. Although Dwight described his meeting with Frankfurter in a breezy, matter-of-fact tone, it was apparently a great humiliation and one that he looked back on with dismay throughout his life, noting more than once that Frankfurter had found his grasp of the technical details of management and plant operation sketchy, that he apparently thought Dwight had not mastered the relevant literature on the subject and was substituting an ideological polemic for careful empirical investigation. Throughout the lunch Dwight could feel the unspoken question: "Is this all these clever journalists know?" Dwight later acknowledged that he had been out of his depth. Frankfurter "argued about the specifics and I didn't come up to scratch. I realized it . . . v. sad for us both—so he eased off."[7]

In November 1936, the Macdonalds left Cambridge for a brief stopover in New York to cast their ballots in the presidential election. Dwight told Fred Dupee that he planned to vote for Earl Browder, the Communist Party candidate. His FBI file (1936 undated) states that "the indices of the New York Office reflect that DWIGHT MACDONALD enrolled under the Communist Party Emblem in New York City in the 1936 Election Registration." Fred Dupee, who actually had joined the Party, was doing organizing work on the docks and editorial work for the *New Masses*. Although he was already irritated by Party pressure on his editing, he was delighted by Dwight's decision. But he expressed regret that Nancy had not come around. Apparently she was going to vote for Norman Thomas, the socialist candidate, who was being supported by the Trotskyists in their brief tactical alliance with the Socialist Party.[8]

In early December Dwight and Nancy left the city, visited Dinsmore Wheeler at the farm, and then headed to the Pittsburgh steel area. They were squired around by a new radical friend, Rose Stein, who wrote for the labor press. Dwight was very much taken with this "small jewess with great qualities of vitality and a refreshing belief in the workers—refreshing because she sees them all the time." Stein took the Macdonalds to Weirton, West Virginia, to take part in a "flying squadron" of the Steel Workers' Organizing Committee who were leafleting a small mill. They expected some kind of a rumpus. Dwight interviewed workers, and attended mass meetings in Aliquippa, Pennsylvania, at a Roumanian Hall decorated with "crude frescos and slavic fraternal signs." The crowded, smoky hall was filled with Old World types "with leathery faces, spiky mustaches . . . fine-looking people, lots of juice, character to them." He met Clinton Golden, the regional director of the SWOC, "a big drawling fellow, a bit on the ministerial side but good at this sort of thing," who made jokes designed to ridicule the steel companies. An older worker told Nancy that where once the cops had harassed the organizers, they were now supporting them. Pittsburgh, Dwight later wrote, was everything a city shouldn't be. It was the product of the "dictatorship of big business . . . a devil take the hindmost capitalism."[9]

Even if Dwight did cast his vote for Browder in early November, he had been won over to the Trotsky position by the end of the year. Dwight had no trouble shifting political gears if events warranted, a trait that often irked his more disciplined friends, who found it difficult to keep up with his fast-changing allegiances. Nancy had a significant impact. Events in Spain were of great importance to her. Communist intrigues against the anarchists and Trotskyist revolutionaries were being reported. Dinsmore Wheeler was constantly reviling the brutal political repression that was taking place in the Soviet Union. News of the first purge trials of Stalin's opponents was reaching the city, and the Trotskyists were actively recruiting intellectuals into their ranks.[10]

It was through Fred Dupee that Dwight met Philip Rahv and William Phillips at the *New Masses*. Although Dupee was still a member of the Party, he was becoming increasingly vexed with the dictatorial approach its officials took to his editorial work. Rahv and Phillips were at that very moment on their way out of the Party's orbit. They had suspended publication of their magazine, the *Partisan Review*, which had been an organ of the John Reed Clubs and supported by the Party. They talked of their hope of reviving the journal as an independent radical journal, and they were in the process of drumming up financial support.

Dupee, who was seeing a lot of Dwight at this time, reported to him on all these matters and suggested that he meet with Rahv and Phillips to discuss the possibility of working with them on the new independent *Partisan Review*. Dwight was immediately enthusiastic. He had been disappointed when the *Miscellany* ended publication. Then he had allowed Luce to dictate to him what his writing obligations were. Here was a chance to do something really important, to become involved in literary affairs and at the same time take an active role in radical politics.

For Dwight, work on *PR* offered an escape from the frustrations of freelance life. He had put together only "two constipated" chapters at best on the steel book. He found the demands of sustained book writing and scholarship too exacting; he did take pleasure in hard-hitting polemical journalism. It stimulated him much more than the long, lonely hours in the library or at his desk spent trying to transform the data into interpretive and interesting prose. Mary McCarthy recalled Dwight's constant talk during these years about doing the book, or not being able to do it. It had become an albatross; he carried the "damn notes" here and there. Among his friends the book had become a "drab joke." As was her way, McCarthy drew upon Dwight's angst in her short story "Portrait of the Intellectual as a Yale Man." It was a problem that plagued Dwight. He was diffident about being a "journalist," even a first-class, literary, intellectual journalist; he could not accept himself, accept his particular talents and limitations, and it never ceased to bother him. He later ruefully remarked that he had never written a book "in cold blood"; the books he did publish were collections of magazine articles.[11]

When Fred Dupee came to tell him that Rahv and Phillips were interested in talking to him, Dwight was excited and flattered. He met them at Phillips's apartment on East 12th Street at nine o'clock in the morning in the waning days of 1936. At that point, Dwight was still a lingering fellow-traveler. Rahv and Phillips were already leaving the Party and they spoke of an independent radical journal in opposition to the Stalinist line of the Party and the popular front in literature and culture. They immediately ganged up on Dwight, countering his defense of the actual role of the Party with detailed accounts of its corrupting effect on all literary and cultural activity. They hit Dwight where he was most vulnerable. At one point, Edna, Phillips's gentle wife, called out from another room, "Leave him alone, leave him alone, leave him alone. Let him breathe." Dwight finally conceded the justice of their arguments and agreed to join the group.

Phillips believed they had converted Dwight, that he might well have joined the Party but for their devastating indictment of it. But it would appear that both Dinsmore Wheeler and Fred Dupee had softened Dwight up so that he was ripe for the arguments of "the boys," as he would soon be calling the scheming cabal of Phillips and Rahv. The very idea of a publishing venture that combined an interest in literature with a strong commitment to radical politics was something Dwight could uniquely appreciate. That it would be an "independent journal" appealed to him. After all, he had chafed under the editorial restraints of the "capitalist" press, which he now identified as proto-fascist. He was enthralled by the hard intelligence of Rahv and Phillips and admired their obvious radical, Marxist, even revolutionary orientation. More important to "the boys" than Dwight's editorial participation was his promise to make every effort to enlist substantial financial support for the project.

Dwight had in mind George L. K. Morris, his old college friend and collaborator on the *Miscellany*. They had bickered of late because Morris's revolutionary commitment did not measure up to Dwight's enthusiasms. Morris had only a tourist's attitude toward the debates Dwight was having with his new radical friends, and he had little patience for or interest in Marxism. But he was very much interested in avant-garde art; he was an abstract painter himself and the founder and spokesman for a group that called itself American Abstract Artists.[12]

Dwight and Fred Dupee enlisted Mary McCarthy as a co-editor to go into the venture with them. Along with Eleanor and Eunice Clark (the latter married to Selden Rodman and the former soon to marry an aide and bodyguard of Trotsky himself), she was one of the Vassar crowd who figured significantly in the literary and radical intellectual circles of the city. McCarthy was a beautiful, witty free spirit who at the moment was winding up a marriage to a young radical actor, Harold Johnsrud. He had introduced her to Communist circles in the theatrical community, so obviously she was to be the theater critic. She was also the lover of Philip Rahv.[13]

An interesting highlight of Dwight's joining this group of sectarian intellectuals was that he, Dupee, McCarthy, and James Burnham, an elegant former Princetonian and serious Marxist theorist, were not Jewish, whereas the ranks of the dissenting left, particularly the Trotskyists in New York, were overwhelmingly Jewish. They came to these struggles out of a tradition of Jewish socialism and revolutionary Marxism—a basically European tradition. Although the younger generation were enamored of American culture, fascinated by its literature, intent on making its classics part of their own tradition and claiming a place in that intellectual milieu, they were, nevertheless, *Jewish* intellectuals with a penchant for disciplined thought, carefully wrought theory, convoluted debate. They were astute, tough street fighters in the argumentative sessions that were their lifeblood. Dwight, with his enthusiasm for debate, was charmed by the aggressiveness of his new associates and impressed with their learning and serious literary interests. He found the Trotskyists genuine intellectuals, unlike the Stalinists. Since most of these new colleagues were faithful believers in an international cosmopolitanism, a world community of intellectuals, Dwight and Nancy, as well as their friends Dupee and McCarthy, all found in their company a political and cultural haven. Their Jewish comrades referred to them as "our distinguished goyim." Dwight was intent on casting off his middle-class Waspish background; a close association with a group of Jewish radicals from Brooklyn and the Bronx was proof of passage.[14]

In the meantime, there was "the book." Dwight and Nancy dutifully went off again to Washington to do research in the Library of Congress and, with an intro from Frankfurter, to interview Justice Louis Brandeis; Dwight found the eighty-year-old justice an over-the-hill old-time Progressive of the Theodore Roosevelt era. Dwight also spent time talking to leaders of the Congress of Industrial Organizations (CIO) in the Washington office.

The Macdonalds were in Washington on Inauguration Day 1937. Neither saw it as anything to celebrate. Dwight accused Roosevelt of leading a reactionary counterrevolutionary movement. Dwight's rhetoric was increasingly militant and sectarian. The Communist Party's newfound support of Roosevelt and liberals infuriated him and pushed him further to the left, which in effect meant toward the Trotskyists.[15] Stalin's rigged show trials, designed to prove the traitorous subversion of his Trotskyist opponents, had a profound impact on radical intellectual circles in New York, where internecine warfare raged between the two radical camps: the Stalinist Communist Party and its fellow-traveler sympathizers versus what was loosely known as the non-Communist left, led, at least theoretically and intellectually, by the Trotsky movement and a variety of freewheeling, anarchistically inclined radicals. Lionel Abel's description of New York as the most interesting part of the Soviet Union in the late thirties was certainly true below 14th Street, in Greenwich Village. It is not surprising that Dwight, accompanied by Nancy, gravitated to this small band who stood against the entire left, since at that point the Communists had become little more than the left wing of the Democratic Party and the liberals were allying with Communists in the struggle against fascism.[16]

It was the trials that seemed to bring about the polarization of the radical scene into two warring camps. For a generation of leftists, the trials were the latest in a series of betrayals that led to a profound disillusionment with Soviet Communism under Stalin. As followers of the Marxist revolutionary movement, they found the horrors of Stalin's methodical liquidation of all opposition a shattering experience. For many, these years marked the beginning of a profound deradicalization. But this was hardly true of Dwight. His flirtation with the popular front had been little more than that, a brief encounter which left no scars. He came to his more serious radical position in a way very similar to that of a character in Mary McCarthy's "Portrait of the Intellectual as a Yale Man," "from the happy center of things, by a pure act of perception." In effect, he began his serious political activism as an anti-Stalinist. One might argue that was the motivating and sustaining force of his radical commitment. Stalinism was to Dwight the latest and most extreme example of arbitrary authority, a nightmarish authority that menaced the entire civilized world.[17]

It was a chance springtime encounter with Mary McCarthy and Margaret Marshall, the literary editor of the *Nation*, that prompted Dwight to read the official Communist Party account of the trials. McCarthy, along with Marshall, took it for granted that the trials were a frame-up. Dwight had had suspicions, but had not imagined the trials to be an absolute fraud. He recalled being incredulous. Shaken, he left them and a few days later bought the official Communist version, *The Case of the Anti-Soviet Trotskyite Center: A Verbatim Report*, published by the People's Commissariat of Justice of the USSR and distributed in New York for a dollar. Dwight found the account the real key to the Stalinist mentality. That the Communists would provide the world with the evidence of their own criminality suggested to him a blind and witless fanaticism. It illustrated clearly how the Stalinists had lost contact with reality, for the transcript could do little but arouse grave doubts in all but the Party faithful.[18]

Malcolm Cowley, the literary editor of the *New Republic*, was one of the Party's few chief apologists with genuine literary credentials. His defense of the trials in that journal prompted Dwight to write a rebuttal and to throw himself into the camp of the Trotskyists, who were at that very moment mounting a defense committee to support the aging exile who had only recently been ridden out of Norway and was living in a suburb of Mexico City. Dwight was delighted to do battle with Cowley. In addition to writing a five-page letter criticizing Cowley's article, he joined the American Committee for the Defense of Leon Trotsky. Dwight's response was not dissimilar to that of other radicals in other times when charged with opposing the State. Guilty? Perhaps, but guilty of what? To get at the heart of the matter, there were two crucial questions. Was Trotsky implicated by the evidence, and did Trotsky's theories of opposition lead inevitably to counterrevolution and support of fascism? Dwight argued that the trials failed to answer either of these questions. The alleged "evidence" was absurd. To believe Trotsky capable of the "hare-brained" acts ascribed to him by Andrei Vyshinsky and Cowley, one would have to believe that Trotsky was in the "last stages of senile dementia."[19]

Dwight's active membership in the Trotsky defense committee and his castigation of Cowley brought him to the center of New York's radical circles and in close contact with what became the hard core of the non-

Communist left and the movers in the Trotskyist camp. By now, both Dwight and Nancy saw themselves as activists, even "revolutionaries." As Mary McCarthy remembers it, Dwight was intoxicated with ideas. The Trotskyist position endorsed his own notion that the capitalist system was in its last rotten stage of decay. He was enthralled by the possibilities. He championed the work of the defense committee. He supported the committee's Commission of Inquiry, headed by the venerable John Dewey, which went to Mexico to take testimony from Trotsky and which completely exonerated him. Through these activities, Dwight rapidly became involved in the exhilarating internal warfare on the left. He, Fred Dupee, and Mary McCarthy attended the Communist-dominated Second Writers Congress to raise a ruckus and challenge Granville Hicks and Malcolm Cowley, leading literary lights supporting the Party line.[20]

In May 1937, Dwight informed Dinsmore Wheeler that the plans for the first issue of the new *Partisan Review* were well under way. From the start, Dwight took great pains to stress the journal's independence and lack of political affiliation. Initially he envisioned no polemics or articles on immediate political and tactical problems. But the tension and ambiguity that developed in trying to walk this independent line were inevitable. Dwight was at that moment becoming more involved in sectarian politics, attending political meetings, organizing writers, participating in debates and conferences, all forcefully against the Communist Party and the popular front. There was no way Dwight could escape partisan identification, and how the new *Partisan Review* could have been expected to avoid a partisan political label is not at all clear. The ambivalent stance of its editors did not clarify the situation. Mike Gold, the Communist litterateur, denounced the enterprise in the *Daily Worker* as "a snake that has shed its skin for Trotsky." In a burst of typical Party slander, Gold denounced *PR*'s editors as "of the same ilk that murdered Kirov, that turned the guns on the backs of Loyalist civilians in Spain and betrayed the army front lines that have been caught red handed in plots with the Gestapo and Japanese militarists to dismember the Soviet Union."[21]

To counteract this kind of assault, the editors planned the first edition to contain several pieces of serious literary criticism. Dwight was assigned the job of wooing Edmund Wilson, and after much anxiety he was successful. In addition to Wilson, the first issue was to have contributions from Wallace Stevens, James T. Farrell, Pablo Picasso, Lionel Abel, Mary McCarthy, Sidney Hook, and Dwight, plus editorial commentary and a section called "Ripostes." When it finally appeared, it also had Delmore Schwartz's soon to be famous short story, chosen by Dwight, "In Dreams Begin Responsibilities," as well as criticism by James Agee.

Dwight labored over his own contribution, an analysis of the *New Yorker* magazine. He wanted to pass the political means test of applied Marxist criticism while at the same time maintaining a level of literary sophistication consistent with the aspirations of his fellow editors and scheduled contributors. His design was to tie the glossy weekly to its class roots but avoid the kind of moral didacticism and vituperation that characterized the *New Masses* in its attacks on the bourgeoisie. Dwight began by charging that the *New Yorker*'s humor was "the humor of the inadequate," reflecting the ruling class's loss of confidence because of its impotence in the late economic crisis. It was an "accurate expression of a decaying social order."

Dwight's piece had the obligatory references to the "class war" and did not neglect a moral indictment: "There is something monstrously inhuman in the deliberate cultivation of the trivial." Dwight couldn't pass up the *New Yorker*'s advertising. What critic has? The glitzy strips down each side of the page formed a cage made of the baubles and beads of capitalist consumer culture, imprisoning the text. The readers, after taking instruction in the arts and literature from the magazine's bland and amiable critics and absorbing the pleasant drolleries of the made-to-order stories, were expected to get out and bring the luxurious goods home—the vintage wines, Paris gowns, movie cameras, golden yo-yos—the things that help the upper classes face their overwhelming problem: "how to get through the day without dying of boredom." Dwight was quick to recognize that the magazine was not simply for the rich. It offered to the middle class, for a small admission fee, a peep into the haut monde. Dwight's piece was well received by Mary McCarthy and Fred Dupee but apparently dismissed as lightweight by Phillips and Rahv.[22]

Dwight had played a large role in recruiting the talent for the first issue, and it was an impressive literary debut. But the main concern throughout the months of preparation remained: What really was the magazine's political stance and what was its actual relationship to the Trotskyist movement? The Communists continued to claim that *PR* was nothing but a Trotskyist rag for anti-Soviet propaganda. Dwight and Mary McCarthy were known as Trot fellow-travelers because of their public endorsement of the defense committee and their acceptance of the verdict of the Dewey committee. Neither was actually a Workers Party member, and surely Mary McCarthy's connection to the Trotsky movement was based on a civil-libertarian stance mixed with high fashion rather than any commitment to a party line. That was true of Dwight also and the other editors to some extent. They sincerely hoped to publish an independent journal, free of ideological or political control and open to diverse views. On the other hand, without surrendering their independence, the editors did want to have a cordial, even collaborative relationship with Trotsky as a literary revolutionist. Trotsky had a similar viewpoint. He wanted to encourage the magazine, but he was suspicious of the editors' expression of "bourgeois individualism." A fascinating correspondence ensued prior to and immediately following the publication of that first independent issue of December 1937. Dwight was assigned the job of preliminary diplomacy, presumably because both Rahv and Phillips wanted to keep a low profile.[23]

Dwight wrote the initial cautious letter to Trotsky informing him that the magazine would be independently Marxist, with its major emphasis on literature, philosophy, and culture. He hoped that Trotsky would contribute essays of literary and cultural criticism. Dwight had read Trotsky's *Literature and Revolution* and was impressed with his prose style and sensitive, intelligent appreciation of fine writing. It was this, combined with Trotsky's revolutionary activism, that attracted Dwight. He was excited by Trotsky's example. It showed that intellectuals, not just men of political power, could make history.[24]

Despite Dwight's flattery, Trotsky remained wary. He would be glad to ally himself with the *Partisan Review* editors in the fight against Stalinism, but Dwight's letter did not offer a clear program based on "certain fundamental principles." The fencing continued. Dwight wrote that the *PR* editors' "conception of the relation of revolutionary literature to revolutionary politics is such that it excludes our taking part in immediate political controversies." He conceded that the editors had political lives and that all were "opponents of Stalinism and committed to a Leninist program of action." But they would not impose their ideas on the literary content of the journal, nor commit it to "immediate political controversies." Because many of their contributors and editors were already allied with Trotsky's movement, they would "continue to be branded Trotskyist," but they would still like him to contribute articles on "culture and literary matters, knowing that your approach . . . fully involves your entire political position." Dwight was offering him a platform but hoping to deny him editorial control or influence. Taking as his model Randolph Bourne, the literary radical of the pre–World War I period, Dwight insisted that intellectuals had the right, the obligation, to be intellectuals who deal with general theory and abstractions and not be compelled to tailor such speculations to fit current controversies.[25]

Dwight had no desire to avoid politics. Still, there was a spongy quality to his letters to Trotsky calling for a fine line between ideological and literary writing, and it was unclear what it might mean in practice. Trotsky continued to maintain his distance, insisting that the journal had to show its hand. He would wait and see.

After the second issue appeared, Trotsky let fly a missile from his fortress hideout in Mexico, a harsh dismissal of the editors and their work. They were educated and intelligent but they simply "had nothing to say." They wanted to be respectable, to be all things to all people and to irritate no one. One can imagine how such a charge hit Dwight, who hated blandness and careerist opportunism more than anything else. Trotsky facetiously remarked that he was not criticizing the content of their ideas, because he had not discerned any ideas. He held out a feeble hope that their vacillating stance would be only temporary, but the tone of his remarks suggested doubt. As for the editors' plan to hold a symposium titled "What Is Alive and Dead in Marxism," it was, he charged, as though they "were beginning history with a clean slate." And whom had they invited? "Political corpses," a group of people with "no capacity for theoretical thinking." Two of the most prominent were Victor Serge and Boris Souvarine, two ex-Communists whose virulent criticism of Stalinism had made them suspicious of the roots of the revolution and the roles of both Lenin and Trotsky. Both were on the verge of raising questions concerning Trotsky's role in crushing the Kronstadt rebellion led by anarchists challenging Bolshevik revolutionary leadership. That the newly revised *Partisan Review* planned to publicize itself by inviting such critics of the validity of Marxist fundamentals did not sit well with "the Old Man." He categorically refused to participate, stating that the editors were not interested in the revolutionary struggle but, on the contrary, were trying to escape to a "small cultural monastery" where they would turn out a "respectable" but harmless "peaceful little magazine."[26]

Trotsky had abandoned the tone of amiable diplomacy and had brought the matter down to the street-fighting arena where the New York sectarians normally did combat. Dwight, the Yale–*Fortune* man as radical, stepped aside and the old pro, Philip Rahv, took over. When it came to this kind of confrontation, Rahv obviously had the experience to engage the master. Rahv's confidence was that of one whose weapons had been honed in the warfare within Party ranks, and he could be harsh and aggressive in debate, but he was also armed with a literary sensibility as keen and subtle as the Old Man's.

The essence of Rahv's rebuttal was that Trotsky had no understanding of the complexities of sectarian politics in New York. Trotsky didn't seem to realize that Stalinism had thrown into question the entire Marxist canon. *Partisan Review*, he pointed out, "was the first anti-Stalinist left journal in the world." It was vulnerable due to its own Stalinist past. It had to feel its way toward viable alliances. As for the symposium, it was simply a way of weeding out alternatives to Marxism not acceptable to genuine revolutionaries. Rahv rebuked Trotsky for his carping criticism and his unwillingness to come to the aid of obvious comrades.[27]

Trotsky took this lecture with more amiability and less impatience than he had showed to Dwight, but he continued to call for a specific program directed against Stalinism, the *New Masses*, and such fellow-traveling journals as the *New Republic* and the *Nation*. "It is impossible to progress without a whip," he wrote to Rahv. "It is necessary to empty this filthy pail of Stalinism to the bottom . . . to destroy their influence on radical thought."[28]

The journal did maintain a relationship with Trotsky, and it was delighted to publish a manifesto purportedly written by Diego Rivera and André Breton, but really by Trotsky, calling for an anti-Stalinist, pro-Socialist International Federation of Independent Revolutionary Art. This was the high point of *Partisan Review*'s direct association with Trotsky. Rahv and Phillips always remained suspicious, cautious, circumspect. They had been burned by their previous commitment to a party discipline and were determined not to take that path again. Dwight, on the other hand, was a seeker, looking for a close radical community, not disciplined but with the comradery that comes with association in a cause. Phillips patronizingly insisted that it was a stage Dwight had to go through. Perhaps, but it was not simply to provide the necessary disillusioning experience that lends political maturity, but more to reinforce his break with his elitist past and to fashion himself as a radical intellectual, one who could participate in making history. At thirty-two, Dwight believed ideas could make a difference and he was intent on participating in the enormous changes that were taking place every day.[29]

4

A Trotskyist Education

WHILE DWIGHT WAS ABSORBED IN PREPARING THE FIRST issue of the new *Partisan Review* in the summer and fall of 1937, he was still struggling to determine what his professional role in life would be. While he remained adamant in his support of the autonomy of art, he was eager to plunge into the political struggle, to make his ideas heard and effective. If he were to be a critic, he seemed to want to be a political and social critic as much as if not more than a literary critic. Already he was a little impatient with the lack of activist political involvement on the part of Philip Rahv, William Phillips, and even Fred Dupee.[1]

As for his own writing, Dwight worried that he might become nothing more than a professional writer of letters to the editor. He could make no progress on "the book" and admitted that he was "paralysed with fright at the very idea of writing such an ambitious book." He realized that to produce a substantial work, he would have to maintain a "cast-iron routine." Though he struggled to keep up his research and writing on the steel industry, he was always drawn to other, immediate writing chores: a series of articles on Time Inc. for the *Nation*, a piece on Pittsburgh for the *Forum*, and more and more short polemics for Trotskyist publications. Immediate engagement had the advantage of taking him away from "the book" and yet could be justified in the name of the urgency of the cause. But by pursuing these short-term projects he never accomplished what he felt was the true work of an intellectual.[2]

Another factor, which Dwight referred to in a casual way but which played a significant part in his decision to abandon "the book," was Nancy's pregnancy in the fall of 1937. A baby boy, Michael, arrived April 8, 1938. Dwight was initially worried about the impact of a child on their life and work. But it was not long before he was bragging about how well he was taking to fatherhood. Having a child in the house "was great sport." Dwight did from the beginning share much of the parenting with Nancy, taking turns getting up to feed the baby his formula and changing diapers. Dwight also shared household duties. He loved to cook and did it regularly.[3]

But domesticity and child-rearing were not the primary concerns of Dwight or Nancy. Nancy was the hardworking and conscientious managing editor of the *Partisan Review*, which was a demanding job, since none of the editors except Dwight took care of the administrative details. Nancy also encouraged Dwight's political enthusiasms and supported him in his efforts to get *PR* to publish a piece on the annual meeting of U.S. Steel. His fellow editors turned down the idea on the grounds that it was hardly appropriate for a magazine devoted to literature and culture. He needed an outlet for the political writing—polemical writing—he wanted to do. James Burnham encouraged him to contribute a column to the Trotskyist Socialist Workers Party's theoretical journal, the *New International*, which Burnham co-edited with Max Shachtman. Dwight jumped at the chance, confirming, for some partisans, the argument that *PR* was also a Trotskyist publication.

Dwight worked very hard on his initial column for the *New International*. Burnham told him that the magazine had a circulation of 4,000 but that many more read it. Dwight was excited to think he would be reaching a more working-class audience. But the Trotskyists' ability to attract the workers was minuscule. With the exception of the contingent in Minneapolis, most Socialist Workers Party (SWP) members and fellow-travelers were to be found among intellectuals and academics concentrated primarily in New York. Dwight did hope to break into actual labor journalism through his friend Rose Stein, but he appreciated the fact that in writing for the *New International* he wouldn't have to water down his arguments.

His column was to be a regular critical analysis of the pundits in the mainstream bourgeois press: Walter Lippmann, Westbrook Pegler,

Dorothy Thompson, Frank Kent, David Lawrence, and Mark Sullivan. Their function, Dwight wrote in his debut column, was to make their readers feel they had a spokesman and at the same time grease the gears of the capitalist system by doing the bidding of the ruling class which hired them. It was a slashing attack on the establishment and its apologists. He dismissed Lippmann's diagnosis as infantile; Pegler, he said, proceeded on the assumption that "ignorance was a guarantee of impartiality." And all the others ran on the fuel of indignation against any collective action or government interference with the masters of the market.[4]

Dwight was pleased with the reaction his column was receiving and sent off some copies to Rose Stein, expecting praise. She replied that the pieces were well written and politically on target, but charged that Dwight was wasting his time. Why devote such efforts to "the ivory tower Marxists"? Who read the journal? True, Trotskyists were scattered all over the world, "but they don't amount to a tinker's damn, nowhere are they sufficiently influential to affect the destiny of a fly—let alone the human race."[5]

Stein urged Dwight to direct his writing more toward the masses—not because "they were good, or honest, or more moral, but because they were *the masses.*" She conceded it might be hard for Dwight to find a place in the labor press, but if he failed he could create his own journal, directed at a genuine working- and middle-class audience. Dwight was irritated by what he saw as Stein's anti-intellectualism and rejection of theory. It was "the usual American philistine attitude . . . the great weakness of American liberalism and the labor movement." Luce exploited this weakness to the hilt. The modern world was an extremely complicated place and could not be understood without "the aid of a lot of abstract and difficult theorizing."[6]

In the spring of 1939, through his Trotskyist connections, Dwight managed to land a column in the *Northwest Organizer*, the official organ of the Trotskyist-dominated Teamster/Chauffeur Joint Council in Minneapolis. Dwight's first piece was titled "Marching to War," a militantly anti-war assault on Roosevelt and his commitment to collective security and defense mobilization. It was also a Randolph Bourne–like critique of academics who were campaigning for defense and militarism. But the editor began to alter Dwight's copy because he was writing above the readership. Among others, Vincent Ray Dunne, one of the famous Dunne brothers, a leader of the Minneapolis Trotskyists and Teamsters, said Dwight's column was "much too highbrow." Since the column garnered no favorable reaction and considerable criticism, it was dropped after running less than two months.[7]

Despite the failure of the Midwestern venture, Dwight was becoming more closely associated with the New York wing of the movement led by Shachtman and Burnham. He was, in effect, their man on the editorial board of the *Partisan Review*. Dwight still refrained from joining the SWP, and he was not encouraged to do so. He had been deliberately recruited as an active fellow-traveling participant in much the same way that the Communist Party of the popular front period recruited well-known writers and celebrities of a liberal persuasion to give it a cultural and intellectual cachet. Having a distinguished Yale graduate, a former *Fortune* writer, a Wasp gentleman, grace the pages of the *New International*, along with his Princeton counterpart Burnham, meant a good deal to the Trotskyists.

Dwight's growing interest in a more militant unionism was further encouraged by a new friendship with a young Trotskyist, B. J. Widick, whom he had met through James Burnham. Widick was a tireless organizer for the Socialist Workers Party, an articulate journalist who had graduated from the University of Akron and wrote for the Akron newspapers and for the Trotskyist *Socialist Appeal*. He persuaded Dwight and Nancy to attend the first national convention of the CIO in Pittsburgh in November 1938. Dwight was developing a strong distaste for reformist labor leadership. Watching the convention confirmed all his suspicions. His notes reflect his scorn for the bland lack of class consciousness that characterized the proceedings. They also reflect his own revolutionary confidence: "Those who are not fanatically militant will go under." The CIO "shows itself to be dead and cut and dried, bureaucratized, unable to oppose fascism effectively or lead a mass movement." Nearly all the protagonists were dismissed with contempt. CIO president John L. Lewis's praise of FDR could have been written by Earl Browder; CIO vice president Sidney Hillman had a "look of intellectual hardness . . . pointed nose, sharp features, cold expression. Speaks in a harsh East Side Jewish accent," a contrast to Philip Murray, head of the Steel Workers' Organizing Committee, whose manner was "soft-spoken, easy-going, discreet."[8]

Shortly after his return, Dwight wrote the introduction to Daniel Guérin's *Fascisme et grand capital*, a Marxist analysis of the relationship of fascism to the corporate ruling class. The introduction was reproduced as a pamphlet, *Fascism and the American Scene*. It mounted a sustained attack on liberal reformers who saw fascism as a maniacal aberration instead of the logical outgrowth of corporate capitalism in decline. As in

Germany and Italy, reformist politicians and labor bureaucrats in this country were rendering the proletariat impotent, Dwight wrote. The popular front was doing its part, through its fawning collaboration with the New Deal and support of defense programs. All were helping to "tame the rank and file," and as a consequence fascist tendencies were expanding all over the country, he warned; fascism could be opposed only by militant working-class resistance.[9]

For all practical purposes, Dwight was a Trotskyist and only cheating the SWP out of its dues. He was following the line with mechanical devotion; his natural skepticism as well as his caustic wit seemed to have vanished. He was particularly disappointed when Guérin's book got no attention except in the Trotskyist press. He wrote John Chamberlain, editor of the *New Republic* and an old Yale acquaintance, asking him whether he could break the conspiracy of silence at that magazine, where, he believed, Malcolm Cowley nixed any reviews of Trotskyist books.[10]

Dwight's service to the cause, plus his putting up $500 toward the publication of Guérin's book and his pamphlet, brought him further into the fold. He was offered a weekly column in the *Socialist Appeal*, the official organ of the Socialist Workers Party.

Dwight's column took on all comers but targeted those on the liberal left, starting with Roosevelt and the "war-dealers" and working leftward through the fellow-travelers and Stalinists. Dwight was now recognized as one of the Trotskyists' official publicists. That he was also an editor of the *Partisan Review* and frequently identified as such in the Trotskyist press points to his service to the movement, which delighted in having a well-known anti-Stalinist intellectual defending their position regularly in their own press.[11]

Dwight's position did have an influence in the editorial offices of the *Partisan Review*. The peak of that influence may be seen in the publication by *PR* of the manifesto of the International Federation of Independent Revolutionary Artists. The purpose of the manifesto was to denounce the corrupting influence of Stalinism and to mobilize artists into the ranks of the opposition. Dwight was keen on getting the journal's support and the editors gave it, despite the manifesto's assertion that "true art is unable not to be revolutionary," which clashed with the modernist editors' promotion of the works of Eliot, Yeats, Pound, and James, hardly four musketeers of revolutionary socialism.

For Dwight the appeal of the Socialist Workers Party was its anti-war stance. He was convinced that the real threat to civilized life was the totalitarian wartime state. It was, as he recalled after the war, his moral revulsion against the war rather than an intellectual conviction about Marxism that drew him. This sensitivity dated back to his studies and commentaries on dictatorships while in Spain in 1935. Dwight was seriously committed to the notion that American intervention in a European war would bring a fascist totalitarianism to America. This was the line of the Socialist Workers Party at the time of the Nazi-Stalin pact and the subsequent invasion of Poland by the Germans and the Russians in the fall and winter of 1939. Dwight and Nancy joined the Socialist Workers Party in September 1939, following the German and Russian invasions. The leadership had continually discouraged their joining. James P. Cannon insisted Dwight would not be happy in the party; he may well have meant that he would not be happy with Dwight as a member. James Burnham and B. J. Widick advised against joining, arguing that it would serve no useful purpose and that he would fit uneasily into the party framework, with its discipline and service obligations. Apparently Cannon's advice only incited Dwight to join. Knowing that the party was soon to be embroiled in a donnybrook, Dwight wanted to get into the fray.[12]

At first his membership was a secret, but when it became known, some in the party or on the fringes thought it was a great coup. Harry Roskolenko saw Dwight as one of the party's great catches. Dwight, he recalled, was gifted, studious, and, coming from the ranks of Time Inc., a living representative of the obvious bankruptcy of capitalism. Dwight had the tweedy appearance of a former Ivy League inmate who had broken out of the bourgeois cage to do battle on the revolutionary turf of the Village. He attended the "Downtown Branch meetings of the party and took part in the endless meetings and debates." His "screech owl" voice was accompanied by a nervous giggle, and when he was really agitated, his somewhat stylish stammer became pronounced and led to an angry sputtering as he tried to express his point.[13]

This new circle was very important to Dwight and Nancy. They had moved in 1939, a year after Michael's birth, to a fourth-floor walk-up in a Federal-style rowhouse on East 10th Street—where Nancy Macdonald has now lived for more than fifty years. A four-room apartment with a closet kitchen and a cozy fireplace, it became a headquarters for literary radicals, a meeting place, even editorial offices. The *Partisan Review* was published out of the apartment in 1940. In moving downtown from the Upper East Side, Dwight and Nancy had moved into a different world, a world they hoped to find drastically removed from the Brearley, Exeter,

Vassar, Yale, Time Inc. orbit, a world of Jewish and foreign intellectuals, a truly cosmopolitan, international community.

It was an exciting time. They were educated, bright, imaginative, just out of their twenties and thoroughly convinced that even in the shadow of war and totalitarian destruction, there was a genuine possibility of revolution. It was a world where words were important, where the thrust of a verbal sword was noticed and made a difference, where taking a stand placed one at the heart of things, near the moving center of history. It was exhilarating, and Dwight's earlier indecision about his future course evaporated. He was intent on gaining a sure grasp of Marxist revolutionary theory so that he could serve in the war against capitalist as well as Stalinist destruction and in defense of a compassionate, civilizing culture.

For all his enthusiasm, however, Dwight must have known that he had come to the movement at the moment it was on the verge of a factional struggle that would lead to a fatal division in its tiny ranks. Accompanying his initial article in the *New International* had been his letter to the editors (Burnham and Shachtman), along with another by Victor Serge, criticizing Trotsky's role in the crushing of the Kronstadt rebellion in 1921, which is often seen as the first great betrayal of the socialist revolution and the beginning of a dictatorial tyranny. Because Dwight had come to the movement as an avowed anti-Stalinist, he was fascinated with all discussions of the root causes of the Stalinist counterrevolution. In his initial Kronstadt criticism he did not debate the tactics of Trotsky or his role in putting down the revolt of the Kronstadt sailors, but he questioned Trotsky's continued refusal to review the circumstances, his shutting off of sincere inquiry. Trotsky, he argued, insisted on attributing Stalinism to economic conditions in Russia and its technical and industrial backwardness. Dwight did not reject this as a partial explanation but wondered whether a "contributory cause may still be found in certain weaknesses in Bolshevist political theory." He felt that Trotsky's defensiveness, his abuse of his critics as part of a "variegated fraternity" engaged in a "charlatan campaign" to slander his anti-Stalinist opposition, was hardly conducive to serious political and historical inquiry designed to avoid the mistakes of the past. Trotsky was incapable of imagining any critic of Kronstadt who didn't have an ax to grind. Dwight was quick to compare such a stance with that of Stalinists toward anyone who questioned the Moscow trials.[14]

Dwight entertained Trotsky's notion that such a rebellion had to be put down because it would only strengthen the counterrevolutionary army already fighting the new socialist republic, but he questioned the brutal mass executions that occurred months after the revolt. He relentlessly brought up the Bolshevik centralization of power and wondered whether mistaken judgments might not have sprung from the very nature of the Bolshevik political organization, which "concentrates power in the hands of a small group of politicians so well insulated (by a hierarchical, bureaucratic, party apparatus) against pressure from the masses that they don't respond to the needs of the masses until too late." Dwight's heresy did not stop there. He conceded that the White counterrevolutionaries would exploit division in the revolutionary ranks, but he asked, "Are not the dangers of an air-tight dictatorship, insulated against mass pressure, even greater?"[15]

There is something bizarre and revealing about Dwight's initial introduction to the Socialist Workers Party. Just at the moment when the factional struggle was getting under way, he became a Trotskyist columnist on the one hand and raised all the major criticisms directed at the movement on the other. Thus a year before joining the party he exposed his deep anti-statism and latent anarchism. It was clear to Trotsky and surely to his hard-core followers that Dwight Macdonald would never be one of them. Even among those who would soon mount a challenge to Trotsky, Dwight was seen as an uncontrollable maverick, unpredictable and irresponsible. Felix Morrow recalled Dwight's amiable charm and his journalistic ability, even though he knew that Dwight looked on him with disdain as a bureaucratic apparatchik. Morrow said that he and others in the Socialist Workers Party knew Dwight could prove useful, but knew he was not a serious, dedicated revolutionary. He simply was "not a Bolshevik." He was a contentious, exhibitionistic intellectual with a flair for the dramatic phrase and gesture, but hardly a candidate for the discipline demanded by party organization.[16]

If Dwight had managed to maintain some of his skepticism concerning the party and its line, he was very serious about its opposition to the war, which he viewed as the duty of an intellectual. This must explain how he could write for the Trotskyist press nearly a year and a half prior to his actually joining the Socialist Workers Party. The irony is that just at the moment he made this decision, the Old Man was fashioning a position that demanded unconditional support of the Soviet Union and an apology for their invasion of Poland and soon to be invasion of Finland. These events, on top of the pact with Hitler, only encouraged a free-for-all questioning of basic Marxist fundamentals on the part of a growing number of American radicals. The "dictatorship of the proletariat," "a degenerate workers' State," "dialectical materialism," the potential of a working-class revolution—all came under the scrutiny of a rapidly deradicalized intellectual community. In the welter of all this, the Socialist Workers Party was still anti-capitalist and anti-war, and for a short while that was enough for Dwight and Nancy. The party may have admitted a Wasp literary figure into their ranks to lend prestige to their increasingly vulnerable position on the war and foreign policy. But they had also taken on an inveterate gadfly, an uncontrollable maverick, a loose cannon that only added to the animosity within their already divided ranks.

Dwight was involved in the factional struggle from the beginning and sided with the minority because of its opposition to Trotsky's demand for unconditional defense of the Soviet Union. Dwight demanded open discussions and debate in the party publications as well as at branch meetings. He encouraged this in the name of achieving some democratic consensus. In his *Socialist Appeal* column he reproduced bizarre passages from the capitalist and Stalinist press and let them speak for themselves. On one occasion he printed the Communist Party line defending the invasion of Finland to show the degree to which Trotsky himself was advocating the Stalinist position. Felix Morrow, the majority editor, deleted the offending material. Dwight had an immediate allergic response to the editing of any of his material. He would demand to know why he had not been informed and what the reason was. If it was not for space, he would not allow material that had been politically censored to appear with his name on it. Shades of his dispute with *Fortune*. The editors of the *Appeal* were promptly supported by Albert Goldman, speaking for the majority-dominated political committee. The *Socialist Appeal* was not "a forum for columnists." It was an organ of agitation and propaganda for the program of the Socialist Workers Party and the Fourth International. Dwight's column served no agitational or propaganda purpose. Trotsky himself joined the fray to accuse Dwight of being unable to understand the nature of Stalinism or the Soviet Union. In this lecture Dwight was elevated to the first rank of oppositionists when Trotsky referred to him as a "Macdonaldist." He suggested that Dwight, who was so quick to see deceit and manipulation in others, should analyze himself and "recognize that he had uncovered his own subconscious."[17]

Dwight returned to his typewriter and wrote nearly a dozen pages of rebuttal, arguing that Trotsky's personal abuse, his reference to Dwight's apprenticeship, his interest in psychoanalysis were all irrelevant: A scribbled marginal note concluded: "As a climax to this hodge-podge of non-sequiturs and rather laborious raillery you uncover the latent threat to the party : 'Macdonaldism.'" There is no question that Dwight took more than a little pleasure in the distinction. But this was not so of his colleagues, who refused to print any of this rebuttal.[18]

The formal split came during the convention of the Socialist Workers Party held in Manhattan in mid-April 1940. The minority faction immediately established their own organization, the Workers Party. They revived the old *Labor Action* magazine to replace the *Socialist Appeal* and they kept control of the *New International*. For the next few weeks after the founding of the Workers Party, Dwight threw himself into the editorial work of getting *Labor Action* going, and he did most of the technical work on the *New International*, since James Burnham was withdrawing from all party activity. Burnham's departure was a real blow, as he had been the theoretical leader of the minority faction along with Max Shachtman, and his developing challenge to Trotsky's analysis of the Soviet state, as well as his repudiation of the dialectic, had an important influence on Shachtman and Dwight as well.[19]

The new Workers Party was not two months old before Dwight was raising hell at downtown branch meetings. He proposed a series of resolutions demanding that the party take a stronger and clearer stand in its denunciation of Soviet foreign policy. He proposed censuring the leadership for failing to react adequately to breaking events on the international scene, and he began work on what he hoped would be a book-length analysis of the nature of the fascist Nazi state, an analysis that rejected Marxist orthodoxy. He fully expected this work to appear first in the Workers Party's publications, but it was turned down with growing hostility from Shachtman, Martin Abern, and other leaders. To make matters even worse, Dwight published an introductory piece in the *Partisan Review* that bluntly criticized Trotsky and his obedient followers for not responding to the new conditions the war had created.

Dwight's article, "National Defense: The Case for Socialism," was prompted by the German blitzkrieg and subsequent fall of France and the occupation of Paris. These events were a devastating blow to all of Western culture, for Paris had been its heart. The speed with which the Reichswehr had crossed the Belgian border into the French capital demanded an entirely new evaluation of this destructive machine. The thrust of

Dwight's review of events was that they could not be explained nor analyzed within a Marxist framework. That inflexible structure didn't allow
its practitioners to recognize a brand-new, mysterious phenomenon. "The
more orthodox Marxists are hesitant to probe too deeply, fearful of finding
in fascism some strange new monster which will not fit into their rigid
categories."

To traditional Marxists, decaying capitalism seemed ripe for socialism.
But the tragedy of the age was the working class's failure to overthrow
capitalism and bring into being the long overdue socialist society.
"Unable to give birth to a natural child, history has whelped the monster,
fascism," which, Dwight wrote, was a hybrid system combining the centralized state power and conscious economic planning of socialism with
some of the more hideous features of decaying capitalism. It was a "black
socialism," a total enigma to orthodox Marxists.[20]

Then Dwight launched into his recently acquired notion that fascism
was not the last stage of a decaying capitalism, but a uniquely new form
of society. Obviously influenced by the Italian ex-Trotskyist Bruno Rizzi,
whose book *The Bureaucratization of the World* had an important impact
on James Burnham and others who were challenging the fundamental
assumptions of Marxism, Dwight was attempting to translate the theory
into a concrete analysis of changing world affairs and to make a defense
of a revised socialism that would confront directly the realities of the present situation. He insisted that it was now clear that the traditional Marxist predictions had proved wrong again. "The gravediggers of capitalism
which Marx and Engels saw" weren't wearing "the overalls of the working
class but the uniform of the Reichswehr."[21]

Dwight's solution remained a revolutionary socialism. There were only
two alternatives: Fight fascism with fascism or fight it with a democratic
socialism. But the greatest illusion was to believe that capitalism could
defend what remained of American democracy by democratic means.
Nazism, he insisted, could be conquered only by another fascist state—
that is, by a total regime imposed by the present ruling class or by a different kind of total state: a socialist working class. The latter would be
more effective, he maintained, because it would not have to worry about
the expropriation of private property, nor maintain large forces of repression, because the ruling class was a small minority.

Surely his comrades could take that hard line, but his conclusion was
beyond the pale. He agreed with those who argued that the rise of a revolutionary socialist movement was unlikely, given the fact that the working

class everywhere was in retreat, without leadership, corrupt and discredited in the Second (socialist) and Third (Stalinist) Internationals. Even his own Trotskyist movement was split by sectarian quarrels. Worse, "Marxism itself, the very fountainhead of all revolutionary science, had been shaken by the failure of its disciples to provide adequate answers in practice and theoretical understanding to the historical developments of the past two decades."[22]

Dwight insisted that the current questioning of Marxist principles was justified, given "the unbroken series of major disasters" the movement had suffered in the last two decades. All of its fundamentals would have to be reexamined with a "cold and skeptical eye." The "revolutionary optimism" expressed in some quarters on the left hardly served the cause of socialism. The only way a genuine intellectual could help the working-class movement was to meet "the stormy years ahead with both skepticism and devotion—a skepticism towards all theories, governments, and social systems, devotion to the revolutionary fight of the masses—only then can we justify ourselves as intellectuals."[23]

There was a Sisyphean quality to Dwight's position that anticipated the vision of Albert Camus, which, ironically, Isaac Deutscher also found in Trotsky: a hardheaded acceptance of failure but a devotion to continue the battle. Dwight accused many of his fellow intellectuals of giving up the struggle for socialism and accepting the "lesser evil" of fascism, a form of society they knew to be "historically bankrupt and without health and without hope."[24]

In 1936 Dwight had wanted out of the Luce organization and had deliberately provoked a confrontation by insisting on introducing Leninist theory into his analyses of the contradictions and failures of the steel corporations and American capitalism. Now he was deliberately going out of his way to challenge the leadership of the Trotskyist sect of which he was a member. He took pleasure in the challenge itself, and he knew very well that they would not let it go unanswered.

C. L. R. James was one of the most attractive, articulate, and learned of the international Trotskyists. A tall, thin man with long, artistic-looking hands that were always in motion, James was a scholar from the West Indies who had joined the Trotsky movement in London. He had written two important books, *World Revolution* (1937), and his seminal work, *The Black Jacobins: Toussaint L'Ouverture and the San Domingo Revolution* (1938). James was a brilliant speaker, perhaps more compelling and impressive than either James Cannon or Max Shachtman. But he was not appreciated by some of the American Trotskyists, who saw his manner as pretentious, intellectual, and aloof. They recognized his intellectual credentials, however, and after a talk he gave before the *Partisan Review* crowd, one member told Shachtman to move over because he had met his master. Dwight frequently quarreled with James, whom he thought "a bit nuts." Hegel was James's god, and Dwight, with typical ingenuousness, told him he could not understand Hegel. To James that was like a novitiate conceding he couldn't understand the virgin birth; one simply could not be a Marxist revolutionary without grasping its philosophical underpinnings. James, writing under his party name Jimmy Johnson, dealt with Dwight initially in the *New International* in an unsigned editorial that did not mention Dwight by name but quoted his "National Defense" piece as typical of the wayward bourgeois who were corrupting the revolutionary movement and were

> rending the heavens with their wails over the defeat of the French army and the extinction of French culture. . . . The fundamental ignorance and stupidity of these learned chatterers are without bounds. The capture or non capture of Paris does not signify either the continuance or extinction of a culture. . . . Intellectuals who moan and slobber over the capture of Paris show no knowledge of either history or culture, but a sure instinct for hiding in the steadily diminishing crevices of bankrupt bourgeois democracy.

James dismissed this "counter-revolutionary squawking" and went on to give a detailed history of the revolutionary class struggle in France, predicting that it would rise again.[25]

Dwight was incensed at being thus bracketed with the proponents of bourgeois democracy, and anonymously. He wrote an endless rebuttal that suggests he was losing his appreciation for brevity, a failing he continually noted in his party opponents. He demanded that his rebuttal be printed in the party's internal bulletin. When it finally appeared, it was accompanied by a letter from James justifying his editorial policy and a statement by Shachtman supporting James. Dwight was exactly where he relished being, in the minority of a minority. Trotsky, Dwight argued, was a great revolutionary because of his concern for culture. James expressed the sort of "philistine proletarianism" that Trotsky had so brilliantly attacked in *Literature and Revolution*. In a footnote, Dwight derided James's assertion that it made not an iota of difference whether the Third

Reich or the Second Republic ruled Paris. In an almost plaintive cry of outrage, Dwight's inherently cultural politics came to the surface:

> What has been the attitude of Hitler and the Nazis to the sort of avant-garde art and letters which has centered in Paris? . . . Do you think that in a Nazified Europe, Picasso and Léger will continue to paint, James Joyce and Gertrude Stein will continue to write, Stravinsky will continue to compose music, Le Corbusier will continue to build houses?

It was the obligation of Marxist revolutionaries "to fight for the preservation of the best of bourgeois civilization as well as for the raising of that civilization to a higher level."

This entire episode illustrated for Dwight the "serious degeneration in the intellectual life" of the Workers Party since the split. The charges and countercharges continued over several issues of the bulletin. Shachtman denounced Dwight and said that the Workers Party had no intention of scrapping its principles the moment some "half-baked notion exploded in its midst."[26]

The last articles and letters Leon Trotsky wrote before he was bludgeoned to death in August 1940 by an assassin wielding an ice ax were largely in response to Dwight's specific criticisms of Marxism and the Socialist Workers, and the sentiments he represented. Trotsky had been sharply irritated by what he referred to as Dwight's "miserable article" and wrote his trusty legal counsel Al Goldman describing Dwight as a plagiarist—simply "Burnham's orphan." He was particularly irritated by Dwight's apparent belief that his ideas were his own discoveries and not common currency among critical theorists. Trotsky's last article, still in his dictaphone at the time of his death, began by describing Dwight's article as "very muddled and stupid," and then he went back over his own previous analyses of Soviet foreign and domestic policy, ridiculing Dwight's charge that he viewed fascism as simply a repetition of Bonapartism. Trotsky was contemptuous of Dwight's call for a "skepticism towards all theories, governments and social systems; devotion to the revolutionary fight of the masses." Skepticism, Trotsky charged, was simply a "preparation for personal desertion." Dwight was not like Burnham, who Trotsky considered an "intellectual snob . . . but a bit stupid." Trotsky predicted that Dwight would abandon the party just as Burnham had, "but possibly because he is a little lazier it will come later."[27]

Trotsky's assault on Dwight's mental capacity was to become somewhat of a legend in later years, turning up in both hostile and friendly portraits of Dwight, usually in an amusing epigrammatic form. Trotsky is alleged to have said or written to an associate, "Every man has a right to be stupid on occasion but comrade Macdonald abuses it." There is some reason to suspect that Dwight himself circulated the alleged story—he frequently referred to it in lectures in later years, and none who quote it ever cite a direct source. In fact, Dwight may well have used the story as a kind of ploy to cancel out the initial shock, even hurt, of the insult and at the same time claim a substantial place as a recognized participant in these weighty dialectical struggles.[28]

Despite the growing rancor between Dwight and the party leadership, he continued to be active, attending meetings, giving lectures, and participating in demonstrations. In late August after Trotsky's assassination, Dwight and several others picketed the Soviet Consulate. Denouncing Stalin as the murderer of their comrade and carrying placards that read "Down with Stalin, partner of Hitler," they raised a considerable ruckus in the dignified East Side neighborhood. Before long, police swooped down on the line and arrested Dwight and four of his comrades. The judge declared their protest unlawful, saying "The streets cannot be made the battleground of political hatreds," and fined them $25. This was Dwight's first arrest and a point of pride as a measure of his commitment, if not to the ideology of the Old Man at least to the radical movement.[29]

Trotsky's murder was a shocking and demoralizing event. Dwight had read his autobiography in the mid-thirties at Nancy's urging, and then many of his other works. He was drawn to Trotsky as the model of the activist intellectual. As Dwight's son Michael recalls, Trotsky may have been one of Dwight's "fantasy characters," a person whose experiences he would have liked to have lived, who had stimulated his imagination and his life. In Trotsky, Dwight admired the fusion of the literary man and the revolutionist. The radical political intensity of these years in New York prompted Dwight to fancy himself an American Trotsky. Because Trotsky had been in exile for so many years, the fight he had waged was largely with his pen. Dwight imagined a similar role for himself: the writer whose work would make a difference, a catalyst for change, at the heart of important events. Trotsky, the working intellectual journalist, was in Dwight's mind a model to be emulated.[30]

In a piece for the *Partisan Review*, "Trotsky Is Dead" (modestly subtitled "An Attempt at an Appreciation"), Dwight devoted much space to Trotsky's intellectuality, his appreciation of good books, his scholarly, reflective bent. Trotsky, he reported with fondness, loved "turning the pages of a freshly printed volume." He was a kindred soul. The man had style, and much of what he wrote was stylistically brilliant. He had a marvelous gift for witty characterization. Sometimes his very brilliance distracted attention from the development of his thought. But the structure was usually there—a beginning, a middle, and an end. Trotsky knew how to write, and for Dwight, such a talent covered a multitude of sins. But it was more than Trotsky's intellectual bookishness and his professionalism that impressed Dwight. Trotsky represented to him what an intellectual should be: He turned his ideas into action. He did things. For Dwight, he was a great teacher. He had taught a lesson that was most important at this moment in the twentieth century: "The fabric of our capitalist society is so rotten it cannot be patched up . . . in world revolution, 'permanent revolution' by the oppressed masses, lies the only hope for mankind."[31]

Although the piece ended by proclaiming that Trotsky's career had helped Dwight to believe "even in times like these in the dignity and capacity of the human race," he could not refrain from raising the old doubts, the current controversies and disputes that were soon to drive him out of the Workers Party. Here, in this funereal oration, he returned to the issue of the bureaucratic and undemocratic features of the party's organization, which could be traced to Lenin. He wished that Trotsky had not accepted so uncritically Lenin's insistence on the vanguard party. Trotsky's pre-1917 criticism of the danger of such undemocratic methods sounded terribly prophetic in 1941. Once again Dwight took the opportunity to cite Trotsky's failure to cope "with the basic changes in politics and economics which have come about since the last war," the inadequacy of his theories, his refusal to reshape basic theory, his inflexible dogma. Again Dwight repudiated the notion of the "degenerate workers' state" and summoned up the "new form of class rule . . . call it bureaucratic collectivism." Dwight was defiantly heretical, publishing again in an offending nonparty publication. He was obviously building the platform from which to deliver his resignation from the Workers Party.[32]

Despite his continued irritation with party discipline, Dwight served as Max Shachtman's publicity director in his campaign for the House of Representatives from the Bronx. Dwight had no quarrel with the immediate political program of the Workers Party, its anti-war stand and its socialist domestic program. But the congressional campaign did not heal the rift between Dwight and his colleagues over party organization, hierarchy, and discipline.[33]

Although Dwight and Nancy were deeply involved in party activity and editing and managing the *Partisan Review*, they always yearned to translate their intellectual and political work into activities that affected people. Because the Workers Party was riddled with factions, Dwight found the *Partisan Review* could be used to create a network of radical consciousness and to promote projects in the cause of fighting Stalinism and fascist repression. In the same issue that carried his appreciation of Trotsky was one by Victor Serge, who, despite his praise for the Old Man, was on the list of backsliders for his criticisms of the undemocratic tendencies in the party. Serge was an old Bolshevik who had been imprisoned by Stalin, then exiled, and was now on the run in France, as the Nazi occupation was imminent. He had written to Dwight asking for help and thus began a major undertaking of both Nancy and Dwight, the effort to protect non-Communist European refugees. The Macdonalds sent Serge money and made valiant efforts to get him, his wife, and his son admitted to the United States as refugees from both the Nazis and the Stalinists. They pointed out that Serge was an eloquent anti-Stalinist, but the State Department did not make such fine discriminations: his records showed that he was a Communist.[34]

Dwight and Nancy were not up against just the Washington bureaucracy; they encountered incompetence, deliberate stalling, and political maneuvering on the part of established refugee committees staffed by liberals, Communists, and fellow-travelers who had little interest in renegade Trotskyists. The Stalinists wanted Trotskyists persecuted and rounded up and were not averse to collaborating with bourgeois governments in that effort. Nancy found the "whole business disgraceful, disgusting, maddening." Nothing but runarounds, delays, botched applications without end. Nevertheless, with extraordinary persistence and effort, they managed to arrange and finance Serge's escape with his family to Mexico. With this invaluable experience they established the *Partisan Review* Fund for European Writers and Artists that operated out of the magazine's office at 45 Astor Place and from their own apartment.[35]

Despite all these activities, perhaps because of some of them, the leadership of the Workers Party remained irritated with and suspicious of the Macdonalds. They were particularly annoyed by Dwight's willingness to transfer his allegiance to the *Partisan Review* when he ran into roadblocks in the party. In November 1940, Max Shachtman demanded that in the future Dwight submit all political articles for the nonparty press to the party's political committee for prior approval. That led to another inter-

minable and heated exchange between Dwight and various members of
the party. As for his own party writings, they were delayed, rejected, and
ignored, which prompted him to turn even more frequently to the *Partisan
Review*.[36]

What it came down to was that Dwight could knuckle under and abandon his dissenting position, he could organize another faction, or he could
resign from the party. The first was out of the question, the second did not
interest him; quitting was the only alternative. Dwight still supported the
party's political program, but since the political committee would not
accept his demands to hold a debate on his case, he submitted his resignation in July 1941, after some soul-searching with friends and associates
in and out of the party. The party had offered excitement and comradeship, and Dwight had felt part of a tight-knit community, despite the rancor and infighting. He expressed regret, because he continued to see the
Workers Party as the only organization that supported his third-camp
position, rejecting capitalist imperialism, fascism, and leftist totalitarianism as embodied in the Stalinist regime. But he could not accept the
inflexible and undemocratic structure and discipline. Characteristically,
at the end of his letter of resignation he could not resist raising the issue
of whether the Leninist party structure bore the responsibility for the
"terrible degeneration of the Russian Revolution." Trotsky had played a
role in the development of that authoritarianism; ever since the split, a
dictatorial atmosphere had enveloped the leadership of his own faction.
Dwight had gone along initially because he had accepted a "sort of lesser-
evil argument." But now he could no longer take the party seriously as a
"revolutionary movement." Its organizational defects were stepping-
stones to Stalinism. The first step for genuine socialist revolutionary
action was to break with the organizational tradition of the Trotsky movement.[37]

5

From Trotsky to Bourne:
War Is the Health of the State

In THE SPRING OF 1939, AS THE EUROPEAN WAR WAS
approaching, Dwight had published a piece in the *Partisan Review* titled
"War and the Intellectuals: Act Two," in which he took his fellow intellectuals to task for gravitating toward the war machine. It is not surprising
that he should have summoned up the vision of that "little sparrow-like
man," the hunchbacked Randolph Bourne. Since his untimely death in
the flu epidemic of 1918, Bourne had become a legend for his "majority
of one" opposition to the First World War. Bourne's line "War is the health
of the State" had an anarchist ring that remains irresistible to those
engaged in individual rebellion against the establishment. Bourne's anarchical distrust of the State struck a responsive chord in Dwight, despite
his own wildly eclectic support of a socialist revolution.[1]

As the likelihood of American involvement in the war increased, it was
entirely appropriate that Dwight chose Bourne's text as the message and
the man himself as a role model. Bourne had opposed the war in
1916–17, when to do so was considered obscenely unpatriotic. He had
sacrificed his writing career to his political commitments, and he had
taken for his targets not the militarists and assorted bellicose reactionaries, but the men of alleged reason and reflection, the liberal intellectuals:
John Dewey, Walter Lippmann, Herbert Croly, and a host of other pro-war
liberals. Croly's *New Republic* magazine had used the big guns of Ameri-

can pragmatism to rationalize the First World War, claiming that American participation would lead to progress and enlightenment.[2]

Dwight singled out a similar cluster of intellectual journalists to attack for abandoning their principles and obligations as intellectuals. Once again, as in 1917, the *New Republic* and Walter Lippmann were included; Lewis Mumford, Archibald MacLeish, and Dorothy Thompson were added. Most horrifying of all was the *Nation*, whose anti-war heritage was being subverted by what Dwight viewed as the mindless sentimentality of its new editor, Freda Kirchwey.[3]

Dwight's piece differed from Bourne's in that it was heavily laden with his recently acquired Marxist terminology. (Dwight's wonderfully lucid, cliché-free prose style took a marked dip during his Trotskyist period.) Where Bourne had railed against an expedient pragmatism and a virulent nationalism, Dwight added, with emphasis, monopoly capitalism. Now that the Senate investigations of the thirties had informed Americans of the role the J. P. Morgan firm had played in influencing U.S. intervention in 1917, Dwight argued, one might have expected even the congenitally idealistic liberal intellectuals to have suspected a "very materialistic nigger in the woodpile" of the current war. With that dubious caveat, Dwight concentrated on the absurdity of contemporary intellectuals presenting themselves as the saviors of Western Civilization from the scourge of fascism. He mocked the idealistic war aims of the intellectuals while pointing to the materialistic and imperialistic goals of the government and Wall Street. Even if one granted the good intentions of Roosevelt and his supporters, the laws of motion of monopoly capitalism worked themselves out with brutal disregard for intentions, Dwight wrote. Already as the "War for Democracy" propaganda was increasing, so was the erosion of democracy. More battleships, lower relief standards. It seemed, Dwight noted with irony, that the "peculiar function of the intellectuals . . . is to idealize imperialist wars when they come . . . and debunk them after they are over."[4]

Dwight's resurrection of Bourne's philosophy failed, partly because it lacked Bourne's critical edge; Dwight was still hewing to the line of the Workers Party, even to the point of giving mild lip service to the "Workers' State" analysis of the Soviet Union. Dwight's piece offered an apocalyptic vision of the catastrophe that would strike unless the masses rose up in resistance:

The next war will be an orgy of slaughter and destruction compared to which the last will seem idyllic. If the masses and their leaders fail to take advantage of the difficulties of their rulers, if they fail to rise up and wipe out forever the whole bloodstained system, the future is black indeed.[5]

There is no reason to doubt the sincerity of Dwight's dire convictions, but it is difficult to accept his hope for a socialist revolution when little evidence of such a possibility existed anywhere in the world. On the contrary, the likelihood diminished with each day. Holding out such hope must have seemed the only way he could maintain his criticism of the "bloodstained system" of capitalism. To settle for half a loaf, supporting imperialistic Britain and ultimately the United States in a war against the fascist or Stalinist variety of totalitarianism, was to give up the fight and make an alliance with the forces he had been battling since the mid-1930s. He was not prepared to throw in the towel, and he had no stomach for the pragmatic tacticians who dictated the policy of the liberal establishment or even of its radical opponents. He was literally calling on faith—an impossible position for a man of deeply skeptical inclinations.

The year between the Stalin–Nazi pact of August 1939 and the assassination of Trotsky in August 1940 had been a terrible one for radicals everywhere. In the spring of 1940 came the invasion and quick fall of France; with the establishment of the fascist government of Pierre Laval, the entire fabric of Western culture was unraveling. Desperately hanging on to his hope for a socialist revolution, Dwight had written that Trotsky was "the one man still living whose name and prestige could have become a rallying point for mass revolution in almost any part of the world." His brutal murder had "sealed the ultimate corruption of the Stalinist regime."[6]

As matters grew more menacing day by day, Dwight, like the Dutch boy with his finger in the dike, was trying to plug the leaks in the intellectual resistance. As one colleague after another not only stepped up their assaults on Marxism but abandoned a radical critique of the capitalist democracies as the only potential source of resistance to Hitler's marauding armies, Dwight became increasingly shrill and judgmental. During the spring of 1940 Stephen Spender, the leftist British poet, contributed a poignant, personal piece to the *Partisan Review*, in which he recalled the marvelously free, libertarian life he had led in the 1920s in Weimar Germany, a life that had been destroyed by the heavy-handed monstrosity of Nazism. Now, full of foreboding and despair and expressing a genuine fear that it might not be long before the British Isles were invaded, Spender declared that under the circumstances he was "on the side of the Chamberlain system against Fascism." He conceded that his motive for swallowing a system he had always been critical of was his selfish fear of "being regimented and losing my personal freedom of action. I carry this feeling too far; in fact, I must admit I carry it to the point of hysteria—i.e., the point where I would really fight." The effectiveness of his confession lay in its concreteness. Spender was not theorizing; he was simply stating his real fears.[7]

One might have expected Dwight to be moved by Spender's tone and infectious honesty. On the contrary, he excoriated Spender's "either/or position" as a classic position liberals struck whenever they felt compelled to justify English or American imperialism in its struggle to crush its German counterpart. Dwight's tendency to equate the capitalist democracies with the Nazi regime was among the more provocative and extreme aspects of his politics and one that infuriated some of his opponents. He did praise Spender for his candor in acknowledging the selfishness of his motives. Most of the bourgeois liberals preferred to rationalize their sellout as a last-ditch defense of the masses against fascist slavery. Dwight scornfully agreed that this "upper-class literary man" might be freer in a Chamberlain system than in a fascist one, but what about a "cook or a bus driver or a coal miner?"

Then came the radical lecture. Spender failed to understand that the freedom he valued was "a class commodity." Spender, who, Dwight noted, had once been an apologist for Stalinism, had now lost his faith and was in full retreat back to an innocuous liberal base, which he had not "left very far behind in actuality." Dwight had no quarrel with Spender's dismissal of pacifism as ineffective and a help to the other side. What incensed him was that this man who claimed to be on the left refused to consider the only real alternative, revolutionary socialism. That was "the only way out of the nightmare into which our age is descending." He raised the hope of a spreading apathy among the masses who were not happy with alternatives offered by either side. "Tomorrow they may find a common revolutionary destiny in that 'Third Camp' whose interests lie with neither of the warring imperialist camps."[8]

Dwight's rhetorical faith in a revolutionary uprising of the masses was (on the surface, at least) religious in its intensity, and his contempt for the backsliding nonbeliever was similar to Trotsky's contempt of Dwight's earlier insistence on "skepticism toward all theories, governments and social systems." The otherworldly belief in a revolution was apparently exempt. It was a peculiarly purist argument that grew increasingly shrill and seemed bizarre, given Dwight's own skeptical nature and his talent for describing concrete conditions. His reflections on the masses in the mid-1930s hardly suggest an unwavering belief in their revolutionary potential. Dwight had developed an apocalyptic vision he would normally have scorned in others.

Spender was quick to note the religiosity of Dwight's criticism. Dwight's refusal to support Churchill as "the lesser evil" and thereby leave room for progressive change aided the cause of a Hitler victory, Spender wrote in a letter to the editor of *PR*. "It's as much as if Christians were to say, 'We cannot support the war unless angels armed with swords swoop down from the sky and destroy the Nazis.'" With bitter irony, Spender feigned humility in the presence of "such an exalted attitude of mind" but in the same breath he cast doubt on its reason. Spender stated that Dwight's repudiation of support for Roosevelt or Churchill, while offering no evidence that a socialist revolution was in the offing, implied that the world was to be punished for not following the editorial policy of the *Partisan Review*. According to Spender, Dwight was advocating a "worse is better" stance, similar to that of the Communist Party in Germany in 1932, hoping that if things didn't go his way the worst would happen. "The patterns of history were apparently not to the liking of Mr. Macdonald, but since they would never be to the liking of any of us . . . we have to choose the lesser evil—or else join the hopeless angels."[9]

The editor always has the last word. Dwight's haughty reply accused Spender of moral and psychological arguments, whereas his own judgment of the situation was arrived at through the process of "objective scientific analysis." Dwight went on to repeat his assertion that a bourgeois capitalist regime couldn't confront fascism successfully and then to remark that Churchill was no different from the Chamberlain Munich men except for his keener awareness of the threat offered by Hitler as another capitalist imperialist competitor. Churchill's speeches as late as 1938 revealed that he "has nothing much against Hitlerism as a social system," Dwight wrote. As for reforming a Churchillian war administration, it was an absurdity, Dwight argued, because the British ruling class would not allow their properties to be socialized without a bitter armed struggle.[10]

Dwight insisted that "lesser evilism" had a history of defeat in Italy, France, and most graphically in Spain, which was another example of the "fatal results of compromising the revolutionary struggle of the masses by giving political support to a democratic capitalist government as a lesser evil to fascism." Revolutionary socialism was the only alternative. If it was not on the immediate horizon, Dwight still was certain of the inevitable breakdown of the present capitalist system in the "coming years" and maintained that the left would have to be prepared to take advantage of revolutionary situations when they did arise.[11]

Dwight was the dominant architect of whatever political position the *Partisan Review* had taken since its resurrection as an independent journal, but his relationships with his fellow editors and even with some of his old friends were difficult as he tried valiantly to hold back the flood tide of deradicalization that was quickening its pace as the war in Europe grew more devastating and threatening to Americans. He was particularly anxious to find a way to gain greater control over the direction of *PR* and limit the influence of his more cautious associates.

In the summer of 1940 he had several discussions with his friend and the magazine's long-suffering angel George L. K. Morris. At first Dwight thought of simply abandoning the project and beginning a magazine of his own. But then he decided he and Nancy could take over *PR* and transform it. He justifiably claimed that they did most of the work in getting the magazine out. Dwight could tolerate co-founder Philip Rahv, for whom he had some respect, if not affection. Rahv's ideas for the magazine were usually practical, and even his "negative Russian pessimism" could be overcome and compensated for by packing the editorial board, which was the crux of Dwight's plan. He proposed to Morris that a way be found to drop Fred Dupee and co-founder William Phillips. Dwight had developed a contempt for Phillips, who, he charged, made no important contribution to the magazine and didn't pull his weight. Dupee's interests were decidedly academic, and, Dwight argued, he contributed little to the journal's vitality or perspective. Being cautious and without serious political commitments or interests, both Dupee and Phillips tended to give their votes to Rahv whenever a conflict arose. Despite his complaining to Morris, Dwight loved arguments, but he didn't like to lose them.[12]

Dwight proposed Clement Greenberg, a new and flamboyant art and cultural critic, and Harold Rosenberg, an equally acerbic critic of art and culture, as replacements. Both were political activists who had been close to the Trotskyist movement. Their main advantages over Dupee and Phillips were that both were brilliant and they could write.[13]

Along with his scheme to dump the two laggards, Dwight presented to

Morris a detailed plan of transformation. He proposed to promote more young and unestablished writers, to publish more experimental prose and poetry, to make much greater contact with readers through forums, and to establish a more receptive letters column. Dwight insisted that the magazine adopt a political line of "uncompromising and caustic criticism of existing social and political systems from the viewpoint of working-class socialism." At the same time, it should be "open-minded" and "unorthodox" and maintain a receptivity to "new ways of fighting fascism from a left-wing point of view."[14]

Morris was not as pliant as Dwight had hoped. He was not happy about undertaking the odious task of "liquidating" Dupee and Phillips, and he had little enthusiasm for Clement Greenberg. Even Harold Rosenberg was not excited about the prospect of working with Greenberg. He considered him a friend but wondered: "Isn't he crotchety, eccentric in judgment, a trivia magnifier, a time-waster?"[15]

Dwight was obviously miffed by Morris's reluctance to make these changes. He again argued for the infusion of new blood in the form of Greenberg and Rosenberg. That was the central issue. Dwight felt his present colleagues were staid, cautious, circumspect, averse to taking risks or strong stands on issues. And they invariably opposed his own ideas. He insisted it was stifling the life of the magazine to drag along the deadwood of Rahv, Phillips, and Dupee. Dwight ended his lament with the threat that he was "unwilling to take on another year . . . as [the] sparkplug in the outfit—of having to plan issues and propose editorial ideas to three colleagues who now seem to be pretty much out of touch with the literary-political field." His co-editors were so timid, he wrote, that they were afraid of "being laughed at if they oppose established opinions." If he stayed on, Dwight warned, Morris could expect much less time from him. He would edit and write only an occasional article. And he insisted that *PR* get a bona fide office; it had of late been published out of his cramped apartment and he had other things to do.[16]

Morris didn't budge. He had heard these complaints and warnings before, and Dwight had not followed through on his threats. The struggles went on, but were relieved a bit when Fred Dupee left for a teaching post at Bard and Dwight managed to get Greenberg to replace him, thus giving him a vote with which to confront "the boys."

The editors announced in the issue of September–October 1940 that come the first of the year, the magazine would change its name. In his proposals for the change, Dwight had listed several possibilities: *Critic*, *Rebel*, *Culture*, and *The Forties*. The first three reflected Dwight's notion of the role the magazine should take. Apparently no consensus could be reached and the peculiarly innocuous *Forties*, which said nothing about the magazine's stance, was the compromise choice. The editors informed the readers that the old name had been appropriate in the thirties but had "led to misunderstandings of the magazine's purpose and character." They felt that the new name was "reasonably attractive and descriptive," but if readers wanted to offer opinions or suggestions they would consider them. The readers did respond, and mostly with scorn: "Too flat, banal and barren of purpose" . . . "Sounds like the shadow of middle age" . . . "Keep the same name." Some wanted *Partisanship in Defense of Culture* and even *Politics*. The response lent support to some of Dwight's contentions, but the magazine simply went on in the old way—and under the old name—without resolving the disputes among its editors.[17]

Dwight thought more and more about getting out from under his burden at *PR*. He was considering several projects: his book on the steel industry, perhaps a book on the labor movement, a prospectus for a book on the political economy of fascism, and an article planned for *PR* on "The End of Capitalism in America."

In addition, the constant stream of Europeans Nancy met through her refugee work, plus Dwight's growing group of writer friends and Village characters, were finding their way to the "regular crossroads" of the Macdonalds' 10th Street apartment. Dwight loved all the coming and going and would shout greetings to visitors down the stairwell of the tenement from his apartment on the top floor. Delmore Schwartz was a frequent visitor. He and Dwight had a mercurial friendship. They took pleasure in throwing each other out of their respective apartments, writing outrageously abusive letters, and then reaffirming their undying friendship. Dwight counted on Schwartz to take a nonconformist stance on the war, but was disappointed; Schwartz's politics were at best erratic. Schwartz, on the other hand, regarded Dwight's enthusiasms and energy as some kind of perpetual motion machine. He felt Dwight had a "pathological excess of energy so great that it triumphed over reality. . . . Dwight was invariably on a phone soliciting signatures to some petition which protested the denial of civil liberties to someone."[18]

Another visitor was the already legendary Joe Gould, a renowned Village character who was allegedly writing *The History of the World*, and had a roomful of notebook manuscripts. Penniless and in poor health, he would make his way, panting, up the four flights and Nancy, out of compassion, would give him a meal, despite his unpleasant tendency to push his nose in her face and sputter and spit when he spoke, probably due to ill-fitting dentures.[19]

But if Dwight, Nancy, and their toddler Mike were firmly ensconced in the intellectual and social life of the Village, Dwight remained obsessed by the notion of intellectual betrayal and backsliding. He wrote to Victor Serge of his disappointment, and Nancy complained of the hysterical "patriotism of intellectuals." Sidney Hook, she reported, had become so extreme that he would soon join the RAF if he could. The only two stalwarts left seemed to be James Farrell and the Columbia University art historian Meyer Schapiro. At the same time, she, like Dwight, was upset at the endless debate and argument without genuine political action. She was finding it extremely difficult to get either help or financial aid in her refugee work.[20]

During the summer of 1941 the Macdonalds found a house on the shore in Miller Place, Long Island. Dwight worked hard on his fascism book and editorial duties, but also spent a good deal of time with the family at the beach. He wrote Serge that he had discovered genuine leisure, something that simply did not exist in the city. Dwight was, however, having a recurrence of the asthma that he had suffered as a boy, which plagued him that summer, keeping him up half the night and making it difficult for him to accomplish much work during the daytime.[21]

Despite Philip Rahv's dismissal of "new form" theories of German fascism, Dwight would not be put off the subject. He not only solicited a piece by James Burnham, a preview of his book *The Managerial Revolution*, but contributed his own long theoretical discussion, "The End of Capitalism in Germany." Dwight's insistence on these rather turgid and detailed technical discussions illustrates the control he had over the political writing in the magazine.[22]

Burnham's piece was a summary of his forthcoming book, which completed his total repudiation of Marxism, ironically in a Marxian manner. The young radical sociologist C. Wright Mills called him the "Marx for the Managers." Burnham's theory of the new class of managers and Dwight's development of "bureaucratic collectivism" reflected the preoccupation of social scientists in the United States and abroad with the increasing centralization of power in the executive branch. Burnham had obvious admiration for the efficiency and competence of this new elite, and he suggested that after assuming power they might well admit a degree of

democracy as a means of maintaining a stable order. So confident was he
of his analysis that he predicted a German victory because of the organizational weakness of the democracies. But in a struggle between fascism and
the Stalinist state, the Soviet Union would be defeated, despite its own
managerialism, due to its industrial backwardness.

Initially Dwight and Burnham were in substantial agreement on the
"emergence of a new non-capitalist and non-socialist form of society."
Dwight wrote Victor Serge about his and Burnham's contributions to the
discussion and their agreement that "fascism is decisively different, economically, from capitalism," noting that such a view was not at all popular in his circle of radical friends. Dwight had a vigorous exchange with
C. Wright Mills over the issue, with Mills insisting that the capitalists
were thriving under Hitler and Nazi collaborators and Dwight responding
that the real policy decisions are made not by the corporate leaders but
by the politicians. "If our view is correct," Dwight wrote Serge, "wholesale revisions of hitherto blindly accepted orthodox theory" were in order.
He also noted that the view they held in common was hardly optimistic.
"But if a thing seems to one to be true, one must accept it no matter how
difficult the consequences."[23]

But as the discussion developed, Dwight parted with Burnham. He
could not accept Burnham's "managers" as the new ruling class. For
Dwight, they were merely the apparatchiks of the system; instead of being
the servants of the capitalist rulers, they were the slaves of their political
masters. Dwight insisted that it was the political rulers who made the
decisions, "responding to non-capitalist needs and demands." Dwight
had returned to his earlier study of the men of action, the dictators he had
surveyed in Majorca in 1935.[24]

Burnham obviously viewed the new managerialism as a progression in
human control over the environment. Dwight, on the other hand, agreed
that the centralization of power among the new class illustrated far
greater efficiency in organization than chaotic bourgeois democracy, but
it was a ruthless and tyrannical control depriving citizens of their freedom
and human potential. At this point Dwight was still insisting that the only
alternative was a socialist centralization of power. Dwight rebuked Burnham for his negativism and sapping of the will to continue the fight for a
more desirable socialist alternative.[25]

This debate was important to Dwight, for he was convinced that the
centralized State was the main enemy, that charismatic political leaders
were the threat. Though in the early 1940s he still insisted on the possi-

bility of a democratic socialist alternative to both fascism and capitalism, which also meant centralization of State power, it would not be long before he was to abandon that hope as an impossibility. There is reason to believe that he maintained that hope solely because it allowed him to continue his opposition to American military intervention and to maintain his criticism against the prevailing capitalist state, which he found vulgar, lethal, and without any civilizing effects. Even at this time, Dwight's repudiation of Marxist theory, particularly the notion that Nazi Germany was the last gasp of a decaying capitalism, was vehement.

The Nazi invasion of Russia in June 1941 made the issue all the more urgent. With this turn of events, the "war party" would and did gain great ground, which lent support to Dwight's notion that imperialist totalitarian nations were now fighting one another in a catastrophic and morally meaningless cause. The move for an alliance with and aid to the Soviet Union was anathema to Dwight, who expected the Stalinists to return to the good graces of the pragmatic Roosevelt administration and its hypocritical liberal supporters.[26]

What was to be done? Dwight insisted the magazine had to take a stand against the surging tide of war hysteria. He teamed up with Clement Greenberg and together they published a program of nonsupport for the Allied struggle. They began by conceding in the very first of their ten propositions that fascism was "less desirable" than democratic capitalism and that a Nazi-dominated Europe meant the end of civilization and a reign of barbarism. But the answer was "not war but revolution." There was, they insisted, only one future for the United States under capitalism: fascism. Only revolutionary socialist mass action could prevail against fascism and save genuine democracy as an alternative to totalitarianism. The call for isolationism was simply a provincial idiocy. On the other hand, support of Roosevelt and Churchill ensured fascism from within and blocked any effective efforts against fascism from without. Again Dwight and Greenberg castigated the intellectual backsliders on the left, singling out Sidney Hook, Max Eastman, and Lewis Corey for their endorsement of the lesser evil. The war supporters presented their position as realistic, but Dwight and Greenberg insisted it was quixotic to think that bourgeois capitalism could defeat the centralized efficiency of the fascist war machine.[27]

The lesser-evil compromise was revealed as a dismal failure by the British Labour Party, which was now nothing more than a servant of capitalist imperialism. They cited George Orwell's "London Letter" to support their criticism of the right-wing movement of the Labour Party, but neglected to mention his assertion that no one believed the war could be stopped by the working class. Dwight and Greenberg insisted that the working class could wage a successful fight against Nazism because its own socialist system could compete with Nazi military methods, war production, and aims—the argument that Dwight had been making since the mid-1930s. A social revolution that would throw a new class, the proletariat, into power would produce even more advanced methods of waging war, precisely because it had no interest in salvaging the inefficient capitalist status quo.[28]

It was also clear to Dwight and Greenberg that Churchill and Roosevelt would never get the willing support of the masses, because they promised them no genuine rewards. Only a program that offered a "real reorganization of society can inspire the peoples." With certainty they claimed that unless a socialist revolution took place, fascism would triumph. Greenberg could not swallow Dwight's refusal to endorse aid to the Soviet Union to meet the challenge of Hitler, but they did agree that the "struggle against the present conduct of the war and for socialism in this country cannot be relaxed one jot regardless of the effect on aid to Russia."[29]

Then the two "revolutionaries" went on to blithely assert that while no organized leadership existed for such a radical policy, such a lack was not fatal. The highly industrial United States had no need for a professional vanguard leadership as in the Soviet Union. Even though there was no organization, party, or movement in place, they as intellectuals could "propagandize such ideas and support such tendencies as lead to the formation of a party or a movement which will take advantage of the 'revolutionary situations' we believe will develop in the future." To work for replacement of the present capitalist governments, denying any support whatever to the Churchill and Roosevelt governments, was all-important. Radical criticism of the capitalist governments and their war had to be unceasing, and it had to be accompanied by positive demands for democratic mass participation in State decisions, referendums, democratization of the Army, equality for Negroes, and working-class control of economic planning, they wrote.

As for the alleged democracies, they had no war aims beyond "kill[ing] enough Germans." Certain that they had a firm grasp of the truth of the situation, Dwight and Greenberg closed with the assertion, "The truth is that the democracies cannot defeat Hitler by force. . . . The only way the conflict can be won in the interests of mankind as a whole is by some method of warfare that will transfer the struggle from the flesh of humanity to its mind. Such a method is offered only by the cause of the socialist revolution."[30]

The interminable theoretical discussions of the nature of fascism, bureaucratic collectivism, and managerial revolution had been irritating to Rahv and Phillips, but this absolutist outburst was too much. It would jeopardize the editors' careers. It would isolate the magazine and leave it without influence and acceptance. All the anger and conflict of the preceding two years came to a head. The decisive split and active warfare among the members of the editorial board were now openly exposed to the readership. Although Rahv did not rise to the bait instantly, when he did, his demolition job was devastatingly persuasive. Citing "moral absolutism," "academic revolutionaryism," "utopianism," "sheer romanticism," and "frivolous irresponsibility," Rahv began by focusing on the arrogant certainty and pretension of his two colleagues. They spoke for no movement, no party, and certainly not for the working class, nor even for any influential grouping of intellectuals, Rahv wrote, and yet they implied they had the backing of the masses and that daily events supported their wild prognosis. The contrary was the case: Hitler's victories had ensured the impossibility of a socialist revolution; if he were now to be challenged and defeated, it could only be achieved by the forces of Western imperialism and Stalin's Red Army, Rahv stated. And the defeat of Germany was the "indispensable precondition of any progressive action in the future."[31]

Macdonald, Rahv noted, had taunted British liberals and leftists for entertaining the foolish hope that Churchill would abdicate in favor of a morale-building anti-fascist government. Rahv wondered how Macdonald could maintain the equally absurd belief that revolutionary changes were likely to occur in Britain or the United States when in fact the labor movements were committed to the war effort. Ironically, all of Dwight's own personal studies of the American labor movement were indictments of the bourgeois sensibility of its leadership. As for repeated claims that this was a different war than the one in 1917 and thus demanded a different analysis, all his colleagues had come up with was a worn-out amalgam of Leninist–Luxemburgian strategies of the last war's radicals. And more to the point, Lenin's "revolutionary defeatism" made sense because he believed that in defeat his country had a chance for revolutionary action. Surely Macdonald and Greenberg would not hold such a preposterous view in 1941.

Macdonald and Greenberg's dismissal of the value of a bourgeois imperialist victory as meaningless ignored the damage this would do to the very structure of fascism the world over and the encouragement it would give to proponents of revolutionary action, said Rahv. As for the argument that the entrance into the war meant the arrival of American fascism, this was one of those "abysmal clichés" that had done "infinite harm to the anti-fascist cause" and encouraged reactionaries like Charles Lindbergh, America First, and native fascists looking for an issue to exploit.

Echoing Stephen Spender's earlier response to Macdonald's revolutionary enthusiasms, Rahv charged that Greenberg and Macdonald's insistence that the Churchill and Roosevelt governments were incapable of mounting an effective war against fascism was wishful thinking. Rahv had a growing respect for the U.S. industrial plant and predicted that the coming American mobilization would "astound the world," which forcefully revealed his rejection of the claim that modern bourgeois capitalism was dead.

What seemed really to drive Rahv to restrained fury was the blithe contention that a socialist revolution could take place quickly and would not be "an especially violent struggle because the ruling class was so discredited by its military incapacity and demoralization to offer much resistance." This seemed to Rahv the height of political frivolity. Would his revolutionary colleagues encourage British workers to engage in sabotage, "thus exposing themselves to mortal danger in order to prepare for a problematical situation that might never arise?"

Finally, Rahv wrote, in order to compensate for the obvious lack of leadership on the left, Macdonald and Greenberg had resorted to the tired rhetoric that such a "leadership must and will be found." This was no more than a ritualistic intonation that had no place in a serious discussion of political reality. Rahv insisted that he was not arguing against "a revolutionary policy in principle" but simply stating that in the absence of a revolutionary movement and certain essential conditions, such a policy was illusory. He could not refrain from ridiculing such an apparent faith in the militancy of the working class. Rahv concluded his piece by saying he did not believe Macdonald, Greenberg, and their friends should rush to join the war party, and even conceded that it was not yet "our war," but that to assume that the salvation of mankind had been entrusted to the intellectual editors of *PR* and they alone knew how to achieve it would be folly.

Dwight and Greenberg replied in the same issue, but their effort was nit-picking, and hardly up to the task, claiming alliance with Luxemburg rather than Lenin, and making other such arcane, but to Dwight utterly decisive, distinctions. He and Greenberg defended their optimism by pointing to the low morale in the Allied army, the poor showing of British and American war industries, Churchill's failure to take advantage of the German involvement on the Eastern Front, widespread strikes (on the left, all strikes are "widespread"), lack of war aims, the fact that the only army able to cope with the Reichswehr was the Red Army (presumably proving the essential necessity of collectivist centralization) in any effective war against fascism.[32]

It was Rahv, they maintained, not themselves, who was the deluded romantic, seduced as he apparently was by Roosevelt's fireside chats. Dwight could not refrain from deriding Rahv's bourgeois caution, which had always irked him. Rahv, he charged, could only support a revolution covered by 5 percent gold bonds and insured by Lloyds against failure; Rahv's approach to history was that of the "homeowning commuter," not the true revolutionary, Dwight's man of action, Lenin in 1910, Hitler in 1925, and presumably Macdonald and Greenberg in 1941, who supported a policy that "looked to the future" and was willing to take risks—after all, "Social change [was] always a gamble."

What was to be done? That remained the question. What in the world was Rahv's program? He had none, so he had put the future in the hands of Roosevelt and Churchill and the status quo. Rahv knew full well "the shabby hypocrisy of [their] war aims." Acceptance meant toleration of the existing capitalist regimes. "Capitalism is *intolerable*," intoned the two horsemen of the apocalypse. Like Macbeth, Rahv, with his reservations, "wants to profit by the crime without committing it."

Dwight had returned to his passionate belief in the role and the obligation of the intellectual, the very raison d'être of the man of ideas, taking a stand as an adversary critic. As he had written in his emulation of Randolph Bourne two years earlier: "In Politics . . . the mask molds the face. You become what you do and say. You don't become what your reservations are." In his mind, Rahv and Phillips and now their journal were giving aid and comfort to the growing legion of betrayers, backsliders, and assorted opportunists. Dwight's position had not only taken on the appearance of unbudgeable intransigence but, as Rahv pointed out, was sustained by an "ultra-metaphysical faith." Pushed to the wall by events,

Dwight had continued to stifle his natural skepticism, for only by holding out hope for the possibility of a working-class socialist revolution could he possibly maintain the vigor, the edge, of his critical contempt for capitalist culture in all its deplorable ramifications.[33]

The debate itself, the continuing discussion, to be carried on in the letters columns and in direct confrontation with every newly backsliding intellectual, every convert to accommodation (worse yet, "patriotism"), became for Dwight the overriding cause of the magazine. This was hardly the view of "the boys," who, with the United States' actual entrance into the war drawing closer, could no longer tolerate such a stance without a struggle for control of editorial policy.

Dwight and Nancy were finding the going increasingly tough, not only in the office but in their personal lives. Their position isolated them, just as Bourne's caused him to be ostracized in 1917. There were rows with former associates on the left, friends, and even family. Nancy wrote Victor Serge of the growing demoralization of the intellectuals and even expressed her own anger, unusual in a person of such quiet equanimity. "My rare outbursts of temper," she reassured herself, "usually had good effects because I don't get angry easily but when I do I have very excellent reasons." She was particularly upset by the return of the Stalinists to respectability following the German invasion of the Soviet Union and complained to Serge that the *Nation* and the *New Republic* had a "tendency to slip back into the Stalinist slime." Nancy often presented herself as not a "political person," but she was not immune to the tensions of international politics as they played themselves out on the battlefields of the New York intellectual community.[34]

Charges were mounting that Dwight's anti-war position was lending support to reactionary forces. He wrote to Serge summing up his current political beliefs: He was in complete agreement concerning the bankruptcy and "fossilization of the old left-wing movements," and he did not grieve his departure from the Workers Party, which was now quibbling about the war. He denied he had departed from the main body of Marx's and Engels's analytical thought and methodology, which, he wrote, "is still valid, living, and incomparably the most illuminating approach to modern history." But he got to the root of his interest when he wrote that Marxism "most needs a theory and practice of liberty."[35]

Dwight was enamored of the psychoanalyst Erich Fromm's attempt to reconcile Marxism with Freudianism and recommended his recently published *Escape from Freedom* to Serge. Dwight was wrestling with his own belief that a truly revolutionary socialist state could combat fascism because of its centralization of power, and with his chronic suspicion of the State as a threat to individual freedom. Since 1935 he had shown contempt for the notion of individual freedom as put forth by freewheeling laissez-faire capitalists; on the other hand, the corruption of socialism at the hands of the Stalinist bureaucrats could not be dismissed. He praised Fromm for raising the important question of how the individual could withstand the demands of conformity imposed by society and its institutions. Authoritarian societies were on the march everywhere in the West as well as in the Soviet Union. In a world where the technical ability existed to end scarcity and provide an economy of plenty for all, a truly libertarian and democratic society was possible. But this had not been achieved in either capitalist or totalitarian regimes, where social neurosis had led to various forms of conformity to escape the burden and the isolation of individual freedom. How to avoid regimentation was a crucial question. Fromm, Dwight noted, felt that much more "radical" changes would have to be made in capitalist-democratic societies than the leaders in political control favored.[36]

Dwight and Nancy greeted the attack on Pearl Harbor and the declaration of full-scale war with equanimity. It came as no surprise; they had simply been waiting for the events that would trigger the crisis. In a letter to Serge Dwight commented on two air-raid alarms in New York that had proved false, almost as though he was anticipating an attack. He was sure there would "undoubtedly be vast changes in American social and political life but just what they would be remained to be seen." He noted that just about everyone, with few exceptions, "is highly indignant at the treacherous and brutal Japanese attack—so much so that for the present at least, everyone seems to have forgotten about the broader questions and issues—war aims, postwar reorganization, etc." Strangely enough, after complaining so much about the pragmatic opportunism of the intellectuals, he now wrote that he thought this unreflective focusing on daily events would soon pass and that before long, "as in England, people will again be thinking of such general questions as war aims, the nature of fascism, etc." There was a hermetic quality about Dwight's intellectual world that enabled him to see things through his own selective filter. For him the nature of fascism was of overriding concern, because he was sure that it transcended the National Socialist regime in Germany or what he referred to as the "Bureaucratic Collectivist" regime in the Soviet Union. Dwight was ripe for the idea of "totalitarianism" encompassing and equating centralization of power on the left and the right. He was certain that mass societies controlled by centralized leadership were the wave of the future, and now that America was in the war, the process of centralization and subsequent dehumanization would accelerate unless that centralized power could be socialized and made human. Radicalism had to have "a theory and practice of liberty."[37]

6

A Majority of One

Dwight's fear that totalitarian regimentation would arise in the United States with the approach of an American declaration of war prompted him to make an aggressive attack on the established literary critic Van Wyck Brooks. Brooks had joined the group of literary patriots and nationalists who wanted artists to take up arms in the struggle. In an intemperate address at Columbia University in September 1941, Brooks had reviled modernist writers for their alienated and adversarial relationship toward their government and society.

Even the divided editors of the *Partisan Review* could agree on resisting these kinds of chauvinistic attacks. Archibald MacLeish, appointed Librarian of Congress in 1939, had triggered similar fears when he indicted a host of literary intellectuals as "irresponsibles" for their skepticism toward the claims of bourgeois democracy and their hesitant response to the demands for military mobilization. Since Dwight had already developed a distrust of MacLeish as a sycophant of the ruling class, it is not surprising that he chose to join the fray by publishing an extensive diatribe titled "Kulturbolshewismus & Mr. Van Wyck Brooks" in the *Partisan Review* of November–December 1941.[1]

Dwight viewed Brooks's speech as "the boldest statement to date" of the "cultural counterrevolution" initiated by MacLeish. Brooks had des-

ignated the naysayers as a "coterie" of "secondary" writers—in Dwight's
paraphrase, doubters, scorners, skeptics, expatriates, highbrows, city
slickers. He might well have been describing the *Partisan Review* circle
and all their modernist heroes. But Dwight didn't leave it at that. Brooks
was not just an old-fashioned philistine defender of vulgar nationalism in
his dotage, he was the "leading mouthpiece for totalitarian values." For
Dwight, ever sensitive to the quirks of ideology, the Brooks–MacLeish
forces were allied with the Stalinist literary front on the one hand and the
fascist totalitarians of Germany on the other—not to mention homegrown
chauvinists. The values Brooks promoted were the same cultural values
of the Stalinists, and they used the specific methods of the Moscow trials
to denigrate their opponents. Brooks's speech smacked of the same
rhetoric as the popular front orators' in their crusade against "formalism"
and demand for "social realism."[2]

Dwight was intent on having Marxism and modernism march together
as political and aesthetic comrades-in-arms against capitalism. But his
heavy emphasis on the totalitarian nature of Brooks's strictures seemed to
his fellow editors, and to some of their prominent contributors, forced and
farfetched. Some, like William Carlos Williams and Henry Miller, simply
did not take such an old "fuddy-duddy" as Brooks seriously. Allen Tate
and Lionel Trilling did consider the issue worth worrying about but were
skeptical of Dwight's political emphasis. Tate agreed with many of
Dwight's charges, even to the point of asserting that nobody who followed
the drift of opinion in the country could fail to see the parallel between
the Brooks-MacLeish school of criticism and the Goebbels-Hitler attack
on modern art. But Tate completely rejected as a form of special pleading
Dwight's claim "that the great writers of our time were exposing the evils
of capitalism." On the contrary, they had written out of a vision of the
eternally universal evils of life. Tate was suspicious of Dwight's tone and
wondered whether after the socialist revolution the alienated modernists
would not appear decadent, a query that the battle-scarred ex-Trotskyist
Dwight Macdonald could understand.[3]

Once again, Dwight had entered that thorny terrain where politics and
art sparred and often collided. While he would resolutely defend the
notion of art free from political coercion of any kind, the suggestion in his
essay that writers such as Proust, Eliot, James, and even Joyce were allies
of revolutionary socialists made him vulnerable to Tate's charge that
Dwight, like Brooks, looked on literature as a branch of politics rather
than as a unique creation in its own right. Dwight's defense of the mod-

ernist writers of the 1920s hinted at a nostalgia that would become increasingly prominent in his critical vision.[4]

Attacking such philistinism made life a bit more tolerable for Dwight at the *Partisan Review* office, since nearly all associated with the magazine and its overwhelmingly young readership could readily agree on the matter. The problem of a common editorial ground remained, however. The January 1942 issue, the first after Pearl Harbor, contained a remarkably innocuous statement informing readers that since the editors were so divided, it was clear that the journal could "have no editorial line on the war." The editors would, of course, feel free to express their own individual opinions. Dwight put his name to this weasel-worded evasion.

Unsurprisingly, this neutral position drew an accusation of editorial cowardice, from a young intellectual street fighter on the rise, one Irving Howe, in the pages of the *New International*. Howe began his attack by contrasting the socialist revolutionary vigor of *PR*'s early issues to the journal's subsequent accommodating acceptance of the prevailing political climate. Obviously under the tutelage of Max Shachtman and Workers Party stalwarts, Howe was earning his spurs by directing his blast at Dwight, as the main political editor, characterizing him as a cowering renegade who in the past would have denounced such wishy-washy expressions of neutrality and detachment. Dwight had joined the ranks of the backsliders he himself had so recently criticized. How could he claim to preserve cultural values while maintaining an alliance with colleagues who supported an imperialist war and with one (an allusion to Rahv) who was openly cynical about the possibility of achieving socialism?[5]

Howe had struck a nerve. Dwight, in a sophistical reply, was forced to deny his earlier revolutionary fervor and to actually argue that the offending editorial statement was "quite in line with the original statement with which *PR*'s career began." He was in an uncomfortable position. He had long wondered how he could continue with Rahv and Phillips and had argued for a much stronger political identification. But here he was, swallowing his pride, accepting the compromise, and arguing for the lesser evil, half a loaf, all the things he detested. To make matters worse, he was on the receiving end of ad hominem attacks on his intellectual integrity. In a brief reply to his defense, the editors of the *New International* settled some old scores and asked pointedly why Dwight didn't "conduct himself in the way of a Randolph Bourne or a John Reed," instead of selling out to fellow editors who were cynical, anti-socialist warmongers.[6]

And then there was Orwell. Dwight was responsible for having brought him to *PR* in the form of periodic dispatches from London. Orwell was cherished in leftist circles for his beautiful evocation of the revolutionary socialist-anarchist struggle in Spain, *Homage to Catalonia*. Dwight loved Orwell, a severe critic of Stalinists, for the enemies he had made. With a lucid expository prose that some found similar to Dwight's and a penchant for saying exactly what he thought, Orwell made a career out of a provocative honesty, which sometimes revealed a profoundly prejudicial attitude toward those he saw as less than real men. However, Orwell, like Spender, began to modify his radical anti-establishment views as the threat of a German invasion of Britain grew. When Orwell began to defend the British government, to work for the BBC, and to attack pacifists, quack socialists, and assorted opponents of resistance to Hitler, his writings became a bone of contention in radical circles.

Dwight liked and admired Orwell, but he had little patience for his recent pro-war writings and his so-called critical support of the British government. Orwell had joined the war parade. In one of his letters, Orwell echoed the Brooks-MacLeish thesis: "If the English people suffered for several years a real weakening of morale, so that fascist nations judged they were decadent . . . it was the intellectual sabotage from the left that was partly responsible." Dwight was irked by what he saw as Orwell's crusading anti-intellectualism, his repeated demands for patriotism, and his abuse of all critics as "negative," "carping," and "irresponsible." Orwell had retreated to a "common-sense philistinism" which, Dwight believed, only supported the totalitarian drift.[7]

Dwight, Nancy, and like-minded friends were frustrated by the new prestige of the Communists and fellow-travelers and the precarious position of critics like themselves, who were increasingly the target of attacks from all sides. It was not long before Dwight found himself in an acrimonious argument with his good friend James Agee over the latter's acceptance of the U.S. propaganda film *Mission to Moscow,* which painted the Soviet Union and even Stalin in a favorable light, poohpoohed the Moscow trials, passed over the Nazi-Stalin pact, and praised the Russians as great defenders of Democracy. Agee asserted, "It was good to see the Soviet Union shown as the one nation during the past decade which not only understood fascism but desired to destroy it and which not only desired peace but had some ideas how it might be preserved and how it would otherwise inevitably be lost." Dwight accused Agee of uninformed double-talk and threw a series of questions at him concerning the role of the Communists in Spain, the Moscow trials, the pact—in short, the litany of Stalinist crimes. But the issue for Agee came down to the fact that the Russians were bearing the brunt of the war against Hitler, and these political details could not gainsay the fact of their struggle and sacrifice. Dwight was isolated, part of a little band of sectarian purists at a time when dialectical discriminations seemed to matter less and less.[8]

But the adversarial position only inspired Dwight. For the July–August 1942 issue of *PR* he solicited an anti-war statement from the anarcho-pacifist Paul Goodman that called on writers, artists, and intellectuals to engage in civil disobedience against the war, to refuse to register for the draft, and to withdraw from the mainstream of American life. Phillips and Rahv found the piece objectionable and wanted it spiked.[9]

Dwight added his own piece, a lengthy challenge to America's war aims. In a devastating critique of a speech by Vice President Henry Wallace, which itself was a gloss on "The American Century," Luce's famous *Life* editorial, Dwight excoriated the administration's self-righteous declarations. Wallace's claim that American might would bring forth the Age of the Common Man was a fatuous absurdity, since the very system he was committed to defend would have to be destroyed. Dwight made an articulate case against a corporate liberal imperialism that might well become fascistic in its foreign policy at the same time it maintained a liberal domestic policy. It would buy off the labor unions by offering them a share in the profits of international exploitation. Anticipating a popular critique by New Left historians in the postwar years, Dwight argued that the "more farsighted members of the business community" saw Roosevelt's interventionist policy as a desirable alternative to fascism: "It would smash the ever menacing Nazi imperialism" and the "United States would emerge with a world position analogous to England's in the last century."[10]

As an indictment of the hypocrisy and ineffectiveness of the administration's stated war aims and its propaganda, Dwight's piece was sharply judgmental, but it was hardly the treasonous polemic Rahv and particularly Phillips accused it of being. Both were decidedly opposed to publishing such criticism on the grounds that it might draw government repression at worst, and placed *PR* in an adversarial position that might jeopardize its future as well as that of its editors. They felt Goodman and Macdonald's double-barreled assault, combined with the attack on Orwell and another article by a British pacifist equating Hitler and Churchill and arguing that "Hitler requires not condemnation but understanding," were too much. Such material was reducing the tone of the journal to that of an extremist rag.[11]

Dwight took the material to his old Trotskyist associate Al Goldman, *PR*'s lawyer, who advised that the criticism presented no legal problem. Dwight wrote to George Morris that Rahv and Phillips claimed to be against his own piece on the grounds that it was not analysis but polemical anti-war propaganda and therefore not editorially acceptable. To Dwight this was a betrayal of an agreement they had reached. The editorial board had insisted that the magazine not take a stand on the war but promised to print opposing positions. From Dwight's perspective, not to print such criticism, not to challenge the State, was a far more dangerous policy.

They were back to square one. It was a fundamental and profound difference of opinion. Dwight saw criticism and conflict as a moral obligation, and his opponents were intent on a "policy of keeping quiet about the war." Dwight told Morris that he was at the crossroads, that a resolution of editorial policy had to be achieved. He felt Phillips and especially Rahv were given to furtive delaying tactics.

Dwight did not see the problem as simply a political disagreement. There was, he felt, a more sordid angle: "The Boys want to build a career on *PR* and hence want to keep it respectable." Phillips was angling for a teaching position and was using the magazine as a stepping-stone, his editorship being his only claim to fame. It was Fred Dupee who had fed Dwight "this charming idea."[12]

Dwight's attribution of selfish career motives arose out of his understanding of his own role as an activist intellectual in a time of crisis. He had made a commitment; his opponents, men of the left, had abandoned all principles and commitments. Of course, it was a good deal more complicated than that. Phillips and Rahv did have a reputation for being manipulative, dissembling, and given to back-stabbing gossip, "always sneaking around corners," as Dwight was soon to put it. And they were much further along on the road to deradicalization, having abandoned Marxism as irrelevant to current political action or activism. Rahv always maintained a respect for its analytical properties, but he and Phillips had already adopted a pragmatic approach to political issues such as the war as not only more sensible and realistic but as serving the interests of the journal and, incidentally, themselves.[13]

Other issues were also involved. Dwight, who frequently described himself as one of the few Wasps in this New York intellectual community, was drawn to his Jewish colleagues, Greenberg, Goodman, Rosenberg et al., in part for their apparent alienation from American society as he had always understood it. Alienation for many of the Jewish writers and intellectuals was a given, made keener by the bourgeois aspirations of their immigrant parents. The second-generation radicals had taken an adversarial stance from childhood. Isaac Rosenfeld compared the Jewish writer or intellectual as an outsider similar to the black. The Jew, he wrote, was a "specialist in alienation," and since nearly all "sensibility—thought, creation, perception—is in exile, alienated from the society," the Jew was particularly suited to the role of critic and interpreter.[14]

Soon many of Dwight's colleagues would be classifying themselves not simply as members of an intellectual cosmopolitan community, but as Jews who identified with the Holocaust victims. Daniel Bell has struck at the heart of the matter in noting the tension between the Jewish intellectuals' faith in the universalism of the Republic of Letters and the horrible particularism of genocide committed against their people. Dwight's own rejection of the prevailing capitalist culture was hardly the same kind of alienation experienced by his Jewish friends and associates. He was *of* American society and culture, if by choice not presently *in* it. They were never in the culture and could not or would not be of it. Dwight could choose to be an outsider because he had the innate confidence of his inherited membership. As his friend Dinsmore Wheeler put it: "Perhaps we were fortunate that we were born Gentiles and citizens of the top-dog country. We can afford to hate objectively."[15]

Dwight may well have sensed this. His own youthful anti-Semitism had long since been curbed, but he was provoked by displays of ethnic chauvinism on the part of Jewish people or anyone else. Sidney Hook has remarked that Rahv and Phillips were not above charging Dwight with an unconscious anti-Semitism "as a kind of bullying weapon" in their arguments with him over Zionism, the government of Israel, the Palestine refugee problem. Hook noted that their accusations had "a quieting effect that was completely out of character." Dwight was quieted because he was sensitively aware of the residue of his Wasp background and his association with people who were still essentially, if sophisticatedly, anti-Semitic. Part of Dwight's cachet with his Wasp family and friends, was his close association with the New York–accented, Jewish intellectual crowd. There is no question that he and others saw a person like Delmore Schwartz as a kind of urban exotic. Association with these Jewish intellectuals positioned Dwight even further away from the scorned capitalist culture in which he had been raised and educated. Now, in this time of crisis, to find his Jewish comrades edging closer and closer to respectability, was only another sign of betrayal. It may well have supported some of the stereotyped notions of pushy Jews, wanting to be accepted. One must recall his accusations against Miriam Isaacs, Dinsmore's girlfriend, for her straining to make herself agreeable, always trying to please and ingratiate herself. Dwight's tone in his complaining letters to Morris is not dissimilar to Lillian Hellman's exasperation at those she charged with acquiescing before McCarthyism to gain acceptance. She denounced the betrayal of her Jewish friends and scolded them as "children of timid immigrants desperately trying to make it." Hellman went on to charge that some found in their anti-Stalinism the excuse for joining those "who should have been their hereditary enemies." The texture, the patterns, the intricate relationship of this Wasp intellectual to these alienated yet highly ambitious Jews, many from out-lands in the Bronx and Brooklyn, formed a complicated weave. However, it must be recorded that Dwight Macdonald's anti-Semitic upbringing, which was not discouraged at Exeter and particularly at Yale, where prejudiced attitudes thrived, and were embraced by his mother and relatives, was never the essence of the man. If anything, its significance lies in his determined rejection of such prejudices and his profound transformation from a dandyish aesthete into one of the more honest and moving critics of the world's decline into barbarism during the war.

Dwight's struggle against genteel racism played an important role in developing his commitment to the notion that an intellectual's identity had to be grounded in a cosmopolitanism that repudiated national chauvinism. This had been exemplified in the circle of anti-Stalinist, Jewish intellectuals he joined after leaving *Fortune*. But his own youthful anti-Semitism and his membership in a larger culture that encouraged and practiced it sharpened his critical sensibilities. He grew increasingly sensitive about the gap between one's public stance and private morality. It encouraged him to constantly scrutinize his own motives and hold himself to high standards of integrity, and to make the same demands of his friends, associates, and society in general.

In August 1942, Dwight, Nancy, and Mike escaped these editorial wars for a month on Nantucket. Dwight wrote Victor Serge with relief that he expected to find it very quiet there. He did feel vindicated about his Wallace article, reporting to Serge that it had been extremely well received, as successful as anything he had ever done. He was also buoyed by the thought that intellectuals were less sure about the war than they had been six months ago. Dwight seemed to take pleasure in his reading of the war news, which he felt supported his charge of Allied ineffectiveness in Egypt, Russia, and the Far East, as well as "the confusion and inefficiency of America's own efforts."[16]

Dwight and Nancy found Nantucket charming, with its eighteenth-century houses and narrow streets, "more like a European village than anything else I've seen." They rented bicycles and took day-long excursions all over the island, with Mike riding behind Dwight in a "funny little basket seat." Dwight noted that the beaches were deserted and that they could swim nude. Mike, he reported, is "especially fond of bathing *au naturel*." But the war intruded even on this island paradise. The Coast Guard patrolled many of the beaches at night because spies had allegedly been set ashore on Long Island. Fred Dupee on a visit to the Macdonalds was startled to be challenged on an evening walk by the hollered demand "Who goes there?" To which the budding poet could only reply, "Me," which apparently satisfied the sentry. Dwight and Nancy enjoyed long evening discussions of politics, high and low, accompanied by pitchers of martinis and pots of steaming clams with Albert and Roberta Wohlstetter. Albert Wohlstetter had been on the fringes of student radicalism while a graduate student at Columbia, but he was already on his way to becoming a well-known analyst of nuclear weapons systems and the transition of peacetime energy into weaponry. Years later they could easily recall Dwight's vehement opinions and flashing wit, including hilarious references to Henry Luce as "Il Luce."[17]

Dwight's earlier prophecies of democratic capitalism's inevitable failure to confront Hitler and his certainty that totalitarian fascism would come to America as the war wore on did not come to fruition. He was now taken with the notion of a liberal corporate internationalism and conceded that the Roosevelt administration was "not fascistic or even developing with any speed in that direction." On the contrary, both the Roosevelt and Churchill governments found their main base in the labor movements. Civil liberties had been preserved to a remarkable degree. Roosevelt was fighting a conservative war, meaning one determined to preserve the status quo. The U.S. government, meaning FDR, was the broker between contending interests in which the "army–big business group" would run the war but the president could still check their more extreme policies. One would have thought that a liberal-labor coalition would fight the conservative drift away from New Deal reformism, but to Dwight's despair, the "lib-labs" had become the most loyal supporters of the war administration and the working class; the only group that might lead a movement for revolutionary change had "allowed its leaders to integrate the trade unions into the structure of these conservative governments." Dwight was particularly incensed by American labor bureaucrats who had not demanded important positions in the war apparatus. Even the relatively enlightened Walter Reuther was scorned as a "boy scout Labor-fakir."[18]

Since his *Fortune* days, Dwight had harbored a deep suspicion of the union bureaucrats. But what distressed him most was his intellectual associates' increasing support of the war. He wrote to Delmore Schwartz, after the most recent ruckus at *PR*, fantasizing about a new magazine where he would be free of the ever vigilant negativism of "the boys." Schwartz replied in his usual fashion with a barrage of amiable insults, jocular ridicule, and genuine affection:

> I always defend you among the academics & the genteel (two of your curse words . . .) by saying: Yes antagonism for its own sake is his appetite and neurosis, and none of his political predictions come true, but he is a master of expository prose. . . . and he opens himself up to all kinds of being and beings, Open House Macdonald ought to be his name.[19]

Normally Dwight appreciated this sort of jovial banter, but his nerves had been worn raw by constant disputes and criticism. "God save me from friends if [you're] typical. . . . Do me a favor, keep silent or join the enemy," he said in an unusually irritated response. For Dwight, so caught up in individual resistance, it had simply become an either/or situation. In an outburst of self-righteous and quite humorless zeal, he admonished:

> My political beliefs, which I take seriously, are not based on mere love of brawling [a charge made frequently by Dwight's friendly critics and usually accepted] and being "different," but on some experience (mostly my years on Fortune), on sympathy for human beings (who are brutalized and oppressed in every way by the social system we have) and an intellectual conviction which I have arrived at by a wide (and still continuing) reading in politics and economics. Goddammit, that nasty crack about my being just a congenital sorehead is the way conservatives have

always and everywhere brushed aside any kind of criticism from a radical standpoint.

Responding point by point to some of Schwartz's criticisms of his editorial judgments as well as his zealous activism, Dwight lashed out:

Why don't you have the guts to speak out in print on anything. As I told you a long time ago (a prediction that has come true) you are ruining your career (not to speak of your moral being as a man) by trimming your sails to prevailing winds, by keeping silent on any hot controversial issues, by excessive diplomacy and hush hush attitude toward all the fakery and shoddiness that's for years been growing so in our whole intellectual atmosphere. I can see how the fact that I insist on talking out on a few such themes might seem to you practically mentally unbalanced. But I assure you the shoe is on the other foot.

Although Dwight closed this missive with an apology for his short temper and asked Schwartz to write to him at the Wheeler farm in Ohio where he was going to spend the Christmas holidays, it is clear that he had put Schwartz in the enemy camp.[20]

In an effort to translate word into the deed and civilize the war, both Nancy and Dwight became involved in the March on Washington Movement to end discrimination in the armed forces led by A. Philip Randolph, the black organizer and president of the Brotherhood of Sleeping Car Porters. Dwight frequently pointed to the blatant racism in the services as exemplifying the hypocrisy of U.S. war propaganda claiming a crusade for democracy. Describing the discrimination as "The War's Greatest Scandal: The Story of Jim Crow in Uniform," he turned out a well-documented (Nancy did the bulk of the research) pamphlet for the MOWM. Detailing the appalling discrimination and indignities inflicted on blacks in all the branches of the service, and the insane waste of resources, it charged that the prevailing Jim Crow system was a direct violation of the Constitution and the draft laws. Since there was no well-organized radical movement against the war, the March on Washington Movement replaced the Workers Party for Dwight as a potential platform from which to confront and oppose the war effort as well as denounce the Communist Party for playing down the race issue and discouraging black protest.[21]

Dwight presented a class analysis, insisting that discrimination against blacks was social and economic, not racial. There was no natural basis for this kind of race prejudice; it was used by the ruling class to divide and exploit workers. A ticklish point was the MOWM's restriction of membership in the organization to blacks on the grounds it was necessary to promote black self-reliance and initiative and to prevent white domination or infiltration by others for their own ends, meaning Communists. Dwight was a vigorous opponent of this policy as a form of black chauvinism or nationalism that was both morally and tactically wrong and injurious in the struggle for genuine equality. To those who blamed the lackluster militancy of the NAACP on its integrated membership, Dwight responded that the NAACP was ineffective because it was liberal, not radical, and was supporting the Roosevelt administration and a war that compromised with racism.[22]

Even though Dwight was not a member, he spoke at a mass civil rights rally April 14, 1943, along with Randolph, Roy Wilkins, the activist lawyer Arthur Garfield Hays, and Dwight's new black friend Wilfred Kerr, co-chair of the Lynn Committee to End Discrimination in the Armed Forces. "Is this the kind of war our leaders claimed it was?" Dwight called out to the audience. "It seems to me the time is ripe and overripe for this country to decide whether it is pouring out blood and treasure to win a war for democracy or a Jim Crow war." Again he declared it was a class issue, rejected any form of black separatism or nationalism, and insisted that blacks join with the white masses to fight for social and economic democracy. He warned that the tide of reaction was spreading and that "Negroes . . . must put up a much stronger fight than they have so far." Dwight pleaded for blacks to take the lead, saying that "large sections of the white population would follow and support their struggle." Anticipating the accusation that pressing the racial issue impeded the war effort, Dwight closed with the warning that the fight would not be easy and that they would be accused by the reactionary press, the Southern Democrats, unscrupulous New Dealers, agents of Stalin's dictatorship, and all the members and fellow-travelers of the Communist Party "of working for Hitler, of sabotaging the war effort, of being defeatists and traitors and God knows what else." The only answer was to throw this question at them: "Is this a war for democracy or not?"[23]

While Dwight and Nancy were absorbed in political activism with the March on Washington Movement, Rahv and Phillips had begun to consider how to gain control over the political content of *PR*, which Dwight had dominated. They were aided by the volatile New York University

philosopher and astute Marxist in retreat, Sidney Hook, who proposed a
series of articles titled "The New Failure of Nerve," about the "flight from
responsibility and a retreat into mysticism and other forms of obscurantism in the face of the terrible catastrophes engulfing the civilized world."
The series opened with great fanfare and a triple-barreled assault by
Hook, John Dewey, and the philosopher of science Ernest Nagel. All concentrated on the repudiation of science and rationalism expressed by a
"motley array of religionists." Enormously sensitive to trends in the wind,
Hook singled out the "frenzy of Kierkegaard," who openly threw over-
board his intelligence to make "leaps of despairing belief which convert
his private devils into transcendent absolutes," and Reinhold Niebuhr,
who more than anyone else had successfully revived the doctrine of original sin as a political tool and who sought "to make God an instrument of
national policy."[24]

Hook's attack on the revival of a religious neo-orthodoxy may have
been just a shade premature for the radical secular community that read
the *Partisan Review*. It was not his critique of the mystics that drew blood,
but his charge that the left also exhibited a failure of nerve that, given its
pretensions, was even more inexcusable than the intellectual conduct of
confused liberals and despairing religionists preaching "a violent return
to a theology without the social gospel." In his challenge to the left, Hook
dismissed the Communist Party as simply the American section of the
Soviet secret police. He directed his attention to the "platonic revolutionaries" who made up the warring sects of the Trotskyist movement and the
ex-sectarians, "romantic revolutionaries," or, more accurately, "bohemian
revolutionists," whose radical views were proclaimed "mainly by half-
sober blusterers at cocktail parties," a sneer that was apparently taken
personally by James T. Farrell. Hook did make a point of exempting
Dwight from his critique by suggesting that at least he proclaimed his
revolutionary intransigence in lively prose for the *Partisan Review*. Hook
always maintained a grudging respect for Dwight's "natural exuberance,"
"uninhibitedness," "engaging and unembarrassed frankness," and willingness to act, as opposed to his colleagues Rahv and Phillips, who
"never took risks, never joined a picket line, did not get involved in any
kind of political life, or made any attempt to win over the general public
to the revolutionary cause." If there had been a revolutionary uprising,
Hook said, they would "probably have gone into hiding until it was all
over."[25]

But if Hook could appreciate Dwight's decent instincts, he thought his message mad. Hook's attack was in the same vein as Rahv's earlier critique of Macdonald and Greenberg's "Ten Propositions Against the War." Their defeatism was a message from "heaven's psychopathic ward," it was "impotent fanaticism"; they were all "theoreticians without practical or theoretical responsibility." Hook was a master at dialectical debate and he put his position very clearly: "If Hitler wins, democratic socialism has no future. If he is defeated, it at least has a chance." It was the socialist left's inability to grasp this simple wisdom that "marks the political insanity of [their] infantile leftism."[26]

Hook's devastatingly effective analysis of the intransigent left's position ridiculed the Trotskyists who argued that Russian workers should indeed support *their* government against the fascist armies but that American and British workers should not. This bizarre stance was the consequence of their stubborn insistence that the Soviet Union, despite Stalinism, remained a workers' state—a degenerate one, to be sure, but still a nonimperialist, noncapitalist workers' state.

During this series of debates on "The Failure of Nerve," Meyer Schapiro, a former Communist Party fellow-traveler and, during the war years, a staunch supporter of the Trotskyist anti-war position took on Hook, whom he denounced as a sellout, and defended the angst of Niebuhr, Kierkegaard, and other religionists of despair as similar to the sensitivity of many modern artists.[27]

This debate deteriorated into an acrimonious exchange of epithets and personal accusations. But the series itself highlighted the profound "shock of recognition" that was occurring throughout the Western liberal and radical community, involving serious philosophic and religious speculations about humankind and its possibilities. The terrible chain of events following the catastrophe of the First World War—the Depression, the Spanish Civil War, the rise of fascism, the Moscow trials, the Nazi–Stalin pact, the second and even more violently destructive world war, with current rumors of mass murders and calculated genocide—all these events dealt a lethal blow to traditional liberal enlightenment thought and contributed greatly to the demise of most Marxian beliefs, to a profound loss of confidence, and to the clouding of a previously optimistic worldview.[28]

The chain of calamities of the twentieth century had forced people to confront the darker forces of human nature, to question the very notion of progress, and above all to challenge the sanguine optimism of the liberal mentality, the "Children of Light," as Reinhold Niebuhr so aptly labeled them. For some Americans, the great spokesmen of the liberal tradition—Holmes, Dewey, Beard, Brandeis—had come to seem irresponsibly optimistic. Daniel Bell has recalled that as the news of the Nazi Holocaust began to unfold, a sense of horror began to take root. The key writers in the intellectual journals of the time "found affinities" with the work of thinkers like Kierkegaard, Niebuhr, and Simone Weil. For Bell and his friends, Niebuhr's *Nature and Destiny of Man* was "a key book." All of them emphasized the element of tragedy. Morton White, the Harvard philosopher, was soon to note that those were the days when Dewey's views were "being replaced by Kierkegaard in places where once Dewey was king." Bell saw his generation as "twice born," finding "its wisdom in pessimism, evil, tragedy, and despair."[29]

Daniel Bell's recollections are particularly helpful when reviewing Dwight's involvement in this debate. On the one hand, Dwight expressed complete confidence in his rejection of the war and continued to call for a socialist revolution. On the other, he rejected Sidney Hook's mechanical faith in rationalism and his defense of the scientific method as applied to contemporary human problems. Dwight was not unreceptive to the skepticism of the younger generation. He had often expressed his own profound skeptical sensibility in his initial notes dealing with the rise of dictatorships in the mid-1930s. Those journals are full of references to "the deep psychic forces" and "dark irrational passions and prides" which were to history as the urge to create something beautiful is to the artist. Dwight had referred to John Dewey and William James and other exponents of nineteenth-century rationalism as the "monsters of reason." At one point Dwight had written, "That this is not a rational world is more borne in on me." And again: "It is virtually impossible to regard men and their institutions with too much skepticism." At that time the Marxists did not escape his scrutiny: "The class struggle is a pre-war concept of 19th-century rationalism which paid too much attention to materialistic factors and too little to the deeper psychic forces."[30]

Dwight could not stomach the religiosity of Reinhold Niebuhr, however, and he could not view his writing as metaphorical. Dwight took detailed notes on several of his books, remarking at one point that he couldn't continue reading because of Niebuhr's "vague categories—'reason' vs. 'nature,' 'sin' vs. 'virtue.'" They seemed to him old-fashioned, meaningless, and carelessly thrown about, reminding him of the "oily oak panelled, stuffy shabby divinity school," a "dreary sapless goody goody milieu one was faced with years ago in college." (Echoes of the hated compulsory chapel!)[31]

Dwight found the obscurantism of Niebuhr and others a disturbing contemporary phenomenon; the trouble with Hook's analytical dismissal of it was his failure to ask why such a panicky retreat into obscurantism should be taking place. Hook's use of debater's logic to stem the tide of irrationality was "like telling a neurotic that his fears were all in his own imagination." The retreat from reality, like neurosis, met a real psychic need, Dwight wrote; it was indeed an unintelligible world, and an unreasonable one. Irrational modes of thought were increasing because "we are in a period of profound social frustration . . . all roads ahead seem to be blocked." The arguments of Hook, Nagel, Dewey were beside the point; they showed far too much "superficial optimism." In view of the crisis facing the world, there was greater reason for a pessimistic, even obscurantist approach. Dewey, for example, seemed to think that if enough children were progressively educated, society would be changed, when it was clear that society had to be changed before children could be progressively educated. It was an absurdity for Hook to believe workers could provide a social purpose for the war without undertaking a social revolution.[32]

Dwight dismissed the Four Freedoms and the Atlantic Charter as hypocritical propaganda refuted by a series of reactionary events: Gandhi jailed and the defeat of the Congress Party in India; in America, regressive taxation, higher food prices, frozen wages, enormous profits, the most conservative Congress since 1933, unions that were simply instruments of government control, with government using the miners' strike as an occasion to further weaken labor, blacks discriminated against as much as ever, and big business overwhelmingly powerful in all aspects of life and culture. Everything was being done to strip the war of purpose, to depoliticize it, make it devoid of content. "As it grinds automatically on, as it spreads and becomes more violent," Dwight wrote, "the conflict becomes less and less meaningful, a vast nightmare in which we are all involved and from which whatever hopes and illusions we may have had have by now leaked out."[33] There was no appeal to the enslaved masses of Europe, merely calculations as to how many bombs it would take to destroy how many acres; no positive ideology, just "opportunistic adaption to the status quo." Political thinkers had abandoned not only the idea of progress but even the idea of consciously attempting to point society's evolution in a desirable direction. There was only "submission to the brute force of events, choices between evils rather than positive programs, a skepticism about basic values and ultimate ends."

Dwight maintained that the eighteenth-century values of the Enlightenment were threatened as never before. They had come into conflict with the development of capitalism, and of course it was the abstract values and not the productive system that gave way. The great scientific and technocratic advances that Enlightenment minds had seen as so hopeful had been used to destroy the very fabric of culture. Whole cities were now leveled with "the most admirable efficiency." Widespread literacy, the goal of all radical and reform-minded activists, had been achieved, "and so the American masses read pulp fiction and listen to soap operas." Technology allowed "a lying and debased official culture" to indoctrinate the masses of the totalitarian states more easily. The State, far from decreasing in power, as was hoped for by all progressive thinkers from Jefferson to Lenin, was now subjugating human beings as the Church had in the Middle Ages.

Dwight offered a fervent Bourne-like denunciation of a technocratic generation in which there was "a hypertrophy of technique, and an atrophy of human consciousness." He declared that only a socialist collectivism could humanize the bureaucratic collectivism that was developing in the world. Despite the extreme weakness of the contemporary working class and the "fossilization of orthodox and sectarian Marxists," the war, with its social upheavals, made for revolutionary opportunities that could be exploited. Dwight argued that those who insisted on "the hopelessness of the socialist cause either show that some other class is more likely to realize progressive values, or else abandon those values." Despite his own deep skepticism and his appreciation of the reasons for despair and retreat into various forms of obscurantism, he returned to a purist line, refusing to accept the notion that a person might maintain a harsh skepticism, even despair, yet continue to struggle for humane values. For Dwight this was no time for a retreat to pre-Marxist precepts or compromise.[34]

Clearly this was Dwight's rhetorical resignation from the *Partisan Review*. If he could not transform it into an activist publication, then he would create one of his own. He would gather his own community around him. Fittingly, he closed this piece on an inspirational note by proclaiming that the revolutionary struggle in itself had a profound effect on human nature. It brought out "virtues and intelligence" latent and unexploited in the masses. To make clear his continuing faith in Marxism, he ended with a quotation from Marx's *German Ideology*, which asserted that to alter mankind on a mass scale, a revolution was necessary, and that in the act of overthrowing the ruling class, the working class could rid itself "of all the muck of the ages and become fitted to found society anew."[35]

There was a ritualistic quality to this last will and testament. It was as if Dwight was girding himself for an individual struggle to find some way to avoid becoming part of a society and a culture he found full of a fakery, shoddiness, and dishonesty that had befouled the intellectual atmosphere.[36] Philip Rahv played down the reader response to Dwight's piece, but Dwight was reassured by the praise he received from his best friend and severest critic, Dinsmore Wheeler. Wheeler hadn't responded to the piece immediately and Dwight had been disappointed, but when the letter of congratulation did finally arrive, Dwight felt he had reached beyond the circle of hard-core political radicals. Wheeler, who was reading a good deal of radical criticism of U.S. foreign policy, wrote that he could think of no other contemporary of their background, upbringing, education, and status "who could bring off such an expression." The article came nearer to "'vindicating' from the point of view of history, of betrayed and murdered mankind, our whole debased generation than anything I know." Surely this was the impulse behind Dwight's declaration: a total rejection of that Wasp ruling class which operated the war machine in its own interests.

Wheeler also appreciated the tone of the piece, which had abandoned the "chilliness, that too objective, case-hardened 'positivism' that is the hallmark of 20th-century intellectualism." He agreed that the *Partisan Review* editors lacked passion—they were "too damned respectable." Wheeler also shared the inspirational, community-creating impulse that carried Dwight along. Dwight had shown the courage to be polemical, had expressed a genuine "feeling of fraternity." After all, a "revolution isn't astro-physical theory." Dwight had shown his awareness that it involved people—men, women, and children—and had infused the piece with this feeling.[37]

Dwight mailed his "Future of Democratic Values" piece to the *Partisan Review* office from a summer retreat in Londonderry, Vermont. It was close to the vacation home of his stalwart supporter Meyer Schapiro. From this vantage point in the Vermont hills Dwight not only finished the article but began the process of finding a solution to the editorial chaos on the magazine. The matter had been brought to a head when George Morris made it clear early in the summer of 1943 that he no longer intended to finance *PR*. Both Dwight and "the boys" now agreed that it simply had to go one way or the other—that either he or they should assume control and the other party should resign. It was agreed that if Rahv and Phillips could come up with the money, they would assume complete control. If they failed to secure the funding, then Dwight and Nancy would take over. Dwight was confident that "the boys" would not find an angel and that he, with the help of Nancy's inheritance added to his own resources remaining from his *Fortune* savings, would take over. He wrote a draft of an announcement informing the readers that the present editorial partnership was coming to an end and that the magazine, in a new form, published and edited by himself, would be forthcoming. In June he wrote to Victor Serge saying that the last issue of the old *Partisan Review* would be the July–August 1943 issue. The disagreements with "the boys" had become so intense that he refused to continue. "Either Rahv would carry it on as an entirely literary magazine or I will carry it on with more emphasis on politics." Dwight insisted on characterizing the split in these somewhat inaccurate terms. Rahv and Phillips were political too; their politics just weren't the same as Dwight's. Dwight added that if he did not gain control of *PR* he would, "with Nancy's help, start in a modest way a completely new publication in the fall."[38]

At the time of this pending split, Dwight wrote Morris a lengthy explanation, citing all the old complaints. Morris was not very sympathetic, replying that he didn't agree that the magazine had become "cliquish." It had its limitations, he said, but it was an intelligent attempt of its kind, and he wanted to stay connected with it. Dwight had been hoping that Morris would announce his own departure when Dwight did. Morris was obviously amused by Dwight's writing of his "interests being with the masses. I suppose this means you want contact with the millions . . . that you want to help them materially. It would certainly be hard to reach them aesthetically on any plane that you could tolerate." But Morris also had his doubts about Dwight's commitment to aestheticism. He picked up on Dwight's concession that his gifts were journalistic. "I have always felt an obstruction between you and aesthetic understanding." However, it was "only an obstruction . . . not a complete impasse; therefore I am distressed to see you retire from the scene for journalistic comments on and for the masses."[39]

Dwight and Nancy were caught completely off guard when the crafty Rahv and Phillips came up with an anonymous angel, one Mrs. Norton, the wife of an Army officer. She wished her support of the magazine to

remain secret because of her husband's position. Rahv and Phillips had
not kept Dwight informed of their negotiations with Mrs. Norton, although
they had been making arrangements for some time before they sprang
the news on their unsuspecting associate. "Well Mrs. Norton has come
through. It's a miracle," Rahv wrote in a tone of obvious satisfaction. She
was a woman of "the finest sensibility, but politically she is an indifferen-
tist more or less." What she liked about the journal was "its sustained
contact with European literary and art ideas."[40]

Dwight was upset by the news, but he put a brave face on it and wrote
a note of congratulations. Although Morris had suggested that Dwight
continue on as a contributor and take his politics elsewhere, Dwight told
Rahv that such an idea was impossible, but that he might like to do "a little writing for PR" if the occasion presented itself.

He did not intend to end with a whimper. Along with his congratulations he enclosed a carefully worded draft of his letter of resignation.
After all, a person who had given five years to such a demanding enterprise had to make it clear precisely why he was quitting. He produced a
derogatory history of the journal's decline from a lively, politically
engaged literary magazine to a cautious, respectable, academic journal
concerned solely with literary criticism. Dwight portrayed his colleagues
as not only abandoning their Marxist socialism but wanting to eliminate
from the journal political discussion of any kind. He opposed this retreat
and called for greater social concern. He flatly asserted that with his
departure the magazine would become "exclusively a literary and cultural
magazine, leaving the thorny field of politics to others"—such as himself.
He announced his intention to start a new magazine that would serve as a
"rallying point" for those intellectuals still concerned with "social and
political issues." It would be a magazine which, while not ignoring cultural matters, would "integrate them with—and, yes, subordinate them
to—analysis of those deeper historical trends of which they are an
expression."[41]

Knowing his opponents, it is somewhat difficult to believe that they
would simply print this disparaging statement and let Dwight have the
last word. At first Phillips simply sent back his draft with some minor
changes. They rejected Dwight's telling the readers what *PR* would
become in his absence; they wanted the statement that it would be exclusively a literary and cultural magazine to read "will become in the main a
cultural magazine." After all, Rahv noted, they intended to deal with
ideas—all sorts of ideas—and "one couldn't deal with ideas without

bringing in questions of ideology and politics." They finally agreed to
state: "*Partisan Review* will devote itself to cultural issues, leaving the
thorny field of politics to others." As an afterthought Dwight in a postscript requested the *PR* subscriber list to solicit for his new magazine.[42]

If Dwight thought he had made a forceful exit, unscathed by retaliatory
barbs, he was shocked to get the proofs of his statement with an accompanying response from the editors. In addition to charging that Dwight had
let himself be carried away by his political passions, they declared he
had sought to convert the magazine to his personal political uses. His politics were still mainly Trotskyist—peppered with his own individual here-
sies; his claim to be the sole Marxist on the editorial board was simply to
identify Marxism with his politics. The editors proudly acknowledged
their role as primarily literary men and repudiated the idea that a magazine of literary sensibility should substitute for a political movement. Noting Dwight's assertion that he would "subordinate" cultural matters to
deeper historical forces, as good dialecticians they translated his "historical forces" as "political interests," and likened his stance to that of Stalinist policy. Resistance to just such political coercion had been the reason for founding the magazine in the first place. They insisted that the
journal's success, and it was doing quite well, was due "to the specific
modulation achieved in combining socialist ideas with a varied literary
and critical content," which would continue to be its policy. They closed
with a sneering reference to Dwight's "self-righteousness and academic
revolutionism, and the incessant repetition of a few choice though all-too-
elementary notions."[43]

In early August Dwight received a copy of the issue containing his resignation and the editorial response. Rahv wrote him a note conceding that
their reply was "pretty sharp," but that Dwight had left them with no
alternative. Dwight's letter had been an attack on the magazine and in
effect it advised the readership to stop taking the journal and wait for his
own. In an irritating parting shot, Rahv again insisted that the journal
would not be "thoroughly de-politicized." There were a thousand and one
ways of slipping political discussions into the magazine—despite their
ostensible agreement with Mrs. Norton.[44]

Dwight was obviously piqued by their stealth in printing a response in
which he had no say, while they had made changes in his letter. He wrote
to complain to Morris, who once again did not offer much comfort. He felt
that of the two letters, Dwight's gave the more distorted picture. All of
Dwight's criticisms were designed to damage *PR*, and Morris could see

why the editors insisted on a rebuttal. After all, there was nothing worse than "a former editor of a periodical denouncing it as safe and stuffy while ignoring its intelligence in a time when intelligence is needed." Morris agreed with Rahv that Dwight had started resigning in 1937 when he joined the staff.[45]

It was a sad parting. With Clement Greenberg in the Army, Dwight had ended with no allies. Matters became even more unpleasant when he and Nancy made a trip down to the city from Vermont to wind up affairs and to get the subscription list he had requested for a circular he planned to send out to announce his own magazine. When they arrived at the tiny office on Astor Place, they found that the locks had been changed. Finally admitted, they were confronted by an adamant Rahv, who refused to give up the list. Rahv felt that Dwight would use it to declare that *PR* was so moribund the readers should switch to Dwight's new magazine. Nancy from her perch atop the radiator could only think how she and Dwight had done nearly all the real work to get the magazine out each month and how she had been, in Dwight's words, "the Unknown Soldier of the little-magazine world." She announced that she would not leave without the subscription list, whereupon Rahv threatened to call the police. The Macdonalds could think of no more fitting turn of events and refused to budge. Rahv saw the absurdity of this situation and finally agreed to surrender the document.[46]

Nancy wrote to Victor Serge describing her and Dwight's shock over the conduct of "the boys," particularly their shrewd tactic of printing a reply to Dwight's letter without informing him. It showed what they had thought of him. But of course the feeling was mutual. Dwight had never had a warm friendship with either Rahv or Phillips, nor had they ever been comrades-in-arms, certainly not politically. Dwight and Rahv respected each other's talents. Dwight believed Rahv to be a sensitive and intelligent literary critic and to have a good editorial sense. Both Rahv and Phillips saw Dwight as a superb journalist, with a lively imagination and an instinct for the immediate crucial issue. In effect, it had been a marriage of convenience, but given the level of political turmoil, it is surprising that the union lasted as long as it did. Murray Kempton recalled that Dwight took with him "a great part of *PR*'s elan and the deficit was never quite repaired. An ineffaceable cast of the dour would always remain" after the Macdonalds' departure. "The note of peevishness that bespeaks the loss of community crept in, asserting a claim on a long-term lease."[47]

If Dwight and Nancy felt a certain sadness over the nature of the parting, they both were glad it was over. Nancy reported to Serge that they were planning to start something new and that they both felt "free and without responsibility. Everybody we know seems to be eager for a new magazine and we have promises of a number of subscribers and even an office helper today in the mails!"[48]

II

MAKING A POLITICS

7

Politics *and the Search for Responsibility*

IT WAS MARVELOUSLY APPROPRIATE THAT DWIGHT AND Nancy would begin their new enterprise in the fall of 1943 in the old Bible House building at 45 Astor Place. The building was a monstrous six-story trapezoidal structure between Eighth and Ninth Streets and Third and Fourth Avenues in the East Village, with an imposing columned entrance on Fourth Avenue. Set into the red stone at the floor was a figure holding a Bible. In fact, the building housed the *Partisan Review*, which probably explains Dwight's ability to negotiate some space there in its warren of tiny transomed offices. One of the building's first tenants had been the colorfully eccentric Horace Greeley, the great liberal editor of the *New York Tribune*, credited with pressuring Lincoln into issuing the Emancipation Proclamation, surely a promising genius loci for *Politics*, as the new magazine was to be called. Just down the hall from the *Politics* office were at least two other radical sects that kept their typewriters and mimeograph machines running regularly. All were apparently godless atheists, as were Dwight and Nancy, but they all had a single-minded vision and deep commitment to make the world a better place, which even the most devout missionary or circuit rider could appreciate.[1]

Dwight's purpose from the moment he left the *Partisan Review* was to find a way to write to and hear from his peers, "to create a center of consciousness on the left, welcoming all varieties of political thought."

During the period of separation from *PR* and the gestation of *Politics*, both Dwight and Nancy were making contacts, projecting their vision in outlines for the new venture. It was to be more than a magazine; it was to be the center of an international community of like-minded people unhappy with the status quo, where they could find a refuge and provide solace for one another and a defense against an increasingly terrifying world. They would dedicate themselves to critical analysis of that world and, even more, to advocating ways and means to encourage a change.[2]

Serious thought was given to the title. Dwight came up with *Left*, the *New Left, Radical Review*, and *Gulliver*, but it was C. Wright Mills, the young sociologist at the University of Maryland by way of Texas and Wisconsin, who pointed out that what they all wanted to do was talk about politics, so why not name it that? And so *Politics* it was.[3]

From the start, Dwight wanted a freewheeling journal that would hew to no ideological line or dogma. It is astonishing just how eclectic his view of the magazine's character was. He wrote to Dinsmore Wheeler that he hoped it would be a combination of the best features of *Time* (presumably he meant the weekly's topicality and up-to-the-minute commentary, and not what he had previously viewed as its neo-fascist slant) and *New Essays*, a revisionist Marxist monthly edited by Paul Mattick and Karl Korsch; it would be critical of Stalinism and committed to a third-camp socialist revolution as absolutely necessary to save the Western world from advancing totalitarianism. It should also be "a three-ring circus," offering literary criticism and a thorough treatment of popular culture in addition to politics. For Dwight, the only consistency would be the unremitting exposure of the hypocrisies of the Allied position in the war, a forceful negativism directed against the status quo of the bureaucratic State and against all the forces of modern life that were contributing to the process of dehumanization, alienation, "thingification." That was Dwight's dominant sensibility, and it is what drew admiring comrades into his orbit.[4]

They were a remarkable group of talented and energetic people, primarily Marxist in their orientation. Dwight's initial statement of purpose claimed that the journal would be Democratic Socialist and that its "predominant intellectual approach [would] be Marxist, in the sense of a method of analysis, not of a body of dogma." While he and his associates were not exclusively Marxist, they were all aggressively anti-Stalinist.

Most were, in the fall of 1943, still advocates of the third-camp position and skeptical of Allied imperialism and foreign policy.[5]

Foremost among the initial group was Nicola Chiaromonte, a refugee from three different countries: Mussolini's Italy, Franco's Spain, where he had fought for the Republic, and Nazi-occupied France. A man of intense moral convictions and acutely critical intelligence, he became Dwight's most influential associate and his closest friend and confidant. Some of Dwight's critics characterized Chiaromonte as Dwight's guru in his search for a pure ideology of anarchism and pacifism. William Phillips is alleged to have remarked that Dwight was desperately looking for a "disciple who would tell him what to think." Dwight and Chiaromonte were, in fact, essentially contrasting temperaments and often disagreed on fundamental issues. Chiaromonte was an almost puritanical Italian of the Savonarola type, given to moody silences. He made great moral demands of himself. He had duties to perform, sacrifices to make, and he was later appalled by the hedonistic pleasure-seeking of modern postwar society. While admiring Chiaromonte's personal dedication and commitment, Dwight himself never had personal self-discipline, always enjoyed immediate gratification, was attracted to a libertarianism in lifestyle, and was given to flamboyant conduct that Chiaromonte sometimes saw as immature and undignified. But there was an anarchist slant to Chiaromonte's thinking that appreciated the spontaneity of good intentions rather than calculated political circumspection and that drew him to Dwight, as it did many others. Some argued it was that spontaneity accompanied by a moral absolutism as to means that led to what they felt was Dwight's irresponsibility.[6]

In addition to contributing the title of the journal, C. Wright Mills played an important role in the early issues; his critical intelligence and his sociological analysis of what later became known as the "power elite" were an integral part of *Politics*. Mills's indictment of the collaborating intellectuals was one of the more incisive pieces published in the early issues of *Politics*. Dwight saw in Mills a young man who, like himself, was looking for trouble, a congenital rebel who distrusted all authority, received verities, and established institutions. Mills was as critical as he was, as alienated and isolated—a natural comrade in opposition; as Dwight recalled later, "misery loves company." (Actually, at the time they met, Dwight was hardly miserable: He relished the outsider stance.) Mills was to his colleagues in the academy what Dwight was to his more cautious colleagues in intellectual journalism. Dwight claimed that Charlie Mills could argue longer and louder about any subject than even he could, but no one else supports that modest concession.[7]

Daniel Bell also played an important early role, giving the early issues of *Politics* their substantive social science in the form of a sophisticated portrait of a liberal, monopolistic, corporate state that absorbed labor into its orbit and ran an economy to serve the interests of the international firms. But it was a world of growing alienation: "People move about in the huge caverns that modern technology has constructed, with little sense of relationship to meaningful events. They live as atomized human beings no longer controlling [their] lives but carried by events."[8]

Such a sensibility of modern alienation helps explain how Bell and Dwight were drawn to each other. Although Bell later rejected the Marxist underpinnings of his analysis of the corporate state, he continued to appreciate Dwight's search for some moral foundation on which to base one's life. Dwight admired Bell both as a professional scholar–journalist and as an intellectual, but he came to see him as a kind of careerist appa-ratchik of the very lib-lab corporate culture he denigrated. This assess-ment was perhaps unfair, but it was hard for Dwight, as an editor of a small magazine, to see clever and able writers go on to other magazines and institutions, leaving him searching for suitable replacements. Besides committing the sin of leaving freelance journalism for the acad-emy, Bell became a serious scholar whose writings were in demand in more lucrative places, and that led to strained relations. Bell liked and admired Dwight, but he was not impressed with his emotional and instan-taneous response to politics and was wont to argue that Dwight, who came latest to and abandoned earliest all political positions, was, as one wag put it, "the floating kidney of the left."[9]

Lewis Coser was another of the initial advisers and contributors. Born in Berlin, he left with Hitler's rise to power in 1933 and moved to Paris, where he participated in a small Trotskyist organization, the Spark, and other left-wing socialist organizations. He arrived in the States just prior to Pearl Harbor. In New York he immediately became involved in social-ist and Workers Party politics. He actually met Dwight through Nancy, because he had sought her out to deliver messages from Spanish refugees living in France. Coser wrote for several of the Trotskyist and socialist publications, and he became an expert on the French radical scene for *Politics*.[10]

Clement Greenberg and Harold Rosenberg had an influence on

Dwight's decision to go ahead with a new magazine, but neither played much of a role after it started, and both became rather critical of Dwight's moral fervor and anti-scientific bias, as well as his meandering political odyssey. Delmore Schwartz was in on the early discussions, and Dwight asked him to do a piece for the first issue. However, he was never driven by politics or *Politics* in the manner of Dwight. He was much more comfortable with "the boys" at *PR*, where he became a desultory editor of sorts.

Paul Goodman, whose convoluted prose style irked Dwight and whose homosexual lifestyle may well have made him a bit uncomfortable, nevertheless became one of the mainstays of the magazine, much to the chagrin of the more disciplined Marxists and social scientists. While Dwight was not totally comfortable with Goodman's libertarianism in practice, he was drawn, at least theoretically, to Goodman's anarchic sensibility. Goodman's rhetoric of sexual freedom clearly attracted Dwight—not the homosexuality, perhaps, but the notion that one should be free to experience genuine eroticism outside a monogamous marriage. Nancy felt that Goodman was brilliant, but his cocky self-assurance and his inordinate self-centeredness strained her tolerant and accommodating good nature.[11]

Other *Politics* colleagues included the historians William Hesseltine, whose criticism of the magazine was of value but whose academic prose was unacceptable to Dwight; Frank Friedel and Kenneth Stamp, two young historian associates of Mills at the University of Maryland; Harold Orlansky, a brilliant anthropology graduate student at Yale; Melvin Lasky, an old Trotskyist associate, now in the Army and later to work as a publicist for the American military government in Berlin.

And then there were the Europeans. At one point Dwight described the magazine as little more than a "transplanted spore of European culture," and another time he referred to *Politics*'s contributors and readers as a little band of immigrants. Andrea Caffi, who wrote under the pseudonym "European," may have been one of the most important. A close friend of Chiaromonte, Caffi was a Russian-Italian intellectual living in France and Italy. Dwight appreciated his "extraordinarily wide range of historical knowledge and imagination" and put him in the company of Albert Camus and Simone Weil, both of whom he published in *Politics*. Dwight was particularly influenced by Caffi's eloquent anarcho-pacifist essays and was helpful in getting them published in America.[12]

Victor Serge, the most eloquent if sometimes anti-Stalinist Bolshevik, became a close adviser to Dwight, and Dwight published him frequently. Peter Gutman (pen name: Peter Meyer), a Czech émigré, wrote frequently for *Politics* as a Marxist critic and became one of Dwight's most forceful opponents as Dwight drew away from his socialist revolutionary position. Niccolo Tucci, an Italian aristocrat and former member of the propaganda wing of the Italian embassy in Washington, wrote a regular column, "Commonnonsense," which contained the most bitter satire and scornful ridicule of practically everything. Tucci's cynical innocence, as one critic called it, irritated the more pragmatic of the intellectuals, but Dwight stuck with him throughout the life of the magazine and played a role in finding him a niche at the *New Yorker*.

Despite this array of extraordinary talent, *Politics* was always Dwight's baby. This was quickly understood by most of the circle. It was often referred to as a "one-man magazine" because Dwight was the sole owner (with ample use of Nancy's inheritance), publisher, editor, layout man, and unquestionably the most prolific contributor. At one point, working over the index to *Politics*, he added to the long list of his signed articles "see also comment, book reviews, discussions to letters reports etc. . . . My God what a lot." But none of it would have happened without Nancy. She had inspired his departure from *Fortune*, encouraged his split with *PR*, led the two-person sit-in over the subscription list, made many of the European contacts through her relief work. Lewis Coser and Peter Gutman knew her first, as did several other contributors. Victor Serge was as much a friend and correspondent of Nancy's as he was of Dwight's. Dwight conceded that without Nancy's talents and energy, the "magazine would have been improbable if not impossible."

If, as some, including Dwight, say, *Politics* was his finest professional achievement, a great deal of the credit, inspiration, and even ideological tone can be traced to Nancy. Her deep moral conscience, her inordinate need and desire to help others according to her well-formed political convictions, gave Dwight, the journal, and those who worked and contributed to it a bedrock of integrity and stability. She kept the books, the calendar, and some semblance of order in the operation. Years later Niccolo Tucci remembered Nancy as "the soul of *Politics*."[13]

The first issue of *Politics* appeared in February 1944; Dwight had hoped to publish in January, but various problems delayed its appearance. Dwight contributed nearly a third of the material in the two-columned, thirty-two-page issue. He stated in forthright terms that he had created the magazine to oppose the war and in the process to create a center of consciousness on the left. He was using the politically devalued term "politics" to revive its genuine meaning. The magazine would be open to all varieties of radical thought, and as critics of the war effort it would measure the weaknesses of the belligerent powers with a yardstick of "basic values." Dwight didn't say "absolute" values, but he felt that there was a moral and ethical standard for critical judgment and evaluation and making that standard clear became one of the major editorial tasks of the journal. As in the past, Dwight once again confronted the recurring question of the relationship between art and politics and stated without equivocation that he would view art, music, literature as historical phenomena. He would also address popular culture, which he maintained had been neglected by American intellectuals.[14]

Dwight also attempted to make clear his own personal political position of the moment. Two important subjects interested him: the nature of the Soviet Union, and what he considered to be the "proper socialist policy in the present war." The Soviet Union was the last stop on the road to totalitarianism, and he found it even more menacing than the National Socialist state because of the superior propaganda of the Stalinists. This horrendous form of "bureaucratic collectivism" would dominate the future if a socialist revolution failed to materialize.

As far as the war itself went, Dwight was resolved to show that it was not a struggle between good and evil or even democracy and totalitarianism, but a conflict between rival forms of imperialism. Any type of "lesser evilism" was political folly and objectively amounted to support of a reactionary postwar government. Translating such a recalcitrant stance into editorial policy required that he continually expose and ridicule every aspect of British–American and Soviet conduct in the war as well as their respective domestic policies. On a broader scale, Dwight saw the war as destroying Western culture. It was literally "the end of Europe." He mourned the bombing of the beautiful old buildings. It was, he wrote, "like living in a house with a maniac who might rip up pictures, burn the books, slash up rugs and furniture at any moment." He cited Rosa Luxemburg for her understanding of the relationship between war and imperialism. World War I, she had written, was

a turning point in the history of imperialism. For the first time the destructive beasts that have been loosed by capitalist Europe over all other parts of the world have sprung with one awful leap into the midst of

European nations. . . . This civilized world has just begun to know that
the fangs of the imperialist beast are deadly.[15]

It was not only European culture and civilization that were being
destroyed. The United States too was becoming brutalized. This was
Dwight's singular and obsessive perspective. As Daniel Bell later
recalled, Dwight and his *Politics* did hold "a unique place in American
intellectual history." It filled a void because "it was the only magazine
that was aware of and insistently kept calling attention to changes that
were taking place in the moral temper," the depths of which were still not
fully realized a decade later. This brutalization was to be Dwight's unrelenting theme. In "The Psychology of Killing," he dissected the technology of the process: The killers are removed and not conscious of the
killing, he wrote; soldiers are manipulated objects, weapons, not human
beings. Dwight ransacked government and military manuals to show that
young men were trained to be efficient at mindless brutality and, worse
yet, to glory in the work. There was an "anatomy of mayhem" whereby
young men were taught to hate total strangers and revere the State's com-
mands. Randolph Bourne's name flits through the pages of *Politics*; an ad
in large boldface type cites that haunting refrain WAR IS THE HEALTH
OF THE STATE!, echoing down through the years and buttressing
Dwight's insistence that the insidious poisons of nationalism, chauvinism,
patriotism, militarism were once again let loose in the land, infecting the
entire society.[16]

As the society was being destroyed, the economy was being transformed into what *Politics* economist Walter Oakes described as the "Permanent War Economy." Daniel Bell added more ammunition by describing the "The Coming Tragedy of American Labor," which had been
co-opted by the promise of a guaranteed annual wage "if they behav[ed]
themselves and accept[ed] an imperialist foreign policy." Dwight prefigured C. Wright Mills's analysis of the power elite when he described a
permanent war economy "organized along democratic/capitalist, not fascist, lines." It was to be a "union of big business and the army, with over-
tures to labor unions in a bourgeois-democratic, rather than fascist framework." This lib-lab treachery was designed to stifle a genuine socialist
revolution. The worst-case scenario was that all "classes will be condi-
tioned to accept, as part of 'The American Way,' military service for youth
and a big army and navy . . . the rulers will make money, keep their power
and prestige, and sidetrack political opposition by first preparing and

finally waging the Third World War." This analysis anticipated New Left revisionism nearly two decades later. Dwight, Daniel Bell, and Walter Oakes had long abandoned socialism by the 1960s, but in 1944 they painted a menacing portrait of a military–industrial complex, built on a co-opted and docile labor force brought into the system to serve what Bell called the coming monopoly state.[17]

Dwight carried on the crusade against all forms of racial prejudice, challenging discrimination in the Army, and he established a column titled "Free and Equal" that drew attention to instances of racism at home and abroad, against blacks and soon Japanese Americans. He also commented on and printed material that can only be described as feminist. In his first and important piece on popular culture, Dwight noted that the heroes of popular culture, those figures who "break out of the mechanized rut of modern life," the Lone Ranger, Tarzan, Judge Hardy, were inevitably men. "When the central figure of a folk saga is a woman, she is likely to be not a *heroine* but a victim. . . . The all-conquering hero is always male." He concluded this commentary with the observation that the "superficiality of the modern 'emancipation' of woman appears when we descend into these profound cultural depths. When heroines are admired, not sympathized with, then we can talk of 'emancipation.'" In a review of an article dealing with sex roles, an unidentified writer argued that "women in American society are an oppressed and an exploited subject caste, like the Negroes, grotesque as it sounds to put it that way." The writer agreed with the author under review that it was essential to get the woman more out of the home and the man more into it—that is, to accomplish "the restoration of the father as a functioning member of the family." Not long after these discussions, Dwight, using the pseudonym Terrence Donaghue, ferret breeder from Staten Island, wrote a letter to the editor to challenge the notion that women were oppressed in America. He quoted the Nazi racialist Alfred Rosenberg to the effect that "the strikingly low cultural level of the American nation is the result of the dominant position of Woman." Dwight often adopted this pseudonym, and "Theodore Dryden," to anonymously challenge, comment on, correct, and frequently criticize an article and often to make fun of his own contributions to *Politics*.[18]

Even more socially avant-garde was Dwight's publication of the poet and writer Robert Duncan's essay "The Homosexual in Society." Duncan, a friend of Allen Ginsberg and other Beats, was forever grateful to Dwight for offering him the opportunity to make a case for homosexuals as equals and to insist that they be allowed the company of humanity at large and not forced into outcast cults. Duncan's argument fit neatly into one of Dwight's most important themes, his hatred of nationalisms, chauvinisms, separate identities:

> It must always be remembered that one's own honesty, one's battle against the inhumanity of his own group (be it against patriotism, against bigotry, against, in this specific case, the homosexual cult) is a battle that cannot be won in the immediate scene. The forces of inhumanity are overwhelming, but only one's continued opposition can make any other order possible.

In the 1940s such an open and frank discussion of homosexuality was an unusual occurrence.[19]

The dehumanization, bureaucratization, and manipulation that Dwight saw as the terrible wave of the future in all of Western society was facilitated by what Dwight chose to call popular culture. To Dwight, this term clearly meant the commercialization of culture for a mass market. He saw this imposition from the top as the vehicle used by tyrannical elites to shape the totalitarian society. His interest in popular or mass culture was always closely tied to his understanding of politics; it was simply another aspect of barbarization. At this point in his evolution, however, he still believed that a genuine "folk culture" could be encouraged in a socialist society, where the old high culture of the upper classes would be replaced by a human culture. He quoted Trotsky to this effect. In the early days of *Politics*, most of the essays under the heading "Popular Culture" were devastating critiques of the pablum dished out to a passive and unthinking mass audience. In an exchange on the subject, Maurice Zolotow charged Dwight with harboring a cranky envy of more popular and successful writers and artists. People like Dwight and the avant-garde he championed were simply out of touch with American society; they were cynics and snobs, Zolotow wrote. The country had had enough of "épater les bourgeois." In a response titled "Lowbrow Thinking," Dwight made the connection between popular culture and the growing political totalitarianism more explicit by charging that the Zolotows were apologists for the status quo as the best of all possible worlds, that they wanted the writer to adjust to a society "organized ever more comprehensively along anti-human lines." Once again the philistines were on the march, led by the Brooks-MacLeish conspiracy, a band of chauvinists demanding a yea-saying nationalist culture. For Dwight, American culture, Allied culture,

Nazi culture, and Stalinist culture all met on the same low ground of vulgar manipulation of the mass mind. He again spoke up for the doubters, the naysayers and rebels who refused to make "peace with all 'the old crap' of respectable, exploitative society with its bishops and bondholders, educated by the soap operas and thermite bombs and fireside chats, and admirals and public relations counsellors."[20]

Dwight and Nancy were gratified by the enthusiastic initial response to *Politics*. They had printed 3,000 copies, which quickly sold out. Since they had not asked the printer to hold the type, copy had to be reset at extra cost; an additional 2,000 were printed and about 1,000 sold. Dwight wrote Victor Serge that subscriptions were pouring in and "all kinds of people were being stimulated to write for it. . . . Everyone seems to feel a terrible moral and intellectual void" in radical and liberal circles that *Politics* could fill.[21]

Dwight's associates and contributors wrote long and constructive criticisms. During the first six months the journal was published, the most consistent criticism was the "negativism" of the editor and most of his contributors, who were masters of the put-down; almost no official or institution of the establishment escaped their sardonic and continued exposure. One George P. Elliot, who held a master's in English from the University of California at Berkeley and was serving on the War Labor Board, wrote that the magazine only "inspired despair." He described *Politics* as representing a group of abnormally sensitive radical intellectuals, who were rejected by (and rejected) their society. Macdonald offered nothing in the way of even "the narrowest channels of action against the disaster" of the war. The magazine's sustained attack on the Soviet Union irked a number of readers, who felt that the constant fulminations against Russia offered little in the way of constructive criticism and only promoted the growing tide of reaction that the magazine feared. Accusing *Politics* of being an appendage to the conservative *Chicago Tribune*, a reader named William Palmer Taylor complained that while the magazine claimed to be diverse in its political approach, only anti-Stalinists seemed to find space to vent their spleen. Many of these critics were appalled by Dwight's growing tendency to equate Stalinism with Nazism.[22]

Since these letters to the editor were representative of a substantial group of critics, Dwight wrote a lengthy response, attempting to make clear the direction *Politics* was taking. The "sequence of left defeats" had led to rampant lesser-evilism, and an acceptance of the half loaf. Those like the *Politics* circle "who insist[ed] on uncompromising radicalism in culture or in politics" appeared to be "foolish utopians," but Dwight declared that he still believed in "the possibility of large-scale progress." The accusation of despair and cynicism was misplaced, he argued; it was the "half-loafers" and pragmatists who were motivated by despair.[23]

Dwight bristled at Elliot's charge of elitism and inability to participate in "public forms of pleasure." In a rather plaintive plea, he insisted: "I go to movies. I read detective stories. I was simply enchanted by the New York World's Fair and went there on every possible occasion." He even went so far as to insist that he thought Bob Hope, Red Skelton, and Jimmy Durante were very funny and that he read and enjoyed comic strips: "Moon Mullins," "Krazy Kat," and above all, E. C. Segar's "Thimble Theatre," starring Popeye and Wimpie. To crown this claim of immersion in the culture of the day, he wrote that he considered Westbrook Pegler the most "brilliant living technician of personal journalism," and claimed that he had been reading Pegler since his days as a sportswriter.

Dwight argued that he had concentrated on the bad qualities of popular culture because they so outweighed the good. He too was grateful for what did "escape the mechanism of commercial exploitation." Even in Hollywood, some life survived, but it was the "inhuman mechanism that [was] historically significant, not the seedlings that sprout in its crevices." It was that image of the machine, totalitarian and on the move, that governed Dwight's view of popular culture.

Dwight denied that *Politics* was driven solely by pity and hate, but he insisted that "outraged protests against the inhumanities" were worth expressing since they were often ignored in the nationalistic and routinized message of most lib-left journals. Barbarism, vulgarity, degrading taste were too easily accepted and accommodated. Society had reached such an extreme of rationalizing lunacy that one could only respond with "a moan or yawp." He had the Marxian text at hand. Marx had written that in grappling with German social conditions, "criticism is no passion of the head; it is the head of passion. . . . It no longer figures as an end in itself, but only as a means. Its essential pathos is indignation, its essential work is denunciation. . . . [There should be] no opportunity for self-deception and resignation."

In defending *Politics* against the accusation of exclusive negativism, Dwight did note the journal's active work against racial discrimination, and he might have also cited his editorials in behalf of rights for women, minorities, and homosexuals. Dwight did concede that it seemed little could be done in America to "advance matters in a radical direction"; however, destroying the illusions and hypocrisies of the lib-labs was useful work. Again the Marxian text: "The disgrace must be made more disgraceful by publishing it, the people must be taught to be startled at their own appearance." To avoid the pending postwar reactionary disaster, Americans had to recognize that "not New Deal II but a Permanent War Economy" was on the horizon; not democracy but an expanded imperialism, not peace but war. The only escape lay in a "revolutionary, basic alteration" in society's economic and social institutions.[24]

Despite the militancy of his rhetoric, inner doubts remained, and Dwight hedged when he argued that even if one had no faith in a revolutionary solution it was a position superior to the prevailing pragmatism, because it "enables one to analyze modern society more accurately and more profoundly." The revolutionary vision he struggled to maintain allowed one to look at social processes more objectively than the expedient realpolitiker could. Did Dwight mean that even if he really didn't have a genuine faith in the possibilities of a democratic socialist revolution, he would continue to view the world as though he did? Was this simply a position from which it was easier to launch criticism? Perhaps, but as his fear of an advancing totalitarianism grew, it would obviously be harder and harder to summon up the rhetoric for an imaginary revolution. Sometimes his sarcastic barbs directed at his former comrades in the Workers Party to puncture their revolutionary pretensions irritated his radical readers and only undermined his own claims to an abiding belief in the potential of a revolutionary movement.[25]

As this extended exchange with Dwight's critics over the stance and direction of the magazine shows, *Politics* could offer a forum for debunking and satirizing what was all too often mistakenly revered. Dwight would continue to solicit, encourage, and promote all new ideas and groups who were interested in the solution to the one great problem of the moment: "how to combine economic collectivism with political democracy."[26]

Nowhere could Dwight's own doubts about the possibility of a successful challenge to the corporate capitalist system and its permanent war economy be seen more clearly than in his continued dismissal of the American labor movement leadership as bourgeois pragmatists completely absorbed into the capitalist system and its consumer culture. Having recently attended the UAW convention in Buffalo, he ridiculed Walter Reuther as too accommodating, a "boyscout labor fakir." In reams of

notes going back to his observations in the mid-thirties, which Dwight
hoped to pull together for a book on the demise of the American labor
movement, he kept up a running barrage on the bourgeoisification of the
CIO. It had evolved, he argued, from a militant army of sit-down strikers
into something almost as respectable as the American Federation of
Labor. Reflecting his own experience in dealing with the leaders of the
steel industry, he concluded that American industrialists had a far keener
sense of class consciousness than American labor leaders or the rank and
file. In effect, Dwight's observations, which had begun with a purely Trotskyist suspicion of trade union leadership, had become a fearful meditation on a potential American fascism emerging out of an alliance of corporate and union leaders against the consumer, the farmer, and the
dispossessed.[27]

In April and July of 1944, Dwight published in the socialist newspaper
a long two-part article titled "'Union Security': Two-Edged Sword." It was
a sustained analysis of the price the unions paid for collaborating with
corporate leaders to gain a measure of security. Throughout the essay,
Dwight drew comparisons with the Nazi state and its control of trade
unions. He closed with the observation that the bourgeoisie in America,
as in Germany, "really has no very deep prejudice in favor of democracy,"
so long as they garner a share of the pie. The focus of Dwight's critique
was once again the all-powerful corporate state which skillfully co-opted
all potential opposition. The unions had simply handed over to government agencies the right to decide all the issues. The image of a creeping
totalitarianism pervaded Dwight's thought.[28]

Word of the horrible consequences of Nazi fascism was reaching
America. In the spring of 1943, Nancy in a letter to Victor Serge had
commented on the Nazi slaughter of the Jews and the deplorable
response of the democracies: "Now there is talk of helping the poor
Jews—but absolutely nothing concrete. They will surely most of them be
killed before the democracies get around to finding some hole in the wall
to put them."[29] News had been continually trickling out of Europe since
June 1942 that unthinkable crimes were taking place all over the Nazi-
occupied territories and in Germany as well. The *Nation*, the *New
Republic*, and *PM* all published reports. The *Nation* of December 19,
1942, ran an outraged editorial under the title "The Murder of a People," which described the killing as so appalling and unprecedented that
it could not be comprehended immediately and had to await the perspective of history. The editors were certain, however, that in time the enor-

mity of the crime would dwarf earlier inhumanities that stretched back through the centuries to the days of the Pharaohs. They briefly recorded the scientific methods of extermination—gas, machine guns, suffocation, starvation, and the introduction of air bubbles into the veins—which spared neither the old nor those too young to walk. They estimated that 2 million had been already exterminated in Poland alone. There was a schedule of killing planned until finally all but a handful of Jews strong enough to be useful as slaves would be put to death. Even the mainstream press, such as the *New York Times* and the *Boston Globe*, were carrying snippets on their back pages by 1943. But practically no discussion of this aspect of the Nazi crimes appeared in *Politics*. It was the lib-lab press that was publishing these horrible accounts, and Dwight suspected them as sources, believing that such reportage tended to provide a blanket justification for everything the Allies did in their war against the fascist barbarians.

The first reference in *Politics* to rampant state murder appeared in Dwight's introduction to a reprint of chapter 37 of Herman Melville's *Redburn*, a moving account of the total indifference on the part of officials, the police, and even citizens of Liverpool, England, to the plight of a mother and her two children literally starving to death in a dank cellar hellhole in one of the slum quarters. Dwight explained that the chapter might serve as a possible model "in its restraint and unadorned reporting for journalists writing of contemporary horrors." Presumably he felt that the dispatches coming out of Germany and Poland were sensationalized and exaggerated. He went on to note that the Melville piece also

> serves to remind us that official murder was not introduced into the modern world by the Nazis. A century ago equally revolting and widespread atrocities were inflicted on millions of beings in peacetime by British capitalism. Theorists of the unique barbarism of the German people, orators dilating on the humanitarian tradition of bourgeois democracy might read Chapter 37 of *Redburn* with profit.[30]

This comparison with the brutalities of the Industrial Revolution reveals Dwight's reluctance to believe the stories of Nazi atrocities as late as the spring of 1944. Despite his fears of totalitarianism, he denied that a unique form of depravity was being practiced wherever the German armies had control. Whereas a mechanistic society in which people were transformed into objects was a growing preoccupation, even an obsession, it remained a political abstraction, grounded in a contextual political understanding. Of course Dwight was hardly alone. Many on the radical left, particularly the Marxist left, had great difficulty confronting the news coming out of Germany. It simply did not conform to their categories, to their theoretical analysis of fascist society. In March 1944, nearly two years after the devastating news first came to America, Dwight was still viewing the genocide as part of the historical continuity of industrial capitalism with its inhuman exploitation and brutalization of an underclass. Whatever knowledge Dwight and many of his associates had was, as Andrea Caffi reported of Europeans well over a year later, "weakened by fits of incredulity."[31]

The early issues of *Politics* were remarkably devoid of reportage or analysis of the Nazi regime's methodical destruction of European Jews. As the horrible reality became increasingly apparent, one might have expected the subject to be a prime target of Dwight's relentless exposures because the atrocities were so illustrative of his worst, perhaps unimaginable, fears concerning the rise of the new totalitarian states. Although Dwight has been justifiably credited with being among the first of the New York intellectuals willing to confront the horrible implications of the Holocaust, his initial response was characteristic and in keeping with the widespread hesitancy on the part of many members of the New York intellectual community to either confront or comprehend the unimaginable dimensions of the state-ordered human destruction. In June 1944, Dwight noted in an editorial comment that the old warning not to buy a pig in a poke didn't seem to hold for the sponsors of committees that had been growing in recent months. He was particularly disapproving of the creation of committees for a Jewish state in Palestine "set up by fascist-revisionist Jews" that were being supported by liberals. A newly established Council for a Democratic Germany was being sponsored by such well-known intellectuals as Reinhold Niebuhr, John Dewey, Alvin Johnson, and the Jewish editor of the *New York Post*, James Wechsler. Dwight endorsed the Council's attacks on the irrational and dangerous Vansittart policy of making the whole German people responsible for Nazism. He also supported the Council's opposition to any attempt to divide Germany; only through the "integration of a democratic Germany" into the Western community of nations, he argued, could a third world war be avoided. This endorsement, however, was followed by suspicions that this new Council was dominated by old Stalinist wheelhorses (Bertolt Brecht, for example), and Dwight posed a series of questions demanding that the Council show its independence from the Moscow line.[32]

Dwight's approach to the issue of Germany and the plight of the Jewish and other victims of Nazism was to view it through his anti-Stalinist lenses. Dwight was answered by the theologian Paul Tillich, who bluntly wondered whether Dwight felt that the Council should adopt the methods of the Dies Committee (later to become the House Un-American Activities Committee) with respect to its membership. Tillich also said that while the Council rejected any Russian form of Vansittartism, it appreciated "the great accomplishments of Soviet Russia in smashing the Nazi War Machine" and would cooperate with all genuine anti-fascists. Dwight rejoined that Tillich was apologizing for the Russian use of collective guilt by throwing in an unnecessary salute to the Russian people. The editor of *Politics* did not consider Stalinists genuine anti-fascists, nor did he think that the Russian people should be singled out for their resistance to the Germans, since England had also been "cruelly ravaged by the Nazis, and her people also fought back with heroism." In neither case was any form of Vansittartism acceptable. As in his earlier exchange with Stephen Spender concerning support of the Churchill government's resistance to the Nazis, Dwight was viewing these extremely emotional matters with a kind of purity and abstractionism that failed to recognize the dimensions of the terrible calamities that were being revealed each day in Europe. Nor did he seem particularly sensitive to the effect of these revelations on his Jewish associates in the New York intellectual community.[33]

In the same issue of *Politics* that rebuked Tillich's Council for its collaboration with Stalinists, Dwight scoffed at Roosevelt's hypocrisy in "shedding crocodile tears over Hitler's Jewish victims," then having the "triple-ply brass" to boast of American generosity in "allowing one thousand (1,000) to enter, provided they stayed within a special 'free port' and get the hell out the minute the war is over."[34]

It was Dwight's contact with the concentration camp survivor Bruno Bettelheim and the reprinting in *Politics* of his seminal essay "Behavior in Extreme Situations" that seems to have jolted Dwight into a genuine awareness that there was something fiendishly unique about the methodical repression of Jews, radicals, and "undesirable" minorities in German-occupied territory. Bettelheim had managed to escape Nazi Germany but only after spending a year (1938–39) in two German concentration camps, Dachau and Buchenwald. Neither was a "death camp" per se but both were illustrative of the meticulously calculated barbarity practiced by the Germans in their effort to transform human beings into objects. Bettelheim, a trained psychologist, took scientifically detailed notes as a means of survival—as a way of maintaining an interest in life in a world of death and human disintegration.[35]

It did not take long for Bettelheim to realize that the camps' purpose was to break the political prisoners' will and make them docile and accepting of the guards' values. The camps were also designed to spread terror among the rest of the population and to provide Gestapo members with a training ground, where they were educated to lose all human emotions and to learn the most effective ways of breaking resistance. In a general way, the concentration camp was deliberately designed as a method of producing changes in the prisoners' personalities and characters, which would make them useful as well as pliant subjects of the Nazi state.

Everything was choreographed from the moment of arrest. That initial shock was followed by the most horribly brutal transportation into the camps and the first terrifying and degrading experiences, during which many perished. Those who survived were then subjected to the slower process of changing the prisoners' lives and, more important, their personalities. The final stage in the transformation was achieved when the prisoner had adapted himself to camp life and was reluctant to leave the camp if the opportunity presented itself. In effect, the camp became home.

Bettelheim dealt with this process as he might have explained the manufacture of a piece of equipment—as a routine assembly-line process. He anticipated what Hannah Arendt would later refer to as the banality of evil, although not so much in his treatment of the SS guards as cogs in the wheels of the machine as in his depiction of the routinization and rationalization of horror. He also employed a class analysis of the different reactions of the victims, pointing out how the perspective of the educated political prisoners was distinguished from that of the common criminals and the determinedly nonpolitical middle-class professionals or white-collar clerks. The latter were the most disoriented because they could never grasp why they were incarcerated. They protested that they had not opposed Hitler and had simply gone about their business. Like Kafkaesque victims, they were sure it was all some horrible mistake of the bureaucracy.

Ingenious methods of torture were applied most horribly in the beginning to break instinctive resistance, Bettelheim reported. Those who fought were immediately killed. That many were able to survive such pain and degradation never ceased to amaze Bettelheim; as a prisoner, he felt that the main task was to safeguard one's ego in such a way that one would be approximately the same person he had been before being deprived of his liberty. The challenge was to remain both alive and unchanged. The final success of the process was achieved when the prisoner did change, identifying with the guards and embracing their values. This was done in countless little ways, by emulating the guards, wearing a piece of a guard's uniform, or sewing and mending their own uniforms so as to resemble those of the guards.

The point of Bettelheim's account, made even more devastating by its detailed, even clinical analysis, was clearly that this mechanical, routinized process was a model for the control of all the citizens in a larger totalitarian society, or, as Bettelheim put it, "the larger concentration camp called greater Germany." The Nazi system was effectively designed to destroy all forms of individualism and independence, to produce a childlike dependency on the will of the leaders of the State. All of this only served to confirm Dwight's long-standing fear of the direction of the modern nation-state systems, of their bureaucratic collectivist control, which he had predicted as the wave of the future as far back as his initial studies of dictatorships in the mid-thirties.

Dwight, a pregnant Nancy, and little Michael were able to escape the somber atmosphere of New York during the terrible months of revelation in the summer of 1944 by taking a cottage on Cape Cod. The Macdonalds had been drawn to the Cape by friends like Mary McCarthy, and they managed to make arrangements to rent a tiny guest house on the property of Polly Boyden, a well-known and well-heeled Cape character who was often well oiled. Both Nancy and Dwight were delighted to be going to the Cape. They had not been happy away from the shore in Vermont the previous summer. Everything about the ocean—the bay, the clams, the beaches, the sailing—attracted them. Dwight wrote with enthusiasm that the neighborhood of Truro and Wellfleet was literally crawling with "literary bohemians." Edmund Wilson lived nearby with Mary McCarthy, although that stormy relationship was about to come to an explosive end. Not far away was Slater Brown, whom Dwight described as "the last of the early 20s gin and poetry crowd."[36]

It was an eventful summer. Nancy got the mumps, which caused con-

siderable consternation because of her pregnancy. In August an unusually violent hurricane uprooted a huge tree on Polly Boyden's lawn close
by their small cottage. This was the occasion of a very long hurricane
party, by the romantic light of hurricane lamps. From the first weeks,
there was a succession of guests and visitors passing through. Nancy and
Dwight were always ready to put up an endless troop of friends and entertain in a very informal way. That summer, Lionel Abel was there, along
with John Berryman and his wife Eileen. They occasionally saw Edmund
Wilson and Mary McCarthy, although Dwight and Wilson were never
close. On one occasion Slater Brown dropped by with three large bluefish
which he pressed upon young Michael. The ten-plus-pound blues were
nearly as large as the boy, and he and Dwight were delighted with them.
Dwight made a special point of living off the local resources of the Cape,
frequently serving his favorite clam pie.[37]

If the paradisiacal atmosphere of the Cape in the summer of 1944
helped ease the tensions of the city and the magazine, Dwight did carry
many of his concerns with him. His associate and best friend, Nicola
Chiaromonte, and his wife Miriam were also on the Cape, and they spent
long evenings discussing the situation in Europe, the terrible events of
the war, and of course the continual reports of the German annihilation of
the European Jews.

Dwight became increasingly absorbed by the question of how such
horrors could have occurred and where to place responsibility. He continued to denounce the attribution of collective guilt. He wrote to Al Goldman, who was serving time in jail as one of the subversive Trotskyists, to
say that he and Nick Chiaromonte had been having some interesting conversations on the Cape concerning the "question of the responsibility of
peoples." To what degree were the German people, as a people, accountable for "the absolutely inconceivably horrible mass executions of millions of Jews in the death camps"? It is clear that Dwight's earlier Marxist
explanation, suggested in his introduction to the *Redburn* chapter, had
been demolished by these devastating reports. Dwight now focused on the
issue of "cultural conditioning." The diabolical Nazis had trained special
groups for special tasks: "the murders are apparently carried out by elite
guards who are systematically brutalized by scientific methods for years
before they become inhuman enough to do this kind of work." Dwight was
also concerned about the apparent "powerlessness of individual citizens
and even of vast groups, before the concentration of State power in a modern 'advanced' society." Given such power, what could or could not be

reasonably expected of the masses in the way of resistance? Chiaromonte had suggested that Dwight make a serious effort to formulate the question and to encourage a number of experts as well as general *Politics* readers to respond.[38]

Al Goldman was an articulate Marxist dialectician, and he argued that condemning the German people as responsible only indicated the "reactionary character" of the present epoch. To lend weight to his observation, he turned to the classic British conservative Edmund Burke, who had written that one could not indict a whole nation for the crimes of a few. One would have to take a poll to determine who favored such a policy, and even then, the conditions that shaped such a position would have to be studied. As for why people didn't revolt, that was such a complex matter, involving so many variables (the overwhelming power of the State, the force of tradition), only the most exceptional combination of circumstances could lead to a revolution. Goldman's conclusion was in keeping with his militant politics. He felt that one could safely assume Hitler's days were numbered, that the masses would overthrow the dictator. Then would be the time to study exactly what factors led to such a mass revolution.[39]

Dwight's work on a long article dealing with all these questions was interrupted in October 1944 by the birth of his son Nicholas. Although Nancy had an easier time than with Michael's birth, the case of mumps she'd contracted during the summer was thought to have contributed to high blood pressure, and she was laid up during the winter of 1944–45. Dwight, who already shared in the housework, took over the entire job. He was looking after the baby and getting Michael off to school, as well as working hard on the article "The Responsibility of Peoples."

In January 1945 Dwight commented editorially on the Roosevelt administration's neglect of the victimized Jews, the deplorable role of Secretary of State Cordell Hull, and the lickspittle apology for such conduct by the *New Leader*. It was Dwight's contention that simply because the secretary was an outspoken anti-Stalinist, the *New Leader* would tolerate any policy, no matter how heinous. In this instance it was doubly offensive, since so many of the *New Leader*'s readers were Jewish. The magazine's praise of Hull's role in helping the Jews flew in the face of the horrible facts. Hull's record was a "cause of shame to every decent American," Dwight wrote; the Jews did have a special claim on the United States for help because they had been marked for death by the Nazis simply for being Jews. "Anyone whose brains have not been addled by the . . . simple-minded Moscow-baiting that the *New Leader* goes in for" knew that Hull was "a narrow-minded, petty-pompous, provincial reactionary" who in fact had done very little for the victims of Nazism. As for his second-in-command, Under Secretary Edward R. Stettinius, he was an unqualified "glad-hander and back-slapper with no ideas of his own." Dwight's especially vitriolic critique of the Moscow baiting of the *New Leader* was somewhat paradoxical, since he was frequently accused of the same kind of political harassment of the Soviet Union by many of the younger readers of *Politics*. It was an issue that was to come increasingly to the fore as America's alliance with the Soviet Union grew strained and the expedient reasons for it declined toward the end of the war.[40]

The Allies' frightful record of indifference and neglect toward the victims of Nazi racism further encouraged Dwight to address the subject of Nazi war crimes and the problem of responsibility, which he felt was shared by all belligerent nations. His lengthy and provocative attempt to confront the matter, "The Responsibility of Peoples," appeared in the March 1945 issue of *Politics*. It opened with a descriptive anecdote of a little French girl who had been tortured by the Nazis. Her mother was relieved that she had not talked before her death and no one in the French Resistance had been exposed. He cited another child, this one Polish, who wrote to a friend: "Now I must say goodbye. Tomorrow mother goes into the gas chamber, and I will be thrown into the well." These and other stories were evidence that "something has happened to the Germans—to some of them, at least; something has happened to Europe—to some of it, at least." Dwight wanted to know what had happened, who was responsible, what it revealed about Western civilization, about the whole system of values. It was the great moral question of the time, and he would continue to return to it. It is not surprising that in the opening descriptive paragraphs, he stressed the methodical efficiency of the killing. It was "rationality and system gone mad." He recounted some of the grimmer details of the "death factories" with their railroad lines bringing in the "raw materials." No wastefulness, pure efficiency, everything used, the byproducts, dentures, hair, spectacles in neat piles. He resorted to what was to become a favorite analogy for conveyor belt dehumanization, the Chicago stockyards. The Nazis had learned well the assembly-line techniques of mass production and modern business organization. "Reality has caught up with Kafka's imagination," Dwight wrote. It was a gruesome parody of Victorian illusions about the virtues of the scientific method. Dwight recalled the boasts of man's capacity to control the environment. "The environment was controlled at Maidenek [a Polish concentration camp]. It was the human beings who ran amok."[41]

One of Dwight's great fears was that the enormity of the Nazis' crimes would be used to justify the inhumanity and destructiveness of the Allies' war record. He was compelled to offer up the obvious examples of war's traditional inhumanity that knew no national boundaries. The Allies' terror bombing of civilian populations, even the Nazis' extermination of large numbers of helpless peoples, had precedents, he observed. Since 1800 the world had seen the annihilation of great numbers of black people by whites in the Congo, the slaughter of aborigines in Australia, the decimation of the American Indians, and the starving and working to death of the lower classes in England during the Industrial Revolution. He could not fail to mention the past fifteen years of forced famine and starvation in Stalinist Russia. The English factory owners and Russian bureaucrats had shown a shocking disregard for human life. The difference was that the Nazis had not disregarded human life; they had paid the closest attention to it and "taken it for the pure disinterested pleasure of it." Hence the German crimes were unique. How so? Partly because of their "intimate individual cruelty," and partly because of the gratuitous character of the worst atrocities. "What has been done [in the past] by other peoples as an unpleasant by-product of the attainment of certain ends has been done by the Germans at Maidenek and Auschwitz as an end in itself."

This was a stumbling block for Dwight. He could not accept the extermination of the Jews as an end one could regard as rational. They could not be seen as a genuine threat to the German nation. Anti-Semitism was one thing; the final solution was indeed another. Genocide was then "neurotic." It was undertaken to "gratify a paranoiac hatred," not for any reason of policy or advantage. What had "previously been done by psychopathic killers had now been done by the rulers and servants of a great modern State." Here was Dwight's nightmare come to fruition, the ultimate result of the process of massification. Having abandoned or certainly modified his earlier understanding of a historical continuity that placed the German crimes on a spectrum of human bestiality, Dwight focused on the irrationality of the ends and sought to discover and explain responsibility.

Unquestionably the fact that both Allied and Soviet spokesmen attributed collective guilt to the German people prompted Dwight to challenge this view. To accept it would mean to condone a brutal retribution as well as "our" own war crimes. It would jeopardize his own critical stance and undermine the basis for his anti-war third-camp position. Dwight insisted that the violent anti-Semitism of the Nazis was not a "people's action." Long before the returns were in, Dwight argued that anti-Semitism may have been widespread, but the rounding up and extermination of Jews was not supported or even known by most Germans. The killing had been carried out by specially trained subgroups not representative of the society at large. In fact, Dwight was quick to argue that the torture and lynching of blacks in the South was much more representative of widespread racism than the deeds of the Nazis in Germany.[42]

Dwight made much of the argument that a particular kind of German had been created, citing Bettelheim's article on the transference of Nazi values to anti-Nazis and even to the very victims of their oppression. In the September 1944 issue of *Politics*, Dwight had exposed an American military manual that deliberately tried to break down the civilized individual's inhibitions against inhuman deeds of warfare. The point of all this was to emphasize the undemocratic, tyrannical nature of modern mass societies where "things happen *to* people." The most menacing aspect of all, he declared, was that modern mass society had become so tightly organized, so rationalized, so routinized that it had the character of a mechanism that ground on without human consciousness or control. The individual citizen in the modern mass state was powerless, misinformed, uninformed, not consulted, and had no role in the decision making that directly affected his life. And worst of all, this trend was developing in all modern industrial nation-states to one degree or another.[43]

Because war had become a permanent institution, Dwight gave several examples of the dehumanization process necessary to the war machine. He quoted a bomber pilot who saw himself as simply "a cog in one hell of a big machine" designed for destruction. The war propagandists of course tried to personalize war. A ship is accidentally blown up at a naval station and a naval spokesman waxes eloquent on the heroism and self-sacrifice of those killed in the blast. They just happened to be in the wrong place at the wrong time: Things happen *to* people. Dwight dismissed collective heroism as well as collective guilt, arguing that some Londoners under the blitz were courageous and others cowards, as is usually the case. Glorifying war by personalizing it was simply another calculating means of manipulating and exploiting the mass mind.

In this treatise on the destructiveness of the modern nation-state, Dwight was not trying to mitigate or apologize for the terrible dimensions of the Nazi genocide. His point was to make the enormity of those crimes even more frightening by suggesting that what caused them was an inevitable development of modern mass organization. If the German people were not responsible for "their" nation's war crimes, the world had already become a far more "complicated and terrifying place" in which "un-understood social forces move men puppet-like to perform terrible acts and in which guilt is at once universal and meaningless." The world, he concluded, "had become just such a place." In short, a totalitarian society had arrived in which the citizenry had been reduced to automatons. Dwight did more than suggest that the Allies were not immune to such comparative atrocities. Again he raised the issues of saturation bombing, the betrayal by the Russians of the Polish resistance, the encouragement of a civil war in Greece in order to restore a reactionary regime in power.

"We have made ourselves the accomplice of the Maidenek butchers by refusing to admit more than a tiny trickle of the Jews of Europe to take refuge inside our borders," he wrote. But then, U.S. citizens, like the German citizens, were misinformed, uninformed, and had not been consulted about these heinous policies. In such a system with its uncontrolled governments, it was no longer the lawbreaker who was to be feared but those who obeyed the laws. Dwight held out some hope that in the United States there existed "an honorable tradition of lawlessness and disrespect for authority" that might muster a rebellion against the garrison state that he and his colleagues saw as part of the permanent war economy.[44]

Essentially, the political focus of "The Responsibility of Peoples" was an anarchic anti-statism. The State had become the criminal force in the world. Dwight applauded all forms of individual and community resistance to the modern State. Only these individual and community forces could cross national barriers and achieve a genuine internationalism. In a pamphlet edition of the article, he added a section titled "The Community of Those Who Endure" denouncing the existing international system designed to keep people apart, and the official class in both government and the military that preached chauvinism and hate. Despite the officials' power, their manipulation, their awesome control of information to serve their ends, there still existed a "community of those who endure" made up of human beings who understood the notion of a common humanity. In a last paragraph of exhortation, Dwight rhetorically queried his readers:

When will those [who] endure . . . the wretchedness and filth and hunger and terror grow weary of so much meaningless suffering? When will they

progress from a Community of Those Who Endure to a Community of
Those Who Will No Longer Endure? When that day comes—*dies irae,
dies illa*—the Responsibility of Peoples will shift from the passive to the
active mood and they will have something really to be responsible for; no
doubt terrible actions often enough, but their own and for *their* ends.[45]

Dwight's most urgent concern was what he deemed to be a change in
the moral temper. So unthinkable was the climate of death and destruction in the Nazi extermination networks that even those involved had no
standards by which to judge human behavior. Dwight quoted a passage
from the work of the young German Jewish scholar and political scientist
Hannah Arendt, who was soon to become a fast friend and collaborator. It
was an exchange between an American correspondent and a death camp
official:

Question: Did you kill people in the camp?
Ans: Yes.
Ques: Did you poison them with gas?
Ans: Yes.
Ques: Did you bury them alive?
Ans: It sometimes happened.
Ques: Did you personally help to kill people?
Ans: Absolutely not. I was only the paymaster of the camp.
Ques: What did you think of what was going on?
Ans: It was bad at first, but we got used to it.
Ques: Do you know the Russians will hang you?
Ans: (Bursting into tears.) Why should they, what have I done?

Dwight echoed that astonishing query: *What have I done?* It was the passivity of the question that was so shocking. The paymaster had gotten
used to the killing. As a defense against the spread of such a calamity,
Dwight adopted the rhetoric of what Arendt later called "fearful anticipation." It was a defense because it produced a criterion for judging the
events of our time: Will it lead to totalitarian rule or will it not?[46]

Dwight's essay met his and Nicola Chiaromonte's expectations, elicit-
ing an immediate outpouring of discussion, praise, denunciation, thoughtful criticism, and additional commentary. On several points Dwight
seemed inconsistent, confused, and uncertain, which hardly separated
him from his critics. His claim that the great mass of Germans were
unaware of the dimensions of the final solution was debatable and

seemed to some unpersuasive. He offered no real evidence to support the claim and soon conceded that the whole world knew a good deal before the summer and fall of 1944, which is when he personally first became aware of it. His discussion of the irrationality of the end (that is, genocide) was challenged by several of his more informed readers, who insisted that once one accepted the racist assumption that Jews were a threat to the Aryan race, everything followed with a frightful rationality.[47]

On the most important point, responsibility, Dwight's stance was unclear and contradictory. His view of what he called "massification," a mass society totally controlled by concentrated power in the state bureaucracy and military, led to the notion of universal irresponsibility, where no resistance was possible. If that were true, then the paymaster's tearful query made sense. In effect, if things were as bad and as far-reaching as Dwight's anticipatory fears suggested, assigning personal responsibility to German citizens would be like blaming people for their sins while insisting on the concept of original sin. It would take on a kind of religiosity that Dwight had repudiated in the work of Reinhold Niebuhr and other obscurantists during the "Failure of Nerve" debates. Jim Cork, a veteran radical and critical reader of *Politics*, wrote in exasperation, "You make the whole damned thing sound like an exercise in Niebuhrian catechismetics, i.e., 'Realize your original sin before you can hope to partake of grace and salvation.'" Niebuhr had been impressed with Dwight's initial article and wrote him to say that he had found it one of the "sanest and profoundest analyses" of the problem of individual and collective guilt that he had read. It was, he felt, a "wholesome antidote to the terrible rot" that had been written on the subject, presumably referring to the wealth of popular material tracing the Holocaust to its origins in the German soul. However, Niebuhr also felt that Dwight had not done full justice to the infinite "varieties and degrees of complicity" and of the possibilities of an "evil regime corrupting even the saner elements of the society." This was of course one of Dwight's great fears.[48]

Dwight's initial exoneration of the German people and his assigning responsibility to the Nazi leadership and their well-trained psychopaths was quickly challenged by readers who claimed it granted too broad a pardon to society at large. Dwight soon backtracked, conceding that he had gone too far in giving the impression that the German people bore no blame of any kind for the Nazis. "On reflection, I think my article stressed the 'No' too strongly, gave too much the impression that the German people have no responsibility of any kind." This was because he dealt with the issue as a moral question, "with the consequent idea of punishment; in that sphere responsibility can only be an individual matter and is . . . related to an individual's freedom of action." He went on to charge that the German people had a political responsibility for Nazism because they had "permitted Hitler to come to power" and they had "endured his rule without revolt." To absolve them of this guilt would be to view them as simply victims, dupes, or slaves. He insisted that he did not believe the masses were simply "inanimate raw material which Führers and demagogues mold at will." At this point in the discussion he asserted that, on the contrary, "they were capable of initiative" and had in fact "intervened on the stage of history with decisive results" and therefore had to be held politically accountable for not intervening.[49] But of course that is exactly what he had initially denied when he wrote that if "the German people are not responsible for their nation's war crimes the world becomes a terrifying place, in which un-understood social forces move men puppet-like to perform terrible acts" and had added that the world was indeed just such a place. It was really a conflict between his anticipatory fears of what the events in Germany signified and his current judgment as to the present state of reality. It was a matter of a political analysis that demanded a transcendent indictment of the danger of totalitarianism spreading across the face of the civilized world and his own judgment as a journalist avoiding ideological or political abstractions and dealing with the concrete situation. This is an important context in which to try and understand political rhetoric. It is the same kind of conflict that can be found in Dwight's insistence on a repudiation of the war and a rhetorical allegiance to a revolutionary socialist third camp which was in conflict with his own observations which informed him that such a socialist revolution was not probable or even possible.

Dwight's attribution of political guilt as distinguished from moral guilt did not clear the muddied waters. It is not surprising that readers of *Politics* quickly turned their attention to what might have been expected from the German working class and the issue of a collective politics. While praising Dwight's article, Jean Malaquais, an articulate Trotskyist, insisted that a "people" don't have a politics. A people are always socially heterogeneous, politically confused, and morally nonexistent. On the other hand, social classes are political because they have specific, well-defined interests. The German working class shared responsibility for bringing Hitler to power. In fact, he insisted, the entire European working class bore some responsibility for both Nazism and Stalinism (an equation that was to become a constant in the pages of *Politics*) because the working-class organizations did not counter the fascists and Stalinists effectively.[50]

It was Nicola Chiaromonte, the originator of the discussion, who offered, nearly two years later, the most menacing conclusion. He wrote from Rome to explain why it was that Europeans were having difficulty taking a charitable attitude toward the Germans.

> The notion that Germans are people like any other people is not rejected, but simply does not work. Through the Germans, Europe committed suicide. But still the Germans did it. The guilt might be general. But still the Germans actually did it, not the French, or the English, or the Italians (even though . . . one might say that Italian Fascism was the ominous beginning). . . . The question is not a historical one—and not even political. The fact is, again and again, that people did not think man could become what he became with Hitler until the Germans showed it was a question of mechanisms, and worked it out completely. This feeling of utter debasement, of having been forced to despair of man, is still present everywhere.[51]

It was during that debate in August 1945 that all of Dwight's nightmarish fears were confirmed by the reality of that mushroom cloud in the far reaches of the Pacific.

8

In a Terrifying World
the Root Is Man

IN THE MIDST OF THE HEATED DISCUSSION HIS "Responsibility" piece ignited, Dwight and Nancy left in June 1945 for a summer of work and play on the Cape with Michael, a precocious seven-year-old, and Nick, not yet a year. Nancy persuaded Virginia Chamberlain, an old friend, to come along to help care for the children. Dwight rented a large two-story house called Four Winds on a bluff overlooking the North Truro fish factory and Provincetown Bay. Railroad tracks ran in back of the house and the Cape Codder made a trip up to Provincetown every two days. This was a major event, and Michael made it a point to be down by the tracks to wave at the engineer and the passengers as the train went by.

It was a busy summer, with many guests and visitors; several friends lived in Wellfleet. Since they were lodging next to the fish house, Dwight made it a point to eat as much seafood as possible. On one occasion Virginia and Dwight worked hard to follow a recipe for squid, but it turned out to be a gastronomical disaster that still serves as a reminder of that fateful summer.[1]

On August 7, 1945, the sun shone brightly on the Cape. It was an unusually clear and beautiful day. The first news of the bomb came when a neighbor appeared at the door and asked whether the Macdonalds knew that we had dropped a bomb on Hiroshima, wiping out the city. Virginia

Chamberlain recalls that they were "all horrified and found it hard to believe." Later in the day they all visited with Mary McCarthy, her brother the actor Kevin McCarthy, and Philip Rahv, who were staying at the Polly Boyden house in Truro, and commenced to argue the pros and cons of what had happened. Most thought it highly immoral. Strangely enough, a niece of General George Patton was present. Dwight had blasted Patton in the cover story that had just gone to press for committing "atrocities of the mind" by giving a warmongering speech to school-children. She was the only one who voiced what quickly became the official defense of such massive destruction: An invasion of the Japanese mainland was in the offing, and it was predicted that millions of American lives would be lost. Dropping the bomb had avoided that. As Virginia Chamberlain recalls, this did not satisfy the little circle of Dwight's friends, and for days a pall hung over the group. A dreadful act of destruction had been committed, and they were convinced it would change their and the world's future.[2] Even though Dwight's politics and his study of the war's destructiveness and his feelings had prepared him for such an event, the news came as the most frightful of his imaginary anticipations.

Dwight quickly called the *Politics* printer to hold the presses so that he could insert a few paragraphs on the bombing to accompany the "Atrocities of the Mind" cover story. His cover editorial began by asserting unequivocally that the bombing of Hiroshima "places 'us,' the defenders of civilization, on a moral level with 'them,' the beasts of Maidenek. And 'we,' the American people, are just as much and just as little responsible for this horror as 'they,' the German people" were for the horrors committed in their name. Dwight argued that the use of this weapon had rendered the concepts of war and progress obsolete. They expressed human aspirations; the bomb expressed only death and destruction. Truman had declared it "the greatest achievement of organized science in history." Dwight agreed, observing, "So much the worse for organized science." He was soon to indict Oppenheimer, Einstein, and other collaborating scientists for toadying to the demands of political leaders. One thing was clear: the futility of modern warfare. In Dwight's view, the bomb was the natural product of the kind of society Americans had created, as representative of the standard of living as electric iceboxes. Those willing to wield such weapons were outcasts from humanity, brutes, not men, and he concluded this agonized outburst with the warning: *"We must 'get' the national State before it gets us. . . ."* The crazy and murderous nature of the kind of society we have created was underlined by the atomic bomb. Every individual . . . had better begin thinking 'dangerous thoughts' about sabotage, resistance, rebellion, and the fraternity of all men everywhere." With some pride in his own characteristic editorial approach, Dwight closed by insisting that the "attitude known as 'negativism' is a good start."[3]

Dwight's unequivocal reaction was virtually a voice in the wilderness; only the tiny group of like-minded souls who surrounded him that summer on the Cape—and the little band of *Politics* readers who were prone to pacifism, conscientious objection, and anarchism—shared his view. At the Cape that summer, Mary McCarthy, Dwight and Nancy, the Chiaromontes, and Niccolo Tucci were the center of a small cadre of writers and critics, political and social commentators who hammered out the outlines of the position they would take in the immediate postwar years. In the next weeks and months they would all spend a good deal of time thinking and writing about the meaning of the bomb.[4]

For Dwight, the bomb was another turning point, one of the most important watersheds in his erratic political odyssey. Niccolo Tucci had already made his own sardonic comment on the meaning of the Allied victory in Europe several months earlier, which Dwight had republished in large type on the cover of the May 1945 issue, accompanied by a grotesque cartoon drawing of skeletons dancing over a barren landscape. Tucci's observations had a horribly prescient quality:

> VICTORY. Comes, lights up the horizon and the hearts, and before you know it, it's gone; you have just the ashes and the dead. . . . Makes me sad for those soldiers who are there, in the line, with ideals all theirs, reserved to the military, "requisitioned for the exclusive use of our boys," and forbidden to everybody else at home or abroad. Their job is that of transforming a torture-chamber into a cemetery; a place of terror and of hope into a place without terror and without hope. VICTORY.[5]

It was this grim vision that shaped Dwight's conception of the final end of the war and of much more. From the beginning, Dwight equated the dropping of the bombs with the Nazi death camps. These atrocities were at that very moment "brutalizing, warping, deadening the human beings who [were] expected to change the world for the better." If Dwight had grown increasingly suspicious of the role of science and technology in the modern world, the creation and delivery of the bomb sealed his certainty that science and modern technology had their own "anti-human" dynamic and that this had proved far more powerful than the apostles of progress, the liberals and the Marxists, had expected. He promised in an aside that a forthcoming article, "The Root Is Man," would deal with the failure of the Marxist vision as well as that of the scientific liberals. The notion that atomic fission could be a force for good or a force for evil was nothing more than an illusory platitude. The history of capitalism and imperialism suggested that it was very difficult to extract the good from the evil, and "already the great imperialisms [were] jockeying for position in World War III."

Again Dwight insisted on making a connection between this inhuman atrocity and the technical efficiencies of the Nazi death camps. Repeating the early rumors of the horrendous potential after effects of radiation, Dwight noted that the men who had produced and those who employed the monstrosity had not known how deadly or prolonged the effects of the radioactive poison would be. Perhaps, he speculated, only among "men like soldiers and scientists trained to think 'objectively' i.e., in terms of means, not ends, could such irresponsibility and callousness be found." It was, Dwight agreed, as some of its proponents claimed: the greatest scientific experiment in history. Indeed it was, "with cities as the laboratories and people as the guinea pigs."[6]

Dwight returned to the question of responsibility, noting that just like the German people, the American people did not know what was being done in their name and had not the slightest chance of stopping it. Here was "the most dramatic illustration to date of the fallacy of 'The Responsibility of Peoples.'" The actual workers in the plants did not know what it was they were working on. "There is something askew with a society in which vast numbers of citizens can be organized to create a horror like The Bomb without even knowing they are doing it."[7]

It was fitting, Dwight thought, that two "democracies," the United States and Great Britain, had developed the bomb and even more appropriate that they were headed by two "colorless mediocrities," Truman and Attlee, "elevated to their positions by the mechanics of the system. All of this emphasizes that perfect automatism, that absolute lack of human consciousness or aims which our society is rapidly achieving." It seemed "the more commonplace the personalities and senseless the institutions, the more grandiose the destruction. It is *Götterdämmerung* without the gods." Dwight was confronting what Hannah Arendt would identify as the banality of evil, a system of destruction as a way of life manned by bland bureaucrats.[8]

Dwight concluded his piece by returning to the problem of immediate responsibility. The scientists received the brunt of his accusations. They had been willing to produce and deploy a weapon whose horrible effects were unknown. Why had they done this? Because they "thought of themselves as specialists, technicians, and not as complete men." Dwight cited the movie image of the "white coat of the scientist" as being as "blood-chilling as Dracula's black cape." If the "scientist's laboratory had acquired in Popular Culture a ghastly atmosphere," Dwight wondered whether it might not be one of "those deep intuitions of the masses. From Frankenstein's laboratory to Maidenek (or, now, to Hanford and Oak Ridge) is not a long journey."

The scientists had all accepted the assignment. There was the rub. They had acted as automatons, doing the State's bidding. The tendency was to think of peoples as responsible and individuals as irresponsible; the reversal of both these conceptions was the necessary first condition for any escape from a decline into barbarism, Dwight wrote. He conceded that "to insist on acting as a responsible individual in a society which reduces the individual to impotence may be foolish, reckless, and ineffectual; or it may be wise, prudent, and effective. But whichever it is, only thus is there a chance of changing our present tragic destiny." Dwight honored those nameless scientists "who were so wisely foolish as to refuse their cooperation on The Bomb! This is 'resistance,' this is 'negativism,' and in it lies our best hope."[9]

The bomb was the final catalyst precipitating Dwight's rejection of Marxism and "scientific socialism." He was now firmly convinced that Marxism was no longer a help but a hindrance to the fundamental aim of *Politics*'s enterprise of seeking out long-range trends and measuring events with a yardstick of basic values. Marxism, he now felt, made it difficult to be radical, to really grasp the root of the matter. To quote Marx, "now the root for mankind is man himself."[10]

Dwight had begun work on his projected essay, "The Root Is Man," early in 1944 and returned to it intermittently for nearly two years. The bomb helped jar it loose, arousing as it did all his guarded reservations concerning the scientific pretensions of Marxism, the scientific method itself, historical inevitability, and the role of the working class in social change. One could trace a skepticism toward many of these issues as far back as his 1935 notes on dictatorships and the modern nation-state. *Politics* had begun to reflect this shift in many ways almost from its first issue. There was an intensified concentration on pacifism as an activist political stance. The politics of conscientious objection was given increasing space. Dwight maintained a wide and growing correspondence with these often religiously committed young men, who were inspired and encouraged by Dwight and constituted a hard core of the journal's devoted readership. Initially he dismissed conscientious objection as ineffective because it isolated the dissident. However, after the dropping of the bomb, when he could no longer consider massive political resistance a possibility, he focused on what the individual could do. Dwight printed with approval a letter from a young reader, Herbert Orloff, who insisted that the bomb had destroyed more than Japanese cities, it had destroyed the fundamentals of the revolutionary socialist movement, but Marxists refused to face up to that reality. Since the ruling class had an all-powerful club, mass military or armed action by the working class was absurd, he wrote. New programs of nonviolence had to be designed; the entire emphasis of socialism had to be shifted from mass action to individual action. The strike, sabotage, boycott, civil disobedience were the only tools left by which to break down a bomb-holding State.[11]

It was almost as though Dwight's Marxist interlude, while sharpening his critical tools, had been a great intellectual burden. Now it could be abandoned and with it all the articles of faith that had accompanied it: the revolutionary potential of the masses, the militancy of a socialist working class, the historicity that promised a socialist future, the entire optimism about Marxism as another example of the nineteenth-century devotion to scientific rationalism and the notion of inevitable human progress. All this could now be openly challenged in the name of a realistic, rebellious, resistant individualism and a thoroughgoing anarchic anti-statism.

In November 1945 while working on "The Root," Dwight published Simone Weil's "*The Iliad*, or The Poem of Force," translated from the French by Mary McCarthy. Weil already had an international reputation in radical and pacifist circles as a person who lived the life she professed. Dwight loved her for her clarity, her ability to probe to the source of a problem, and her hatred of all forms of oppression. She became for him a saint of individualistic activism. Thus he was delighted to publish her treatise on the evils of power and coercion. Her essay on *The Iliad* was an eloquent interpretation of the classic as a paean to force, a force that ultimately enslaves and destroys those who resort to it. Force was that power that turns those who are subject to it into things, Weil wrote. It turns man into a thing in the most literal sense; it makes a corpse of him. Reading the poem from this perspective, one sees the continual contrast between a humane and peaceful world and a world of violence, enslavement, and death. Force crushes its victims, but it also intoxicates its possessors and ultimately brings them too to ruin; both those who use it and those who endure it are turned into stone. Force "effaces all conceptions of purpose or goal, including even its own war aims."[12]

Weil's explication was a restatement of the themes that preoccupied Dwight and emerged in *Politics* as his revolutionary socialist rhetoric eroded and as he became more interested in anarchic pacifism. He was pleased as well as astonished by its reception. He wrote to one of his correspondents that he thought it a great piece of political writing, which dealt with "the moral questions implicit in the terrible events one reads in every day's newspaper"; the only people who failed to grasp why such an article had a place in a political journal were, "and I think this profoundly significant, all of them Marxists." To a Marxist, Dwight observed, "an analysis of human behavior from an ethical point of view is just not 'serious'—even smacks a little of religion." It is true that Hans Gerth, the very epitome of the Marxist-oriented social scientist, was appalled by the applause the essay received and charged that it represented a drift toward aestheticism and reflected the "political retreat of a homeless radical intelligentsia."[13]

Gerth's statement confirmed Dwight's perception that two opposing camps were developing among the little band of *Politics* readers and contributors. One was made up of what Dwight saw as the "traditional liberal left whose philosophy is scientific, materialist in general and Marxian and Deweyan in particular." These he labeled "Progressives." The other camp, whom he labeled "radicals," were "as yet an amorphous group which includes many of the contributors (and the editor) who have lost their scientific materialist faith but still consider themselves socialists and are groping for a more adequate philosophy."[14]

Since it was indeed Dwight's magazine, and since he was hardly one to shy away from division and conflict, he announced in November 1945 that *Politics* would host a series of Friday-night discussion meetings in the barnlike hall known as the Stuyvesant Casino on Second Avenue on the Lower East Side. He led off the series with a truncated version of his pending piece, "The Root Is Man," exposing for all to see what he described as his un-Marxist and unscientific ideas. The hall was packed with over 350 people. The Marxist radicals, many of them Trotskyists, were there as a claque, as were a vociferous group of anarchists. Dwight, in his high-pitched stammer and with an occasional giggle, delivered what amounted to a dismissal of Marxism as an obsolete holdover of nineteenth-century materialism that had proved inaccurate as a prophetic instrument and was not even reliable as a method of analysis. At the same time, he dismissed the pretensions of the scientific humanists for their unjustified optimism concerning scientific progress. This was greeted by many of the partisans with hoots, jeers, and polemical, sometimes abusive, criticisms of Dwight and the *Politics* circle. Dwight felt that the meeting "degenerated into a battle of political stump speeches between the Trotskyists and the Anarchists, producing more heat than light." He blamed it partly on his own paper for attempting more than could be developed adequately and his reluctance to hog the floor during the discussion period.[15]

Many *Politics* readers were unsympathetic to what they saw as Dwight's religiosity, his vague moral stance, and his wholesale repudiation of not only Marxism but the rationalist tradition. To the novelist Calder Willingham, Dwight had become one of the obscurantists he had formerly criticized in the *Partisan Review.* Despite the criticism, the discussions continued. Lewis Coser somewhat mollified the Marxists with an upbeat account of the potential resistance he saw building in France. This was followed by another provocative baiting session conducted by Nicola Chiaromonte on why "Socialism Should Be Utopian," designed as a repudiation of Marx's famous critique of utopian socialists and an insistence on a moral and ethical approach to political action. Dwight expected "fireworks" and he was not disappointed.[16]

It was true that Dwight did seem bent on narrowing his constituency by assaulting nearly every potential ally on the left. He dismissed the labor movement as simply the co-opted handmaiden of corporate capitalism and intricately dependent on the permanent war economy. The lib-labs were apologists for war and Allied imperialism, not excluding the Russians; in effect, they were "totalitarian liberals." The socialists continued to support the war effort and to put forth Norman Thomas as their candidate for president, a man who in Dwight's view was indistinguishable from the bourgeois liberals. The Trotskyists were hidebound Marxist ideologues given to Leninist authoritarianism. The British Labour Party was in the same camp as the American lib-labs and socialists and had completely sold out to British attempts to save as much of the empire as possible. Negativism remained Dwight's single weapon, a purity in *Politics* and in politics, too, that had its comforts but offered little in the way of genuine political activism.

Seymour Martin Lipset, a young socialist reporter, wrote a forceful critique of the political ineffectiveness of the journal in a letter to Dwight, saying that *Politics*'s contributors and readers were little more than "smart aleck masochists on the left" who took pleasure in proving that "everything is hopeless" and spent their time jeering at those who were still trying to build a socialist movement. Dwight, as the editor, did not contradict this despairing message. "Damn it Dwight, whether you realize it or not you are playing a reactionary role while editing the best magazine the left has ever had."[17]

Dwight did indeed want to do something. He took pleasure in being the head of a like-minded group. Lipset had referred to the "Macdonaldites . . . hundreds, perhaps thousands of them." Andrea Caffi had written from Paris to praise *Politics* as "a lighted torch around which to rally." Dwight wanted to assume such a leadership role. He was obviously flattered by such attributions and must have still harbored that old Trotsky image of the intellectual as man of action. He had declared early in March 1945 that *Politics* could serve in reestablishing socialism as a moral idea. He was certain the "human spirit was tougher, more resilient and tenacious than the more mechanical varieties of bourgeois and Marxist thinking." While he did not think it worthwhile to look to existing Marxist parties as a vehicle for resistance, he did assert that the "main thing was simply not to put up with things as they are."[18]

The October 1945 cover of *Politics* carried in heavy black type the announcement: "Here's ONE Thing We Can Do! An Appeal to Readers of 'Politics.'" It was a plea for aid to the starving people of Europe. The American government had destroyed Europe's industrial apparatus but had been "callous and hypocritical" in providing aid. Dwight printed firsthand accounts of the terrible plight of the Europeans and how the United States government had cut off lend-lease abruptly and deliberately weakened the United Nations Relief and Rehabilitation Administration. What our government would not do could be done by concerned individual Americans.

Dwight then outlined what became known as the *Politics* Package Project. Bundles of food, clothing, and other essentials would be gathered and sent off to the beleaguered intellectuals of war-torn Europe. Both Dwight and Nancy considered this project a direct refutation of the charges of ineffective negativism. They also saw it as deeply political in tone and in practice. It was an organizing tool to draw together small groups of like-minded radicals. It was anarchic in its anti-government stance. The idea from the start was to create an international community of independent radical intellectuals who refused to accept the leadership of the imperialist capitalist democracy and were veteran foes of Stalinist tyranny. So political was the project that when an editor of the socialist *Call* accused Dwight of stealing the idea from them and then refusing to join in collaboration, Dwight explained that he had chosen to set up his own plan because "the readers of *Politics* and its friends overseas" were "presumably a somewhat homogenous group." Many of them were personally known to Nancy, who had never stopped working with refugees and displaced persons and had initiated the package project among personal friends in the summer of 1945.[19]

The *Politics* office was the center around which a great many in the group pivoted. Danny Rosenblatt, a young man from Detroit who had lost his job with the maritime union and was a regular guest at the Macdonalds' on the Cape, vividly recalls his first encounter with Dwight in the Dickensian warren at the Bible House. Dwight was picking over some cast-off garments from the package project—bundles of old clothes to be sent to Europe. Dwight had apparently just put on a pair of old trousers with a hole in the crotch. "One of his balls was hanging out," Danny recalled.

I was so impressed with his appearance, with the tiny ratty office, the slanted floors caving in, books scattered everywhere, old newspapers piled in the corners . . . mountains of clippings. It was absolutely wonderful. . . . There was something so honest, so genuine, so idiosyncratic. He seemed to have such an open direct approach to everything. It was just like the magazine.[20]

The project was scorned by the Marxist wing of the *Politics* readership. Sam Horn, a staunch socialist, wrote that the idea resembled "an attempt to bail out the ocean with a thimble." It helped relieve stricken American consciences, but not the suffering in Europe. But Dwight's critics soon understood the project as a personal action on the part of the Macdonalds and a reflection of their anarcho-pacifism.

In the December 1945 issue of *Politics*, Dwight published an eleven-page supplement titled "Starvation: America's Christmas Gift to the European People." It was an exhaustive account of the entire relief situation and an unequivocal condemnation of United States policy, which, he said, had resulted in mass starvation, particularly in Germany and Eastern Europe. He quoted a Quaker report that asserted, "No child born in Germany this year will survive the coming winter. Only half the children aged less than three years will survive." The American people were apathetic about the situation while consuming meat at a rate many times that of the rest of the world. He labeled Truman "the happy hypocrite" who repeatedly issued pious statements about America's concern while approving cuts in the food allowances. Dwight quoted reports of American soldiers in Europe living high on the hog, with post exchanges loaded with provisions, while their neighbors were starving.

As for the peace itself, at Potsdam the victors had drawn up "Genghis Khan terms for defeated Germany." Dwight quoted the British *Economist*'s denunciation of the peace as the Hitlerian finish to a war against Hitlerism: The settlement would not last a decade, and when it broke down there would be "nothing but the razor edge balance of international anarchy between civilization and the atomic bomb." Dwight was particularly horrified by the accounts of starvation and violence in eastern Germany under Soviet control. He quoted a British major's remarks as reported in the London tabloid the *Daily Mirror*:

> The greatest horror in modern history is taking place in Eastern Germany. Many millions of German people have been ejected on to the roads . . . are dying by the thousand . . . from starvation, dysentery and exhaustion. Even a cursory visit to the hospitals in Berlin . . . is an experience which would make the sights in the concentration camps appear normal.

There was something insensitive as well as revealing about Dwight's willingness to quote a British military officer in a cheap London tabloid equating the postwar trauma and dislocation of civilian populations with the methodical destruction of Jews and other "undesirables" in the German concentration camps. But conditions in eastern Germany, were indeed dire, and the Russians were hardly able or willing to offer much in the way of aid. Dwight saw the Russians as being chiefly responsible for the horrors. But his explanation was astonishing. He wrote that such barbarism was to be expected, since "the Russians are, as is well known, a primitive, semi-Asiatic people whose national traditions and present totalitarian government combine to place a low estimate on human life." What was not to be expected was the Allies' tolerance of such behavior.

That they had allowed such mass expulsions of people made them also responsible for the millions of Germans who would surely perish.[21]

Dwight began to dwell on the theme of the subhuman quality of the Russians. It became his firm belief that "something monstrous had developed in the USSR." He began printing accounts of Russian atrocities immediately following the end of the war. In the October 1945 issue Dwight published a letter from an American sergeant who helped run a small camp in Belgium where groups of Russian displaced persons and then German war prisoners were successively quartered. Dwight titled the piece "500 Red Army Men," and in his introduction noted that the sergeant's "close observation of the Russians gives one an intimate sense of the kind of human beings" the Soviet Union was turning out. It also gave "a decidedly 'unofficial' view of the Germans," presumably meaning that the prevailing official view was one of prejudicial denigration and slander to justify Carthaginian peace plans.[22]

The sergeant's letter characterized the Russians in great detail as ignorant, crude, primitive peasant types, dirty as well as uninformed, who cared only for their creature comforts. They had no knowledge of their own history or culture and essentially no interest in it. By contrast, the German prisoners of war who occupied the same quarters after the Russians left were clean, orderly, well educated, concerned about the quality of their environment, and ingenious at making life more pleasant for themselves. Interested in literature and culture, they even went so far as to put on theatrical productions that were intelligent and entertaining. For *Politics* readers who might fail to get the point, the sergeant summed up the implications of his observations:

As long as the Russian was eating and sleeping, regardless of how sloppy his conditions might have been, that's as far as his mind traveled. . . . Now then, without any further material, one can draw a great difference between the Russian and German regimes on a comparative basis of what those regimes did to the individual. Now assuming, as we have been the entire length of the war, that German Fascism is a thing to be avoided and fought against, and considering the above evidence, which is only partially given at that, What do you think the Russian form of Government must be?[23]

This contribution was a bit much for at least two *Politics* readers. One, who signed his letter with a pseudonym, Balticus, conceded that one could not build a better world on hatred of the Germans, but that to try to prove that the Germans were better than others was going too far, even for a good international socialist. This critic was particularly scornful of the sergeant's praise for German proficiency in organization. He conceded that the Germans were indeed "specialists in organizing a regimented life," but noted that one should not forget those "for whom that proficiency meant death." Referring to Dwight's earlier defense of the German people, Balticus also wondered where the German underground movement was, if the German people were so opposed to the Nazis and did not condone their methods? Underground resistance existed all over Europe. The Italian workers killed Mussolini and shot every fascist they could get their hands on. "Had one Nazi died at the hands of the Germans since V.E. Day?" He concluded that Europe during the war had been one vast concentration camp and unfortunately every German soldier was a guard in the camp. Dwight did not, as was often his practice, respond to this biting criticism of his editorial judgment. The poet and essayist Edouard Roditi complained of the unreasonable comparison of Russian peasants with the far more educated, urban, and industrialized workers and peasants of the Balkans and both Eastern and Western Europe. "It is a cultural rather than a political difference," Roditi argued. "The differences between Russian and German prisoners proves nothing about the values, as processes of Nazism and Stalinism." In a marginal note Dwight dismissed Roditi as a "typical pro-Sovietist. They justify the Soviet regime as uniquely progressive; and they justify its results because you can't expect it to reshape people."[24]

There is something in the total political configuration of Dwight's position at this time, prior to the cold war, which, when viewed as a whole, is troubling. No one can quarrel, and few in his circle did, with his genuine attempt to oppose the tide of hate, prejudice, and vengeance that follows a war and was surely inevitable immediately after Germany's defeat. This was especially to be expected given the searing accounts and photos of the liberated death camps. It is true that American chauvinism was on the march and the ghastly material was being exploited for political and foreign policy purposes. It was in Dwight's nature and in his understanding of his professional obligations to react against this exploitation. As for his anti-Russian perspective, it must be recalled that there had been, in his view and that of most of his associates, a nauseating pro-Russian sentimentality during the war that ignored and apologized for the terrible realities of the Russian state. Again Dwight's notion of his professional obligation was to set the record straight.[25]

Nevertheless, to suggest that those reports coming out of the death camps were sensationalized, exploited, and then, by implication, exaggerated, to insist that the invading Russian army was made up of monsters who committed crimes on a par with the atrocities committed in the camps, to argue repeatedly that Stalinism was not just equal to Nazi totalitarianism but a worse danger, simply because there were those on the left who refused to recognize it as such—all of this, combined with an increasing solicitude for the German "people," was more than merely contrary to the prevailing wisdom; it showed an insensitivity to the terrible shock many of Dwight's own Jewish associates and friends had experienced. If he was politically responsive to the situation, he was unresponsive in the very areas that one had reason to expect Dwight Macdonald to appreciate. To plead the case of the German people so eloquently may well have had a genuine and justified political rationale, but since discussion and analysis of the plight of the European Jews was not a major preoccupation of the journal, the overall impression one gets from this material is unsettling. While Dwight denounced the notion of collective guilt as it applied to the Germans, he had no trouble discussing collectively the kind of human beings the Soviet state produced. Nor was there ever any discussion or analysis of Russia's role in the war or its cost in human lives, which far exceeded all of the Allied casualties put together. Dwight saw the Soviet Union as the great menace to Western civilization. But the Russians could, and did with some justification, argue that it was their defeat of the Germans on the Eastern Front that had saved that Western civilization. Surely that argument had as much weight as his suggestion that the raping and pillaging Russian soldiers were one of a kind with the SS and the Gestapo. When Dwight did print descriptions of Allied rape he never suggested that it equaled Nazi war crimes, it was simply another example of the terrible brutality of war. It would seem that Dwight's intellectual concerns, his political interests and even his principles were intact but that his moral sensitivity, which counted for so much, was encased in the very kind of Waspish certitude he deplored.[26]

An example of the insensitivity may be seen in an exchange Dwight had with one Arthur Steig, a *Politics* reader and occasional contributor. It grew out of a passage in a review by Nicola Chiaromonte of an article by Salwyn Schapiro which charged the nineteenth-century French journalist Pierre-Joseph Proudhon with being an anti-Semite and a "harbinger of fascism." In his defense of Proudhon's having frequently made an ethnic connection between Jews and international finance, Chiaromonte argued that such a connection "was not, after all, altogether arbitrary and without foundation." Steig, incensed by the use of this old saw, wrote Dwight an impassioned letter suggesting that it was just this kind of stereotyping that had led ultimately to the gas chambers.[27]

Dwight responded in a lecturing tone, asking whether Steig was familiar with such sentiments in the writings of Marx and Engels. He repeated the Chiaromonte "Jew-capital amalgam" and argued that the connection had "some justification" historically. Anyway, Proudhon's statement would be similar to "someone saying they detested Scotsmen." It was true that the Jewish stereotype had turned into "a monstrous cold-blooded and pathological lie," but Steig had to view the issue historically: Jewish bankers had indeed played a "big role in an earlier stage of capitalism and thus had been a force for evil." He closed by telling Steig to keep his shirt on. Talking about the murder of Jews as resulting from Chiaromonte's defense of Proudhon was "just adolescent."[28]

Steig was not to be put off. He conceded he had not read Marx or Engels on the subject, but he was familiar enough with the "demonology of anti-Semitism to understand the background of the 'Jew/banker'" connection. He felt that scholarship had long since dismissed that trite slander and in fact most Jews were simply "poor bastards." Then, applying Dwight's logic, he wondered that if it proved to be true that Jews played as much of a role in banking as they played in film production then presumably the "cold-blooded lie would become an attitude with some justification."[29]

Dwight did not temper his continuing attack on the Marxist left and particularly those lib-labs who continued to make excuses for the atrocities committed by the Soviet state. In December 1945 he initiated a series titled "New Roads" to criticize the "dominant ideology on the left which is largely Marxism . . . and to suggest and speculate on new approaches to the central problem: how to advance toward a society which shall be humanly satisfying?" The initial contributions were pieces by Will Herberg, already well into his odyssey from left to right, Andrea Caffi, and Paul Goodman.

This triple-barreled assault, expressing an individualist, anarchic, pacifist, and libertarian politics, a politics of feeling and sensibility, was indeed the main thrust of the Macdonald wing of the movement. Goodman's article presented his libertarian sexual argument that throwing off society's oppression would free people from their "anxious submissiveness to authority." He was particularly intent on liberating small children and adolescents from Victorian taboos so that they could escape the personal patterns of coercion and authority that might reemerge no matter what form political change might take.[30]

When one "Constant Reader" wrote in to object to the "imbecilic and confused ex-Marxist philosophical anarchistic rubbish by idiots like *Goodman!*", Dwight defended Goodman's piece as a "serious and competent contribution." Calder Willingham also complained of Goodman's "gaseous discharges" and unintelligible prose. Dwight himself had trouble with Goodman's syntax, but there is no question he endorsed the libertarian message. To the charge that he did not allow Marxists sufficient space to answer, Dwight said the pages were always open to their criticism but that perhaps the inadequacies of their position explained their present "paralysis."[31]

Unintimidated by growing criticism, Nicola Chiaromonte offered a dismissal of the "scientific pretensions" of Marxism, the dialectic, the vision of a world free from class conflict, the entire litany. In addition, Chiaromonte maintained that Marxists scoffed at the "ideal of justice" and jeered at the ideas they were supposed to uphold because they considered ideas to be instruments to gain their ends. Like Dwight, Chiaromonte insisted the only way of looking at socialism was with a utopian vision. On this score, Chiaromonte reflected a Camus-like sensibility, especially when he stated that a "socialist ideal is not firmly grounded if it does not accept at the very start the possibility of nonrealization as one of its fundamental possibilities." This was a deliberate attempt to express, in contrast to the optimism of the lib-labs and the Marxists, the "radical pessimism," which Reinhold Niebuhr was to call the Nerve of Failure.[32]

The New Roaders had come on strong, filling up the pages of *Politics* with relentlessly earnest philosophical and moral speculation to the point of tediousness. Dwight soon complained of the "rarified atmosphere of the journal." He even felt relieved when the critical opposition began to flood the magazine with angry retorts and a stream of abuse and often ridicule.[33]

Even prior to Dwight's own contribution to the series, the Marxist readership was complaining of the drastic direction of the magazine. The first lengthy assault was delivered by one Virgil Vogel, a leader of the Young

People's Socialist League in Chicago and a contributor to *Politics*. He expressed dismay over the rising "cult of Marx baiters" who were beginning to fill the pages of *Politics*. The intellectuals, obvious summer soldiers, were in retreat, he said. From that point on, a stream of criticism from the left flowed in.[34]

Feeling pressured by his critics, Dwight continued to put off publication of his own essay month after month. He apologized at one point, conceding that the delay was due to "incompetence, dilatoriness," and his unhappiness concerning the conclusions being forced on him by reality. It finally appeared in April 1946. Dwight modestly spread "THE ROOT IS MAN BY DWIGHT MACDONALD" across the cover of the magazine, accompanied by a smoky photograph of industrial pollution by Walker Evans. Dwight began with Marx's epigram "To be radical is to grasp the matter by its root: Now the root for mankind is man himself."

With learned references to earlier theories of bureaucratic collectivism as the abstract alternative to a capitalist or socialist world order, Dwight returned to Trotsky's revealing assertion that if the proletariat failed to fulfill its revolutionary mission, then "nothing else would remain except openly to recognize that the socialist program based on the internal contradictions of capitalist society ended as a utopia." It would then be "self-evident that a new minimum program would be required—for the defense of the interests of the slaves of the totalitarian bureaucratic society."

There had been no revolution. It was time for leftists to recognize reality and quit looking at the present as though it were the "entrance hall to the spacious palace of the future." Trotsky's deadline had already been reached. All the Marxists offered was a limited program of defense for the slaves of totalitarianism. Why not, Dwight asked, base one's socialism on utopian aspirations? Why not simply "begin with what we living human beings *want*, what we think and feel is *good*? And then see how we can come closest to it, instead of looking to historical processes as a justification of our socialism?" Dwight denounced the deterministic blueprints of history which were the foundation of Marxist belief systems. (This came almost word for word from his mid-1930s critique of Marxism.) How did one determine goals? By simply starting from one's personal interests and feelings and working from the individual to society rather than the reverse. The ethical would come not from history but "from absolute and non-historical values such as Truth and Justice." Those who wanted the certainty of a preordained text and catechism would not, he warned, be happy with his analysis. He promised only a personal explanation of why the Marxian approach to socialism no longer satisfied *him*.[35]

It is quite clear, as he had conceded in an exchange with James Farrell, that Dwight had never been a "real Marxist." He insisted that this had hardly been a secret when he was in the Trotsky movement. "The Root" was simply his open, final disavowal of any Marxist residue. Having pronounced Marxism obsolete, Dwight returned to his early skepticism and came up with two new political categories by which to distinguish the controversy absorbing the American and Western left community. That community was divided between Progressives and Radicals. The former thought more in terms of historical processes than of moral values. They believe there was an answer to all social questions through the application of the scientific method. They applauded man's control over his environment as inevitably good and progressive and insisted that the use of science and technology for evil and inhumane ends was an aberration. For Dwight, this bland, fatuous optimism characterized the left from the Communists, Trotskyists, and European Socialists to the New Dealers, by way of the lib-labs, British Labour Party, and other assorted knee-jerk sentimentalists and hold-overs from the days of the popular front.[36]

Dwight's alternative category, radical, applied necessarily to a minority of anarchists, conscientious objectors, pacifists, and renegade Marxists like himself. He had already conceded that he really didn't deserve the Marxist label. Once again came the doctrine of pessimist radicalism, with its repudiation of the notion of an inevitable progress as part of any historical process. Skepticism was an indispensable guide toward the good life. Dwight expressed grave doubts about mastery of the environment and favored what he understood as the adaptation of technology to mankind rather than the reverse. Dwight denied that he and his supporters, or comrades (some already referred to them as Macdonaldites, as had Trotsky) rejected the scientific method. He simply limited its sphere and argued that its alleged benefits were considerably less than was often assumed. Finally he argued that the firmest ground to struggle for human liberation was "the ground not of History but of those non-historical Absolute Values (Truth, Love, Justice) which the Marxists [had] made unfashionable among socialists."[37]

"The progressive makes history the center of his ideology. The radical puts man there," Dwight might have said at the center of his theology, for that captures the transcendent tone, if not purpose, of his defense of

"absolute values." There was something strikingly similar to Reinhold
Niebuhr's essay, *The Children of Darkness and the Children of Light,* in
Dwight's own division between Progressives and Radicals, and there was
the same claim to see the world objectively, realistically, and not through
the lenses of a preordained historical process. For Dwight, the Progressives' optimism toward human nature and its ultimate perfectibility was
similar to Niebuhr's pollyannish Children of Light's underestimation of
the Children of Darkness. In their skepticism toward the beneficence of
science, the Radicals were aware of the unpredictable and "tragic element in man's fate . . . in any conceivable kind of society."

To the charge that he had succumbed to a newfound religiosity, Dwight
conceded only that his stance was non-materialistic and non-scientific,
but denied any belief or interest in a transcendental authority. He conceded that his radical view cold be compatible with the religiosity of a
Will Herberg or some of the pacifists like D. S. Savage, but there was certainly no necessary connection.[38]

On the issue of authority for his absolute values, Dwight was entering a
thicket of philosophical debate. In limiting science, Dwight urged that it
was important in its own sphere but that was not the sphere of values
which were outside of scientific purview. Values and ethics simply could
not be demonstrated in scientific terms. They fell into the sphere of art
and morality. Being concerned with an ethically superior kind of world,
socialism needed to establish standards of values and ethics no matter
how much its growth might be shaped by historical processes. It must be
primarily concerned with the sphere of human, personal interests—that
is to say, "The Root Is Man!!!"

Dwight conceded that as an editor of *Politics* he had maintained the
nineteenth-century notion that man's liberation came through scientific
progress. However, his experience editing *Politics,* as well as the tragic
world events of the past few years, had caused him to abandon Marxist
optimism. The crisis simply went beyond the contradictions of capitalism
imagined by the Marxists. The brutality and irrationality of the West's
social institutions defied belief and life was now dominated by warfare of
a ferocity and on a scale unprecedented in history. Evenhanded in his
condemnation, Dwight listed horrors that Attila cold not have matched:
the extermination of the Jews, the gulag of Soviet tyranny, the "'atomiza-
tion' of the residents of Hiroshima and Nagasaki." All of this had forced
him to stop and rethink and to try to come to grips with this strange and
terrifying world.[39] Dwight's statement was another example of the "shock

of recognition" that had been experienced by so many intellectuals during those horrendous years. Clearly stated, his introduction was the very best example of his superb ability to bring the issue together, to pose the questions that were on so many thoughtful minds, in effect to confront the world he lived in and attempt to find a way to cope and to seek a more humane life.

Dwight's next section reflected a tone of anarchic libertarianism that frequently brought the anti-state left into agreement with the nineteenth-century Jeffersonian liberals. There is a conservative cast to libertarian individualism, and the Soviet Union became Dwight's frightening image of the wave of the future—the sinister trends of nationalism, chauvinism, and the all powerful warfare state reached their "paranoiac pitch" in the Soviet Union, "where imperialist policy is more aggressive than anywhere else on earth, where 180 million lived in a combination barracks and munitions plant over which flew the red banner of a Marxist revolution."[40]

Since it seemed more and more apparent that there existed no mass base for a genuine socialist movement, Marxist revolutionaries had turned to the State and to all of the odious forms of statism, bureaucratic organization, and statistical measures; in short they developed a dehumanization process that was more menacing than the iniquities of industrial capitalism they opposed.[41]

What was most appalling to Dwight was that there seemed to be no viable counter movement to check this tendency. At least the nineteenth-century laissez-faire liberals had been genuine and stuck to their principles. The contemporary politics of the liberal left spectrum remained uncritical of rampant statism. Thus all that currently existed on the left was a banal clash between the totalitarian heirs of Bolshevism and the hapless sons of intellectual fathers, the reformist socialist.[42]

Dwight raked over the Marxist failings that he had been listing for several years. But his major argument concerned the lack of any ethical foundation in contemporary Marxism and its sneering dismissal of values. Dwight insisted that Marx had had the same values as present-day radicals, anarchists, and utopian socialists. Marx had written eloquently about the unfettered development of the individual, stating that society existed for the individual. The liberated society of the future would be free from all forms of coercion (the "withering away of the state"). It was the subsequent Marxian belief in inevitable progress, in evolution, in deterministic history that had led to the current statism.

How one determined one's values and established their validity was left for the second installment of the essay, which was scheduled for the following issue of *Politics*. When it failed to appear, Dwight told his readers it was not the kind of piece that could be hurried. There was no June issue, and Dwight's promised answers to many questions were once again put off.[43]

In June Dwight, Nancy, and the boys headed once more to the Cape. This time they rented a house in Wellfleet owned by the artist Ed Dickinson. It was on the bay side of the Cape, which made it easier to take the kids swimming—the water was warmer and much calmer. The house was a classically simple Cape Cod saltbox right on the marsh water at high tide. When the great tide receded, it exposed vast muddy sand flats on which the family could walk out and dig hard-shell clams (quahogs to the natives) by feeling for them with their bare feet or using a clam rake. Dwight also had an opportunity to renew an old interest in sailing, because the owner had left a sailboat with the cottage.

The social life was hectic. Every evening there was either a dinner or at least a cocktail party. By the end of August, Dwight was a little worse for wear. He likened the round of partying to an athletic performance. The usual friends were nearby. The Chiaromontes were in a lakeside cottage. Dwight and Nancy saw a good deal of the John Berrymans, who stayed with the Macdonalds for part of the summer. While Mary McCarthy was in Newport with her then-husband Bowden Broadwater, Edmund Wilson and his son by Mary, Reuel, were in Wellfleet. McCarthy had sold Dwight her Wellfleet heap of a car for one dollar. One day on the street of the village, Wilson accused Dwight of driving his car. Dwight replied that he had the bill of sale and that he was the proud owner of the humble vehicle. If Dwight did not see much of Wilson, whom he found hurried and constrained whenever they met, Reuel was a friend of Michael's and the two boys spent the summer exploring the flats and wooded areas, clamming, and spying on the adults.

One evening Dwight and Nancy had dinner with John Dos Passos. Dwight found it dull and painful. Having abandoned any pretense of radicalism, Dos Passos had become a staunch Republican on his way to the *National Review*. Dwight found him neurotically "scared of Russia and Communism." On another evening he had dinner at Anna and Norman Matson's with the émigré painter George Grosz. Grosz got drunk and spent the better part of the evening abusing socialism and art, praising Hollywood and money-making, obviously baiting Dwight, who of course bit. Dwight reported to Mary McCarthy that in both Dos Passos and Grosz "you see the wreckage of the leftism of the 20s; they're lost without a movement; memento mori."[44]

Dwight had begun to be a leader for this community of writers, artists, and assorted bohemians, always arranging picnics, clam digs, and get-togethers of one kind or another. At one point he compared himself to a social director at a Borscht Belt hotel. He remarked to Chiaromonte that the summer had become "fantastically sociable . . . people dropping in all the time from nearby houses, to swim, play ping pong or from more foreign parts to talk and drink." Dwight jokingly complained of the amiability of the summer people and wished there were a few mundane stockbrokers and wholesale grocers from Montclair, New Jersey, to lessen the temptations to socialize. Dwight and Nancy were beginning to create the community of like-minded (if incredibly argumentative) souls, the retreat "from the rotten world" that Dwight had dreamed about with his friend Dinsmore Wheeler twenty years earlier. It was in effect an oasis, which became the title of Mary McCarthy's perceptive satire of the group a few years later.

When Mike was eleven and obsessed with baseball, Dwight was one of the chief organizers of a Sunday morning softball game in which fathers and sons and an occasional mother and daughter participated. It soon attracted a vociferous audience of kibitzers. The rule was that there should be no limit to the number of players and often as many as twenty played on a team. It became such a renowned event that after a few summers people from as far south as Hyannis were demanding to be allowed to play.

Dwight played second base. He would stand behind the bag with his long FDR cigarette holder hanging out of his mouth and little Nicky hanging on to his pants legs, delivering a running commentary, shouting directions interrupted by a cackling, giggling laugh. Norman Mailer often came from Provincetown to play shortstop. Arthur Schlesinger, Jr., and his neighbor Edwin O'Connor, the novelist, played first base. Cyrus Rembar, Mailer's friend and lawyer, took the game so seriously he illegally slid into second and broke his collarbone. Charles King, who later ran the Wiltwyck School for Wayward Boys, was the single black participant.[45]

Like so many of these events, what began as amusement for the kids quickly became dominated by adults. Following the games, there would often be large cooperative beach parties, frequently organized by Dwight. These affairs ran well into the evening, with songs and storytelling. John

Berryman would on request give his great recital of "The Monkey's Paw," done in a straightforward way, not hammed up for the kids, and therefore all the more terrifying.[46]

There was also a good deal of nude bathing on the lonely stretches of ocean beach between Wellfleet and Truro. Nancy vividly recalled one occasion when Freda Utley, the staunch anti-Communist political writer, was arguing with Dwight clad only in a "small bra to hold her hearing aid." Some recall that Dwight and Nancy had cocktail parties in the early 1950s where nudity prevailed. Adele Mailer remembers arriving at one such occasion and approaching the drinkers, all of whom were nude. "All these intellectuals, the whole bunch. It was so cute." Even her husband Norman was embarrassed by a naked woman's pubic hair at eye level as he sat in a chair holding a drink. Ann Birstein recalls the first time she met Dwight. She and her then husband, Alfred Kazin, drove up to the Macdonalds' cottage on Slough Pond and found Nancy reading on the sun deck. She told them to walk down to the water's edge where Dwight was taking a dip. When they called to him he "arose out of the water like a great bird, stark naked."[47]

So well known was Dwight's practice of nudity that Saul Bellow created a hilarious episode in his satirical put-down of Dwight as the "light-weight" intellectual, Orlando Higgins, in his novel, *Humboldt's Gift*. He describes Higgins at a summer gathering engaged in an animated discussion of the McCarthy hearings. Higgins and his partner in conversation are sitting stark naked at either end of a log. His "penis which lay before him on the water-smoothed wood expressed all the fluctuations of his interest." While Higgins gave his views with much puffing and "a neighing stammer, his genital went back and forth like the slide of a trombone." Bellow ended the passage: "You could never feel unfriendly toward a man of whom you kept such a memory."[48]

The nudity was a part of Dwight's libertarianism and quest for a new sexual freedom. Mary McCarthy looked back on the nudity with some astonishment. "They even had them [nude cocktail parties] indoors . . . with middle-aged people. It sends horrors down your spine." When Mary questioned Nancy about it, Nancy turned the question around: "Isn't that rather superficial?" Her brother Selden Rodman and his friends had all practiced nudity at his place in Oakland, New Jersey, in the 1930s. Nudity was an expression of the anti-bourgeois bohemianism that both Dwight and Nancy affected. It was a break with convention, with restraint; it suggested genuine liberation from what they both saw as a conformist, even repressive, society. Apparently it was upsetting to young Michael, perhaps even traumatizing, since he recalls no nudity at parties and concedes that he may well have blocked it out. On one occasion Dwight reported to Mary McCarthy under the heading of "Cape Gossip" that the Truro police force had arrived on the beach unexpectedly and told the bathers that nudity was illegal and that the next time they would be fined $20—"the stinkers."[49]

Nude or clothed, cocktail parties, which turned into potluck dinner parties, were a fixture of beach life. Conversation, argument, debate was Dwight's strong suit. No one enjoyed it more. He was disappointed if the talk was small talk. He wanted an issue to be pursued and he worked tirelessly to see that some controversial topic would engage the company. Eileen Simpson remembers a party where Dwight turned to her and said "'Eileen, we're not discussing anything.' Everyone was talking, but to Dwight we ought to be talking about some serious subject."[50]

Theatrical in his response, Dwight was always quick to put down any idea he considered absurd. "If you believe that you'd believe anything. Come off it." His voice would rise and fall, and his "high-pitched, shaking-away, giggling-wheezing laugh," as his son Michael described it, would punctuate his argument—invariably with good humor but always on the attack. Sometimes while listening he'd "give a quizzical, owl's look, head mildly cocked to one side, waiting for the humbuggery . . . when he could swoop down." But even the swooping wasn't fearsome. It was "campy fun as he pounced." Winning the argument was important to him, but at the same time, he would readily concede defeat when convinced by his opponent, which was not so unusual.

Sometimes with his sons, however, he became impatient and less than sympathetic and would mock their attempt to engage in adult discussion. Dwight was not interested in children as such and invariably treated young people as respectfully as adults or simply ignored them if they were too young to be included in the grown-ups' conversation. He seemed to genuinely like young people, took great pleasure in his students when he was on the visiting professorship circuit in his later years but small children bored him. However, later on he greatly enjoyed his two grandsons, Ethan and Zachary.[51]

Dwight would often be deliberately contrary just to raise the heat of the discussion. On one occasion he baited John Berryman by saying he didn't like Robert Lowell, a close friend of Berryman and a poet for whom Berryman had great admiration. Soon they were shouting at one another. The following spring while on a lecture trip Dwight wrote Nancy that he had stopped in to see Lowell who was poet in residence at the Library of Congress, and had found him very friendly and cordial: "How could I have baited John so maliciously at the Cape on the subject of Lowell. I'd never even met him. *Must* stop this sort of judgment on people as you have often advised."[52] On another occasion Dwight and Berryman had such a heated dispute over the nature of unconscious that Berryman walked off into the darkened ocean. "The icy water shocked him back into his senses," but not before he had scared the wits out of the assemblage.[53]

One evening Dwight had a row with one of the local "Stalinoids"; the next day when he visited the man's garden, which was open and available to the community (one left money in a box), a hastily fashioned sign greeted him: "No Trotskyists may take vegetables from the Garden."[54]

These disputes were no doubt fueled by alcohol. Dwight seemed to have a limitless capacity for bourbon, as well as before-dinner martinis, and most of his friends drank pretty heavily too. Some argumentative evenings ended with partygoers driving into the gullies that bordered the sand roads winding through the piny woods. Trapped in the sand, cars would be stranded and forced to wait for rescue until the following morning when sobriety returned.[55]

Totally absorbed in the summer's activities, Dwight did practically no writing or even reading. To Mary McCarthy he blamed it on his spending too much time "gawking at nature" and the fact that there was "almost no intellectual life—too many painters and musicians (why don't any sociologists and philosophers ever come to the Cape?)" McCarthy, a very disciplined writer who set aside her mornings for work even when she had houseguests, faulted Dwight for his lack of rigor and thought he was wasting his enormous talents. At prep school and college he had shown unusual diligence in his study habits, carefully ordering his time and mastering extensive reading lists. It may well be that Dwight's drinking did affect his work. Most of his friends and associates did not think of Dwight as having a "drinking problem." While there were obviously incidents where Dwight was tight and overly flamboyant, even tactless during the verbal jousting at parties, he wasn't regarded as someone who couldn't handle his liquor or let it interfere with his work. Nevertheless, the endless round of cocktail parties and beach picnics must have taken their toll.

But even at the Cape, Dwight had been unable to escape the heat of the "New Roads" controversy and the avalanche of criticism directed at his initial "Root" essay. This put considerable pressure on him as he struggled with the more philosophical exercise of the second installment. He became increasingly anxious and irritable. At one point Nancy wrote Andrea Delacort, one of their Paris contributors, that Dwight was "hardly human" as he tried to complete the piece and that she was "going to have to find his root when he finally gets finished."[56]

The long-awaited last section appeared in the July 1946 issue. Dwight began defensively by conceding that he was taking up this part of the argument with reluctance, because he "had no philosophical training" and didn't feel at home in the field. Nevertheless, in a world of such over-specialization there was room for amateurs. The present course of society was catastrophic, he wrote. Questions long considered closed deserved to be thought out again: determinism versus free will, materialism versus idealism, the concept of progress, the basis for making value judgments, the precise usefulness of science to serve human ends, the very nature of mankind—all these age-old questions deserved rethinking.

As promised, Part Two presented Dwight's thoughts on the relationship between the scientific method and the making of value judgments. It was an act of personal will, because the first part of the essay and the implications of "The Responsibility of Peoples" piece were so dark and foreboding as to suggest that the possibility of resisting the tide of encroaching totalitarianism was nonexistent. But Dwight insisted that there was a dimension of individual free will that could not be controlled, manipulated, shaped, or stifled by the power of the State.

Dwight declared that there was no scientific method for determining values. Drawing on Henry James,he gave an eloquent defense of subjectivity and the validity of ethical and aesthetic standards that can not be validated by scientific measurements. The crux of Dwight's argument was his assertion that "each man's values come from intuitions which are peculiar to himself and yet . . . also strike common chords that vibrate respondingly in other people's consciences. This is what ethical teachers have always done; it is the only way we have ever learned anything essential about ethics or communicated our discoveries to others."[57]

Such a notion seemed mystical only because far too much weight had been given to the authority of science. Not anticipating postmodern criticism, Dwight wrote with confidence that most would agree one could not determine whether a poem was good or bad aesthetically by using science as a guide, but that few would claim there were no standards for making aesthetic discriminations. Obviously influenced by the existential vision from Europe that had penetrated *Politics*, Dwight, like Albert Camus, denied any transcendent crutches: "The 'trick' in living seems to me to reject all complete and well-rounded solutions and to live in a continued state of tension and contradiction which reflects the real nature of existence."

Dwight was a seeker, but he had concluded that the human condition was one of uncertainty. His radical pessimism was appreciated by younger critics like Daniel Bell, who were quick to see that although he was of an older generation he was part of a revolt against the dogmatic certainty of Marxist radicalism and the smug complacency of previous "progressive thinkers." Dwight was skeptical, disenchanted, and reflective, and this gave his analysis a modern postwar tone that was in tune with a younger generation of *Politics* readers.[58]

In a later edition of the essay, Dwight titled the second installment "Toward a New Radicalism," to refute the charges of passivity and reaction leveled by his critics. He completely rejected the notion of mass political action. Replying to a young Trotskyist *Politics* reader from Ceylon just as he was completing the last section of the essay, Dwight wrote: "It is my opinion that serious political activity toward socialism is not possible on a party or mass basis and that we must begin again in a much more modest and directly personal way." He recommended that his reader steer clear of the Trotskyist Fourth International.[59]

But how could one translate this personal, subjective, individualistic ethical, and moral stance into effective political action? By action Dwight meant action to "bring about a free classless, warless, humanly satisfying society—in short, just about the opposite kind of society to the one we have now." Always defensive before the charges of religiosity, mysticism, escapism, and scholasticism, Dwight turned to the anarchist, pacifist, libertarian themes that had increasingly found a voice in the pages of *Politics*. The kind of society desired was one whose only aim, justification, and principle was the "full development of each individual and the removal of all social bars to his complete and immediate satisfaction in his work, his leisure, his sex life and all other aspects of his nature."[60]

Dwight's vision had clearly turned inward. He now saw solutions in terms of small units, little communities in which direct, personal human relationships could be achieved. It must be a small unit society in all aspects, economic, political, social. Marxism always addressed a mass society and thus had based itself on the very cause of alienation which it thought it was combatting. To the chagrin of many of his radical readers,

Dwight quoted with approval Marx's own anti-statist theories of alienation from *German Ideology*, and lamented the fact that he had not made these ideas the cornerstone of his thinking rather than choosing to "waste years on economic analysis which today has only historical interest." What was worse, the attempt to replace capitalism with collectivism had turned out to be the cause of twentieth-century alienation.[61]

Dwight had returned to the Hedonists of his school days, to his Brook Farm ideal of a communal settlement of like-minded people living and working together. In the last pages of "Root II," he drew upon his youthful bohemian fantasies as a model of individual escape from the menacing dehumanization that so characterized the centrally controlled mass societies of the mid-twentieth century. (He was soon to write in *Politics* that living in a city as large as New York made it impossible to respond to others as human beings. They had to be passed by as "indifferently as ants pass each other in the corridors of an ant hill.") But in addition to a close-knit personal comradeship, the members of this ideal community would also engage in direct political action to combat Jim Crow, and to promote pacifism and other common goals of the group. In a passage that prompted much jeering among his critics, Dwight wrote that members of the community would preach their ideals, "or, if you prefer, make propaganda—by word and by deed, in the varied everyday contacts of the group members with their fellow men (as, trade union meetings, parent-teachers associations, committees for 'worthy causes,' cocktail parties etc.)"[62]

The reference to cocktail parties, symbolizing what became known as the "radical chic" of the New York celebrity set and well-meaning bourgeois leftists, brought Dwight unmerciful ridicule, and he later conceded that he had been more assiduous at imbibing at cocktail parties than making propaganda at them. He was being unduly hard on himself, however, for no one was more intent in turning a casual social occasion into an intense political discussion. It is entirely possible that Dwight felt a serious discussion at a cocktail party was more worthwhile and productive than the formal debates at the monotonous meetings he had attended.[63]

Much of what Dwight wrote in these last pages of the essay reflected his own personal, even psychological, feelings at the time. He had written to his friend Arthur Wiser that while Wiser and his wife were working at enriching their own internal lives as pacifists, his own life had become far too busy, unreflective, "externalized to an almost frightening extent." He complained of feeling dominated by outer compulsions although there was an appearance of free will and choice. People should be happy and should be able to "satisfy their spontaneous needs here and now. . . . All ideas which required the sacrifice of the present in favor of the future should be looked on with suspicion."

Indeed, the emotional intensity of the last pages, the almost confessional tone, not only anticipated the sensibility that would capture the imagination of a generation of young people in the 1960s, it deeply reflected Dwight's personal fears and disenchantments with his own life. His relationship with Nancy was intellectually and politically fulfilling; they did indeed love and admire each other and were genuine friends. But it was emotionally and apparently sexually unsatisfying for both of them. Dwight's personal and political confidence was waning, as was his pleasure in the demanding work of producing the journal. The work was wearing him down. "The Root Is Man" was a document that criss-crossed the boundaries between Dwight's public life as an intellectual journalist and his personal life as a human being trying to find some satisfaction in a world grown increasingly complex and difficult to understand.

It was the openly untheoretical, personal tone of "Root" that most irked his critics. It was to be expected that the Marxists and the scientific and philosophically sophisticated would find his piece wanting. Younger Marxist radicals such as the UAW organizer Frank Marquart, Irving Howe, and Lewis Coser wrote lengthy critiques accusing Dwight of abandoning a class analysis, of abandoning all hope in the organized working class and becoming, as Howe put it, "the 13th disciple." Coser tracked what he saw as the somersaults of these political roués who "travel from Marx to Reich, from Trotsky to Tolstoy, from Lovestone to Berdaieve. . . . they invent God when they lose their party cards." Dwight, who had leveled similar criticism against his own associates when working on the *Partisan Review*, was on the receiving end now. Younger members of the leftist cadres looked to him as a summer soldier fleeing the radical struggle for the safety of bourgeois purity. But Dwight remained convinced that a genuine radicalism had to start with individual human beings and their needs and not with the rigid ideological abstractions of class.[64]

It was probably harder for Dwight to take the criticism of Nicola Chiaromonte, since he had been the most influential in promoting Dwight's criticism of Marxism and adoption of an increasingly personal anarcho-pacifism. Chiaromonte endorsed Dwight's repudiation of Marxism, but he was disappointed with the philosophical vagueness and inconsistencies of the second part of the essay concerning ethical values and standards. He wrote from Paris to say that Part Two had left him "exceedingly confused." Chiaromonte felt that Dwight's general principles, the inapplicability of scientific measurement to the question of good and evil, and Dwight's appeal to an ethical rather than historical base were not necessarily related to such practical proposals as the rejection of violence, anti-statism, the establishment of "psychological communities." One didn't have to disbelieve in the scientific method to refuse to serve in the Army.

Chiaromonte was uncomfortable with Dwight's separation of values from the world of reason, so that "the question of good and bad might have nothing to do with the question of true and false." While Dwight had dismissed the assumption of values that were not acted upon, he had hardly redemonstrated his own absolutes: nonviolence, or anarchism. The tone of Dwight's argument obviously unsettled the highly disciplined and philosophically sophisticated Italian. It pointed up the differences in their temperaments. Chiaromonte took exception to what he interpreted as a hedonistic self-indulgence in Dwight. Dwight had written with suspicion of all ideologies "which require the sacrifice of the present in favor of the future." From Chiaromonte's point of view, this amounted to saying he didn't give a damn about moral ideas. For Chiaromonte, morality was nothing at all if it was not giving up something in the present in favor not only of the future but even of the purely ideal. Chiaromonte expressed astonishment at the naïveté of an article attempting to go back to the ethical foundations of social action. It was almost as if his student had simply misunderstood and taken the repudiation of Marxist rigidity and demand for individual self-determination as license for a freewheeling libertarianism. But Dwight's notion that one should not base one's politics on the sacrifice of the present for a better future was meant to be akin to Arthur Koestler's claim that the brutally totalitarian Soviet state had always based its demands for sacrifice on such an argument. Perhaps more to the point was Dwight's own personal need to somehow break away from what he saw as the external restraints and limitations of his own life. Both he and Chiaromonte were soon to become involved in serious extramarital affairs; Chiaromonte's disciplined sense of morality dictated the way he dealt with the disruption in his marriage, while Dwight followed the path of individual freedom and satisfaction.[65]

While Chiaromonte took Dwight's piece seriously, others, even among the *Politics* circle, found the entire "New Roads" and Dwight's lengthy

contribution in particular to be both tedious and naive. Even Niccolo
Tucci, whom Dwight constantly defended against charges of frivolity,
recalled that many were embarrassed by the concluding section's unsophisticated handling of standard philosophical questions. Mary McCarthy
wrote to complain of the ethereal quality of the New Roads discussions
and Dwight's persistent tenacity in keeping the tiresome argument going.
One Gwynne Nettler charged that Dwight had joined the ranks of the
intellectual whifflebirds. A whifflebird, he wrote, was a fabulous creature
that "flies backwards in ever decreasing circles until it flies up its own
ass hole and becomes extinct." He saw only three alternatives for Dwight:
suicide, religion, or ethical hedonism. He felt that at the moment religion
of the Quaker variety seemed most likely.[66]

Dwight had undertaken a large challenge, one for which he had little
disciplined training. But he was a superb journalist, and the strength of
his statement lay in his ability to bring together the questions, the anxieties, the deep-seated fears of his contemporaries and express them in a
vivid and enduring language that intelligent, literate laymen could understand. Later a student of Dwight's political thought dismissed his efforts
as little more than a period piece. But that was a serious underestimation,
because Dwight drew distinctions, provided definitions, posed questions
that remained at the root of intellectual discussion for several decades.[67]

9

A Political Desert Without Hope?

NEITHER DWIGHT NOR NANCY LOOKED WITH ANY enthusiasm toward their return to the city as the summer of 1946 drew to a close. Nancy was so involved with the package project that she felt she had no private life autumn through spring, and she did not appreciate returning to the "grimy, gloomy, weary city" after such a wonderfully relaxing summer. Spending long afternoons in Ed Dickinson's studio, practicing the piano while he worked eternally on his fine painting "Ruins at Daphne," which still hangs in the Metropolitan, she wondered why she and Dwight continued to live in New York. Dwight was loath to leave off "exploring the dunes, bays, moors [and] ocean breaks" surrounding the Dickinson cottage.

They both were born in Manhattan and remained there all their lives, so it seems odd that Dwight and Nancy seldom expressed any enthusiasm for the city as a place to live, to explore, with its exciting pace, the street life, the colorful neighborhoods, or even as a place to work. Dwight maintained a love-hate feeling for the intellectual life. He harbored the characteristic Manhattan provincialism that sees the rest of the world as the boondocks, but New York as a vibrant, pulsing town that can lift the spirits of a returning citizen appears to have failed to charm them.[1]

They stayed in Wellfleet as long as possible, until they felt compelled to return late in September. It was like a day of reckoning. Once again Dwight confronted the issue: What is to be done? Having abandoned Marxism and even the notion of a socialist revolution of any kind, his only alternative was to fall back on his support of pacifism as a politics of resistance and opposition. It was an anti-state position that was congenial to his negativist criticism. The pages of *Politics* carried endless discussions of pacifist politics and conscientious objections to all forms of state coercion.[2]

The thrust of Dwight's position was aggressive anarchism as a means of challenging the State. "My kind of pacifism," he wrote, "may be called 'non-violent resistance' or even better, 'friendly resistance.'" Dwight became a member of the War Resisters League and the more militant Peacemakers. Driven by the notion that one had to *do something*, he lent support to all forms of civil disobedience and argued at one point that if the Jews had adopted such a tactic against the Nazis, they would have made it far more difficult, perhaps even impossible, for the Nazis to herd them by the millions into death camps.[3]

In February 1947 Dwight was one of the main speakers at a public protest where some four or five hundred youths burned their draft cards. He insisted that it was essential for citizens to resist the war-making state. Pacifism, he argued, could be "a way of actively struggling against injustice and inhumanity."[4]

Dwight hoped for the creation of small groups of pacifist resisters who would band together to resist the war-makers. His entire identification with pacifism fit neatly into his personal anarchist inclinations and his continued notion of communards who would transform their ideas into effective political action while at the same time transforming their own lives in the process. His wide correspondence with militant pacifists and jailed conscientious objectors revived his interest in pacifism as a form of self-transformation. He was fascinated by the experiments among the pacifists with communal living.

His friend the pacifist Art Wiser wrote him an article on one such community, but Dwight complained of its too dry statistical approach, its academic, bureaucratic jargon. Dwight wanted to know more about the human, psychological problems—the personal dimension. "What do you get out of it as individuals? Do you ever get drunk? Do you paint, write poetry, get spiritual illuminations?" He wondered how this experiment compared with Brook Farm, thinking back to his youthful enthusiasm for a communal retreat. "What are your relations to the local folk? How about sex?"[5]

For Dwight, anarcho-pacifism was hardly confined to a militant political agenda. He was drawn to it for ways to live his own life, to find more personal fulfillment, to escape the pressures, the lack of time for reflection in the rat race of urban existence. With the loss of community had come a loss of any shared values, and Dwight sensed this loss even within the *Politics* circle. It was growing difficult to communicate with other intellectuals, so rigid were the political boundaries, so harsh the polemical abuse, so weak the sense of comradeship. He could hardly remember a time when "there had been so little contact between intellectuals and so little concern about establishing it."[6]

Once again the pressure of deadlines and the stress of editorial decision making led Dwight to drink heavily. After a run-in about a cover drawing for *Politics* by George Barbarow, Dwight wrote a note apologizing for being tight and thus "obstinate and wrong-headed." Dwight lamented what he saw as the narrow perspective of *Politics*. He had been behaving like a "specialized idiot." He had gone to only one concert and not a single play or art show, and had read only one or two novels in the past year. "That's a dog's life and something should be done about it."[7]

Nancy was also unsettled and pressured by the package project, which she administered almost single-handedly. Her life was also burdened by taking the major role in caring for the children. Dwight no longer shared in the daily supervision of the boys. Nancy was a permissive parent, which added to the hectic state of their domestic life. She believed children should be allowed to do anything they felt like doing that wasn't suicidal. The apartment, still the four-room walk-up on East 10th Street, was forever cluttered with books, magazines, newspapers, children's toys, and the never-ending packages to Europe. There was, as Nancy complained, little order, but it was "cozy if nothing else."[8]

Except for the more relaxed times on the Cape, neither Dwight or Nancy seemed content with their personal relationship. While Dwight was unquestionably devoted to Nancy, he was not inspired to great passion by her. For her part, Nancy did not feel from him the warmth and affection she needed and believed she deserved. It was hardly an unusual situation. In his early forties, Dwight saw his life drifting by. After twelve years of marriage, of intellectual and political comradeship that had literally built a small but loyal community, a mutual feeling of dissatisfaction had set in.[9]

It is not surprising given these feelings that Dwight evidenced more than a passing interest in the sexual liberation politics expressed fre-

quently in the magazine by Paul Goodman and also in the highly controversial organistic therapy of Wilhelm Reich. Both seemed to dovetail into
Dwight's anarcho-pacifist, communal interests, and he was delighted to
provide space in *Politics* for their ideas. He published several pieces by
Goodman advocating the need for a much freer sexual life than was practiced even in the bohemian circles below 14th Street. A long essay by
Goodman on recent revisions of Freud in which he singled out Reich for
special praise caused considerable controversy. Goodman argued that
Reich did not capitulate to a therapy whose goal was the "smooth running
of the social machine as it exists." This was a message that resonated
with Dwight's frequent lamentation of being trapped in the vast social
mechanism. The thrust of Goodman's argument was that even the revisionists of Freud such as Karen Horney and Erich Fromm instructed the
alienated patient in the art of conforming to the larger demands of society.
While this was a relatively common leftist challenge, Goodman irritated
Marxist readers by chiding them for dismissing the revolutionary political
implications of sexual libertarianism.

In a piece appropriately titled "The Barricade and the Bedroom,"
C. Wright Mills and Patricia Salter, colleagues at the Bureau of Applied
Social Research, dismissed Goodman's "gonad theory of revolution."
Ironically, they saw Goodman's message, with its stress on natural
instinct, as a reactionary appeal to biology and a neglect of the social and
historical forces shaping personality and conduct. But since Dwight was
also turning away from Marxist social science, he could appreciate the
anarchic individualism that was at the heart of Goodman's argument.[10]

Despite Goodman's impenetrable prose style, he was imaginative and
asked important questions, and Dwight was taken with his critique of
"sociolatry," the worship of fitting the individual into the mechanized
system. Dwight thought the constant criticism he received for giving
Goodman a platform was prompted by Goodman's insistence that sex was
a "primary factor in our lives." Although this was not a revelation, Goodman was clinically specific about it, which made people uncomfortable
and even angry. As usual, Dwight wanted something more than abstract
theorizing; he wanted some kind of personal statement, and he was open
to all imaginative efforts. Goodman, he argued, carried ideas to their
rational conclusion and speculated freely on every possible and impossible occasion. Dwight conceded that he was naturally prejudiced in
Goodman's favor, since his own thinking "had been influenced by Goodman's ideas, where it had not come independently to conclusions similar

to his." Dwight saw Goodman's sensitivity to the needs of individual freedom a genuinely "poetic insight into the problem of revolutionary action much more practical than the work of Norman Thomas or Max Shachtman." In his view, Goodman got down to human cases "without the usual palaver that really gets under people's skin." Dwight was quick to defend his publishing of Goodman. To one reader who complained that Dwight published "confused, ex-Marxist philosophical anarchist rubbish by idiots like this Goodman" and wondered whether Dwight failed to read the material he printed, Dwight replied that he read everything that appeared in the magazine and that Goodman's contributions were "serious and competent." Nancy too had a selective respect for Goodman's ideas, conceding that despite his murky writing style and cocky, self-centered manner, he often had "something to say" and was "brilliant in spots."[11]

To Dwight, the criticism from the Marxists and other social scientists was evidence that Goodman and Reich were undoubtedly on to something. This was further endorsed by the Food and Drug Administration's harassment of Reich for being a charlatan. Reich's outlaw status undoubtedly made him all the more interesting to Dwight. Reich's repudiation of the taboo against masturbation, his claim that to live fully one must have sexual satisfaction regularly and to achieve such satisfaction one must not feel obliged to remain monogamous—all these ideas had appeal.

Dwight and Nancy were seeking a new bohemian radicalism of self-fulfillment. Paul Goodman, Wilhelm Reich, Alfred Kinsey, and A. S. Neill were of keen interest to them because all challenged prevailing restrictions and repressions. Dwight was struck by the connection between sexual repression ("the vast mechanism of repressive society") and totalitarianism. In a letter to one of Reich's students and future biographer, Dwight noted that Orwell's *1984* had independently arrived at many of Reich's concepts. Dwight felt that one of Orwell's main themes was the "relation of sexual deprivation to totalitarianism." Dwight pointed out that the spokesman for the New Order says: "The sex instinct will be eradicated. . . . We shall abolish the orgasm."[12]

Although Dwight pursued his interest in Reich and solicited a long article on him for *Politics*, he was very dubious concerning the claims Reich made for his orgone accumulator, a machine designed to free one's sexual energies. Dwight spent a good deal of time with friends who had become converts and was convinced that Reich was either on to something big or was self-deceived on a grand scale. At one point Dwight went down to Isaac Rosenfeld's loft on Hester Street and spent forty minutes in Rosenfeld's orgone box. He denied feeling the meaningful twitches and tingling his friend claimed. But Dwight did concede that he felt much hotter, adding with a grin that he would have felt hotter if he had simply remained in a phone booth for forty minutes.[13]

Part of Dwight's passing interest in Reichian theory may have been due to its claim to be an escape from the "blows of history" experienced by many on the left. As one critic has observed, Reich offered "a war of liberation against the residual puritanism and production-oriented austerities of American life."[14] Another critic observed: "In the gloom of the Cold War years intellectuals whose historicism had been shaken faced the choice of either accommodating themselves to a prosperous anti-Communist society or taking a stand directly on what Mailer, citing Reich, called 'the rebellious imperatives of the self.'" For some, Reichianism became a substitute for radical politics.[15]

Confronting these political doubts became more difficult, and Dwight grew more depressed with the political situation as well as with his own personal life. The increasingly erratic publication of *Politics* reflected Dwight's unsettled state. There was no February issue in 1947, and the next issue combined March and April. Dwight seemed to be losing his direction and perhaps some of his motivation. Even as he became more engaged in militant pacifism, his doubts as to its political effectiveness intensified. While defending resistance to the draft and the public burning of draft cards, he conceded at the same time that such tactics were open to many practical objections. He did not know whether the demonstrations were effective, but he embraced them because they were the only actions he believed directly expressed his opposition to conscription. "A beginning must be made somewhere," and Dwight hoped that these actions would lead to "more effective ways to arouse people against the violence and killing which have become the most prominent features of the age we live in."[16]

What really weakened Dwight's commitment to pacifism and nonviolent direct action was his gradual acceptance of the notion that Stalinist Russia was led by men of a different breed, men who had abandoned the basic values of the West. Passive resistance would have absolutely no effect on these men. A *Politics* reader and an employee of the Civil Administration of the Military Command in Germany, Kurt Glasser, had sent Dwight the reports of Soviet savagery in Berlin. Glasser had written to say that his experiences in Germany had forced him to abandon his own commitment to a religious pacifism as a philosophy of action, because religious pacifism rested on the assumption that even one's enemy held the same basic ideals as oneself. This could work against the British, "even possibly vis-à-vis the Germans." But the Stalinists had placed a minus sign in front of the entire value system. Pacifism would have precisely the opposite effect from that intended: "They have no basic moral identity with us that we can appeal to."[17]

Dwight, who had been entertaining these ideas for some time, replied to Glasser that he too had been bothered recently by the realization that Russia put a minus sign where "we of the West put a plus sign" and that such a drastic dichotomy in values would make a decisive "tactical difference." Dwight's phrase "we of the West" marked the kind of national or even regional identification that he had always repudiated. In the early days of the war he had seen the Nazi occupation of France as a blow to Western civilization, but even then he could not identify with the Allied opposition. He was still reluctant to abandon his pacifist resistance, because it was the only dissenting position from which he could maintain his critical perspective against the American state. But his growing fear of the Stalinist state as a menace to Western civilization obviously surpassed his fear of national socialism and began to cloud his critical vision.

Soon after his exchange with Glasser, Dwight published a lead article in *Politics* on "The Russian Culture Purge," which revealed, even more than "the aggressive character of their foreign policy," the dimensions of "their war on the West." The cultural purge announced "a break with Western culture." After citing outrageous attacks on the arts in Russia and the establishment of total control over their creation and production, Dwight conceded that commercial capitalism "perverts art or makes its practice more difficult, but totalitarianism simply liquidates it. At least in the capitalist West there were crannies in which the artist could survive and there were conflicting forces of which he can take advantage." In the ghastly present period, the very "contradictions of capitalism," that bourgeois anarchy generations of Marxists had railed at, now turned out to have their advantages. In Russia there were "no crannies, no contradictions, no conflicting forces . . . there is only one culture, one conception of art, one criterion of taste and achievement." There was simply no freedom for the artist; control was total.[18]

In that same issue, Dwight printed Kurt Glasser's document reporting rape and pillage by the rank and file of the Red Army in Germany. He introduced the piece by asserting that there was "no better indication of the inhuman, perhaps sub-human" types being produced by the present "Russian experiment." Dwight made note of the "less brutal but equally corrupt and insensitive behavior of our own 'boys' in Europe which might serve as an eye-opener to those who foolishly talk of American Democracy."[19]

It was a knifelike line to follow, to be aware of the menace of international Stalinism and at the same time avoid a blanket endorsement of the Western opposition. Even as Dwight denounced the bestiality of Stalinist Russia, he became embroiled in a vitriolic controversy with the editors of the *Partisan Review* as to which journal was sufficiently critical of the Soviet threat. In the heated exchanges that filled their letters column, the *PR* editors singled out Dwight and *Politics* as illustrative of the kind of passive religiosity that refused to face the "realities" of the confrontation between the East and West and the need to support American foreign policy. Dwight accused the *PR* editors of taking a position that supported "power plays by the U.S. State Department, backed up by the atom bomb and the biggest navy and air force in the world." Still claiming a socialist perspective, he criticized the military solution as in no way an advance toward socialist goals. Despite his constant nagging doubts about both pacifism and socialism, Dwight was still trying to remain true to his third-camp position. But denial that there were only two choices did not quell Dwight's nagging doubts. He was growing increasingly contemptuous of the "Stalinoids" who dismissed the very notion of a Soviet threat. In Dwight's view, these deluded apologists, Henry Wallace, Ralph Ingersoll (his old boss at *Fortune*, now editor of the left-wing tabloid *PM*), Max Lerner, Freda Kirchwey, and Frederick Schuman, were "effectively spreading the Stalinist infection." Here Dwight was anticipating the cold war liberal anti-Communism and the kind of terminology one of its chief practitioners, Arthur Schlesinger, Jr., used in referring to Carey McWilliams as "the Typhoid Mary of the left, bearing the germs of the [Stalinist] infection even if not suffering obviously from the disease."[20]

The cold war polarities were making Dwight's position very difficult to maintain. During the shattering spring of 1947, he returned to the problem in an editorial discussion of the Truman Doctrine, the presidential declaration of a cold war commitment to oppose any further Communist expansion. In the immediate instance it concerned Greece and Turkey, which the British could no longer afford to defend. Dwight, who had enthusiastically defended the rebellion in Greece and dismissed the threat of a Communist takeover, was now far more worried about Soviet expansion. Throughout his political analysis he remained critical of U.S. policies, but he felt obliged to concede that life in America, despite the terrible power of the State, "had not reached by far the totalitarian horror that Russia has." Dwight was beginning to adopt the rituals of the cold war, which demanded some declaration of purity before embarking on any discussion. Even if he did concede that there were only two alternatives, the USA or the USSR, he found it impossible from the standpoint of his own values to choose either. The Truman Doctrine's purpose was to "prop up the present reactionary governments of those countries so as to contain the expansion of Soviet imperialism."

As a counter to Russian imperialism, he concluded, the doctrine had its points as an effective means of confronting the Stalinist state. Using all the prevailing arguments of some of the most conservative critics of Roosevelt, Dwight charged that Teheran and Yalta had been a policy of Munich-style appeasement. Even Truman at Potsdam had failed to effectively hinder Stalin's drive toward world power. Nevertheless, the Truman Doctrine was at best only a means of postponing World War III. As a means of "helping free people maintain their free institutions" it could only be described as "grotesque." Both Greece and Turkey were ruled by "viciously reactionary regimes." The Truman Doctrine was really more of a competitor than an opponent of the Kremlin Doctrine.

What was the alternative to such a dismal choice between two war-making states? To choose either was to deny all progressive values, to deliberately turn away from the goals Dwight had long sought. In a footnote he referred to the third-camp position of revolution by the masses in the last war as the last flicker of Marxian hope; it had turned out to "be tragically non-existent." The question remained: Was there any third alternative now?

Dwight was not prepared to throw in the towel and declare his allegiance to the West. That seemed to be the terrible choice large-scale politics presented to the radical—truly a desperate situation. Perhaps on a small, individual scale, however, an individual's own thoughts and actions could make a difference. Perhaps small seeds of resistance might be planted that would later produce larger changes. Taking his cue from Jane Addams, who defended the "subjective" rewards of commitment, Dwight argued that small acts of resistance were rewarding to the individual in themselves. But these personal rewards were scant comfort because on the world scale, "politics is a desert without hope."[21]

Dwight's own individual act was to throw himself into the rapidly developing presidential campaign by conducting a personal crusade against Progressive Party candidate Henry Wallace. This was a way of challenging the lib-labs, fellow-travelers, and assorted "Stalinoids" he saw as refusing to acknowledge the danger of an aggressive Soviet totalitarianism seeking world domination. He completed a brutally critical portrait, "Henry Wallace: The Man and the Myth," which appeared in two very long installments in *Politics* in the spring of 1947 and was then published as a small book during the campaign year of 1948. It was a deliberate hatchet job which, in addition to tearing Wallace apart, disparaged Democrats, Republicans, New Dealers, Trotskyists, almost any political organization. But of course it had the virtue of Dwight's crisp, biting prose. It was, as Arthur Schlesinger wrote with delight in the *New York Times*, "etched in acid." Unrelenting in its sarcasm, it portrayed Wallace as an empty shell, a "phony who pretends to be something he is not." Dwight insisted he had a moral and literary obligation to oppose a fraud for ethical and aesthetic reasons. His hostility was stimulated all the more because Wallace, he was certain, only pretended to stand for the very things Dwight valued, "peace, democracy and human freedom," when in fact the man was "an instrument of Russian Foreign Policy, really an agent of the enemy and thus all the more dangerous."[22]

The book was intended to demolish the myths about Wallace: (1) that he was a well-meaning idealist; in fact, Dwight insisted, he was a "realpolitiker"; (2) that he was a man of integrity; Dwight called him a backtracker and a liar who would never defend principle if it threatened his position in the power structure; (3) that he was a man of moral courage; he had invariably folded when challenged and was easily intimidated by businessmen and Communists, Dwight wrote; (4) that he was a dreamer, a visionary; he was a conscious political operator and was willing to collaborate with the most ruthless of political manipulators, the Communists; (5) that Wallace was rigid, even doctrinaire in his insistence on holding to principle: in fact, Dwight wrote, Wallace was "not a man of principle but of principles—all of them, all together, all at once"—in effect two-faced, contradictory, expedient; (6) that Wallace fought privilege and injustice; in fact, Dwight wrote, it was only banal rhetoric. He called the rhetoric "Wallese, a debased provincial dialect," an impersonal idiom used in Wallaceland: "a region of perpetual fogs, caused by the warm winds of the liberal Gulf Stream coming in contact

with the Soviet glacier." In Wallese one compulsively refers to the
"freedom loving" and "forward looking," as opposed to "reactionaries"
and "red-baiters." (Dwight proudly claimed membership in the latter
group.)[23]

What was most interesting about this assault is that it was aimed not
just at Wallace but at all liberals, defined as those not identified with
what had become known as the "non-Communist left." In the early
1930s, hard-core Communists reserved their greatest hatred for the liberals and Democratic Socialists, whom they labeled "social fascists." They
hardly bothered with dyed-in-the-wool reactionaries or even the overt fascists. The same may be said of leftists in general: They found establishment liberals the most odious and dangerous of the American political
breed. Dwight did not escape this syndrome. In his book there is almost
no specific indictment of the open, card-carrying members of the Communist Party. It is the fellow-travelers, the dupes, the sentimental apologists for the Soviet Union under Stalin that receive the brunt of his contempt. Dwight prefigured the provocative charge made by Irving Kristol
during the debate over Senator McCarthy that American liberals had
indeed come late to "the facts of life about Stalinism." However, he
seemed pleased that the confrontation between the USA and the USSR
had speeded up the recognition on the part of liberal groups like the
Americans for Democratic Action. He resorted to the extremist language
of those he had only recently indicted as infected with Stalinphobia by
insisting that Wallace could be likened to earlier pro-Nazis and was,
indeed, "an American Fifth Columnist."[24]

Dwight actually expressed irritation because Wallace, while on a trip
abroad, "assailed his own country's policies and praised with equal
warmth those of Britain and Russia." Of course it was not the criticism
but the praise that rankled, failing as it did to meet Dwight's demands
for an across-the-board negativism when treating the two belligerent
forces. Throughout, Dwight managed to remain harshly critical of nearly
all U.S. policy. He withheld support for the Truman Doctrine because it
supported reactionary regimes. In addition, his book was full of anti-
Roosevelt remarks. He ridiculed Wallace's ineptitude as a leader with
the casual statement "Unlike Hitler, Roosevelt and other modern demagogues, Wallace is not adept at manipulating physical crowds." It was
this sort of thing that made the Americans for Democratic Action shy
away from distributing his book, even though they relished his smear of
Wallace. In the late 1940s the liberals were certain that New Deal liber-

alism would live forever once the Communist treachery had been eliminated from American life. Their notion was that their anti-Communist,
"vital center" liberalism was the normal political climate in America.
They were tragically mistaken, and it was a costly error in political
analysis. From any historical perspective, one might well make the case
that the decade of the 1920s was a far more representative era, even
more so than the New Deal thirties. The focus of centrist liberals on
Communism, their perception of the extreme left as the major threat to
liberal democracy, was a gross exaggeration. They underestimated the
power of the right, to their great regret during the second half of the
century.

So sweeping were Dwight's attacks on liberals, labor, and the New
Dealers that his publisher, James Henle, pleaded with him to cut his sarcastic and sneering remarks about World War II as a "crusade for Democracy" and his ridicule of Wallace's rhetorical banalities in support for the
war. Henle found it hard to stomach Dwight's unrelenting assault on anything related to Roosevelt, the New Deal, or the war, and to accept
Dwight's overly sympathetic and generous apology for Herbert Hoover,
whom Dwight described as a nonpolitical man of "good intentions." There
was in this eccentric analysis a strong leftist libertarianism that appreciated the anti-statist rhetoric of Hoover as a man of principle, as contrasted with the unprincipled pragmatists and men of expediency who led
and staffed the New Deal. As was always the case, Dwight was irked by
any editorial suggestions based on matters of opinion and he refused to
remove the sympathetic reference to Hoover. He did cut some of the more
blatant anti–World War II material.[25]

But even with this continued criticism of New Deal liberalism, particularly as it was carried over into the Truman administration, Dwight's tone
toward contemporary foreign policy was much more muted than in the
past. By adopting the phrase "We of the West," an identification that his
earlier cosmopolitanism and anti-capitalism would never have allowed,
Dwight was gradually and reluctantly beginning to speak the language
of the cold war and by doing so ignoring his own earlier warning that
a single-minded anti-Stalinism was dangerous, since its practitioner
inevitably came to support the U.S. State Department. He had not
reached that point, but his Wallace book was certainly an important step
in that direction and his ultimate abandonment of resistance and his
choice of the West.

Even Chiaromonte worried about Dwight's tendency to see the issues of the cold war in stark terms of good and evil. He addressed Dwight's praise of the people of Berlin for sticking their necks out on the side of the grey democracies as against the "black totalitarians" by replying that the problem was just this "Koestlerian notion that one must choose the grey as opposed to the black. There is a grey side to Stalinism, too, if we start going in for nuances." What the moment called for, Chiaromonte insisted, was a strong stand on principles, not expediency and compromise. All of a sudden Dwight was on the receiving end of a lecture against his old anathema, the lesser evil. It was the very "greyness" of democracy, Chiaromonte insisted, that caused it to be so discredited. Preferring American democracy to Stalinism should not mean that one accepted both Jeffersonian freedoms and Southern racism. It meant that we simply "recognize the permanence in American political life of the principle of individual freedom." Choosing the grey was an impossible step, as long as one "believes in the reality of the human conscience, which as long as it exists, can submit to the Lesser Evil (no philosophy is needed for that) but not *choose it.*"[26]

Chiaromonte really brought Dwight up short. Dwight could not accept his assertion that there was "a grey side to Stalinism." He replied almost angrily: "Is there really a grey side to Stalinism and WAS there really a grey side to Nazism?" Dwight insisted that since the thirties the world had witnessed the growth of "totalitarian political formations" which "really ARE total . . . and that totality is an evil one." There were no redeeming virtues in the Stalinist state. "It can easily be shown that Stalinism is all black," and if the Russian people got anything good out of it, it was only when they evaded or resisted the system.[27]

Once again Dwight felt compelled to compare it with the American system, which was mixed. For example, he could publish *Politics* because Nancy had some capital and the State did not interfere. There was some space to use the system to promote some good. In the United States, total control did not exist. A worker could change jobs and could have some security if he obeyed "certain written and well-known laws." He was driven to the conclusion that "the East really *is* black as compared to the West's grey." As an internationalist, Dwight said, he would love to see as much grey in the East as possible, and he pleaded with Chiaromonte to help him continue to maintain his independent stance.[28]

Chiaromonte replied that regardless of the evils of Stalinism, it was dangerous to talk about those evils "without trying to show to what extent Stalin's foreign policy is, if not justified, at least made intelligi-

ble by the conduct of his adversaries." He insisted that if the situation
looked bleak, the blame could not be placed solely on Stalin. Since the
only interest of both countries was pure power politics, creating a world
of "geo-political gangsterism, Stalin concluded he should be the tough-
est gangster of them all." Certainly on the political level "the guilt must
be shared by both parties." If the United States was, as Dwight insisted,
"less brutal, inhumane and stupid, it was America's responsibility to
find a reasonable answer to the beastly situation." Chiaromonte astutely
pointed out that if Dwight kept insisting that "the Soviet State is
unadulterated evil you admit in principle the use of unadulterated force
against it." Dwight could not accept that logic. He had maintained his
pacifism during the war even though he had seen the Third Reich as
totally evil. The only part of this argument that Dwight could swallow
was its critique of the United States' aggressive foreign policy. That was
important, but for the moment he was primarily concerned with making
a case for the total depravity of Stalin and the Soviet totalitarian
society.[29]

While one can grasp Dwight's extreme repudiation of the Stalinist
state, there was something perplexing in the vehemence of his attack on
Henry Wallace and the energy he expended to crusade against Wallace's
presidential ambitions. At a time when an open-ended debate would
surely have been a good thing, Dwight was making every effort to silence
one of the main sources of opposition to the establishment cold war position. He had in fact allied himself, at least in part, with the Schlesingers
and Wechslers who were part and parcel of the very statist lib-lab politics he had condemned for so long. The cutting edge of his criticism
under the circumstances could not but be blunted. The self-proclaimed
"dupe," I. F. Stone, who conceded the weakness of Wallace and the Progressive Party, was still determined to vote for Wallace as a "protest
against the Cold War, high prices and hysteria." During this period
Dwight dismissed Stone as a "Stalinoid," but it would seem that Stone's
position was in many ways vintage Macdonald politics. Surely Dwight
was practicing what he had previously labeled an "obsessive" or excessive anti-Communism and in doing so ignoring legitimate criticism of
American foreign policy. Despite Wallace's naïveté about the Soviet
Union, his arguments challenging the Truman Doctrine, his criticism of
an administration that thought almost solely in military terms, and Wallace's own bipolar worldview were indeed worthy of discussion and
debate and not a "red-baiting" dismissal. On the other hand, his apolo-

getics for the Soviet Union and his refusal to condemn the brutal and totally irresponsible Soviet police state were in Dwight's opinion the worst kind of immoral politics. Dwight, despite his doubts, was still trying to maintain that aspect of the third-camp position that made him a harsh critic of both United States policy and the Soviet Union. Congenitally unable to tolerate political cant, Dwight felt a moral obligation to expose Wallace's sentimental hypocrisy. It may not have been the most astute of political decisions, but it was honest.[30]

In that spring of 1947 Dwight had written that on the world scale, politics was "a desert without hope." This suggested the relief he felt when he and Nancy packed up the kids and lit out for the Cape in mid-May. Both were exhausted, depressed, and hopeful that life in the more relaxed beach community, among friends, could shore up their flagging spirits and relationship. Both had begun to experiment with occasional infidelities, which neither took great pains to hide. Nancy had initiated the practice and Dwight had not shown much concern or even interest; he had been too absorbed in his own political angst. Apparently at the Cape he hoped to find some fulfillment, personally, socially, sexually.[31]

They were once again in the Dickinson house on the marshes looking out on a vast sand-flat cove. As in the previous summer, they were quickly caught up in a round of parties, picnics, and serious clamming expeditions.

On July 24, while Dwight was preparing a meal in the kitchen and Nancy was practicing piano in Ed Dickinson's studio, the kerosene space heater caught fire. The flames quickly ignited a curtain and then the woodwork, and by the time the Wellfleet volunteer fire department had put out the blaze the bathroom had been badly damaged and the attic room completely burned out. Both Nancy's and Mike's clothes were lost. Dwight had scampered around throwing *Politics* manuscripts and materials out the window, as well as the revisions for his book on Wallace. Because the house was a primitive structure and had no electricity, they continued living in the undamaged part for the rest of the summer and well into the fall. The Dickinsons were upset about the damage and did not invite the Macdonalds back the following year.[32]

Around the time of the fire Dwight had become involved in a summer romance with the wife of one of his old college classmates. It consisted of hurried embraces in bedrooms at parties, meetings across the vegetable bins at the First National, and swims and sails together on the pretext of entertaining the children. Dwight recorded it all on his trusty typewriter in a kind of flirtation calendar. "Breasts in loose bathing top . . . brown legs . . . Nancy and kids in PTown . . . took a chance on a brief kiss at the car . . . love, scandal, must stop . . . I love you . . . I love you." Dwight wrote her poems and she played the harp for him. At this point Dwight was making some effort to be discreet, but he did confide in John Berryman when the poet told Dwight that he was having an affair with one of his students and that it was causing him great consternation. Berryman noted in his diary, "Dwight's despair like mine, Nancy's like Eileen's." Driven by desire but marred by fear as to what it might do to his marriage and his children, it was an unsettling experience for Dwight.[33]

In addition to attending to the problems and complications brought about by the fire in the house and the fire in his heart, Dwight was also having great difficulty getting out the long-awaited French edition of *Politics*. Despite considerable squabbling over the production and publication, due to Lionel Abel's refusal to meet deadlines, the issue was received with much praise. In it appeared articles by Albert Camus, Simone de Beauvoir, John-Paul Sartre, and a strong endorsement of the Soviet state by Maurice Merleau-Ponty. Dwight still appreciated the third-camp stance implicit in Sartre's existentialism at this time, but he was terribly suspicious of Sartre's theories of engagement. They seemed to force art into a collaboration with leftist politics. To Dwight, Sartre's literature of engagement smacked too much of the Brooks–MacLeish thesis he had railed against in the early days of the war. It was not long before he was censuring Sartre and his followers for being far too soft on Stalinism, which, he said, exposed their ignorance of politics. They had come late to politics, knew little of the popular front or the Moscow trials, and thus had little understanding of what "Stalin's new order really means." But despite his disagreements with Sartre and Merleau-Ponty, Dwight was delighted with the positive reception the issue met with among New York intellectuals.[34]

Once again Dwight and Nancy delayed their departure from the Cape as long as possible; again the thought of returning to the city depressed them both. Dwight wrote to Clifton Bennett, a young *Politics* staffer, that he wondered "why people live in cities. At least cities like New York." At a *Partisan Review* party for Simone de Beauvoir, when Beauvoir asked him what life was like in New York he responded bitterly that living there was like "living in a concentration camp . . . an unpleasant necessity."[35]

Even as a young man, Dwight had seen the city as antithetical to the notion of community. After visits to the Wheeler family farm, he often lamented his return to the city. In a short editorial piece he limned the dehumanizing nature of New York, relating anecdotes that illustrated the lack of human connection: "no eye contact" has always been the street-wise warning to the city-bound hick. He ended this latest lament with the biblical passage on the Good Samaritan.[36]

In the past, after the initial culture shock the pressing round of editor-ial duties seemed to revive their spirits. This was not so in the fall of 1947. Both Dwight and Nancy were suffering from depression, brought on undoubtedly by the fear that their marriage was failing but exacerbated by the rise of cold war tensions and their feeling that an effective politics seemed increasingly less likely.

Not long after his return, Dwight wrote to Melvin Lasky that he was in "a dismal mood . . . everything looks stale and black." The daily papers were unreadable, he said, and he saw himself going through an "emo-tional and spiritual change. . . . I feel low as hell." Everything seemed such a great effort. The magazine was already three weeks past deadline. Dwight felt that for the first time he understood what it meant "to a reli-gious person to lose his faith." He noted that his depression was odd since the magazine, despite its tardy appearances, was doing well, and he looked forward to the publication of two books, the one on Wallace for Vanguard Press and *The Root Is Man*, which he thought would be pub-lished by Houghton Mifflin. He wondered about the patterns of one's life and told Lasky that he intended to read Freud again.[37]

Dwight was clearly trying to break out of the doldrums of his marriage. Nancy, in order to maintain the relationship and, of course, the social and political collaboration, apparently tolerated the situation a little better than Dwight did. In an exchange with a *Politics* reader, William Rickel, Dwight responded to Rickel's assertion that personal experience was the source of ethical values and knowledge rather than study, reading, and history. "Maybe so—indeed probably so—but that doesn't get me very far, since it is 'personal experience' which is the basis, largely, of my pre-sent state of gloom about things."[38]

In a deep funk, apparently not brightened by the summer flirtation which had developed into a brief affair, Dwight quite typically felt obliged to inform his readers that he was seriously considering packing up *Poli-tics*. In November he sent a letter to all the subscribers, which began by noting the lateness of the current issue. The last issue had been

July–August 1947; the next issue did not appear until February 1948. Dwight wrote that despite what had become "a long and dishonorable tradition of appearing late," the present lag was due to the fact that he was "feeling stale, tired, disheartened and . . . demoralized." He had considered an indefinite suspension, but there was such an adverse reaction, far stronger than he anticipated, that he thought perhaps the "magazine did serve some useful function." Furthermore, there was no other magazine he would prefer to write for.[39]

Dwight analyzed the causes of his despair: the ever bleaker political outlook, his own "growing sense of ignorance," and the unrelenting "psychological demands of a one-man magazine." The latter had been stimulating initially but had of late become "simply demands." He went on to say that little could be done about the grim political situation, but that if he changed the magazine to a quarterly and found some suitable people to serve on an active editorial board, he might make a real effort to continue publication. He promised that the next issue would appear in January (it didn't come out until February), that it would be larger (seventy-two pages instead of forty-seven), and that it would cost more (75 cents instead of 50).

The response to Dwight's letter was astonishing. A great number of readers wrote urging him to do everything possible to keep the journal alive. Many of them expressed their own despair over the state of international and domestic politics, but they urged Dwight to find a way to continue. Some were lighting out for the territories, fleeing the city, and invited Dwight to come along. "To us," one wrote, "the day that *Politics* arrives is a good day . . . we are always cheered by the air of obvious sincerity and good will."[40] *Time* magazine carried a story on Dwight's letter, accompanied by a photo of a suspendered Dwight at his desk with his FDR cigarette holder clenched in his teeth.[41] Some well-known conservatives were attracted to the magazine for its anti-statist attitudes. Henry Regnery wrote Dwight to praise his demolition job on Wallace and to ask whether he was interested in doing a pamphlet on generals in politics, or on American liberalism. Given Dwight's consistent attack on lib-labs, it is not surprising that the journal had some appeal for members of the libertarian right. Regnery was the publisher of Albert Jay Nock, and Dwight resembled that curmudgeon in many ways, including the sparkling prose style.[42]

Dwight was buoyed by the response of his readers. The same issue of *Politics* carried the results of a readership questionnaire that had been created by C. Wright Mills. After looking at the earliest responses, Dwight had concluded that the readers were sharing his feelings of depression and had also lost interest in political questions, turning instead to cultural issues. When the poll was actually completed, however, the tallies suggested that he had simply been projecting his own malaise onto the results. On the contrary, the readers wanted more political articles. A full 37 percent wanted more analysis of contemporary political issues and only 7 percent wanted more cultural criticism. Nor did the overwhelmingly young readership (most were under thirty-five) accept his denigration of Henry Wallace. While some doubted Wallace's political acumen, they did not suspect his moral intent. Since Dwight felt that Wallace's foggy sentimentality was so extreme as to be politically immoral, this rejection only served to depress him all the more, because he consistently feared the Stalinoids were most effective in their appeal to idealistic but naive young people.[43]

If Dwight's personal and political life was troubled, Nancy too was suffering from dissatisfaction and unfulfillment. She was considering psychoanalysis. In 1948 she began seeing a Dr. Peterson-Krag, with whom she felt she was making progress. It was not long before she was urging a reluctant Dwight to also seek therapy. But matters had deteriorated too badly; Dwight and Nancy had begun to go their separate ways.[44]

Throughout the winter Dwight continued to complain of depression and inability to work. "The world situation is so bad and so hopeless," he wrote a friend, that it was pointless to comment on it. How inextricably his political despair was tied to his personal crisis can be seen in repeated references to a "trapped life. . . . We human beings seemed to be trapped in a vast social mechanism which has become so complex and which works so fatally according to a logic of its own that it is simply impossible any more for the human consciousness to understand it or to control it." Dwight found "a certain bitter pleasure in accepting this reality." Truth, he wrote, "is perhaps the one ethical value which even the presidents and dictators cannot deprive us of."[45] To the editor of the French journal *La Révolution proletarienne*, he wrote of having "lost [his] last illusions" and being able to adopt a more stoical attitude.[46] Dwight wrote another friend that he was reading Kierkegaard. Surely he was not seeking to find "faith on the far side of despair," as the Danish clergyman had promised, but perhaps just a way of accepting more easily the weakness and frailty of humanity. Dwight was also attracted to Kierkegaard's prose style. He felt the theologian had "solved the problem of writing about the abstract, metaphysical

ideas in an informal, personal style, better than any other writer." During
the previous spring when Dwight had editorially appraised the Truman
Doctrine, he had declared that international politics presented "a desert
without hope." Now he seemed to be looking for a way an individual could
simply cope with the bleak situation.[47]

10

Goodbye to Utopia

THERE WAS SOME CAUSE FOR OPTIMISM IN A DARK AGE, Dwight wrote, in Harry Truman's upset victory over Thomas E. Dewey, the former governor of New York, in the 1948 election. The triumph of Truman actually gave Dwight little to be excited about, but he took pleasure in the resounding defeat of the Wallace campaign. Dwight also relished the discomfiture of the "smart boys" of the press, notably those of *Time* and *Life*, who, he reported, had to call Sidney Hook requesting enlightenment on the electoral catastrophe. While grateful for these small favors, Dwight had no real preference between the smug "homunculus," Dewey, and the Democratic hack, Truman, since both stood for "what goes on anyway," which is precisely what he opposed. He also dismissed the Trotskyist candidate (a "midget totalitarian") as well as Norman Thomas, whose respectable brand of socialism was indistinguishable from contemporary liberalism. Having in effect dismissed all the candidates, Dwight expressed delight in the poor voter turnout. Half the eligible electorate had failed to go to the polls. While the nonvoters were obviously not principled anarchists, he saluted them as "47 million brother anarchists under the skin" and hailed the high degree of "civic irresponsibility." It was the low turnout that was the cause for optimism in a dark age.[1]

Dwight still had his purity intact, if no really viable politics. He seemed to be growing a bit more accustomed to this seat on the sidelines, a return to the detached elite he had belonged to as a member of the Hedonists at Exeter. He wrote friends that he was feeling better, making time to read and plan new writing projects. He was particularly thrilled that none other than T. S. Eliot had acknowledged in the preface to his *Notes Towards the Definition of Culture* his debt to Dwight Macdonald for his article "A Theory of Popular Culture," which had appeared in the first issue of *Politics*.[2]

It seems clear that Dwight wanted to do more essay writing on literature and culture, leaving the political struggles and the endless theorizing to others. He applied to the Yaddo writers' retreat and was accepted. Besides wanting to turn "The Root Is Man" into a book, he had other projects in mind. But what of *Politics*? In November he talked of more issues, but by December he wrote Nicola Chiaromonte that he would bring out two quarterly issues in the winter and spring of 1949 and then suspend publication until the fall. Dwight wanted to keep his summer free for reading and writing. He planned a fund-raising program and would attempt to recruit an *active* editorial group to help share the editing burden: Mary McCarthy, Hannah Arendt, and Isaac Rosenfeld. He hoped that C. Wright Mills would agree to come in and, of course, Chiaromonte would be the magazine's European correspondent and liaison with the various independent radical groups. With some hope of returning to a more literary and reflective life, Dwight's spirits were brighter than they had been.[3]

At the turn of the year Dwight felt better about *Politics* and its future, and even about his other writing. Nancy was still very busy with her psychoanalysis; she went four times a week. She also played the piano, continued to administer the *Politics* package project, and tried, along with Dwight, to keep up with her reading and social life. Dwight felt that the two boys were flourishing. Michael was just about to be admitted to an experimental school. He had attended Bank Street for three years and then the Downtown Community School. Dwight had not been happy with these previous schools and complained that the "intellectual sinew and crusading spirit had gone out of the progressive education movement." As might be expected, Dwight had quickly become embroiled in conflicts with the "Stalinists and fellow-travelers" who, he claimed, dominated the parent-teacher meetings. At one such affair he denounced the Community School administration for inviting as speakers "such banal apologists for Stalinism as Leo Huberman and Corliss Lamont." He was immediately denounced from the floor as a "Trotskyist." Dwight denied the

charge, arguing that only a Stalinist would have used such archaic terminology and insisting that he "had not been a Trotskyist for years and was
at present rather anti-Trotskyist in outlook."[4] Strangely enough, although
both Nancy and Dwight showed a strong interest in their son's formal education, they did not encourage Michael to read in any particular direction
and did not devote time to his education at home. They seldom read to
him, nor was he encouraged to develop literary tastes. He remembers
reading little but comic books as a child and not being introduced to serious literature until he entered Exeter.

At this time Dwight and Nancy were undergoing some financial strain.
Dwight was two years late in paying the tuition bills for Michael, resulting
in dunning letters and angry responses. He wrote Chiaromonte at one
point that Nancy had tallied up their expenses against their income and
they were horrified to discover that they had been spending on normal
living alone about twice what they had reckoned. They had only been
able to do it because Nancy had gotten "a fat and unexpected inheritance" and Dwight still had some income from money invested while at
Fortune.[5]

Dwight did fret a lot about money and was always talking about getting
"money-writing jobs". During this period he complained to Chiaromonte
that his whole life and work had been "oriented in a non-money (in fact
anti-money) direction" and that he couldn't seem to get pointed the other
way. He now planned some money-writing projects and even Nancy, who
had heretofore done only God's work, was also seeking gainful employment. They'd had some thoughts of going to Europe, ostensibly to consider publishing *Politics* from Paris or London, but that would have to be
put off. He rather dramatically pictured himself to Chiaromonte as having
"dropped from the ranks of the comfortably off into those of the hard up."
It was, he added, "a little frightening, also, of course stimulating."[6]

If Dwight was worrying where his next meal was coming from and
hunting up paying work, he still had time to become involved in another
nonpaying activity even more demanding of his time than fighting the
Stalinist domination of the PTA. In the spring of 1949 he threw himself
into the struggle against the Soviet peace offensive by becoming a partisan opponent of the National Council of the Arts, Sciences and Professions, a well-known liberal-leftist group seeking accommodation in the
cold war but viewed as little more than a Communist front organization.
The Council had issued a call for a Cultural and Scientific Conference for
World Peace to be held at the Waldorf-Astoria Hotel in New York on the

weekend of March 25–27. Anti-Stalinists immediately traced the conference's origins to the World Conference of Intellectuals for Peace and
Freedom in Wroclaw, Poland, the previous August. That meeting was
identified as a Soviet-organized and Communist-dominated propaganda
affair that had launched "violent and scurrilous attacks on the United
States."

At the initial meeting of the Friends of Russian Freedom on March 4,
Sidney Hook, after being elected chair, reported on the plans for the conference at the Waldorf. Denouncing it as a pro-Soviet event that had excluded
all but pro-Communists, fellow-travelers, and naive dupes, he called for the
formation of a committee to "expose and counteract the conference." Plans
were made to organize a countermeeting on the same evening as the conference at which liberal-leftist anti-Communists would reveal the conference
for what it was—an aggressive, Communist-sponsored propaganda event
designed to exploit the natural desire for peace among the politically unsophisticated that made up a large proportion of the remnants of the Progressive Party and liberal and pacifist organizations.

Dwight's involvement with this protest grew out of his earlier association with Nicola Chiaromonte and Mary McCarthy in the short-lived
Europe American Groups, an effort to rally and aid independent radicals
in Europe, who would take a third-camp stand against the militarism and
belligerence of both the East and the West. At the start, they collaborated
with the *Partisan Review* editors and Sidney Hook, who became increasingly impatient with what they felt was a lackluster defense of the United
States as leader of the free world. The Hook-*PR* faction was not at all
happy with the anarcho-pacifist line of Macdonald, McCarthy, and
Chiaromonte that tended to creep into the organizational literature and
the kind of groups contacted. It seems clear from the vantage point of the
post-1960s that the Europe America Groups movement was a part of the
"non-Communist left" movement that eventually became a tool of the
Central Intelligence Agency's covert operations in Western Europe. Certainly the EAG's manifesto labeling Stalinism the main threat to Western
cultural freedom had a "proximity to official thinking." The movements of
Sidney Hook at the time, his trips to Paris, his consultations with CIA
director Walter Bedell Smith in 1950, suggest that the agency did indeed
want to penetrate organizations like the Europe America Groups and
direct them toward a pro-America policy rather than see them operate as
independent radicals not allied with either camp.[7]

Despite his suspicions of Hook and the *PR* "boys," Dwight's own fears about the aggressive intent of the Soviets and his waning belief in anarcho-pacifism rendered his involvement in the Europe America Group somewhat halfhearted. Chiaromonte warned Dwight as in the black and grey argument that "nothing could be more disastrous for the Groups as to become involved in the support of official policies, or to let people think that they have become so." At the time, he had urged Dwight and Nancy to hold out against Hook and "the boys." Chiaromonte felt that anti-Stalinism had become anti-Russia and was only disaffecting independent leftist radicals. "What the boys and Hook do in the last analysis, is not to say that they are happy about the State Department, but that finally they are prepared to yield to American raison d'Etat as against the Russian." This was not political analysis but a "preordained act of conformism and a very unconstructive one, from, precisely, the democratic point of view." Given the later shock of so many intellectuals upon discovering that in fact the CIA was funding their conferences and publications, this early sensitivity to the problem is revealing. There is evidence that penetration, manipulation, and covert use of political organizations by government agencies were always feared and taken seriously.[8]

But Dwight had moved closer to Hook and even joined an interim group known as the Friends of Russian Freedom, which proposed a "Russian Institute" to serve Russian displaced persons. It too was to be a counteraction against the Soviet propaganda blitzes. Given these earlier collaborations, Dwight was ready to join with Hook in opposing the upcoming Waldorf conference. The initial meeting of the counteraction against the conference was held at the Macdonalds' apartment, and it was at this meeting, crystallized and led by Sidney Hook, that the organization Americans for Intellectual Freedom was created. Hook had applied to the National Council to deliver a paper on "Science, Culture and Peace," and several members of the program committee had no objections. Hook's proposed paper defended the thesis that there were no national truths in science, nor were there class or party truths, and international peace was being undermined by those who held that these specific truths existed. Harlow Shapley, a well-known Harvard professor of astronomy and veteran champion of leftist causes, as chairman of the conference apparently overruled the board members and denied Hooks' application. On the eve of the conference, Hook, with a *Herald Tribune* reporter in tow, attempted to force a confrontation with Shapley. Shapley, however, managed to lock Hook out of his suite of offices high up in the Waldorf. These initial skirmishes heightened the enthusiasm with which the counterdemonstrators approached the event.[9]

During their initial preparations for the protest, Mary McCarthy and Dwight informed Hook and his trusty lieutenants Arnold Beichman, formerly of *PM*, and Merlyn Pitzele, a former Trotskyist who was an editor of *Business Week*, that they were going to apply for delegate passes to the session on Writers and Publishers and would interrogate the Russian visitors and their stooges from the floor. Hook and his associates were much amused by such naïveté. They ridiculed the idea, saying there was simply no possibility the Stalinists would admit the likes of these two ex-Trotskyists into the meeting. But Dwight, Mary McCarthy, Robert Lowell, and Elizabeth Hardwick ignored this warning and went around to the conference planning office at the Hotel Seymour and presented their credentials as genuine writers and publishers. They were told to return for their delegate passes the next day. McCarthy called Hook to tell him what had happened. Still cocksure of his analysis, he laughed and assured McCarthy that no passes would be forthcoming, that he knew the ways of the Stalinists.

When Hook learned that they actually had gotten their passes, with no embarrassment whatever he insisted that they undergo a training session by himself and his associates on just what to do once they gained admittance to the hall. In a tiny room in the Waldorf "some goon from the Intellectual Freedom Committee," as McCarthy recalled, told them they would be hazed as soon as they tried to speak. They would have to take chains and ropes with them and fasten themselves to their chairs in order to make their inevitable removal more difficult. They would also have to have mimeographed copies of their planned remarks to give to the press as they were being thrown out of the hall. Dwight and McCarthy thought Hook's cronies were crazy. Dwight broke into his squeaky giggle at such absurdities. They certainly had no intention of trooping into the conference carrying chains and ropes. They did agree, however, to write up their planned questions and remarks. McCarthy and Hardwick were apparently intimidated enough by these dire predictions to arm themselves with umbrellas as a weapon against expulsion and as a way of demanding recognition.[10]

This small band of warriors were among the very first to enter the meeting room, which was, appropriately, the Starlight Roof, a place designed for gaudy extravaganzas. They grabbed seats in the second row. When Louis Untermeyer called the meeting to order, McCarthy and

Hardwick, expecting to be ignored, banged their umbrellas to get attention. Untermeyer could not have been more courteous. He asked them to await the question-and-answer period when they would b. given their two-minute turns. McCarthy was red-faced with embarrassment at her own rudeness and even more irritated at Hook for causing her to expect shabby treatment. Untermeyer took a quick rebuke from Robert Lowell because Untermeyer in his introductory remarks had referred to Hook as a "dirty word." He quickly agreed to emend that to "a four-letter word," and went on to introduce the panel members. They included Agnes Smedley, W. E. B. Du Bois, Howard Fast, Norman Mailer, F. O. Matthiessen, Stalin Prize novelist P. A. Pavlenko, the composer Dimitri Shostakovich, who had been unmercifully disciplined by Stalin, and A. A. Fadeyev, secretary of the Union of Soviet Writers, one of the more notorious of the Stalinist apparatchiks, whom Dwight had already pilloried in *Politics*.

As Dwight sat listening to three long and somewhat boring speeches by panelists, he compared Shostakovich with Fadeyev. The composer was "pale, slight and sensitive looking, hunched over, tense, withdrawn, unsmiling—a tragic and heart-rending figure." In contrast, Fadeyev was "a big, bulky, square-shouldered man, with a ruddy, fleshy, big-jawed face." He had a cold and wooden expression, and he appeared to Dwight more like a plainclothes detective than a writer. After the initial speeches the floor was thrown open for questions and Dwight was quickly recognized. He began by asking Fadeyev why he was there, since he had no standing as a writer of any consequence—why wasn't Russia represented instead by Boris Pasternak, Isaac Babel, Ivan Katayev, and several other renowned and serious authors? All had been harassed, persecuted, or jailed by the Soviet government for opposition to the regime. Dwight asked specifically just where six of these famous writers were. Wasn't it true that Fadeyev himself had revised his own work to conform to the Party line, and had he not attacked the United States at the Wroclaw Conference, saying its vaunted capitalist culture was little more than a St. Vitus dance? Dwight ended by asking how this sort of thing advanced the world toward peace.

Fadeyev responded with equivocal and evasive answers offered in an aggressive and sneering tone, which drew hearty applause from the partisan audience. Mary McCarthy got the floor next and directed her questions at F. O. Matthiessen. The mild-mannered Harvard English professor had been one of the intellectual luminaries of Henry Wallace's campaign, and in his brief remarks he described Thoreau and Emerson, Whitman and Melville as the Henry Wallaces of their day. McCarthy asked whether the good professor thought that Fadeyev had adequately responded to Dwight and whether he thought Emerson and Thoreau could organize liberty and practice civil disobedience in the Soviet Union today? Matthiessen, appearing to Dwight "pale, harassed, defensive," and who had spoken "clumsily," responded that while he did not know the background to Macdonald's question, he thought Fadeyev had met the question head on. As for whether Emerson and Thoreau could exist in the Soviet Union, he conceded that they might not be able to, but such matters had to be seen in their historical context. The nineteenth century was a more individualistic era and today societies were more collective. Then, in what Dwight deemed the "non-sequitur of the year," Matthiessen observed that he didn't know whether Lenin could live in the United States today.[11]

All the protesters in the hall, including Robert Lowell, Jean Malaquais, and Nicholas Nabokov, got their opportunity to challenge the panelists at this and other sessions. Dwight wrote at some length of the pathetic figure of Shostakovich, bullied and browbeaten by the Soviet state, who was embarrassed by Robert Lowell's query as to whether the state's criticism of his work had been helpful. He managed to reply in halting Russian, which was translated by Fadeyev, who "rolled out the Russian in vigorous periods: The criticism does me much good: it helps bring my music forward."

Dwight thought the high point of the session was the anguished statement of Norman Mailer, the much acclaimed author of *The Naked and the Dead*, who was ostensibly a supporter of the conference. He had been a Wallace backer and was obviously seen as being in the ranks of the "Stalinoids" and fellow-travelers. *Life* magazine in a cover story on the meeting described Mailer among the "dupes and fellow-travelers," along with Arthur Miller, Langston Hughes, Albert Einstein, Dorothy Parker, Mark Van Doren, and a host of other celebrities from the arts and sciences. When Mailer rose to speak, he quickly made it known that he came as a "Trojan horse" and proceeded to denounce both the East and the West. In a "most moving and sincere manner," Mailer expressed profound skepticism about "peace conferences" because they wouldn't do any good. Taking the Trotskyist line of his political mentor, the French writer and activist Jean Malaquais, Mailer insisted that only with the arrival of "a decent equitable socialism" could peace be achieved. The United States and the Soviet Union were both moving toward state capitalism, for which there was no future. Sounding much like Dwight had only a short time before, Mailer saw the citizens of both states "caught in a mechanism which is steadily grinding on to produce war." He apologized for being so pessimistic, but he claimed to be "only a writer, not a political man," and all a writer could do was tell the truth. Given the few genuine writers in the conference camp, Mailer's defection was an obvious disappointment. Dwight felt a certain compassion for the Stalinists, so weak and abandoned did they appear.

One observer from the Eastern bloc was Czeslaw Milosz, who was already well on his journey away from Communism in his native Poland. As a member of the Polish delegation he was charmed by the spectacle of Dwight and Mary McCarthy raising hell with apparatchiks like Fadeyev. He found the gathering "awful" and felt certain that half the Communists and fellow-travelers in the hall were working for the American security agencies. Mary McCarthy wrote that it was Milosz's "first exposure to the democratic left and he just fell in love with us." He must have, for several years later he wrote a glowing sketch of Dwight and *Politics* as the perfect embodiment of an American type: "the completely free man capable of making decisions at all times about all things strictly according to his personal moral judgement." Milosz linked Dwight's extraordinary self-confidence to an American tradition of individualism and self-reliance embodied in the writings of Emerson, Thoreau, Melville, and Whitman.[12]

Dwight was intrigued by the constant turmoil throughout the conference on the sidewalk in front of the hotel. Picket lines marched before the entrances denouncing the Commie-run meeting. Representatives from the American Legion and Catholic and patriotic societies booed Dwight and his group as much as they did the other delegates because, as Dwight presumed, "their hatred was directed against all alien-appearing intellectuals." Hook's cadre of Americans for Intellectual Freedom, which had not attended any of the sessions, held a countermeeting at Freedom House just off Fifth Avenue on West 40th Street, on Saturday afternoon when Dwight and his group were doing battle on the Starlight Roof. Arthur Schlesinger, Jr., the young Harvard professor and recent Pulitzer Prize winner, Morris Ernst, an ACLU lawyer, and George Counts, the ex-radical and educator from Teachers College, all gave speeches denouncing the Stalinist propaganda. Through much of the excitement, Nancy, in the tradition of a good Trotskyist trouper, with Michael in tow, attempted to sell copies of *Politics* to the people coming and going at both conferences. Typically, Dwight carried a batch of *Politics* into the Waldorf-Astoria reception, where he handed them out to the members.[13]

The entire episode ended in an unexpected fashion. After the panel session in which Dwight and Mary McCarthy asked their questions, McCarthy got into an argument with Howard Fast, the well-known Communist Party publicist and leader of the American delegation. He invited her and her comrades to attend a reception of the National Council of the Arts, Sciences, and Professions at the rather shabby Hotel Seymour. That evening they all gathered in a grubby, smoke-filled room drinking bar Scotch at 50 cents a highball and listening to boring speeches. Dwight was transported back to his Trotskyist days: "no showmanship, no fun, no dash—all very dutiful, worthy and abstract." But despite his boredom he did get a much different impression of the Stalinists than he had previously gleaned from his reading of the Communist press. He found it possible to communicate with them, since, he was astonished to discover, they shared a common culture and political background. They read many of the same books, went to the same art shows, foreign films, held the "same convictions in favor of the (American) underdog—the Negroes, the Jews, the economically underprivileged"—and against the Catholic hierarchy and the U.S. State Department. Dwight and even Hook had denounced the State Department for denying visas to several delegates from Communist countries. Dwight found that he had more in common with these assorted fellow-travelers than with the right-wing pickets out in front of the hotel. He found himself engaged in easy argument and discussion with the National Council of the Arts, Sciences, and Professions delegates and was taken with Howard Fast's geniality. Perhaps more important, however, was Dwight's sudden perception of the weakness and ineffectiveness of the Stalinists. They were far less dangerous than he had expected, partly because they belonged to the culture while their politics (as in the case of the Trotskyists) were so "disconnected from the rest of their lives and personal interests." They seemed to act out of duty more than spontaneous pleasure. Of course, there was one important stumbling block: their apparent sentiment in favor of the present Soviet system. It remained, Dwight observed, the "keystone of their faith" and of their "whole ethical and mental existence." But even its deep-rooted psychological character was not invincible. There were inner forces of commonality that worked on the side of the anti-Stalinists.[14]

This recognition that these Party members, fellow-travelers, and assorted dupes were in fact human beings and not unapproachable was inconsistent with some of the more extreme assertions Dwight had made in the pages of *Politics*. One reader joshed Dwight about his "scotch-drinking camaraderie" with Howard Fast. He likened Dwight's meeting with Fast to two Colonel Blimps encountering one another after a long stretch in the bush who spent their time talking about the natives. Dwight was tickled by the characterization and quickly conceded that as a former Marxist it was "something like the old school tie." Having once been a part of the Marxist movement, he had a sense that "one belonged to a club" no matter how far one had departed from the cause.[15]

Another critic, Harry Frankfurt, pointed out a contradiction in *Politics*. In a previous issue Dwight had argued that among Stalinist Communists "the basic premises, the very concepts of reality have become so different that there is simply no communication possible, and it is either ignorance or hypocrisy to assert that there is." Yet in his encounter with Fast and the Stalinists at the conference, he recognized a common culture and found that communication was possible. Dwight immediately conceded the contradiction and explained that the note denying the possibility of communication had been written a month before his report on the conference and he had, as editor, simply failed to reconcile the two in proof. He added that if he had done so he would have noted that on a personal and direct conversation level, provided discussion of the nature of the Soviet Union was avoided, it was possible to have a rational conversation. "The premises from which both sides start are similar (same values as to art, culture, human ends but the . . . dogmas are flatly contradictory." The Stalinoids think that the USSR is "the palladium of those values and aspirations. We think it is the nemesis of same." He had to conclude that what he had discovered was that one could talk to a Stalinist about everything but the "main point. Thus it might be possible to so hem them in on secondary issues as to shake their confidence in the main point." In effect, communication with them was then a worthwhile activity. However, Dwight added that one implication of his report on the meeting with Fast was wrong if it led one to believe that people like Howard Fast were amenable to such influences.

But he also saw Fast's geniality as a tactical nuance. "U.S. commies" were considering a new popular front. He did not mean to include Fast among the friendly rank and file with whom genuine communication up to a point was possible and worthwhile. Fast, he charged, was "a professional cynic and liar in the service of the USSR." For Dwight, Fast was one of the "new men," an amoral, hard-boiled apparatchik, not a simple, well-meaning dupe.[16]

The most important consequence of the conference was Dwight's realization that the Communist movement in America by the spring of 1949 was already in pathetic retreat. The tide had turned. The movement had suffered a stunning defeat in the election of 1948, its ranks were decimated, and American liberals and progressives were abandoning whatever allegiances they may have had to it. The labor movement was busily purging the Communists from its ranks. In the academy, suspected reds were hounded and harassed not only by state and local investigating committees but also by fearful college administrators who were collaborating with the FBI and congressional agencies to make life miserable for dissenting professorial types who did not declare their undying hatred of the Communist left. In effect, the Waldorf Conference was one of the last great efforts of American Communists and their fellow-travelers to engage in the political dialogue. After the conference, the story is primarily one of a beleaguered remnant suffering abuse and isolation. In a country like the United States, which maintains a measure of political pluralism, the absence of a viable left encourages those in the political center to move to the right, and the defense of civil liberties and the maintenance of an adversary stance becomes all the more difficult before the demands for conformity and consensus.[17]

Dwight seemed well aware of these political developments during the spring of 1949 but it did not offset his estrangement from whatever existing forms of dissent and political opposition lingered on. In fact, while the preparations for the Waldorf Conference were taking place, he had become involved in a bizarre incident concerning the writers' colony, Yaddo, just outside Saratoga Springs, New York. Dwight had been accepted for admission to transform "The Root Is Man" into a book to be published by Houghton Mifflin. However, in February he had a conversation with Elizabeth Hardwick, who urged him to reject his admission. It seems that Robert Lowell and Hardwick had been influenced by charges in the press against the writer Agnes Smedley, the radical enthusiast of Chinese Communism, who had stayed at Yaddo as a friend and guest of its director, Elizabeth Ames. Amid sensational reports of an FBI investigation and lurid tales of spy networks, Lowell and Hardwick concluded that Ames was using Yaddo as some kind of a haven for Communist subversives and conspirators, even though Agnes Smedley, under constant harassment, had left the premises.

These events occurred just prior to one of Robert Lowell's breakdowns and, according to one close observer, his sickness "required a conspiracy." The only explanation for Elizabeth Hardwick's participation was her need to placate Lowell, whom she intended to marry, and the hysterical tenor of the times. At a meeting of Yaddo's board of trustees on March 1, Lowell and Hardwick charged Ames with subversion and asked that she be "unconditionally fired." The meeting was held in the Yaddo garage in an "atmosphere strangely comparable to that of a purge trial," according to Eleanor Clark, a former Yaddo guest. As part of their campaign to rid Yaddo of Ames and her fellow-traveling friends, Hardwick contacted prospective applicants, urging them to boycott the colony until the board had dismissed the offending conspirator. Although little was offered in the way of supporting evidence, Dwight immediately wrote Ames a rude and blistering letter declining his appointment and condemning her for turning the retreat into a "center for pro-Soviet propaganda." That a figure like Agnes Smedley, a "lifelong apologist for communism and the Soviet Union," should make Yaddo her winter headquarters through the efforts of Elizabeth Ames was a scandal. It was even more shocking that Ames defended Smedley as just "another nice old Jeffersonian (or maybe agrarian) democrat." This suggested that the Communists had a "strategic behind-the-scenes position at Yaddo." Dwight insisted that he did not oppose pro-Communists being invited to Yaddo, but Smedley was beyond the pale, since she "had no connection with cultural life and activity." Dwight assured Ames that he knew how Communists worked: Once they got a foothold, they would ruthlessly exploit their advantage to "reward the faithful and knife the heterodox." Yaddo's mission had been perverted, and until the mess had been cleaned up Dwight would have nothing to do with the place.[18]

This shoddy affair reveals the climate of panic and hysteria that gripped the country in 1949. The Communist coup in Czechoslovakia, the imminent victory of the Chinese Communists, the Berlin blockade, the Soviet detonation of an atomic bomb, a series of congressional investigations of subversion, the dramatic Alger Hiss–Whittaker Chambers confrontation, the Smith Act trials of the leaders of the Communist Party, and the House Un-American Activities Committee (HUAC) Hollywood investigations had dominated the headlines for a year and created an atmosphere of fear and suspicion.

Liberals and leftists succumbed to the prevailing hysteria to a surprising degree. Mary McCarthy recalled feeling quite convinced at the time that Yaddo might actually be a Communist front. Dwight was ultimately brought to his senses by the organized resistance of a group of former Yaddo guests—Harvey Breit, John Cheever, Alfred Kazin, Kappo Phelan—led by Eleanor Clark, the sister of Eunice Clark (Eunice was the former wife of Selden Rodman, Dwight's brother-in-law). The Clark group sent out mailings demolishing the case against Ames. Pointing out that there was no evidence of any political means test at the colony, they denounced the accusations as preposterous. Conforming to what was becoming an ironclad rule, they too made a declaration of their own anti-Communist purity, in the course of insisting that the defamation arose "from a frame of mind that represents grave danger both to civil liberties and the freedom necessary for the arts."[19]

Dwight's genuine position on the real evils of Stalinist Russia, combined with his loss of faith in pacifism, working-class militancy, and the possibility of a socialist revolution, were leading him into a political dead end, where he found it increasingly difficult to maintain a critical stance toward the establishment's foreign policy and American culture vis-à-vis the Soviet Union. Dwight's independence, his self-proclaimed negativism, his refusal to accept any kind of nationalist loyalty had marked his political vision and sustained his political life. It was not a matter of betrayal of commitment: he had simply arrived through his own painful analysis to a point where he had no viable political position other than the "lesser evil." For him it was a discouraging dilemma. Even as he continued to identify with a radical, or at least dissenting, tradition, and still felt himself to be a member of an alienated elite in opposition to American nationalism, imperialism, and mass culture, he was, even if inadvertently, coming to support the maintenance of American power abroad and established institutions at home.

It was not until September 1953, when McCarthyism had begun to subside, that Dwight belatedly wrote an apology to Elizabeth Ames as part of a letter recommending the appointment of a friend to Yaddo. Dwight admitted he should have written long ago and apologized for his "rude and what's worse foolish and unfounded letter" of accusation. He felt that he was "usually fairly objective and level-headed, but that time I must admit that my political prejudices ran away with me and I accepted a flimsy case . . . on spectacularly insufficient evidence." He confessed that he had long ago realized he had been "sold a bill of goods." In a gracious response, Ames pointedly noted that if "great numbers of my friends had not rallied to my defense," she would have been even more dismayed than she was, for she "would have had to wonder what our future in this country could be if a majority of intellectuals could be so easily misled." She expressed appreciation for his apology, but her observation surely got to the root of the failure of American liberalism in those years. As one historian has written, "There were so many victims [of McCarthyism] because there were so few defenders." It is now clear that the red scare that transcended as well as preceded McCarthyism was not a right-wing or even partisan Republican phenomenon. It ran across the entire political spectrum, and some of those most vigorously involved were members of the liberal-radical left, who adopted tactics that undermined and even betrayed their most cherished values and beliefs.[20]

During that spring of 1949 the New York intellectual community was rife with conflict and animosities. Nothing provoked more serious dispute than the award of the first Bollingen Prize for Poetry to Ezra Pound, which was announced February 20, 1949, by the Fellows of the Library of Congress in American Letters. This panel of distinguished judges determined that Pound's *Pisan Cantos* was the best book of poetry published in the United States during the year 1948. Having served as a propagandist for the Italian fascist government during the war, Pound was in St. Elizabeths hospital for the criminally insane in Bethesda, Maryland, and under indictment for treason. He was also a notorious bigot and anti-Semite, and the *Cantos* contained ample evidence of his virulent prejudices. At the same time, he was one of the most highly esteemed of modernist poets, the initial patron of both T. S. Eliot and James Joyce and revered by the postwar generation of poets such as Robert Lowell, John Berryman, Allen Tate, Robert Penn Warren, and many others. The Fellows, who included Eliot, Lowell, Tate, W. H. Auden, and Katherine Anne Porter, anticipating the criticism that their choice was bound to arouse, justified it on the grounds that the only legitimate criterion for judgment was poetic achievement. Any other moral or political consideration "would in principle deny the validity of that objective perception of value on which civilized society must rest."[21]

This abstract defense of the absolute autonomy of art in the case of such a bigot as Pound was bound to cause ill feelings in an intellectual community largely made up of second-generation Jews. Jewish cosmopolitans such as Philip Rahv, William Phillips, Delmore Schwartz, and Irving Howe could not but appreciate the negative vote cast by Karl Shapiro, one of the Library Fellows. On the other hand, Tate, Lowell, Berryman, not to mention Eliot, dismissed Jewish objections as chauvinist, provincial, a betrayal of their obligatory devotion to the autonomy of art. Allen Tate went so far as to lecture Shapiro, arguing that Shapiro should vote for the award in order to offset the common anti-Semitic charge that "Jews always act as Jews and not as free human beings"; in the popular parlance of the day, they determined all values by asking the "simple question is it good for the Jews?" Tate was so incensed by an equivocal editorial in the *Partisan Review* that he challenged the writer, William Barrett, to a duel, which was in keeping with his own devotion to Southern aristocracy. Dwight, who had no love for Barrett, was equally incensed by what he saw as a mealy-mouthed and equivocal editorial in which it was difficult to discern on which side Barrett and *Partisan Review* stood. Shades of earlier battles with "the boys."[22]

A deep strain of anti-Semitism ran through modern literature. It could be found in Henry James and T. S. Eliot and could be traced to Dickens and even Shakespeare. Most of the Jewish intellectuals had long since made their peace with that as part of their passage into the cosmopolitan world of twentieth-century culture. But this was 1949, and the revelations of the Holocaust, an unavoidable preoccupation of an entire generation of Jews and many gentiles as well, could hardly be ignored when judging a man like Pound. To recognize his technical craft as a poet was one thing; to honor him was quite another. As Daniel Bell wrote in explaining his own vote against another award for Pound to be given by the Academy of Arts and Letters: "One cannot escape the meaning of the Holocaust." What, Bell wondered, does one say to those who allied themselves with the murderers?[23]

Clement Greenberg wrote in the *Partisan Review* that awarding such a prize to Pound literally made him "physically afraid." (For this confession he was placed in the category of "Jewish chauvinism" by Nicola Chiaromonte.) Irving Howe was another who could not stomach paying homage to such a man. While a champion of modernism, he felt that as Jews "we had our own distinctive outlooks and interests to defend. To render Pound honor a few years after word of the Holocaust reached us was unbearable." It was not, Howe thought, "anti-Semitism, not at all, but it's a kind of bleached narrowness of spirit."[24]

Typically Dwight did not shy away from this minefield. He had chafed for some time at what he felt was his Jewish comrades' betrayal of their allegiance to the international and cosmopolitan intellectual community. He cleverly took this opportunity to denounce that betrayal, while at the same time making the issue another chapter in the developing cold war polarity. There was an ironic contradiction here. Dwight admonished Jews for succumbing to a Jewish nationalist identity, while he was bent on favorably comparing the United States with the horror of the Soviet Union. Claiming incompetence to judge, Dwight made absolutely no case for Pound's poetry. What he unequivocally defended was the principle on which the decision was made: the validity of an objective rather than an ideological perception of value. The judges' courage, despite their claim of having avoided all political considerations, was, in Dwight's view, "the best political statement made in this country for some time . . . the brightest political act in a dark period." Pound's *Pisan Cantos* were full of detestable social and racial prejudices. Nevertheless, a committee of eminently qualified poets had determined that the literary quality of Pound's work was deserving of the prize. Here was a committee appointed by a high official of the United States, the Librarian of Congress, giving a prize to a man accused of treason. "There were," Dwight wrote, "few other countries and certainly none east of the Elbe where this could happen." Dwight took pride as an American in knowing that his country had "preserved a society free and 'open' enough for it to happen." In addition to this defense of the autonomy of art, the recognition of Pound as a poet as distinct from Pound the anti-Semite fascist, Dwight took the occasion to insist that the committee's choice demonstrated that the United States stood for civilization. In the United States there was a conception of humanity where "clear distinctions were made between the value of a person's work and the value of their politics. The horror of the Russian system was that it reduced the individual to one aspect, the political." Dwight argued that the liberals and the radicals who opposed the award "were bothered by the very thing that was healthiest politically about it: the fact that Pound's treason and fascism were not taken into account when honoring him as a poet."[25]

There was an ironic twist in Dwight's argument. Robert Hillyer, a well-known critic for the *Saturday Review* whom Dwight and other supporters of the award saw as a philistine of the first order, had denounced the award as representing an esoteric cult of "New Critics" notorious for their anti-American, anti-democratic elitism. Dwight, in an equally nationalist outburst, was defending the award as evidence of the United States' position as the chief defender of the free world. If Dwight had simply rested his case on the autonomy of art, it would have been consistent with his long-standing criticism of those who called for a politically acceptable content. But his insistence on seeing the award as a blow for freedom as

opposed to totalitarianism was difficult for some of his readers and colleagues to take, and understandably so, since Dwight had always shown
contempt for any kind of nationalistic self-congratulation. The further
irony is that Dwight could hardly see Lowell, Tate, and Eliot as spokesmen for a libertarian freedom or any of the politics of personal liberation
with which he identified himself.[26]

Dwight's defense of the Bollingen judges drew considerable praise.
Felix Giovanelli, the translator and Dwight's political comrade, wrote to
congratulate him for his "graceful gesture" of support. The editor of *Poetry*
magazine, Hayden Carruth, praised Dwight's editorial as "by far the best
thing that had been published on [the] matter." Allen Tate thought well of
Dwight's support and wrote a note of praise. Dwight was pleased by the
generally favorable response of the readership and took a perverse pleasure in an extremely "vituperative, hot-tempered and abusive note" from
Pound, which engendered in Dwight an "instant, personal liking" for the
man. Pound had written to say that Macdonald was "certainly one of the
most ignorant apes that ever reached a typewriter." He assumed from
Dwight's editorial comments that he "lied from ignorance rather than malice." Dwight responded with a one-liner: "My Dear Pound: As one ape to
another, what's eating you?" Dwight was proud to show this letter to the *PR*
boys. For Dwight to have been called stupid by Trotsky and an ape by
Pound was no mean accomplishment.[27]

At the very time Dwight was paying his homage to the judges of the
Bollingen award, he became involved in another related issue that caused
no little resentment among some of his Jewish friends. At one point in his
supplementary account of the Waldorf Conference which appeared in the
same Winter 1949 issue of *Politics* supporting the Bollingen judges,
Dwight indelicately described a claque of the delegates as "youngish,
Jewish, male" trade union organizer types, who bombarded the meeting
with petitions and announcements. Dwight's description of them as "Jewish" drew a threatening rebuke from Clement Greenberg, who was known
for throwing his weight around. Irving Howe, who had only recently
described the "return to Judaism" of people like Isaac Rosenfeld, Leslie
Fiedler, and Irving Kristol as a "sickeningly sentimental business," nevertheless found Dwight's seemingly gratuitous observation stupid and a
mistake, if not anti-Semitic. Dwight was upset at what he considered an
onion-skinned reaction by his Jewish friends and associates, which he
saw as all of a piece with their opposition to the Pound award. He wrote a

complaining letter to Chiaromonte about an "upsurge in Jewish nationalism," with an increasing "super sensitivity and a super aggressiveness on
the part of Jews about their Jewishness." He saw an irony in the fact that
the very man he had been baiting at the Waldorf Conference, Fadeyev,
had only recently been exposed as the one most responsible for "carrying
out the purge of Jewish (read cosmopolitan—apparently the euphemism
used by the Politburo) writers inside the USSR." Chiaromonte, who saw
eye to eye with Dwight on this issue and only a year earlier had been
rebuked for his defense of Proudhon's anti-Semitism, wrote a reassuring
letter saying that he, his wife Miriam, and even Lionel Abel did not think
that Dwight's "reference to Jewish-looking people in the Waldorf piece
was in any way nasty."[28]

The problem caused tension at social gatherings. Dwight noted that
one could not joke about Jews anymore. He loved to repeat Delmore
Schwartz's comment "After six months, Israel will change its name to Irving." Now telling that joke would be considered anti-Semitic. This at the
expense of marvelous Jewish humor, wry and self-deprecating. Dwight
collected all sorts of incidents of what he considered an aggressive Jewish
nationalism. To Dwight this was a "retrogressive" rejection of cosmopolitanism, a revival of a kind of "primitive clannishness." As for the explanation that the Holocaust had demanded that Jews rethink the meaning of
their existence in the Diaspora and their relationship to non-Jews, Dwight
considered it "an irrational response, a neurotic response to a neurotic
crime." He wrote in his notes that he understood it, but that didn't make
it rational.

Dwight denied the assertion made by Jews that anti-Semitism was
increasing in America. He saw no evidence of it, only "an excessive sensibility" developing among Jews. Why, he wondered, were so many of his
friends writing about strictly Jewish matters in *Commentary?* Why were
such bright and able people devoting so much time to "uninteresting
exercises in Jewish culture"? It was all the more regrettable, he said,
because there was not "a great fund of intelligence over here and the Jews
had an abnormally big part of it." He cited recent articles on strictly Jewish themes by Irving Howe, Daniel Bell, Isaac Rosenfeld, Harold Rosenberg, Paul Goodman, and Hannah Arendt. In his notes, Dwight argued
that Jewish culture, like his own Scottish culture, simply could not bear
the weight of such scrutiny; that it was a provincial, second-rate culture
in interest and importance and could not measure up to first-rate cultures
such as the French and German. "Does the extermination of Jews make

the Jewish culture one whit more interesting?" he asked. In the past, Dwight noted, "Jews made contributions to culture not as Jews but as Frenchmen (Proust), Germans (Heine), or mostly as Europeans (Marx, Freud)." In the thirties and forties, the Jews played a leading role in the avant-garde culture, but now there seemed to be a withdrawal, a return to a "ghetto culture," a turning away from subjects that appealed to Jewish and non-Jewish intellectuals alike. He predicted the intellectual decline of *Commentary* if it continued to allow such ethnocentric articles to dominate its pages.[29]

Dwight was specifically anti-Zionist, in keeping with his anti-statist attitudes, and he expected his Jewish associates to be the same. He was personally annoyed by what he saw as the assertion of a Jewish monopoly on the notion of alienation. Bell and Howe wrote of alienation as though it were wholly Jewish. But Dwight had always taken pride in his own alienation from the prevailing mass culture of the United States. It seemed to him that his former comrades were excluding him from their own alienated ranks.

There was a tension here that transcended political differences. Dwight had worked for over a decade in a predominantly Jewish intellectual environment. Sometimes he claimed to be a "non-Jewish Jew: I have spent my whole adult life in NYC radical-intellectual circles . . . have long proclaimed myself an 'Honorary Jew' and Red Rosa has long been my favorite revolutionist." On the other hand, he often felt like an outsider. He had been a kind of apprentice not only to Marxist radicalism but to the vigorous mode of argumentative dialogue, New York Jewish humor, the way of speaking only in questions—"This is a magazine?" "That is a political position?" He had worked hard to become one of "the boys," but he always felt the distance. In his brief memoirs of these years he noted the head start the Jewish intellectuals had. They had marched in May Day parades when they were in short pants; he had not really discovered Marx until he was thirty. Looking back on the experience later in life, he recognized that even in the Trotskyist movement he was "never involved psychologically" the way Rahv and Phillips had been. "I was always a tourist; I could get out anytime." For a talk he gave in 1970 on being a New York intellectual and the anti-Semitism that people directed toward Jewish New York intellectuals, Dwight wrote in his notes of the distinction between Nazi anti-Semitism that saw Jews as inferior and a populist anti-Semitism that resented them because they were superior.[30]

Sometimes the issue did turn nasty. During a small party at the Rahvs' apartment, Dwight got into a boozy argument over the issue of the new State of Israel and its treatment of Arab refugees. Philip Rahv complained of the anti-Semitic tone of an article by William Arrowsmith criticizing the *Partisan Review*'s defense of Israel in the *Hudson Review*. Dwight denied the charge or conceded its truth only in a minor degree and then turned on the two editors: "You Jews are no better than anyone else." He delivered a rather drunken diatribe on their obligation as Jews to show some concern about the Arabs, threatening to start a committee to send help to the refugees and damning the Israel government for its lack of concern. At one point when William Phillips made some rejoinder, Dwight replied, "Oh, you sharpshooters. You shift gears too quick for me—I can't keep up with you." Phillips took this to be anti-Semitic, a suggestion that Dwight was attributing their facile footwork to some kind of "Jewish trait." When he later reviewed the incident, Dwight thought he was referring to his past experience with "the boys" during his squabbles over editorial policy.

As Dwight was leaving, the mild-mannered Phillips, in an effort at reconciliation, remarked: "Well, anyway, Dwight, I do admire you for being so frank and honest, and for not sneaking around corners." To which Dwight replied: "Well, anyway, Will, I don't admire you for not being frank and honest, and for always sneaking around corners. But you can't help it—you were born that way." Phillips later called Dwight and expressed his shock at what he took to be anti-Semitism, adding that he did not feel Dwight's "extreme drunkenness" was an excuse. Dwight wrote in notes about the incident that the comment about "the boys" not being able to help themselves might indeed have been anti-Semitism. However, he also thought it had simply been prompted by his irritation at Phillips's "usual, sanctimonious, solemn-sweet manner" and that he was indicting him as himself rather than as a Jew. "Consciously," he wrote, he "did not think of Jews as being especially sneaky or unfrank. But of course there may be unconscious anti-Semitism." Looking at his notes in 1972, Dwight wrote a marginal note suggesting that Will and Edna Phillips were right, that his phrase "you were born that way" sounded like "you were born Jewish."[31]

There is something poignant and admirable about Dwight's continuing self-analysis and attempt to confront his deepest feelings. It also helps explain the intensity and seriousness with which he approached such issues as the award of the Bollingen prize to Ezra Pound, his anger over Israel's treatment of Arab refugees, and the bitter controversy he was soon to become embroiled in over Hannah Arendt's *Eichmann in Jerusalem.* These episodes that dotted his career have importance beyond his rejection of anti-Semitism. He rejected the notion that the historic contribution of liberalism was to separate politics from morality and insisted that the most significant task was to figure out how to relate morality and politics. Dwight's struggle with his past is related to his status as one of the first of American intellectuals to directly confront the moral horrors of the war and the frightful implications of the Holocaust. During these difficult years, Dwight devoted much effort to finding a path to some kind of human sanity. He continued to insist that political morality must "emphasize the emotions, the imagination, the moral feelings, the primacy of the individual human being once more." He recruited an astonishing community of writers and critics who posed in his magazine the crucial and enduring questions of modern existence. Just how do people lose their humanness? How are they made into things? What is the process? "How may we tell good from evil?" Under his direction, *Politics* took up the cause of the victims of all oppression, defending the rights of blacks, women, homosexuals, conscientious objectors at a time when these issues were either beyond the pale or ignored. All this was part of an ongoing effort to learn how he could "live lovingly, truthfully and pleasurably" and, as a good journalist, pass the word on.

While Dwight was struggling with the residue of prejudice from his past, which his intellect and good sense rejected, he was also carrying on a continuing debate over domestic and international politics. That Winter 1949 issue of *Politics* contained a long and closely argued critique of the entire direction of Dwight's politics and *Politics* from an attentive subscriber, which prompted Dwight to rethink just how far he had removed himself from a radical approach to American life.

The letter of criticism was written by Robert Bone, a young radical drawn to Dwight's anarcho-pacifism. Expecting so much from Dwight and the magazine, he was keenly disappointed by what he saw as Dwight's rampant deradicalization. In Bone's reading, there was a deep watershed that by the spring of 1949 divided the new *Politics* from the old. Dwight's "emotional hostility" to Stalinism had caused him to abandon his third-camp position and, as a result, the critical edge that had defined the magazine. The shift was evident in the language Dwight used now. He had adopted the East-West polarity of the mainstream press. Bone was particularly chagrined by Dwight's use of the phrase "We of the West" in his comparisons of the United States with the Soviet Union. Bone asked rhetorically, "Who of the West? Acheson, Franco, the U.S. Military, and Cotton Exporters?" Did Dwight feel solidarity with them? Bone, who was at the moment under indictment for refusing to register for the draft, took a dim view of the U.S. government's persecution of nonregistrants; whereas Dwight had once expressed condemnation, he was now writing justifications.

Bone also took on what he saw as Dwight's willingness to collaborate with Sidney Hook and the Americans for Intellectual Freedom group. Why had Dwight not attacked Hook's support of the "purging of CPers from all teaching positions"? Surely Communists were no more given to distortion in the classroom than Catholics or "the average bourgeois professor." Bone insisted that the "ruling-class domination of education" was far more threatening than a few isolated Communists.

Nor could Bone, like so many other critics of the Bollingen award, accept Dwight's separation of Pound the fascist from Pound the poet. He maintained that on this score Dwight was inconsistent. In the same issue of *Politics* that carried the Bollingen editorial, Dwight praised Norbert Wiener, the M.I.T. mathematician who a few years earlier had publicly refused to give technical information to a colleague working on guided missiles, for acting as a man first and a scientist second and thereby accepting a relationship between ethics, morality, and intellectual discipline. And as for Dwight's sentimental closing columns in the Winter 1949 issue, praising such "uncommon people" as the Kuwait Oil Company, which had donated $100,000 for the relief of Arab refugees, and the governor of the Leeward Islands, for treating his "colored" population with respect and giving scholarships for poor students out of his own pocket, Bone demanded to know how such drivel could appear in a radical magazine unless it was a "deliberate kind of softening-up process preparatory to accepting a philosophy of 'reformed' power politics." The very fact that Dwight had listed himself on the cover of the Waldorf Conference supplement as an *American* intellectual in battle with *Soviet* intellectuals confirmed his acceptance of American nationalism. Bone noted that in the past Dwight found the designation "American" repugnant. Bone closed by reminding Dwight that it was the role of a radical to expose the forces of total war, State worship, mass industry, and mass culture, not to argue that one nation had a monopoly on these evils. Dwight was literally urging the destruction of an enemy nation rather than taking a critical stance and declaring that the deep ills of modern civilization knew no national boundaries. To soften criticism of the United States in order to emphasize the ills of the Soviet Union, Bone charged, only fostered a complacency that was making the nation resemble its opponent.[32]

Bone's long critique arrived toward the end of April, and Dwight waited to reply after he had left the city for the summer and had time to think things through. In the middle of June he began his response, which took nearly a month to complete. It was a painful reassessment, undertaken to clear his own mind as much as to enlighten his critic. He began by saying that Bone's letter had hit home. It was true, Dwight conceded, he was changing his politics. Initially he had said to hell with both war camps, even while arguing that the Nazis were qualitatively different and much worse than the Allies. He had opposed support of the Allies because he genuinely believed a "revolutionary third camp" was a "historical possibility." He could no longer maintain this revolutionary socialist point of view because no third camp had materialized, and the chance that one would emerge seemed negligible. Pacifism still had a hold on him, but he had recently resigned from both the Peacemakers and the War Resisters League. He had tried to publicly explain his dilemma earlier in *Politics*.

Dwight also conceded that *Politics* had not printed much about the evils of Nazi Germany versus the virtues of the United States, whereas it had harped on this theme when the Soviet Union replaced the Nazis as the great peril to civilization. Once again he claimed that *Politics* readers, being on the left, needed no warnings about Nazism, but many still cherished illusions about the Stalinist regime. They continued to see it, erroneously in Dwight's view, as much less threatening to socialist and pacifist values. Therefore Stalinism was far more dangerous. But even more important, Dwight no longer felt the American people were going to make a revolution, a genuine "break with ancient evil and hoary iniquity," which he had once hoped they would. He frankly confessed: "I no longer see any political (or . . . historical) reality in such utopian, all-or-nothing doctrines as revolutionary socialism or pacifism." In addition, Dwight acknowledged that he had become very conscious "of how much better things (are) ordered in this country so far as the interests of the great mass of human beings are concerned" in comparison with the conditions in Russia. In effect, Dwight was now taking the position of Stephen Spender in 1941, which Dwight had denounced as a sellout, although the Russian threat in 1949 seemed far less imminent than the notion of a German invasion of the British Isles in 1941. Dwight went on in a confessional tone: "I have been forced by the logic of events to abandon this ultimatist Utopian perspective, and to think in much more limited, empirical, here and now terms."

With great reluctance, he said, he had joined his early half-a-loaf opponents. Without a "faith" (for that is what it had been) in an alternative position, which Dwight had clung to despite much skepticism, he now felt forced to support U.S. policy. He could not argue for America's withdrawal from Berlin and the "consequent giving over of the Berliners to the terror and repression of the Soviet system."

So exhausting was this repudiation of his entire political past that Dwight put the letter aside for nearly a month, finally returning to it July 11. In the second part he summed up his earlier remarks by stating explicitly that he had "lost . . . *faith* [emphasis added] in any general and radical improvement in modern society" by either Marxian socialism or a "pacifist persuasion and ethical example." He would now look at politics only in immediate pragmatic terms. With this vision he came to the despairing conclusion that not only were socialism and pacifism "here and now quite bad for people," but that pacifists instinctively underrated the threat of Stalinism and denied the argument that it was any worse than the political systems of the United States and Western Europe. Pacifists refused to recognize it as a system of "total human degradation which is to the old tyrannies . . . as the Atom Bomb is to the gatling gun." Dwight quickly dismissed the pacifist power of persuasion and brotherly love. The new totalitarian regimes of the Nazis and the Stalinists were impervious to such appeals. They used the "utmost of brutal force" and "psychological pressures" to reduce their victims to the kind of zombies that Orwell described in *1984*.

Dwight felt that Bone's worldview was so different from his own that there was no way he could communicate the urgency of the situation. He accused Bone of representing an intellectual and moral absolutism that blinded him to the nature of the new totalitarian systems. Pacifist and socialist thought had not "kept up with modern history, had not evolved, developed, produced new theories which are capable of coping with the post–Hitler-Stalin world."[33]

At the same time, Dwight had the overwhelming feeling that the individual was dwarfed by world events. During the same period that he was making this assessment of a personal political deradicalization, he wrote to a student publication at the University of Chicago lamenting the powerlessness of the average citizen of any nation-state: "The individual 'citizen' (what a mockery!) has about the same chance of determining his own fate as a hog dangling by one foot from the conveyor belt of a Chicago packing plant." He noted the terrible irony of contemporary events wherein the U.S. Army, "a most reactionary organization, whose purpose is mass slaughter," had now become, with the airlift, the protector of the popularly elected government of Berlin against the threat of liquidation by the Russians. Only a few years before, they had firebombed Dresden and killed hundreds of thousands. There was, Dwight wrote, "indeed a logic to both actions, but it is not a human, not a rational or ethical logic. It is rather the logic of a social mechanism which has grown so powerful that human beings have become simply its instruments." This continued to be Dwight's great theme: dehumanization, the mechanization of society. Despite his tilt toward the West, he took little comfort in the rapid development of the national security state.[34]

The total candor of Dwight's political retreat and his insistence on facing up to the consequences of his conclusions cannot be dismissed. He was genuinely trying to see the situation as the evidence presented itself and not shirk the logic of the argument. In effect, he had simply lost whatever faith he had in a revolutionary alternative. In the past he clung to that hope, sometimes almost as a way of maintaining his adversarial stance. But now, as he had stated so plainly on the cover of *Politics*, he was trapped by the terrible dilemma of having to make a choice between undesirable alternatives. He was, as he was soon to announce publicly, "Choosing the West," but not without regret, and even despair, that his active political life was drawing to a close. The critical ideal, the adversarial stance was the spark behind Dwight's editorial being. As an intellectual one could not adopt a party line or establishment position. Dwight had believed deeply in the power of ideas, but now his own political ideas, centering around socialism, pacifism, and anarcho-individualism seemed dated, ineffective, even dangerous. As he later was to write "then [in the period of 1935–45] we believed in revolution, now we don't."[35]

Dwight was tormented by the need to formulate his ideas, to explain to friends, colleagues, comrades the dimensions of his disillusionment. On May 19 he gave a talk declaring "Goodbye to Utopia" at one of the "Packages Abroad" discussion meetings at the Rand School auditorium. It was a part of a fund-raising series of anti-Soviet addresses delivered by such anti-Stalinist luminaries as Bertram Wolfe, the author of *Three Who Made a Revolution*, the China scholar Karl Wittfogel, and Arthur Schlesinger, Jr., all of whom were to find their way into the American Committee for Cultural Freedom, a bastion of cold-war intellectual activity.

Repeating his view that both socialism and pacifism no longer provided any solution to the menace of Soviet expansion, Dwight described himself as at best a "pacifist fellow-traveler." He confessed that he had always been a fellow-traveler of all the movements he had briefly allied with. He always had doubts and hesitations when in the Trotsky movement and only joined out of a "moral revulsion" against the war rather than any intellectual Marxist convictions. The same was true of his pacifism. It had been prompted by the deep shock to his ethical values caused by the atomic bombing of Japan rather than by any real belief in it as a political doctrine.[36]

Dwight was probing his real political motivations. He now confessed that he was a "hopeless empiricist," which he had been ashamed to admit because it was considered a kind of leprosy in the Trotskyist movement. Now it was "Goodbye to Utopia," goodbye to attempts to find any "consistent systematic pattern for understanding reality." He gave a hearty welcome to a kind of political thinking that assumed "reality is contradictory, subtle, complicated and ALIVE." Dwight had been reading Alfred North Whitehead's *Science and the Modern World*, which he judged to be "one of the most important books of our century" and essential to any understanding of the ideas that governed the age. He quoted Whitehead's warning that to be mentally in an ideological groove is to live contemplating a given set of abstractions. The groove prevents "straying across country." There was nothing Dwight liked better than leaving the groove and straying across country, tacking and zigzagging as he went. And nothing irritated his comrades more than his refusal to stay in the groove. In a fit of annoyance at Dwight's dilettantism, C. Wright Mills had called him the "Peter Pan of the Left." Now Dwight flaunted his repudiation of ideological theory. In a letter to one young African student of a Marxist persuasion, he wrote: "Personally I feel that history or theories can often be sterile, while people's own personal experiences seldom are. Our western culture puts too much emphasis on big scientific concepts and not enough on human experience."[37]

Just a year earlier, Dwight had published extracts from Alexander Herzen's *My Past and Thoughts* because Herzen rejected grand theory. In an introduction to the selections, Dwight wrote that men a century later were in much the same state of mind as Herzen had been after the failed revolution of 1848: "Despair and doubt ravage us, the Marxian dream has turned into the Russian nightmare and we can appreciate the unsystematic, skeptical, free thinking approach of Herzen." Here Dwight, the radical, the seeker, the enthusiast, embraced the political vision that became known as "the end of ideology." But he did not join forces with those who endorsed the status quo and began to celebrate American life as the best of all possible worlds. If Marx had been "our man in the thirties," Dwight argued, Herzen was more relevant today: "When mankind seems to be at an impasse, such a thinker, precisely because he is free, uncommitted, is better able to make us aware of our real situation."[38]

Dwight also acknowledged his agreement with the premise that "the really terrifying enemy of human values today is not the traditionalist, the conservative (say, Herbert Hoover or T. S. Eliot) but rather the Utopian (say, Henry Wallace.)" Considering how mild-mannered and ineffective Wallace had proved to be, this was pretty heavy hyperbole designed to ruffle the lib-lab leftists in his audience. Utopias when carried out "seem to lead almost fatally to the reverse" of what they sought to achieve. Calling on his thirties commentary on Swift and the Houyhnhnms, he concluded that a society which denies evil and the irrational denies life itself. When good becomes universal, he now argued, there is no longer room for life, since "life means Evil, Conflict, Unhappiness, Diversity and Irrationality as well as Good, Harmony, Happiness, Uniformity and Reason." If mankind "created and accepted a world in which there was neither economic exploitation (Marx) nor sexual neurosis (Freud), the result would also be a world without art, religion or culture as we know it." Dwight Macdonald may well have been educated at Exeter and in the Trotskyist movement, but after roughly a decade as at least a fellow-traveler of radical and Marxist movements, there is little doubt as to which school dominated his thoughts. He was now repeating the conclusions he had come to shortly before he began his adventurous journey from Luce to Lenin, and it is apparent that not much of that journey's ideological message had stuck. Without resorting to the terminology, Dwight realized that he was getting "very close to religion: Original Sin, the Imperfectability of Man, etc.," an acceptance of a kind of Niebuhrian view of man's frailty which utopians refused to entertain. That still made him uncomfortable, so he made an analogy to sport, where only opposition, error, the irrational make the game interesting.

Dwight still clung to the existential position he had adopted in the early 1940s. One had to act on the basis of possibilities even if the probabilities were slim. To do otherwise was to court inertia. So he now spoke for what he facetiously called a new kind of utopia: "namely that of regarding any situation as to some extent open." For some reason, however, he did not apply this to the pacifist proposition of brotherly love and persuasion. But he did argue that it was important to avoid either-or positions such as "We must go to war with Russia or lie down and be conquered." With remarkable prescience, he argued that there were many other possibilities: the Russian system itself might break down without war. As for American foreign policy, he continued to support the Marshall Plan simply because it sustained starving peoples. The United States should give to Europe useful commodities and technological know-how and the liberal democratic tradition. What it should take from Europe was culture—arts and letters and the art of enjoying life. The United States should make Marshall Plan aid dependent on free elections, independent trade unions, progressive income taxes, democratically elected presidents to replace kings, minimum wages, and social security laws. The result of such a program would be the weakening of Communist-led guerrilla movements. In effect, Dwight was now solidly in the camp of the Americans for Democratic Action, the liberal impulse of the Central Intelligence Agency, all of the once hated lib-labs. He now called for more intervention, not less, in European affairs, more exporting of American goods and ideas abroad. As for imports, he called for the immediate opening of the gates to greater immigration. It would be "the most effective political counteroffensive against the Commintern and it would make America a more humanly enjoyable place to live." In this address, which may have elicited some raised eyebrows if not hoots and hollers from the Rand School audience, Dwight seemed bent on snatching some measure of political satisfaction from the jaws of defeat and disillusionment that had characterized his dilemma.[39]

This was a painful exercise for him; his honesty demanded that he follow the thread of his thought to its logical conclusions, but there was no joy in the writing, no humor, no pleasure. It was indeed a "Goodbye to Utopia" and a good deal more as well.

11

The Breakup of Our Camp

WHEN TRYING TO DESCRIBE WHAT WAS HAPPENING to the New York intellectual community in the late 1940s, Irving Howe recalled the poignant title of a collection of stories written by Paul Goodman, "The Break-up of Our Camp." The title story described the end of a small community, the breakup of a summer camp; in a very different context, Dwight and others were experiencing a similar fragmentation and imminent death, of an American, or at least New York, intelligentsia. Dwight sensed in 1949 that an era was passing; it left him feeling adrift and, as he wrote to Chiaromonte, a bit frightened. From the last issue of *Politics*, with its ad on the last page for his formal "Goodbye to Utopia," Dwight was saying goodbye to much of his past, goodbye to *Politics* and politics, to the bohemian Village community, to his wife Nancy. The five years starting with 1949, which led him through two love affairs, divorce, and remarriage, were years of turmoil, indecision, and at times a coming apart. During this period three major themes run through the narrative of his life: a traumatic personal crisis at once exciting and depressing; the abandonment of his earlier radical form of politics and the great difficulty of finding some substitute for a way to live a satisfying and moral existence; and finally, his need to fashion a new career, to become "a money writer" who could make a living with his pen. The personal, the political, the professional, all the dimensions of his life, were completely disrupted, and he went through spells of disturbing self-doubt and anxiety, but at the same time he felt the exhilaration of a new life in the making.

When Dwight left for the Cape in June 1949, he planned to make a break with his hectic life, to take stock of his career, his marriage, the way he lived. He complained to Chiaromonte that he had become little more than a switchboard, with no time to reflect. The whirl was often fun and sometimes stimulating, but he wanted to "become a writer and a reader again." The rather large house of Norman and Anna Matson that he was renting in the Wellfleet woods had a separate study in a wooded glade and he planned to set aside some time each day there for reading, thinking, and pondering his future.[1]

On "Escape Cod," after bidding "Goodbye to Utopia," Dwight hoped he could create his own new anti-utopia: an anarchistic, unplanned, spontaneous freedom, a dismissal of bourgeois rules; the prime impulse would be self-expression, self-awareness, even self-indulgence—a reckless flouting of convention.[2]

The first step toward making his escape a real possibility was to give up *Politics*. He wrote a letter to subscribers that suggested he didn't know when, if ever, the next issue would appear.[3] There was vague talk of recruiting an editorial board to help with the publication, but nothing definite at all. Intellectually and emotionally, Dwight had decided to abandon the journal. He would be free to pursue his personal life; it was the first and necessary step in the painful break with Nancy. *Politics* had been their raison d'être. Ending their collaboration on the journal would mean the beginning of the end of their marriage.

Dwight was seeking personal liberation through a deliberate attempt to violate the normal restraints of married life. His reading in Reich and Goodman was not simply a desire to understand their psychology. He was searching. The affair with the wife of the former college classmate that had begun on the Cape in 1947 had continued sporadically during the year, but had ended in 1948. It seems clear that Dwight was hoping for another liaison this summer, and he was not disappointed.

The Macdonalds had been summering on the Cape for five years. There had developed a close-knit community of congenial couples with relatively similar lifestyles and bohemian inclinations. Affairs were tolerated; it was not unusual for men and women to make known their interest in other people's spouses. One frequent guest of the Macdonalds recalled: "We were all very promiscuous. Everybody bedded down with everybody else." That recollection was probably exaggerated by nostalgia, but there is evidence that such was the case and that the community was plagued by family crises and broken marriages.

What made these relations even more complicated was that everyone knew everyone else. In Dwight's case the women he became involved with knew each other and Nancy very well. They were constantly socializing at the casual get-togethers in the afternoon on the isolated beaches, the cooperative cocktail and dinner parties, and the nighttime beach parties, with a roaring fire, skinny dipping, and petting in the dunes. So when Dwight began the serious pursuit of a longtime member of the community, Joan Colebrook, an unhappily married Australian woman with three children, it became common knowledge.[4]

What was so thrilling for Dwight about this affair was his feeling that he stood on the threshold of really changing his life. With this woman, he felt he had somehow discovered the talisman that would allow him to escape what he had come to see as his routinized, unimaginative, monogamous relationship—a relationship based, to be sure, on love, genuine affection, and respect, but lacking any spark, any "magic," as he put it.[5]

His view of the new relationship was one of innocent awe. From adolescence he had been troubled by sexual insecurities, including fears of homosexual tendencies, which still concerned him at forty-three. They haunted his dreams. His sexual life, from his early interest in Dot Wheeler through the turmoil of love and rejection during his years at Yale and then in New York while working for *Fortune*, represented to him a record of failure. It might be said, and he clearly thought, that he had married the intelligent, sensible, and sensitive Nancy Rodman out of a desire to escape sexual passion. Looking back on his life, he felt he had been deeply afraid of women and had been unable to believe a woman could love him. Eileen Simpson, John Berryman's wife at the time, recalls that when she first met Dwight in the early forties she thought of him as not interested in women as women. He seemed so cerebral. Women were attracted by his intellectual energy and when they made their interest and admiration known, he was surprised, she said. While it is true that he wrote of a sexual interest in Nancy early in their marriage, he now saw most of the years as bland, constrained, and unexciting for both of them. He confided to his journal, "Something should happen in a good marriage; nothing happened in mine."[6]

Nancy's own perceptions were not very different from Dwight's. She had hardly found the marital bed an exciting place, either, but she was not so compelled to experience the physical delights of sexuality at the expense of her deeply valued relationship with Dwight as a partner, a friend, an energetic collaborator. What she missed was emotional warmth, not just sexual heat but a personal connection—a loving feeling for her that transcended intellectual respect and at the same time was not simply physical attraction. Eunice Clark Jessup, one of the Vassar group, who was briefly married to Nancy's brother Selden Rodman, in a revealing and perhaps unfair analysis remembered the "performance-oriented" Wasp approach to lovemaking: "getting it up, getting it out, getting it off."[7]

Dwight was keenly aware of what he took to be his own failures as a lover. He often referred to it in his private journals, in an almost Hemingwayesque image, as "the wound." He spoke of his inability "to make life interesting emotionally" for women. "Women," he wrote in a piece of self-analysis in the midst of his affair in 1951, "want to feel a RELATION with a man above all. . . . An emotional not just a sexual relationship." Dwight felt that his tendency in the past had been "to shy off from an emotional relation, to keep them at a distance" and then to want to "go directly from the impersonal, intellectual plane to the sexual plane."[8]

Joan Colebrook was an attractive woman with more sexual experience than Dwight and more physical energy than Nancy. She was able to open up a new world of sensuality to Dwight. So dramatic was the experience that Dwight felt he was being introduced to sex for the first time. She was also able to make him feel that he was "attractive to (some) women and for the first time [he] felt a responding emotional and sexual attraction."[9]

So enthralled was Dwight by this new sexual pleasure that he made little effort to hide his obsessive desire to see his lover, even to the point of dropping Nancy off at the cottage and spending the night with Joan, when it was clear that Nancy knew and was terribly hurt. Once when Nancy took umbrage at his blatant behavior, he brusquely replied, "Would it be better if I slept with her without your knowing it?" Nancy replied with more acidity than usual: "It is one thing to know that there are cockroaches in the bathroom, but it's another and worse thing to find one on one's toothbrush." Thinking back on these sorry episodes later in therapy, Dwight was contrite and guilt-ridden over his "callous and crude disregard of Nancy's feelings."[10]

Dwight's belated discovery of sex reminded Joan of a "college kid's discovery of atheism." She found his wild recklessness and devil-may-care attitude immature and irresponsible. It was not long before she began to suspect that she really had a different agenda than he did. Her feelings for Dwight were genuine; she was keenly attracted to his energy, his stimulating intellectuality. She felt that he had a fine literary talent and a bright future in the American intellectual community. But she had already experienced one failed marriage, and her second was in the process of dissolution. She was interested in a measure of security, stability, responsibility. She did want, as Dwight came to understand, a loving, sensitive, intelligent husband, a free spirit like Dwight, but a creative man with more orderly habits, more personal discipline than he seemed capable of during this heady emotional awakening. She frequently reprimanded him for acting like a child, to which he responded that that was exactly what he wanted to do—break away from stultifying discipline.[11]

By the end of the summer and into the fall the love affair had become all-absorbing and so full of turmoil that Dwight could think of little else. Both he and Joan were racked with anxiety about the course they were pursuing. She had confessed to her husband that she was having an affair. Dwight had made the same clear to Nancy. But his angst transcended his lingering fears about destroying his family. He was also concerned about the direction of his professional life. Could he make a living as a writer? Ever since his experience with *Fortune*, he had been torn between his need, even desire, to do commercial "money writing" and his long-held hope, or perhaps fantasy, of undertaking serious, intellectual scholarship, "to write a book in cold blood." He was determined to make known his desire to switch from noncommercial intellectual journalism to freelance writing for a popular market. The first step was to announce the definite end of *Politics*. Dwight sent to each subscriber in October a notice of the magazine's suspension. This time he conceded that it was not entirely a question of finances. Under an understated separate heading, *Personal*, he publicly admitted that he had failed to work up the promised prospectus for a new *Politics*. The uninteresting and depressing political situation made him reluctant to continue to shoulder the burdens of putting out the magazine. His own values and interests were rapidly changing, and he wanted to spend more time on his own writing and less time tinkering, as an editor, with the work of others.[12]

Dwight had been asking around for freelance writing jobs. He landed an assignment to do a generalized critique of the *New York Times* for Max Ascoli's *Reporter*. He predictably lambasted the paper as poorly edited, poorly written, and poorly designed. It was not edited, "it just happens once a day." He wrote a surprisingly appreciative review of Peter Viereck's *Conservatism Revisited: The Revolt Against Revolt* for his friend

Robert Hatch at the *New Republic*. He liked the book because it was so hard on liberals, and he even accepted Viereck's call for more law and order in world affairs. In the fall of 1949 he did a major *New Yorker* profile on Richard Weil, Jr., the forty-two-year-old president of Macy's New York store. He was delighted to receive $3,000 for the two-part article. "I am a rich man," he exclaimed to Joan Colebrook. Despite his excitement, he could not shake a gnawing unease. Somehow writing for money invariably seemed to leave him unsatisfied. While editing *Politics*, he had written Nicola Chiaromonte that working for UNESCO, a job Chiaromonte detested, was "no worse than writing for the *New Yorker*." Now he wrote Chiaromonte about his *Times* piece, the *New Yorker* profile, the *New Republic* review. After each was a dollar figure. But he felt demoralized, stale and disoriented. His writing habits were still "disorderly, slack," he was still too "self-indulgent," and he had yet to find a way to get down to some serious writing.[13]

He did reveal to Chiaromonte what was really on his mind. The continuing love affair was "wonderful and satisfying, but full of frustrations, difficulties, dangers, guilt, complications." He was writing pages of self-analysis, long letters, journal entries, all in an effort to describe and sort out his feelings as he went through the whirlwind of his illicit romance. It was a remarkably thorough record that was in keeping with his compulsive need to analyze, measure, record, and judge. It also revealed his instinctual need as a writer to put things into words, to capture the experience and make it come alive again. What emerged was a painful portrait of a man suffering through what a later generation would label a midlife crisis. Plagued by serious depressions and at times an almost debilitating self-doubt, he became unmercifully demanding of attention and suspicious of his own and his lover's motives.[14]

Being in love was a full-time occupation, absorbing all of Dwight's energy, intellectual as well as physical. He became enamored of Henri Bergson and insisted that Joan read him. (If Joan introduced him to satisfying lovemaking, he tried to instruct her in the ways of the intellect.) For Dwight, Bergson's message in the *Introduction to Metaphysics* was that modern man should trust intuition. It was the only recourse, and was the "typical method of the poet, the artist, and dear, of the woman." Dwight was drawn to Bergson because he saw the scientific mentality as deterministic and the poet mentality as exemplifying free will.[15]

This close connection between his public stance and his private struggle is also evident in his contribution to a *Partisan Review* symposium

inquiring into the religious interests of intellectuals. Dwight responded
with his usual dismissal of organized religion and the notion of a transcendent force. He remarked that rather than great theories, he was interested in the "small questions":

> What is a good life? How do we know what's good and what's bad? How
> do people really live and feel and think in their everyday lives? What
> are the most important human needs—taking myself as that part of the
> universe I know best, or at least have been most closely associated with
> as a starting point? How can they be satisfied best here and now? Who
> am I? How can I live lovingly, truthfully, pleasurably?[16]

These open, innocent, even naive queries could have been copied from
his Exeter and Yale notebooks. It was as though he had decided to go
back to his youth and try to find a way to start out again.

Although Dwight's plans to end his marriage and start a new life with
Joan occupied his mind in the fall and winter of 1949–50, his professional concerns were also paramount. Dwight learned that to do "money
writing" effectively, it was better not to get too involved. He confided to
Chiaromonte that he had just discovered something any idiot knows: that
in commercial writing, one should avoid subjects one has much background in or has strong feelings about. The result—the Macy piece—was
"lovely" and the financial satisfaction was considerable. He found that he
now talked about his work in terms of the size of the commissions.
Dwight's departure from *Politics* and his pending departure from Nancy
were extracting a considerable price in self-esteem and jealousy of those
who had a passion for their subjects.[17]

Somehow Dwight hoped to write for money and at the same time expe-
rience the old enthusiasms. It was for this reason that he bombarded his
New Yorker editor, William Shawn, with profile ideas and other potentially
important projects: Earl Brown, a new black councilman in New York;
Ben Davis, the black Communist Party functionary; Cord Meyer, the
World Federalist soon to be in the employ of the CIA; John Dewey; James
T. Farrell. Shawn showed little enthusiasm. Always lurking in the wings
were portraits of successful executives. If he couldn't get Shawn to bite,
he would end up doing the president of Metropolitan Life. Ah yes, but
there is Wilhelm Reich. Too kooky. Well, what about A. J. Muste? Too
obscure. By the spring of 1950 Dwight had centered on Roger Baldwin,

the founding head of the American Civil Liberties Union. He persuaded Shawn that the piece could be a modern history of civil liberties in America. Shawn liked the idea and thought it appropriate becau .e the junior senator from Wisconsin, Joseph McCarthy, had only recently made a splash in the press with stories of Communist subversion in high places, particularly in the Eastern-educated establishment. A piece on Baldwin might pose an interesting confrontation.[18]

Dwight was excited about the Baldwin project, but quickly found it much more ambitious than he had imagined. Baldwin proved to be a fascinating personality to unravel. Dwight appreciated the complexities and quickly discovered that his portrait would have its "dark shades, but would not be without its highlights," a Rembrandtesque "chiaroscuro-like" effect.

As Dwight got into the research, he began to see that Baldwin's life raised all the issues of his own and that the profile would have to be large in scope. Shawn was reluctant to guarantee him the space he envisioned. Dwight complained to Chiaromonte that he was being forced to leave out "almost everything." Dwight focused on the way he could use Baldwin to describe how liberals went Stalinoid in the thirties. But he had to explain so much to the *New Yorker* audience that *Politics* readers would have taken for granted. Now he had to popularize, even trivialize the subject. The history of civil liberties since 1920 was important to Dwight, his own values and convictions centered around these issues, but he wasn't writing for himself and his comrades—he had to follow the standard formula of a *New Yorker* profile. Tailoring the piece to that formula while meeting the meticulous demands of Baldwin's enormous ego proved to be a difficult and frustrating task.[19]

Politics was part of the problem. Baldwin had been far more of a pragmatist and compromiser than Dwight had initially understood. By 1950, however, Dwight was attacking the ultimatists who demanded purity, and was showing an appreciation for those who recognized reality and strove to achieve half a loaf. But he frequently admitted that he was baffled by the complexities of world affairs. Contemporary life in America was far more complicated than he used to think. In a vein similar to his endorsement of Viereck's conservative demands for stability and order, Dwight was critical of "Heretic[s], Rebel[s], Iconoclast[s] and the other incorruptible Prometheuses of our age" who reject "in toto bourgeois (i.e., post-1790 Western) society and culture," and thus destroy too much. Moving toward a stance of anarchic libertarianism, he insisted to Dinsmore

Wheeler that one must "make one's own culture, one's own ethical code."
But unless one was a Gandhi or a Tolstoy, one would probably make a
botch of it and come out as barbarous as the nearest AT&T executive. He
agreed with Wheeler that he had come a long way in his thinking. He had
indeed changed, he admitted, "and also improved since those dear dogmatic, dead days."[20]

Dwight and family were back on the Cape in the summer of 1950, but
even its whirl of social activity failed to lessen the tension of his life.
Dwight and Nancy had rented a prefabricated cabin that an old Wellfleet
veteran had had shipped from Georgia and placed on the dunes above the
ocean beaches and near Slough Pond close to the Truro-Wellfleet border.
It was a beautiful spot, a short walk through the woods to the breathtaking
dunes overlooking the wide and often barren ocean beach. Nancy thought
it the best place they ever rented; they managed to buy it two years later
and it remains Nancy's cottage.

As in the past, even the tranquillity of the Cape was ruffled by the
latest eruption of the cold war, which had turned hot in June when North
Korea invaded South Korea. Despite Dwight's confusion as to just where
he stood politically, he lost little time in coming to the defense of Truman's decision to intervene with American troops. Nancy was astonished
by the vehemence of his arguments with a local pacifist. She confided to
their young friend Danny Rosenblatt that she was a bit nonplussed
because Dwight "was smashing away at all the arguments he used to use.
God knows where I stand." Nancy did not simply follow in Dwight's footsteps politically or any other way. She was more restrained about repudiating her own past anti-statist and pacifist loyalties. One friend recalls
that while she did not come on strong, she did question Dwight and ask
him to explain himself carefully, and he spent a good part of the summer
doing just that in sandlot debates from Wellfleet to Provincetown and
back.[21]

With his usual gusto, Dwight set about baiting critics of U.S. policy.
One day Clark Foreman, an active member of the Progressive Party, a
strong supporter of Henry Wallace, and the director of the Emergency
Civil Liberties Union, which was labeled a Communist front for its legal
defense of Communist Party members tried under the Smith Act, appeared
on the beach as a guest of Robin and Gloria Lanier, regular summer residents and friends of Dwight and Nancy. Dwight was quick to denounce
him vociferously, shouting out across a wide stretch of sand that Foreman

was a fellow-traveler and Stalinist apologist. At a large evening beach party he continued the confrontation. Ralph Manheim, his old associate and translator for *Politics*, was also astonished at Dwight's enthusiastic support of Truman's police action. When Manheim ran into Nicola Chiaromonte in Paris, he reported that Dwight had more or less gone off the deep end and was a "100 percent Trumanite." Dwight denied that was the case, since he had strong disagreements with much of Truman's general foreign policy: the defense of Formosa, the refusal to seat Chinese Communists in the UN, and the failure to devise any kind of a program that might appeal to the Asian masses. It was always Dwight's argument that the United States should oppose Stalinist Russia, militarily when necessary, but always politically; the United States should counter Soviet propaganda from the left and lend support to leftist opponents of Communism rather than always lining up with the right-wing opponents. This was the remnant of his third-camp position—he was still maintaining the hope of a viable left in Europe capable of opposing Stalinism. It was also the policy of cold war liberals collaborating with the Central Intelligence Agency. But through it all Dwight resisted joining the celebration of American life. He complained to Chiaromonte of the "whole political line of Truman, the failure to compete with the commies in freeing, or pretending to in their case, masses from landlords and top dogs."[22]

Dwight had written very little that summer except for a damning attack on Louis Fischer's biography of Gandhi. Fischer, he declared, failed to deal with the most important aspect of Gandhi's life, his attempt to "apply moral values to practical politics." Since that was Dwight's constant theme, the review represented his continuing effort to discover how to live his life as well as how to conduct his politics and carry on a worthwhile profession.[23]

He did complete another piece for the *Partisan Review*, which apparently was never published. It was an extended review of Bruno Bettelheim's *Love Is Not Enough: The Treatment of Emotionally Disturbed Children*. Dwight had published Bettelheim in *Politics* on the behavior of concentration camp inmates and had genuine admiration for the man. It is clear from the draft of his review that he had a keen interest in the subject and found much provocative material in Bettelheim's treatment of it. Dwight took Bettelheim to task for a kind of Victorian prudery. He felt that sex was glossed over, and when Bettelheim declared that at the school for disturbed children he had founded in Chicago, "undressing and bathing have their critical moments because of their potentials for erotic stimulation," Dwight insisted that one didn't have to be a Reichian to find this "too defensive." It was indicative of the modern habit of "taking as 'problems' such natural pleasures and opportunities as sex, religion, marriage, eating, bringing up children and even art and leisure." Even if one granted that "sex" in such a school might pose a problem, he wanted to know more than Bettelheim was willing to reveal. Masturbation was treated with some tolerance, but Bettelheim made only passing reference to homosexuality, a subject that still haunted Dwight's own psyche and appeared as a recurring element in his dreams.

But even more revealing were Dwight's own generalizations which he felt Bettelheim, to the book's detriment, failed to make. He concluded that a child's natural, healthy development was encouraged if its needs were satisfied by the loving care of its parents, and if the outside world seemed friendly, the home orderly, and the child could spontaneously imitate and identify with the parents' values. Bettelheim's residential school ostensibly sought to provide an environment where the child could "literally make a fresh start in life," in effect experiencing a kind of "second childhood that takes him step by step through the stages of development that were blocked by his natural environment."[24] What is fascinating about Dwight's observations is the way they reflected his own self-analysis. He felt that he had had a distorting childhood, that he had been deprived of genuine emotional and sexual development. He had missed out on important stages. When Joan Colebrook accused him of acting like a child, immaturely and irresponsibly, he responded, "I have got to go through a real expression now of immaturity and irresponsibility if I am ever to become really mature and responsible." He claimed that for years he had led a disciplined, monogamous, hardworking life but on a false basis, and that it had begun to fall apart. He wanted to become self-disciplined and productive again, but there had to be some inner change and growth first. These thoughts had probably arisen in Dwight's mind when he read Bettelheim, or perhaps his reading of Bettelheim had given him insight into his own psychic needs. He too, he insisted, was in need of a second childhood that would take him step by step through the stages of development.[25]

Although Dwight had written on psychiatry for *Fortune* with a sympathetic attitude, he was skeptical of its practical value. Nancy had been in psychoanalysis for nearly a year and a half and most of Dwight's comments about it were to complain about the drain on their resources. However, his own personal problems finally led him to consider seeking therapy. He felt that "it had done Nancy a lot of good"; she was more stable and independent and had more insight into their ways of feeling and acting. Nancy had urged Dwight to get help, and she made an agreement with him to delay any decisions concerning their marriage till after he had consulted a counselor.[26]

Not long after they returned from the Cape, he broached the issue of a separation with Nancy. He felt that her analysis had made her better able to cope with the possibility of divorce. Again she insisted that before a decision was made he should have some talks with a psychiatrist. He agreed, feeling that the "creative paralysis and depression" he had lately experienced were excessive even considering the turmoil in his life. In early November 1950 he agreed to have a talk with Nancy's therapist, Dr. Peterson-Krag, and also with another psychiatrist recommended by Delmore Schwartz. But Dwight was still resistant and he told Joan, who was also encouraging him to see a therapist, that he would not go into regular psychoanalysis because he "did not have enough faith, or hope or charity about the results."[27]

It was not simply Dwight's emotional life that concerned him; he was becoming obsessed with his failure to complete the Baldwin profile. He began to tie all his ailments together, complaining of an "existentialist angst." He kept chipping away at the damned thing but he made as much progress as Sisyphus. "Is it I can't produce because I am unhappy or I am unhappy because I can't produce?" he wondered.

Dwight's erratic and frenetic state, his inability to make any decisions, his insistence that he could not leave Nancy were driving Joan Colebrook to distraction. She had her own concerns. She had separated from her husband and was financially strapped. As the Christmas holiday season approached, she decided she had to flee to Mexico, a more accommodating climate where living would be cheaper and where she could get some of her own serious writing work done. Joan had already written a brief history of her native Australia and two novels by the time she met Dwight. He gave her much encouragement and helped get her essays published in the *New Yorker*. Some of these essays would be published, in 1967, in a widely acclaimed study of women in prison titled *The Cross of Lassitude: Portraits of Five Delinquents*. She hoped that Dwight would follow after her, and live openly with her in keeping with his professions of freedom and rejection of bourgeois convention. He could not betray his agreement with Nancy, however. "It would mean her shipwreck . . . I simply cannot abandon her, she is dear to me and I love her, in another way and the kids—OH God!"[28]

When Joan left for Mexico, Dwight quickly rationalized that they really were not meant for each other after all. Lengthy, involuted explanations were carefully jotted down, all suggesting that she was not the free spirit he was looking for. She was looking for another stable, successful husband as much as a lover. He made peace with their separation and even concluded that it was perhaps for the best.

These rationalizations revived Dwight's energies; he completed a 153-page manuscript on Baldwin and submitted it to Shawn at the *New Yorker*. Finishing up the Baldwin profile freed his mind enough for him to finally make the decision to make an appointment with Max Gruenthal, the psychiatrist recommended by Delmore Schwartz. Gruenthal was a German-born analyst who had received his medical degree from the University of Berlin in 1916 and had emigrated to the United States to escape the Nazis. Dwight described him to his friend Gertrude Norman as an "eclectic" psychiatrist of the "hurry-up school, uses Freud and everything else." Gruenthal had told Dwight to expect results after eight to twelve sessions. After four sessions Dwight was optimistic. He was having the most vivid and interesting dreams, which he carefully typed up and then discussed with Gruenthal.[29]

In his correspondence with Joan Colebrook, Dwight had already begun the process of self-analysis. Now under the care and direction of the analyst he undertook a serious study of his dreams and also of his past. He requested letters from Dot Wheeler, Dinsmore's sister, and from Dinsmore too. He went over his schoolboy correspondence and notebooks in an effort to find out who he was. It was the first time he had ever done such a thing and, with the help of the sessions with Gruenthal, he saw "a pattern emerging." The analyst combined "extreme tact with extreme persistence in questioning, and insistence on concrete, non-evasive answers." Dwight was confident that it would not be long before he had "some suspicion" of his true self. He looked upon the project as a "whodunit." Since childhood he had kept every scrap of personal writing. He now rummaged through it and composed summaries of his letters to his previous girlfriends as well as to Nancy. Dwight literally undertook a research job on his own past emotional life, his relationships with women, his homosexual anxieties, his fears of failure professionally, lack of close friendships, problems of self-confidence and self-esteem. On this project he had no writer's block; the pages flowed in a torrent.[30]

He kept Joan closely informed of his progress. Gruenthal's efforts had put him into a much better mood because the doctor was forcing Dwight "to face up to things, to confront myself and my relations'.ips. I think there's hope for me and us." But he also suggested to Nancy that the therapy might resolve *their* problems. He believed that once through the process with Gruenthal he would be a free man, able to make decisions and take hold of both his personal and professional life. By April, after a half dozen sessions, Dwight thought he was "getting on" to himself.[31]

In mid-April 1951, with the Baldwin profile accepted by Shawn and needing only severe editing, with Mike scheduled for Exeter in the fall if he managed to pass the entrance exams, and with Dwight's analysis beginning to give him a sense of control and confidence, he summoned up the courage to once again approach Nancy about ending their marriage. They went out to dinner in the Village and carried on their discussion in several local bars ("good places to talk privately, better than at home—nothing more anonymous, private than a public place"). They more or less agreed to try a separation during the summer. Dwight felt he had reached a turning point in his life; he felt a sense of pending freedom. His mood of excitement and confidence affected his attitude toward his work. He was busily cutting the Baldwin profile. At the same time, he was solicited to write a review of Hannah Arendt's *Origins of Totalitarianism*, the most important political book in years, to be followed by a three-part series on totalitarianism for the *New Leader*.[32]

Dwight responded to *The Origins of Totalitarianism* because it confirmed many of his own ideas. There was, he declared, "a remarkable resemblance" between Arendt's basic theory of the unique nature of the totalitarian system and his own ideas as expressed in his earlier writings on Soviet bureaucracy. Now with the cold war at fever pitch, they did not have quite the same adversarial punch that they had when written from his third-camp position. His repeated equation of Stalinism with Nazism, his constant argument that Stalinism was more menacing due to the technological advances in manipulating a mass society, tended to provide cold warriors with ammunition. As in the past, Dwight made those on the lib-lab left his main targets. He was quick to pick up Arendt's accusation that the "temptation to explain away the intrinsically incredible by means of liberal rationalizations" was a form of "liberal wheedling." He charged that the twentieth-century "irrational irruptions of . . . anti-humanity" had literally made the whole "rational-utilitarian way of looking at life obsolete." The Nazis and the Communists (Dwight had given up making any distinction between Stalinists and Communists) had completely shattered the "progressive-materialistic world view that went back to the Encyclopedists of the enlightenment."

Hannah Arendt's great strength was her willingness to "go against the mainstream of scientific-materialistic thought," still running strongly in various academic and leftist circles. Arendt, like Simone Weil, had the courage to face the fact that "our society is rotting and that Nazi-Soviet totalitarians" were the extreme expression of this rot. Their liberal-progressive opponents were also infected. A fundamentally new way of thinking (and above all, feeling) was essential if Western society was to escape destruction. What Dwight really found fascinating and convincing was Arendt's sardonic notion that the totalitarian man was the dull burgher next door, the stolid "bourgeois family man." That struck an important note in Dwight's own understanding of the menacing quality of totalitarian systems. This image of the perfectly normal neighbor transformed into the horrific instrument of human destruction brought the issue out of history and into the everyday world of all citizens of Western society. It aroused in Dwight that old fear of massification transcending the Nazis and even the Stalinists and becoming a way of life for modern man.

It was this notion that enabled Dwight to transfer his interests from the explicitly political concerns of the cold war to the broader, all-inclusive world of mass culture. Confused by international politics, finding it complicated and beyond his grasp, and with his dissidence losing its bite, he was able to redirect his adversarial criticism toward mass culture, denouncing it as being as destructive to civilized life as international Stalinism. Just when so many well-known intellectuals were on the threshold of embracing American culture and celebrating life in the United States as the best of all possible worlds, Dwight saw a way to continue to see himself as "against the American grain." It would be an ironic stance for a "radical" to take, because it became a defense of modernist tradition and an aging avant-garde that was no longer on the cutting edge of aesthetic and artistic experimentation.

In a letter to Nicola Chiaromonte about the stalemate in Korea, Dwight directly connected the threat of Soviet totalitarianism with the problem of the "disease of mass culture in America":

If the U.S. doesn't or cannot change its mass culture (movies, radio, sports cult, comics, television, slick magazines) it will lose the war against the USSR. Americans have been made into permanent adolescents by advertising, mass culture—uncritical, herdminded, pleasure-loving, concerned about trivia of materialistic living, scared of death, sex, old age—friendship is sending Xmas cards, sex is the wet dream of those chromium-plated Hollywood glamor girls, death—is not. . . . The happy ending is de rigueur in Hollywood, but there is no such thing in real life—everybody's life had an UNhappy ending, namely death. . . . Anyway we have become relaxed, immersed in a warm bath, perverted to attach high values to trivial things like baseball or football (kids' games really), and we just don't function when we get out into the big cold world where poverty, the mere struggle for existence, is important, and where some of the people are grown ups.[33]

Arendt's analysis of totalitarianism pointed directly to the menace of mass culture, that is, the mass manipulation of human beings in the capitalist West. Neither the failures of American democracy nor the crass commercialism of American culture could be equated with Stalinist tyranny, but conceding this much did not negate the challenge of a consumer culture that was increasingly decadent and destructive of human values and aspirations. This overriding fear led Dwight to concentrate on the parts of Arendt's work that seemed relevant to mass society. Totalitarianism, Dwight extrapolated, "begins and ends with masses . . . men reduced to uniform statistical units." The mass man is produced by modern industrial capitalism, which is uninterested in human concerns. "The masses are the chips, the slag, the superfluous raw material. Chips and slag are, presumably, content to be such. But men are not." Despite totalitarianism's "absolute lunacy," there was a method to its madness. Quoting Arendt, Dwight agreed that the aim of the totalitarians was "nothing less than 'the transformation of human nature itself,' reducing human beings to animals or rather to bags of reflexes." This was the secret of totalitarians; they created a dream world of total delusion that provided a haven for alienated modern man who could not tolerate the burden of finding a code of human conduct within himself, who in a world without faith in God or reason needed a crutch. Again quoting Arendt, Dwight wrote:

Before the alternative of facing the anarchic growth and total arbitrariness of decay or bowing down before the most rigid fantastically fictitious consistency of an ideology, the masses probably will always choose the latter and be ready to pay for it with individual sacrifices—and this not because they are stupid or wicked, but because in the general disaster, this escape grants them a minimum of self-respect.

Throughout the piece there is a tone of foreboding that the rational Western mind simply can't grasp the true nature of the menace. Dwight summoned up Kafka and finally his old standby Lewis Carroll, warning that Stalinist totalitarianism was "bent on reshaping all of humanity . . . to prove that the totalitarian dream world is the reality and the real world the illusion." In a footnote he translated a passage from Carroll comparing the Communists to Humpty Dumpty, who when challenged by Alice for using words any way he felt like responded: "The question is, which is to be master—that's all." Dwight added,

> And indeed, the communists have demonstrated that that *is* all, that 'peace' means 'war' that 'people's democracy' means 'anti-popular repression,' etc. They have made real what Kafka and Carroll imagined. The Castle has taken shape as the Kremlin, millions of people have been condemned to death like K in *The Trial,* without even being able to discover what they are accused of, and the Red Queen, the Ugly Duchess and Humpty Dumpty have materialized as the rulers of great nations.

Satisfied with the *New Leader* series, Dwight felt that he had broken out of his despondency. He was spending a good deal of time with Dr. Gruenthal, as well as preparing for the sessions with summaries of his dreams and past journals. With so much going on, he was not prepared to join Joan in Mexico, and for the moment the trial separation from Nancy was put off. Dwight and the family returned to Slough Pond in June 1951 and he became immediately absorbed in dealing with the cantankerous Roger Baldwin, who was not at all happy with the draft of Dwight's portrait of him for the *New Yorker*. Neither was his second wife, who wrote to ask that Dwight eliminate reference to her two children by a previous marriage—and also to their practice of nude bathing on Martha's Vineyard. Dwight agreed, but he thought cutting the reference to nude bathing was prudery. It was one of the few aspects about the Baldwins that he found attractive.

Dwight depicted Roger Baldwin as a canny operator motivated by idealism combined with a healthy dose of cynicism and shrewd calculation. Baldwin was hardly pleased with this mixed verdict. When Dwight described him as deceptive, "a master of the backstairs finagling that seems to be inevitable in committee work," Baldwin was piqued and began demanding revisions. To Baldwin, but not to Dwight, Baldwin's practice of keeping items off meeting agendas to "prevent minority decisions" was not unethical. In the margin of the complaining letter Dwight wrote "I think it is."[34]

However, the real thrust of Dwight's piece was its portrayal of Baldwin as a classic example of the liberal leftist who while critical of civil rights violations in the United States apologized for the Soviet Union in the thirties and lent support to Communist front organizations under the rubric of fighting fascism. Ammunition for this attack on the liberal left's gullibility came from Baldwin's own article of September 1934 in the propaganda sheet *Soviet Russia Today,* where he argued that the Russian dictatorship was necessary because the Soviet Union was surrounded by reactionary enemies.

Dwight's dredging up this material on Baldwin's flirtation with Soviet apologists, not so dissimilar to Dwight and Nancy's own flirtation with the Communist Party during the same year, did raise an interesting and always difficult problem. Joseph McCarthy was exploiting the virulent anti-Communist sentiment that had been unleashed during the Truman administration. Was an assault on Baldwin for being sympathetic to Communism wise, when many liberal organizations, such as the American Civil Liberties Union, were under attack? Baldwin was understandably irked by this apparent "red-baiting" of him when he was fighting for the rights of unions to exist and defending freedom of speech everywhere in the United States. Dwight ended his piece by arguing that, ironically, the ACLU had become tamed, respectable, and really part of the establishment by the fifties. He saw no connection between that evolution and attacks on Baldwin very similar to his own. A later historian of the ACLU described how the right wing in America had seized upon Baldwin's *Soviet Russia Today* article to tarnish the organization's and its founder's reputation.[35]

As the weeks went by, the correspondence became testy. Baldwin claimed it was libelous to call anyone a Communist, Communist sympathizer, fellow-traveler, or anarchist without proof. He was annoyed by such a slant, especially from a person who was writing from a "professedly friendly standpoint." Dwight scrawled in the margin "Not so—an objective standpoint." Baldwin felt that the profile presented a man without moral principle or stable social philosophy, a lover of power for its prestige, a tightwad and petty tyrant. At one point Baldwin made a trip to

Wellfleet and an accommodation was reached. They parted on the "friendliest terms," Dwight wrote Shawn, "at least as friendly as one can be with an SOB like Baldwin." Dwight said he felt as though he had negotiated a settlement with U.S. Steel.[36]

Dwight would not budge on the fellow-traveling point, however. Baldwin notified Dwight that he had turned the matter over to his lawyers. The issue remained unresolved and the publication date put off.[37]

The threats from Baldwin and the circumspection of Shawn only added to the confusion of Dwight's life. He wanted to move on to other things, but the Baldwin imbroglio just wouldn't end. And the crisis of his personal life had reached a climax. With Joan Colebrook in Mexico, the affair had cooled considerably, and Dwight's sessions with Gruenthal encouraged him to pursue his own needs, sexual and emotional. At one point in June he took a trip to New York and wrote Nancy that he was making real progress in his sessions with the analyst: "I'm becoming more selfish by the minute." Dwight thanked her for persuading him to seek treatment. But Nancy must have been chagrined, for while Dwight may have been in the process of cutting his ties with Joan, he was also determined to end his marriage.[38]

During the previous summer Dwight had had a brief flirtation with Gloria Lanier, wife of Robin Lanier. They had been regulars in the summer community for several years. After a beach picnic Dwight and Gloria had danced together and he felt "a sense of communion and pleasure and delight" that astonished him. He was a notoriously awkward dancer, but somehow she had made him feel lithe and competent. He later recorded in his journal that at that moment he knew it would be good in bed with her, but he was still attached to Joan. Characteristically, Dwight wrote Joan about the incident; he was compelled to disclose all his actions and innermost thoughts, even if it meant hurt to his confidante. Although both he and Joan made light of it and claimed to see Gloria as foolish in her forward approach to Dwight, he nevertheless returned to the event several times in his correspondence and journal.[39]

Now, in the summer of 1951, Gloria's own marriage was not going well; her husband was actually seeing a good deal of Nancy Macdonald. Late in July Dwight wandered off from a night beach party with Gloria, and before he knew it he was involved in another serious summer romance.

For the rest of the summer, Dwight pursued Gloria with as much enthusiasm and determination as he had pursued Joan two years before.

In September, while still at the Cape, he wrote to Joan of his new affair. "Faithfulness is just not in me," he wrote with some pride. Surely to Joan's irritation, he commented that he and Gloria as "man and woman" were "still childlike because we both, unlike you, in our 20s and 30s missed out on sex pretty much and we are now making up for lost time." Dwight took up the theme of personal liberation. After conceding that this was a terrible time for Nancy, he remarked that the whole idea of monogamy and marriage depressed him. "The only responsibilities, duties, obligations I will accept are those I want to, those I choose to, freely and spontaneously."[40] Marriage, he felt, was unnatural, while fatherhood or motherhood did seem natural.

In his journal he pursued this "childhood" image, portraying himself and Gloria as having a "wholehearted" relationship. They thought only of the moment. Comparing himself and Gloria with Nancy and Joan, he wrote: "We burn our bridges, they always keep a line of retreat open; we give, they bargain; we think we can do anything, can get away with anything, they know they cannot." Often the two couples would be together at parties and Gloria would wrap her leg around Dwight's under the dinner table or deliberately brush her lips against his cheek. Sometimes this was so open and brazen that both Nancy and Robin were hurt.[41]

This affair, carried on into the fall in the city, gave Dwight a big boost out of the lethargy he had been experiencing. During the last days on the Cape, he renewed with enthusiasm his struggle on the political front against the tendency of fellow-travelers and lib-labs to play down the menace of Soviet totalitarianism.

In the circles Dwight moved in, the liberal left was well represented, and he took pleasure in challenging what he saw as their self-righteous self-pleading. He would point to a man like the historian Henry Steele Commager as a muddleheaded liberal—"a particularly naive and obtuse (simple-minded, clumsy) type of routine liberal." He targeted I. F. Stone as a fellow-traveler, had no use for Carey McWilliams of the *Nation*, and was never hesitant to confront and red-bait when the occasion arose.[42] And yet Dwight never defended government action, opposed every form of repression of Communists, and did write and speak out against McCarthy. He collected a good deal of material on the senator and always championed any move to denounce him. Nevertheless, there is something to the notion that his choice of targets, while legitimate and honest, represented a political stance that did not present much of a challenge to the Truman and Eisenhower administrations. Many other writers who were former Trotskyists and otherwise on the non-Communist left also gave a certain degree of aid and comfort to the right. Mary McCarthy in her 1952 novel *The Groves of Academe* chose to make a self-serving liberal professor her target; Robert Warshow found it important to decry the sycophantic, sentimental tastelessness of the Rosenbergs not long before their execution; Leslie Fiedler did much the same in an attack on American liberalism for its deluded tolerance of the Communist movement at home and abroad.[43]

But Dwight would not have seen himself as serving the camp of nationalists and patriots wanting to repress the rights of freedom of thought in the name of national security. He regarded writing and saying what he believed, at the time he believed it, his obligation as an intellectual. Perhaps in some fundamental way, Dwight was not "political," as some of his critics charged, for he refused to tailor his published thoughts to fit an effective political agenda. Ideas and principles were what was important to Dwight, not the politics—nor the historical context. Still, there was something misplaced in singling out a man like Henry Steele Commager for criticism in the very midst of the McCarthy era, a time when the liberal defense of civil rights was both weak and compromised. It may help explain why Dwight attracted some men of the far right like Henry Regnery, the conservative publisher and McCarthy supporter (he wanted to send Dwight to Europe on an assignment, but Dwight, pressured by both Arthur Schlesinger and Mary McCarthy, declined, despite his great desire to go, because of Regnery's close association with Senator McCarthy); Frank Hanighen of the right-wing journal *Human Events*; and T. C. Kirkpatrick, the managing editor of *Counter-attack*, a private exposé sheet that listed Communist sympathizers and deliberately contributed to their dismissals from jobs in private industry as well as the federal government. Dwight wrote Kirkpatrick as late as 1948 that he "wholly approved of using information and publicity to fight the Commies."[44]

While Dwight continued to participate in the cold war battles on the left, he also continued his analysis with Dr. Gruenthal and his exciting if disquieting love affair with Gloria Lanier. The constant tension and strain of the affair was interfering with his work. Apologizing to Chiaromonte for not having written, he blamed his usual "summer state of analphabetic and solipsistic enjoyment especially in August and September," the months when the affair had begun. Dwight reported that he had done almost nothing but "bird-watching, beach-walking, beach-lying, bathing and enjoying parties, conversations etc." He wanted to confide in his friend, but he found it terribly difficult. He did tell Chiaromonte that his unproductivity was due to a new affair. "It's the most equal such relationship I've ever experienced, only pleasure, delight, spontaneity, harmony on both sides, god knows how it will end or IF it will end." There were, he felt, "great disadvantages, practical ones anyway, to postponing sex to the age of 45. On the other hand, certainly better late than never."[45]

Dwight spoke of reviving *Politics* or bringing out a new magazine, but the trouble was that his personal preoccupations "often blocked his view of History." Hannah Arendt became so exasperated by this excuse that she exclaimed: "Really, Dwight, why in the world can't you have *both*? Lovemaking can't possibly take up ALL your time, or even most of it. So for Christ's sake get down to work in the rest of your time! Plenty of other people have done it." Dwight promised Chiaromonte he would return to work and if he failed he would "consult Dr. Gruenthal, who fixes you up for work as well as for love."[46]

Nancy had encouraged Dwight to see an analyst because she thought it might resolve his sexual fears and allow him to have affairs without taking them so seriously as to consider leaving her and the children. But Dwight was determined to end the marriage. Without *Politics*, there was no longer a foundation for living together. While Nancy devoted considerable time to the European refugee problem, Dwight's interests were mostly stimulated at social affairs, in cocktail conversations and arguments with his old leftist friends and associates. As he acknowledged to Chiaromonte, he no longer spent much time poring over his morning *New York Times,* and clipped very little. He wondered whether this was because he was "escaping from history," or was it because history seemed to have "settled into a jog-trot in which repetition is the keynote." It was like a ballet done by an "unimaginative choreographer. . . . History seemed to be 'marking time' and statesmen 'going through the motions.'" His reading was more focused on literary and cultural materials. He took a sudden interest in *The Letters of Ezra Pound*, whom he had come to admire as a poet and writer. The collection of Pound's letters was the most "exhilarating book" he had read in a long time. "Pound was a Man, that is, Somebody, who had guts and wit and humor and a humble respect and love for poetry, also generosity towards other poets and literary people (quite different from the dog-eat-dog atmosphere of NYC literary politics)," which Dwight conceded he "touched only tangentially—thank god."[47]

Perhaps this shift from the political to the literary can best be seen in Dwight's extended review of his old friend and *Politics* associate C. Wright Mills's important book, *White Collar*. Mills opened with a vivid image of the middle-level clerical worker as an ineffective, passive automaton: "The white collar man is the hero as victim, the small creature who is acted upon but who does not act, who works along unnoticed in somebody's office or store, never talking loud, never talking back, never taking a stand." It was similar to Dwight's own characterization of mass man. Influenced by Hannah Arendt, Mills, like Dwight, was concerned with this herd of bored, routinized, dehumanized, replaceable parts in some vast mechanism. Superfluous, having no control over their lives, and feeling no joy, these were people obviously ripe for totalitarianism. One would have expected Dwight to be sympathetic to Mills's book. But instead he honed his hatchet.

After announcing that he had found the book boring to the point of unreadability, he charged Mills with mishandling his data, filling the text with cluttered, irrelevant, and miscellaneous information, failing to complete thoughts, and resorting to grandiose generalizations with little substantial supporting evidence. He expressed surprise and disappointment, because he considered Mills "the least academic professor" he knew, but the book had all the failures he associated with scholarly writing. Dwight claimed he could tolerate some abstract words in a sentence, but when they were multiplied interminably without being anchored to some concrete word his attention wandered. By plucking phrases from the text, he diminished Mills's work without addressing his argument. Mills's may not have been the most sprightly prose, although taken as a whole it was far more imaginative and the style superior to most sociology written in America. One might have thought Dwight would at least discuss Mills's analysis of the co-option and taming of labor, but he ignored the substance of Mills's weighty book.[48] Daniel Bell, a close friend of Mills's in these years, though often in disagreement with him, recalled that he and the historian Richard Hofstadter felt Dwight had been frivolous and unfair. "You don't take a work which has a serious impact on American sociology and spear it on the basis of a clumsy phrase. It wounded Mills enormously. I don't think Mills ever recovered from it. He constantly referred to it."[49]

Before Dwight was through, his critique had become almost ad hominem.

Mills was "a propagandist rather than a thinker"; he wasn't disinterested enough. In a remarkable example of psychological projection, Dwight accused Mills of exploiting ideas rather than respecting them. Mills had little interest in the current political system because he knew of no way to affect it. Then, adopting the collective "we," Dwight observed:

> Almost everybody, masses and intellectuals alike, feel ineffectual in politics . . . but we intellectuals suffer a further frustration: can we understand politics and history any more, can we fit them into any conceptual frame, can we still believe that we can find the theoretical key that will lay bare the real forces that shape history—indeed, can we believe there *is* such a key at all?

To this rhetorical question Dwight answered no. "The liberals and the Marxists . . . had their keys, but they didn't fit the locks. Mills recognizes this, but he has no alternative theory or explanation of why things are as they are." Attributing to Mills his own frequently expressed confusion, he charged that Mills "feels modern society is just not understandable, he feels helpless and confused, as, for that matter, I do myself. Only Mills won't admit it." Dwight accused Mills of having manipulated sociologese, abstractions, irrelevant quotations, and name dropping so as to disguise the fact that "he hasn't anything to say. I wish he hadn't done it."

It was an astonishing and somewhat brutal performance that revealed Dwight's own deep-seated feelings of intellectual inadequacy, his admitted inability to wade through the heavy texts of the social science literature, his increasing closet absorption in the very mass culture he condemned, his gravitation toward the middlebrow *New Yorker*, his failure to write a book "in cold blood" himself. Mills must have posed a threat to Dwight on some unconscious level, although Dwight was not usually given to those kinds of career anxieties. Mills was a strong believer in "the Critical Ideal": he was grappling with a theory, desperately trying to devise a radical position against the brutalities of the postwar capitalist industrial system. Mills had an advanced degree and was connected to a prestigious academic institution (Columbia University); he was involved with the younger generation of politically active intellectuals. Dwight, on the other hand, was drifting and unconnected. In fear and anxiety he lashed out at his friend with the one weapon he could always rely on: his devastating talent for verbal invective. He may have been only an honorary member of that street-fighting, dog-eat-dog

New York intellectual community, but he had obviously not forgotten the tools of the trade.[50]

With a breezy insensitivity, Dwight wrote to Mills that at first, finding the book so bad, he had considered turning the reviewing assignment down. Instead, he explained, he had decided to say what he thought, since if the roles had been reversed he would have wanted Charlie to do the same. He closed by saying that he respected Mills's work and that this book was simply not up to snuff; he was harsh because he had anticipated so much more.[51]

Mills responded in a restrained way, considering the hurt he felt. Addressing his letter to "Dear Old Pal," he wrote that of course Dwight should have accepted the assignment and that he understood there was "nothing personal in it (but only fast irresponsible reading)." He was most upset by the fact that he could learn nothing constructive from Dwight's criticism and by Dwight's failure to discuss or acknowledge the "new facts" he had spent seven years digging up. All he had left Mills to do was "close up shop."[52]

Dwight replied that he didn't inform the readers about the new material and ideas the book contained "because honest-to-god I didn't know what they were." He had gotten so lost in the details that the book was just "one big blur." As for "constructive" criticism, he did not think it was his job to write a "How-To-Do-It review"; his job was to evaluate the work, to praise the good and damn the bad.[53]

Although the exchange was amicable enough, the friendship was badly damaged. When Norman Thomas asked Mills to participate in a panel discussion with Dwight, Mills wrote that there could be "no discussion of any public interest" between the two of them. The "stuff" Mills thought was important "was simply a blur" to Dwight. Mills told Thomas that he was still committed to empirical research, whereas Dwight had "made a fetish of confusion and drift, which is now his charming style." He suggested that Thomas read "Dwight's nonsense on Arendt," referring presumably to Dwight's assault on liberal and Marxist attempts to explain an irrational world in rational terms. Mills did say that having heard Dwight speak he did not think it would be difficult to "beat him up" intellectually, but he had no interest in doing that. He was interested in working out a set of problems and discussing them with those he shared some possibility of communication with. He closed by informing Thomas that he was sending a copy of the letter to Dwight, even though he was "no part of his intellectual or political world." Across the top of the copy, Mills scrawled "Dwight: So to hell with guys like you." All in all, it was a rather sad falling-out and one that symbolized Dwight's journey away from the oppositional kind of dissent he had so long been associated with.[54]

This episode was part of the stress Dwight was undergoing as he withdrew from Nancy and his past. He was not seeing or communicating with his old leftist comrades, and he was taking up a new life in a different part of town and with a different social set. Gloria was not a writer, not an intellectual. She was, or at least she appeared to be to Dwight, a straightforward, open, sensual person with a kind of breathless innocence. As he undiplomatically confessed to Joan, his affair with Gloria was "all so different" from what he had had with her. That had been an affair between two strong-willed, opinionated people from very different sexual and intellectual backgrounds, with turfs to protect, demands to be made, and accommodations to be agreed upon. With Gloria there seemed only harmony, pleasure, common enjoyment, and a mutual feeling that both were finding the sexual satisfaction and emotional fulfillment they had been deprived of. "For the first time in my life I have an equal relationship," Dwight confided to his journal. He compared their relationship to all his past ones in which one partner held back and the lovemaking and relations were the product of bargains and negotiations, tactics, obligations, debts, threats, and intimidation. Often, he recalled, one would try to prevent the other from acting as he or she wanted to act, try to subjugate the other, to force the other by external means to conform to one's wishes and needs, to use and exploit another for one's own needs. "This is the sin against the holy ghost in my book," Dwight wrote in his journal. For Dwight, using another person was unethical, and so he managed to translate his affair into a moral adventure of satisfying emotional, sexual, and even intellectual integrity. Despite Dwight's abstract theorizing, what really attracted him to Gloria was her ability to make him feel, for the first time in his life, fully adequate, fully a man, confident and secure that she desired and enjoyed him as much as he did her.[55]

By November 1951 the affair had reached such a critical point that Dwight initiated a meeting with Robin Lanier to discuss "the Situation." He wanted to assure Robin that he was not a philanderer, that he was serious and not simply "playing around." He also wanted to refute Nancy's charge that he and Gloria were in control, calling the shots, while she and Robin were helplessly suffering. On the contrary, they were, he insisted, all in the same boat, the victims of powerful forces. In this way he thought he could establish a bond with Robin and ultimately make it easier for him to accept the affair. Dwight recorded in his journal that both Gloria and Nancy had remarked on his facile ability to rationalize almost any situation. He acknowledged that they had a point.

Both Dwight and Robin agreed that the situation could not drag on interminably. Dwight said he felt that his affair with Joan had done just that. When Dwight reported back to Nancy that Robin had been polite and even noble in expressing his desire to see that Gloria did what was necessary for her happiness, Nancy ridiculed Robin's feigned nobility, and snorted that if Gloria made a decision to live with Dwight, there would be nothing he could do about it. So much for Nancy's passive acceptance of "the Situation."[56]

So open and intent were both Gloria and Dwight in pursuing their own desires that they informed their spouses they wanted to go away for a week together and expected them to look after their respective children. Nancy was outraged and hurt because Dwight had chosen to do this the very week of their wedding anniversary, which he had simply forgotten, as he often did. For years he had insisted that birthdays, even Christmas observations were bourgeois. Robin Lanier's reaction to the proposal was so violent that Gloria became "hurt, frightened and disoriented." She asked Dwight to "cooperate for a while in a platonic relationship" until she had mended some fences.

They did take the trip a week later and drove through Pennsylvania and Ohio, staying at country inns and acting out with "reckless spontaneity" all Dwight's fantasies about what a genuine loving relationship was. The trip was badly planned; everything went wrong. There were no reservations, the car broke down, and they kept getting lost. Gloria couldn't read maps or find the hotel. They ended up in a ratty place with no heat. On the way home they got stuck in a horrendous traffic jam in an exhaust-filled Holland Tunnel. Yet it was a marvelous lark, a joyous shared adventure: "You know, dear Gloria, we haven't yet, since we first really knew each other that night on Horseleech pond . . . almost four months ago, we haven't yet had a quarrel, a row, not even a small one, nor have I, even for a moment, felt any serious division from you."[57]

Throughout the holiday season Dwight and Gloria spent days together in bars, museums, library reading rooms, galleries, restaurants, tramping the city in the well-worn tradition of lovers, delighted with themselves and almost everything they saw. Dwight wrote Mary McCarthy shortly before Christmas, "I have become happy." But he apparently did not tell her the reasons. Dwight always considered her one of his closest friends, but they did not confide in each other. He did record in his journal that his happiness was "because of Gloria." She had transformed him, and he had the concrete evidence. He was standing on a subway train and a little man, in thick glasses, offered him a seat. "Most unusual for a man to offer another man a seat." At the market an old Italian asked him whether he was going to use the parsley to decorate a dish, and another elderly woman who was next told the butcher to wait on Dwight. Dwight wondered at all these expressions of friendliness; he was convinced it was because his

> happiness was showing, that when one is in love, it gives one an aura, a mane, and that others want to get in contact with it; for indeed when one is happily "in love," the precious stuff spills over onto other people, one loves everybody, wishes everybody well, wants to give to everybody, to joyfully yield to them what they require, just as with the loved one; so love of one person is a symbol of love of all persons.

At a party given by the public relations expert Benjamin Sonnenberg for the poet Karl Shapiro, Sonnenberg's wife, Hilda, whom Dwight had not seen since the 1930s, commented, "You look bigger and better, and happier." He responded, "Yes, I am happy." "You didn't used to be," she said. "You always had a cold or something, and you looked strained and old. Now you look younger than you did fifteen years ago."[58]

Dwight's glow of warmth and happiness carried him through the winter and spring of 1952. But a resolution had to be reached, and although he and Gloria were committed to spending their lives together, both had terrible bouts of guilt and anxiety. He had pretty much made it clear to Nancy that he intended to leave, and he sometimes urged Gloria to be more forceful with Robin. He felt they could not continue to put things off as he had done with Joan. Gloria promised she would indeed leave Robin when summer came and go off with the children somewhere. While he chastised her for stalling, Dwight was having the same doubts and anxieties. The thought of leaving his own family was liberating one moment and terrifying the next. Often as he prepared to make the break, he would stare at the familiar surroundings. The whole apartment, with its "big dilapidated sofa," the ill-ordered books, he would no longer awake to each day. At some near time he would no longer sit at his desk and look at the pockmarked pink wall, scarred from wayward flings at a dartboard used to raise funds for a "Trotskyite" affair years ago. The file cabinet he would never use again. He would no longer see Nicky's crayon drawings—"I won't be sitting here EVER EVER EVER AGAIN."

Even though he had long been resigned to the fact that their marriage didn't work for Nancy or himself, "Still it is a death, something is about to die—the marriage of Nancy and Dwight Macdonald expired at its home after a long illness." It was hard for him to think about moving out, clearing the desk and bureau, taking his books ("and what were 'his' and 'her' books?") and leaving, going away not for a weekend or a week or a month or even a year but just going away for good. And then what of the kids? Could they "split the atom of a marriage"—could they divide their love and their need for love in two pieces and give one to him and one to Nancy? Or, he wondered, did Nancy and he exist as one for the kids, and that if split, did "both die for them, a little . . . a lot . . . all." No, Dwight was not above the fray, as Nancy had once charged, he was suffering; Gloria was suffering too. Nancy and Robin were suffering. And the kids would suffer too.[59]

III

CRITICISM AS A SUBSTITUTE FOR POLITICS

12

Making a New Career: From Politics to Culture

In THE SPRING OF 1952, DESPITE THE TENSIONS OF HIS personal life, Dwight did not remain aloof from the heated skirmishes of cold war controversies. He continued his campaign against the threat to culture and civilization posed by Stalinist Russia and its apologists and fellow-travelers at home and abroad. The threat of what had become known as "McCarthyism" was also very real to Dwight. He had worried about an uncontrolled anti-Communism that adopted the methods of its enemy.

By 1952 he was part of a dissenting faction within the American Committee for Cultural Freedom. The Committee was affiliated with the Congress for Cultural Freedom (CCF), founded in Berlin in 1950 by Arthur Koestler, Sidney Hook, James Burnham, and Melvin Lasky, among others, and it remained the most active organization of what was loosely called the non-Communist left. The CCF, with headquarters in Paris, was designed to counter the propaganda campaigns of the Stalinists; it published magazines and held conferences throughout the world. From the beginning there was a struggle within the organization. One faction, which included Francois Bondy, Michael Josselson, and Nicholas Nabokov, favored promoting leftist organizations that were hostile to the Communists, but also insisted on a critical attitude toward Western imperialism, capitalism, and U.S. domination. Their ideological opponents were the more belligerent ex-Communists, who were stoutly pro-American, more aggressively anti-Communist, and deeply suspicious of radicals attempting to remain neutral between the Soviet Union and the United States. They were led by Koestler, Burnham, Hook, and Lasky. It was much the same struggle that had been going on since the end of the war in the Europe America Groups and the Friends of Russian Freedom. Suspicion was widespread among many on the left that a covert connection existed between the Congress for Cultural Freedom and the U.S. State Department. This was a sensitive issue among European intellectuals, and the accusation was made constantly by Parisian leftists. In 1952 the liberal non-Communist left wing had managed to remain powerful in CCF headquarters in Paris. The membership of its U.S. affiliate, the American Committee for Cultural Freedom, had shifted considerably to the right, however; the organization had given a platform to adversaries of the left like Max Eastman, Karl Wittfogel, and before long Whittaker Chambers. The thrust of the Committee was distinctly antagonistic toward the Paris office of the Congress for Cultural Freedom, which was still attempting to fight Stalinism abroad from a liberal-leftist position. The American Committee sought a united front against Communism through an alliance of the non-Communist left with anti-Communist American conservatives.[1]

While Dwight's anti-Stalinism remained strong, the belligerence of the Committee and its equivocation concerning civil liberties worried him and several other liberal-leftist members of the Committee. Dwight and the journalist Richard Rovere fought for a general condemnation of McCarthy and McCarthyism. Much to the chagrin of the conservative members of the Committee, they argued that Communism was no longer an internal political or cultural threat. The real menace to American values was the current witch-hunt atmosphere fostered by demagogues like Senators McCarthy and McCarran. Dwight proposed that the Committee publish a pamphlet exposing and condemning the "abuses of anti-Communism." To make it all the more effective, he suggested, it should be written "by veteran red-baiters," presumably meaning to include himself.[2]

Now the Committee split into three factions: the liberal anti-McCarthyism faction; the outright conservative defenders of McCarthy plus those who were at least tolerant of the senator's activities, if less than enchanted by his clumsy methods; and the pragmatists in the middle who wanted to keep the coalition of anti-Communist liberal, conservative, and even reactionary cold warriors together at all costs. The conservative faction was incensed by the liberals' dismissal of domestic Communism as no longer a threat. Karl Wittfogel was so angered by Dwight's resolution that he called him a Stalinist apologist. Dwight and Rovere insisted that Joseph McCarthy was "the central figure" leading an "ignorant vigilantism." Dwight conceded Daniel Bell's point that the campaign against McCarthy was largely sponsored by Communists and fellow-travelers and a Committee statement condemning the senator would only give aid and comfort to the enemy. But of course the real rub, as Bell confessed, was his stance as the "congenital centrist," a peacemaker trying to keep the broad base of the Committee together. Dwight insisted, however, that it was immoral, unethical, and poor tactics as well to pay attention to or be influenced by the political position of others. McCarthyism threatened cultural freedom, and the Committee had an absolute obligation to oppose it.[3]

At a meeting chaired by Sidney Hook on April 25, 1952, the Committee soundly defeated Dwight's resolution and accepted a weak and watered-down compromise proposed by Bell which attacked "unprincipled and exaggerated innuendos and downright lies as techniques of political discourse." Dwight ended his notes on these skirmishes with the observation that if the *Nation* was the stronghold of anti–anti-Communism, the American Committee for Cultural Freedom had become the stronghold of anti–anti-McCarthyism. He characterized their meetings with a gloss on a well-known bit of verse by Hughes Mearns, which he proposed as a theme song:

> *As I was going up the stair*
> *I met a man who wasn't there.*
> *He wasn't there again today.*
> *I wish to God he would go away.*

He suggested that they follow the advice of one observer and "rename the committee The American Committee for Cultural Freedom—in Russia."[4]

While Dwight was engaged in these struggles in the inner councils of the ACCF, he was attempting to live up to his own ideals and principles. In a review of some books on Simone Weil he noted that "the curse of the modern intellectual" was the "division between thinking and living," the intellectual's apparent inability or refusal to "practice what he believes."[5] He followed this with *New Yorker* profiles of Weil and Dorothy Day, both of whom served as role models of the committed life. Both made heroic efforts to live what they believed. They took risks, they dared to fail and even to make themselves look ridiculous in their efforts. Their persona was that of a modern "holy fool," which Dwight may well have been crafting for himself.[6]

These pieces were followed by a hilariously provocative defense of William F. Buckley's *God and Man at Yale*. Buckley, a young postgraduate teaching Spanish at Yale, had published this attack on the university administration for tolerating a faculty in which professors teaching atheistic socialism thrived. Dwight was not interested in Buckley's denunciation of academic freedom and his calling for alumni control; he was attracted to Buckley for the enemies he had made. Anyone who reviled college administrators couldn't be all bad, despite the outcry of wounded liberals who predicted the return of the Inquisition. Buckley was articulate, "brisk, brash, indecorous." He had all the earmarks of the campus radical: clear-eyed, grim-lipped, shaggy-haired, "an intellectual who would rather argue than eat"—in short, a man after Dwight's own heart. The fact that he was also a reactionary who called for religious indoctrination and the firing of professors who failed to toe the line was of little concern, since in twentieth-century America such a goal had about as much possibility of being realized as Dwight's own youthful faith in a workers' rebellion. Buckley had exposed the discrepancy between the creed and deed, between the stated ideology of the university administration and the educational practice of the institution. The administrators paid lip service to Christian values and free-market capitalism as a way of extracting donations from mossback alumni, but Yale, like the rest of the society, was not very religious or free-enterprising. Instead of conceding that the university reflected the predominant culture of the nation it served, the administration was caught flat-footed and reacted with all "the grace of an elephant cornered by a mouse."

Dwight took particular joy in seeing the pompous Harvard dean McGeorge Bundy bested in an acrimonious exchange on the subject in the *Atlantic*. Dwight described Bundy's contribution as an "apoplectic denunciation" which, while seeming to have the weight of the argument on its side, was so stylistically cumbersome that the two-fisted kid outshuffled him. Although Dwight was to have several other encounters with

Buckley, some unfriendly, he admitted that he "could not help liking Buckley" and was delighted to visit with him in later years on his estate in Stamford, Connecticut, and to go out sailing in the Sound. "He's not a bad fellow, really," he assured the skeptical Gloria, "just has bad ideas."[7]

In February 1952 Dwight, seeking some security and stability in his work life, had begun to negotiate seriously with William Shawn, recently named editor-in-chief, to become a staff writer for the *New Yorker*. Dwight had met some of the regular staffers at the magazine's headquarters on 43rd Street and found them amiable and interesting. Shawn, a shy and reticent man, immediately liked him, particularly his good humor. He felt that Dwight and longtime staff writer Geoffrey Hellman "had something in common." He wasn't sure just what it was, but "it had something to do with humor. They had a certain kind of humor" that he always enjoyed. Hellman had brought Dwight to Shawn's attention and was supportive of his desire to join the staff.[8]

In the spring of 1952 Dwight was given a contract, plus a phone and a desk in a long narrow office which he quickly filled with books, papers, and the inevitable manila folders overflowing with clippings. It was not long before he'd managed to turn his *New Yorker* office into the kind of cluttered den he had maintained at *Politics*. Dwight became so much a part of the place that Shawn allowed him to keep the office nearly ten years after he had quit writing for the magazine. "It gave him a place to go and we liked to have him around."[9]

Dwight's first great success for the *New Yorker*, which did much to establish his reputation as a world-class demolition expert, was his exhaustive review of the fifty-four-volume set of the Great Books of the Western World, with its accompanying Syntopicon, the two-volume index to the Great Ideas in the Great Books. This massive collection of presumably "the best that had been thought and said" throughout Western civilization's history had been selected by Robert Maynard Hutchins, formerly the chancellor of the University of Chicago, and Mortimer J. Adler, who had once taught the philosophy of law at Chicago and was at the time the director of the Ford Foundation–financed Institute for Philosophical Research. The publisher of this extraordinary venture was the Encyclopedia Britannica, jointly owned by Senator William Benton of Connecticut and the University of Chicago. In effect, this was a mass-culture entrepreneurial operation fueled by all the technology and marketing sophistication that Dwight condemned as being responsible for spreading spurious culture and driving out the genuine article.[10]

Adler had long been on Dwight's list of aggressive philistines. In the second issue of *Politics* he had reviewed Adler's 1944 book *How to Think About War and Peace* thus:

> No one who wants to understand the meaning of this war, and the problems that will confront us afterwards, can afford to read this little book. It is true that, as Clifton P. Fadiman writes in his introduction, the book is by no means easy reading. But for all its dull writing, *How to Think About War and Peace* is superficial. Mr. Adler once wrote a book called *How to Read a Book*. He should now read one called *How to Write a Book*.[11]

Dwight had Adler well within his sights when he took on the Great Books. His review, thirteen columns between the strip ads, was itself a discussion of a cultural commodity:

> In its massiveness, its technological elaboration, its fetish of The Great, and its attempt to treat systematically and with scientific precision materials for which the method is inappropriate, Dr. Adler's set of books is a typical expression of the religion of culture that appeals to the American academic mentality. And the claims its creators make are a typical expression of the American advertising psyche. The way to put over a two-million-dollar cultural project is, it seems, to make it appear as pompous as possible.

Dwight then proceeded to go through the project in a methodical way, attacking most of the selections as impenetrable and the translations as deplorable. He dissected the promotional hype—an easy target. He had some juicy material. Clifton Fadiman, who had replaced William Lyon Phelps in Dwight's stable of galloping philistines, had given an after-dinner testimonial to the Great Work. The editors, Fadiman intoned, had taken upon themselves "'the burden of preserving, as did the monks of early Christendom, through another darkening . . . age, the visions, the laughter, the ideas, the deep cries of anguish, the great eurekas of revelation that make up our patent to the title of civilized man' (applause)." Dwight pointed out that nearly all the works were currently in print. Remembering his own youthful collecting, he argued that it was much more fun—and cheaper, too—to buy the books separately. Sets were "monotonous and depressing; books, like people, look better out of uniform. It bothers me to see *Tristam Shandy* dressed like the *Summa Theo-*

logica. Milton should be tall and dignified, with wide margins; Montaigne smaller, graceful, intimate; Adam Smith clear and prosaic." One can sense the delight Dwight took in this flight of bibliographic fashion design. Rabelais looked grotesque in a double-column textbook format, but such a dull, stultifying layout was "admirably suited to Dr. Adler's Syntopicon."[12]

The rest of his review was devoted to demolishing Adler's formidable guide to the Great Ideas in the Great Books. It would take the "patience of Job and the leisure of Sardanapalus" to plow through the plethora of references, the worthless effort of hundreds of scholars and graduate-student slaves marching under the orders of Dr. Adler, who mistakenly imagined himself as the great "codifier and systematizer of Western Culture."

Dwight summed up the whole project:

> Its aim is hieratic rather than practical—not to make the books accessible to the public . . . but to fix the canon of the Sacred Texts by printing them in a special edition. Simply issuing a list would have been enough if practicality were the only consideration, but a list can easily be revised, and it lacks the totemistic force of a five-foot, hundred-pound array of books.[13]

Dwight received a congratulatory note from William Shawn for a "remarkably sane and witty piece of writing." Alfred Kazin, often a critic of Dwight and not particularly friendly toward him, wrote to praise his piece as "one of the most exciting and really useful critical articles I have read in ages." He felt that Dwight's earlier piece on Dorothy Day had yielded a bit to the *New Yorker* assembly line, "but the Adler is just a triumph." Saul Bellow remembers the review as an important piece of cultural criticism made all the more effective by Dwight's caustic wit.[14]

Dwight was definitely a defender of the canon of the Sacred Texts himself; it was the grading and the huckstering hype of the Great Books operation that repelled him. It was a perfect example of counterfeit culture foisted on the populace for profit. Dwight diligently wrote the publishers for promotional material and collected the unctuous advertisements. One brochure assured Mr. Macdonald that the set would enable him to "talk intelligently, and stimulate [him] to greater mental activity." The project exploited people's insecurities, attempting to make them feel they needed the set in their living room in order to create the proper appearance.

After seeing this pitiless dissection, Shawn agreed to give Dwight almost unlimited space for other devastating assaults on what Dwight interpreted as a wave of cultural barbarism overtaking the entire Western world. It was an important piece to Dwight, because it showed him he could fashion a career as a serious cultural critic and avoid the blandness and the cool circumspection of *New Yorker* formula writing. It might be the way back to serious work as well as a way to make a living and embark on a new life that would be intellectually satisfying and would also be of service.

Inspired by his success, Dwight briefly entertained the hope, or perhaps it was by now a fantasy, of a revived *Politics*, to be undertaken chiefly through the efforts of Mary McCarthy, helped by Arthur Schlesinger, Jr., Richard Rovere, and Hannah Arendt. But when Dwight shied away from the real organizing work, McCarthy took over the task of raising funds for a magazine to be called *Critic*. It was to be a journal that would directly confront the tide of conformity threatening political and intellectual activity. She put a good deal of time into the project, sent out a prospectus, and went to work to raise money. She never did manage to secure more than $55,000, however, and they had determined that $100,000 would be necessary for the kind of magazine they envisioned.[15]

The failure of the *Critic* project was a further indication of "the break-up of our camp." The core "New York intellectual" group had been made up of young writers who had lived on the margin, were connected through their alienation from the mainstream, their radical anti-Stalinism, their devotion to the avant-garde, modernism, and a freelance stance toward the world. It had been on the decline since the last years of the 1940s, fragmenting because its radicalism was severely shaken and its old faiths destroyed. New prosperity and many new career opportunities had also arrived in the early fifties. The former contributors to *Politics*, *Partisan Review*, and other little magazines were now finding assignments with large-circulation magazines, such as the *Reporter* at ten cents a word or even the *New Republic*, which was paying more. Or they were offered visiting lectureships and professorships at well-known universities. Grants and fellowships were more available.

Irving Howe recalled that Dwight, with his keen antennae, sensed that the old world was disappearing and that he had better dig in at the *New Yorker*, despite some misgivings and his occasional impatience with its editorial management and discipline. It would not be long before Howe would get *Dissent* off the ground as a remarkably enduring social democratic voice. *Dissent* editors frequently asked Dwight to contribute, but even their revisionist Marxism was not of much interest to him. The magazine always seemed too social-science oriented, almost academically disciplined. Dwight found it boring and wrote only in response to articles criticizing him. He continued to drift away from his radical, bohemian, intellectual past and toward the uptown world of the *New Yorker*.[16]

Dwight wrote a surprisingly unfriendly review for the *Reporter* of Simone de Beauvoir's *The Second Sex*, which he labeled a "black book of the male terror," provincial in its perspective and biased and distorted in many of its judgments. But the close of the review revealed his own emotional investment in the subject. He quoted a long and eloquent passage he felt showed insight into the possibility of a man–woman relationship that could be one of "equality in difference." Beauvoir argued that such a development could occur if in "love, affection, sensuality" a woman succeeded in "overcoming her passivity" and established "reciprocity with her partner." The ever present differences between the eroticism of men and that of women did create "insoluble problems as long as there [was] a 'battle of the sexes.'" But they could be overcome when a woman found both desire and respect in her partner. "If he lusts after her flesh while recognizing her freedom, she feels herself to be essential," her integrity remains unimpaired even in her submission to which she freely consents.[17]

It is little wonder that Dwight appreciated these ideas, for they echoed his journal entry about finding with Gloria, for the first time in his life, a real sense of equality in a loving and sexual relationship. He had denounced the "give-and-take" arguments of Nancy and their Wellfleet neighbors, the Jenckses, who had defined marriage as bargaining and compromising between contending partners; in Dwight's view this was the old "battle of the sexes," demanding a victor and a vanquished. Beauvoir's moving passage about the possibilities of a genuinely reciprocal relationship mirrored his enthusiasm. Unfortunately, he wrote, this incisive and meaningful passage was "contradicted by the bulk of this immense and deformed work."[18]

As the summer of 1953 approached, Dwight looked forward to the Cape with mixed feelings. He had a bundle of work laid out. An updated article on mass culture scheduled to appear in *Diogenes*, a Ford Foundation cultural publication, was in proofs, which he needed to correct. He had begun a very ambitious project, a review of the Revised Standard Version of the Bible, and he was also mapping out a profile of Alfred H. Barr, Jr., the director of collections at the Museum of Modern Art, both for the *New Yorker*. Both were demanding subjects that called for considerable research. He had amassed reams of material, and he hoped to get work done at the Slough Pond cottage. So he went with some trepidation, because he knew he was seldom disciplined and productive at the Cape. It had also become the site of so much emotional tension that he looked upon its woods and beaches with a mixture of fear and guilt. His relationship with Nancy was now so strained that they deliberately planned to spend much of the summer apart.[19]

The Laniers were not on the Cape that summer, so Dwight had to arrange to see Gloria on visits to the city and to Connecticut, where she was spending some time with friends. While at the Cape he wrote her often of how quiet his life was. The days were good for work and thinking. He didn't see anyone until dinner time. Daphne and Geoffrey Hellman were there, and he saw them quite often. There were the "inevitable Jenckses." He did see Joan Colebrook, who had returned after her long stay in Mexico, but found her quite wanting when contrasted with Gloria.[20]

Dwight's relationship with his boys during these trying years of the decline of the marriage was often troubled, particularly with Michael, whom he felt was deliberately difficult, volatile, and bent on gaining attention by causing as much turmoil as possible. Two years earlier, just before Mike was scheduled to go to Exeter, Dwight recorded in his journal that the boy wouldn't miss him. He had become "rude, sullen, aggressive, mean to me. Uses whatever power to frustrate me he has." Mike was even more antagonistic toward his younger brother, so much so that the boys would often have very little to do with each other. One of Dwight's problems that had shown itself in his adult relations was a reserved emotional distance. There was plenty of jocular banter, but he seldom made any display of affection except in extreme situations or crises. Virginia Chamberlain, who spent so much time living with them in the cottage on Slough Pond, had noted that Dwight never gave Nancy or the boys a hug or a kiss. Both sons came to resent the practice of addressing their parents as Dwight and Nancy instead of Dad and Mother, which was a part of their "liberated," "progressive" approach to child rearing.

Dwight was aware of this distance. When he had lived briefly with Geoffrey Hellman the summer before while Nancy and the boys were in

Europe, he noted that they got along fine but that there was always a barrier he could not break down. To Nicola Chiaromonte, whom he described as his closest friend, he did write about his love affairs and the demise of his marriage, and after some time, Chiaromonte wrote a far more open and personal letter about his own marital woes. But even with Chiaromonte Dwight hesitated to really confide his innermost thoughts. Dwight was intellectually open, honest, candid to a fault, but he could not overcome a sense of deep separation from others, and found it difficult to express affection and love. Dwight wanted close confidences, but he was awkward and ill at ease in achieving them. His friends, associates, and professional colleagues had affection and admiration for him, but many admitted that they did not really know him in an intimate way.[21]

However, the sessions with Dr. Gruenthal, the diary entries, and the analysis of his dreams had encouraged him to try to overcome this distance during that summer of 1953 and to spend time with Mike and Nicky. He wrote Gloria that he had never been closer to them, that he was constantly delighted with their company and with their newfound enthusiasm for reading and their desire to please him. Nick, almost nine, had just finished reading a biography of Buffalo Bill and another of Sir Walter Raleigh. Dwight thought that showed a range of interest equal to his own. Nick was just beginning Thackeray's *Rose and the Ring* and constantly interrupted Dwight to announce what page he was on. Fifteen-year-old Mike's intellectual interests had been awakened since entering Exeter and he had grown out of his obsession with sports and comic books. He had just finished Norman Mailer's political novel *Barbary Shore* and at Dwight's urging was trying to wade through Trotsky's autobiography. When Mike heard that summer that Mary Day, Gloria's older daughter, had read *War and Peace*, he asked his mother to bring up their copy on her next trip from the city. That summer was also when Mike informed Dwight that his first intellectual hero was Edmund Wilson, because "he reads and learns so much." Dwight agreed that Wilson's mastering Hebrew in order to read the Old Testament was impressive. It was not the sort of thing he could discipline himself to do, even while researching the Bible revisions. He noted rather wistfully that he was in the number two spot after Wilson because Mike had gathered that Dwight wasn't as much a scholar.[22]

Although Dwight seemed more at ease with Michael that summer, their relationship continued to be a stormy one, with frequent arguments that turned into full-scale rows upsetting to both of them. Dwight had a way of teasing Mike, of even showing a contempt for what he saw as his son's intellectual pretensions and his aping of Dwight's argumentative confidence. Michael felt his father could never, would never, concede a point. He knew Dwight had a reputation for recognizing a mistake and changing his mind, but saw him as incapable of making such a concession to his son.

One very close friend said the relationship between Michael and Dwight seemed at times unusually competitive. Michael was intelligent, but as a young person growing up in a family where the father was such a larger-than-life figure, he could not really compete with Dwight. Nevertheless, he tried. He had the energy and intelligence, if not the same infectious enthusiasm or verbal skills that could mix insult and acerbic aggressiveness with humor and amiability. Mike wanted to be like Dwight, whom he greatly admired. In his attempts to emulate Dwight, he would say something awkward, naive, or misinformed, and Dwight, instead of overlooking it or responding gently, would pursue the issue, not let it drop, crowd the boy, and later the young man, into a corner, often with guests watching. When Mike would inflate himself, which he did in an effort to gain attention, Dwight would cut him down to size. This was his son, yet he felt compelled to humiliate him. There was in Dwight a blind insensitivity that astonished many of his closest friends and people who loved and admired him. He was a caring, thoughtful person who took no pleasure in hurting those he loved, yet he seemed unaware of the intimidating intensity of his verbal aggression. For Dwight, argument was the elixir of life; the subject did not matter, and he could and would, for the sake of the argument, take any side. One woman who observed Dwight thought that he was "orally sadistic." Mike later recalled his father's notorious lack of tact, usually displayed during an argument and frequently after a few drinks. He wondered whether it was an "unconscious sadism." The first time Danny Rosenblatt saw Michael Macdonald on the Cape remains an indelible image for him. Mike was standing on a dune, high above the sea, shouting down to his father: "Fuck you, Dwight." It was perhaps the most rebellious thing he could think to do, since Dwight seldom used obscenities.[23]

Michael, like his father, was insecure, lacked confidence, desperately needed a greater display of love and affection and real support and reassurance. A psychologist who gave Mike a Rorschach test when he was ten noted that he strove for perfection in everything he did, and invariably found himself wanting. She thought his perfectionism stemmed from a "generalized feeling of inadequacy." While he was very adult in much of his behavior, in some ways he seemed much younger than his age. The tester thought he had been given too much freedom as a child and had been forced to make too many decisions. What he needed was more routine, more rules, and more boundaries. Neither Dwight nor Nancy had the temperament nor the inclination to provide more structure and discipline. Those who were guests at Macdonald parties invariably remember Mike and Nicky crawling and later wandering around in the post-midnight hours amid the adult revelry. Even at ten, the psychologist felt, Michael had "a certain amount of resentment against his parents, probably for failing to give him the support and reassurance he needs."[24]

If Dwight's relationship with his son Michael was turbulent, with periods of comradeship interrupted by acrimonious bickering, his relationship with Nicholas during his early years was one of much mutual pleasure. Nicholas had a sweet and accommodating disposition. At Dalton, the prestigious private school in Manhattan, he was a good student, popular with his classmates and teachers. Prior to the breakup of his parents' marriage, he was a gregarious and outgoing youngster. The separation and divorce were difficult for him. He did not use the word *divorce* for years. He felt that Dwight had betrayed and abandoned him and over the years he developed a quiet, sometimes intense, resentment toward his father. That summer, fearing very much what the pending breakup would mean to the boys, Dwight went out of his way to spend time with them.

There were the usual beach walks, picnics, softball games. For Mike and for many of the young people, as well as adults, in the community, Dwight Macdonald *was* the Cape. He was the organizer of activities children participated in. He treated young people as adults. They loved to josh and tease with him. Mary Day Lanier remembers him as incredibly popular with the kids, "full of fun." She and her friends delighted in rushing up and calling him a "dirty Stalinist," which induced him to chase them down the beach or into the water.[25]

Despite the fact that Dwight felt closer to his boys than ever before, his unresolved situation continued to be upsetting. There were gnawing doubts and recriminating letters because Gloria was not taking any decisive steps to break with Robin. Dwight experienced bouts of deep depression, making late-night drunken phone calls to berate her for not having sympathy for him alone on the Cape. These spells were often followed by exuberant rejuvenation. His spirits soared when he received a call from William Shawn acclaiming his review of the Revised Standard Version of the Bible "a masterpiece, as good as the Great Books review and being rushed into print." Dwight was ecstatic: "Maybe I can make a go of money-writing after all." Forgetting all his complaints about the *New Yorker* formula, he wrote Gloria with enthusiasm that he now "felt at home" there. Confident, for the moment, that he could make a good living as a writer, he closed his note with a burst of optimism: "Let us both be strong, and you paint and i write and WE WILL COME THROUGH!!!!"[26]

When Dwight had proposed this extended review to Shawn, he argued that although its revisers and publishers claimed the revision preserved "the timeless beauty of the King James version, actually it changes it a lot . . . for the worse." He saw the new translations as a trend undertaken on the grounds that the King James was "archaic and hard to read." Dwight noted with disdain that Michael, even at Exeter, was using a revised version and "it reads horribly." He had told Shawn he would like to make a "*reactionary* case for sticking to the King James, on literary grounds and on grounds of our cultural traditions—the King James is part of our language, has shaped the expressions we use still, and revising it is like getting your mother's face lifted."[27]

This would be another way for Dwight to pursue his cause: protecting the culture from barbarism, resisting the encroachment of mass culture, which was bent on stripping the language of its verve, excitement, traditions in the name of accessibility. Dwight did enough homework to offer a brief and fast-moving summary of Bible history from the Venerable Bede translation of 735 through all the subsequent versions and translations. In addition to this history of the Bible, the first part of his essay offered an eloquent appreciation of the King James version as having all the "genius of a period when style was the common property of educated men rather than an individual achievement. It was written between Shakespeare and Milton, when Englishmen were using words more passionately, richly, vigorously, wittily and sublimely than ever before or since." He compared it to a Gothic cathedral, built over time and representing the "collective expression of a culture." To tamper with such a monument was a violation of the sacredness of art and perhaps incidentally of religion too.[28]

To make the Bible readable and accessible to the modern tin ear meant to deliberately "flatten out, tone down, convert into tepid expository prose what in the KJV is wild, full of awe, poetic, and passionate." What hurt the most was the deliberate attack on "style," which was most precious to Dwight. Style, he insisted, was not mere decoration. Form in a work of art like the KJV cannot be separated from content; this was fundamental to

Dwight's aesthetic. For him, the Bible was in a class with other works of high culture, such as *Ulysses* and *The Wasteland*; they were indeed difficult, because they demanded more from the reader. That was the price one should pay for genuine artistic quality. It was unfortunately a price the revisers were unwilling to pay and so destroyed an artistic monument.

In a review of *Against the American Grain*, the collection in which the piece was reprinted, Steven Marcus complained that nowhere did Dwight "evince the slightest awareness that for several millions of people the Bible is the living Word of God, or that such things as salvation and immortality are abiding realities for the largest part of mankind." Dwight mourned the passing of the King James version because it was a great literary monument "to which, because it also happens to have a religious function, practically everybody, no matter how unliterary or meagerly educated, was at some time exposed." This dismissive afterthought, which sees the Bible as mattering no more than *Love's Labour's Lost*, revealed Dwight to be a mere aesthete rather than a literary or cultural critic, Marcus charged. Dwight certainly had no interest in religion and was deliberately making an aesthetic judgment. But was Marcus claiming that to make the Bible more accessible was furthering true religion? If at the heart of religion lies mystery and if the simplified version reduced the complexity and the mystery, perhaps the revisionists really undermined the power of true religion. It does not appear that believers found Dwight's critique offensive.[29]

Dwight did not care whether the Bible was accessible to the great masses or not. He adroitly compared the revisers' work to the bombing of Dresden. Allied bombers had tried to spare the great monuments of Europe, but "military necessity" often compelled their destruction. Reading the revised work was like walking through a city that had just suffered saturation bombing. The eye searches for familiar landmarks. "Is this gone? Does that still survive? Surely they might have spared *that*!" Dwight had found a way to transform his fear of Stalinism into a fear of an encroaching cultural totalitarianism, which, like Stalinism, threatened an entire way of life. In effect, he saw his work as a moral obligation. He was taking a stand.

Dwight's persona as a custodian of the culture was being put together block by block during the year 1952–53. First there had been the Great Books, then the Bible piece and Shawn's declaration of a masterpiece. Now to complete the foundation was the *Diogenes* publication of his

"Theory of Mass Culture," an update of "A Theory of Popular Culture," which had appeared in the first issue of *Politics*.

In the initial piece Dwight assigned the blame for the low level of popular culture to the entrepreneurial hucksters who deliberately exploited the masses with the cheap production of inferior goods for a profit. This flood of trash, he wrote, was made possible by the development of mass democracy, mass education, and the development of electronic mass media: radio and movies. He distinguished popular culture from genuine folk culture: Folk art grew from below, whereas popular culture was imposed from above. The piece, written while Dwight was still close to Trotskyism, bristled with references to the "ruling class" which hired technicians to manipulate the needs of the masses in order to make profits for them. This mass-produced material threatened not only folk culture but high culture as well.

Now, influenced by his current reading of Albert Jay Nock's *Memoirs of a Superfluous Man*, Dwight adopted Nock's use of Gresham's law as it applied to culture: Cheap goods drive out superior goods and the worst drives out even the bad. Popular culture competed with high culture, infected it, and eroded its distinction. Academicism was a kind of spurious high culture, Dwight wrote; it masqueraded as high culture but it was nothing more than a popular culture for the elite. An insidious merger of the two cultures was taking place, he warned. Popular culture was taking on some of the aspects of high culture; high culture and the avant-garde culture, which were one and the same, were increasingly watered down. He named names throughout the piece, citing Archibald MacLeish, Walter Lippmann, Dorothy Thompson, even the *New Yorker* as purveyors of this "Kitsch."[30]

What was so menacing was that the masses were being exploited culturally as well as economically, and this was having a devastating effect on socialist ideas. "The deadening and warping effect of long exposure to movies, pulp magazines and radio can hardly be overestimated." Dwight ended the *Politics* piece by arguing that the whole problem of popular culture involved one's conception of the role of the common people in modern history. He cited Ortega y Gasset and T. S. Eliot as aristocratic elitists who would restore the old walls between a high culture for the few and a folk art for the masses. Dwight saw this as reactionary and politically unlikely. He ended on an optimistic and radical note: Rather than blame the masses for their low taste, it would be more accurate to see the rise of popular culture as the product of "a peculiar historical situation" largely created by the "persistence of class exploitation" through the century of the common man, with its mass education and the expansion of democracy. The exploitation of democracy had to end if culture was to recover its health. Dwight closed the piece with this optimistic declaration:

> Since my own convictions are democratic, I believe that the trouble with the revolt of the masses is that it has not been rebellious enough, just as the trouble with Popular Culture is that it has not been popular enough. The standard by which to measure the popular culture is not the old aristocratic High Culture, but rather a potential new *human* culture, in Trotsky's phrase, which for the first time in history has a chance of superseding the *class* cultures of the present and past.

At the start of the *Politics* venture, in 1944, Dwight still believed that out of the carnage of the war a rebellion was possible. In the summer of 1952, seeking to return to full-time writing, he had begun to rework the essay, removing its explicit and provocative radical message and making it into a piece of cultural criticism with decidedly conservative implications. The change in terminology was significant. He replaced the term "popular culture" with the more pejorative "mass culture." The new version was based on the *Politics* article, but instead of the radical terminology of "class rulers" and "manipulated masses," it referred to "the mob" and, significantly, no longer suggested that the masses might be integrated into high culture. The new piece discussed the mass culture of Soviet society, which was purveyed not for entertainment, as in America, but for propaganda and pedagogy in the interest of their "ruling class." No longer applying that term to American manufacturers of popular culture, Dwight now labeled them "The Lords of Kitsch."

As for the reciprocal degradation brought on by the merging of high and low culture, he developed a much more extensive indictment of cultural homogenization, the wiping out of all values, standards, and distinctions. "Academicism" was still his term for spurious, middlebrow culture that masqueraded as high culture and contributed to the "spreading ooze which threatens to engulf everything." He noted the creation of "adultized children and infantilized adults": Children had access to adult materials, and great masses of adults spent their time reading comic books and watching Westerns. Everything was going from bad to worse—from Sherlock Holmes to Sam Spade and then to Mike Hammer. It was a

journey from rationalism to primitive violence. Heroes were no longer
men of talent but celebrities who got a break, Dwight observed. Science
was no longer thought of as a boon to mankind. On the contrary, it meant
Frankenstein and Hiroshima. Dwight wondered whether there was a
"popular suspicion, perhaps only half conscious, that the nineteenth-
century trust in science" was a mistake like the nineteenth-century trust
in popular education. Again he presented a notion of Albert Jay Nock's,
that mass education only ensured a larger audience for propaganda and
cultural exploitation.

Even the *New Yorker* short story took its lumps. It was formula writing,
without individual distinction, smooth, "suggesting drama and sentiment
without ever being crude enough to actually create it." Dwight stressed
the bland homogenization, the mixing of glossy advertisements with coverage of violence, wars, atomic theory.

Dwight again alluded to Ortega y Gasset and T. S. Eliot's desire to
bring the vulgar masses under aristocratic control. But in this new piece
Dwight pitted those writers against the Marxists and liberals who saw the
masses as intrinsically healthy but duped and victimized by the Lords of
Kitsch. The Marxists believed that if the masses were offered good stuff,
they would respond and the level of mass culture would rise. Dwight
found both arguments unpersuasive. He maintained that mass culture or
culture on a mass scale could never be any good. Genuine culture
demands individuality and a sense of community, whereas the man in the
mass loses his identity, his humanity, his sense of community. The mass
man, a term Dwight would not have used in his 1944 piece, was now seen
as that basic ingredient of totalitarian systems: lonely, isolated, he is part
of what the sociologist David Riesman called "the lonely crowd." The
scale of mass societies was simply too big and too inhuman for genuine
culture, folk or high.

Of the two positions, the conservative versus the radical or liberal,
Dwight now sided openly with the conservative. A classless democratic
culture was an impossibility. "All great cultures of the past were elite cultures," he noted. But politically even that stance was without much meaning. While there was little hope for the revival of an elite culture, there
was equally little hope for an improved mass culture. He concluded that
the future was dark. "Human beings have been caught up in the inex-
orable workings of a mechanism that forces them . . . into its own pattern." Mass culture would probably get worse. Dwight offered no way of
avoiding the "spreading ooze."[31]

Dwight had been doing some reading in the growing literature on the subject and cited Max Horkheimer, Theodor Adorno, and Leo Lowenthal of the Frankfurt School, but it does not appear that they influenced his analysis very much. He had been writing about mass culture, particularly movies, the popular press, and Hollywood, for a long time and had not changed his views significantly since his Yale days. A Marxist rhetoric did briefly mark his *Politics* piece, but that had been abandoned.

Dwight was now confident that he had found his own formula, a form of negative cultural criticism, informed, serious but done with wit, verve, and humor. The positive reaction to these pieces gave him the incentive to get back to work on his profile of Alfred Barr of the Museum of Modern Art and an essay review of the Polish poet and ex-Communist Czeslaw Milosz, both of which he had wanted to complete during the summer of 1953 but had failed to stick to. Once back in the city, however, things did not improve as he had hoped. Angst and despair abruptly overtook him. He was seeing Gloria at every opportunity, but she was also absorbed in her own family affairs. By September Dwight was in a serious state of anxiety.

In a detailed self-analysis, prepared for renewed visits to Dr. Gruenthal, Dwight wondered why he shied away from "making money by writing so as to support my own children whom I love and [Gloria's] children whom I also love. Why this not only shrinking from responsibility but also a positive horror of it." He then launched into a list of resentments. He resented Gloria's stalling *and* his own tolerance of it, because it made him think that he "had been as cowardly and weak as she." Making the final break with their spouses had come to seem an act of personal heroism. Gloria's constant excuse that she had to think of her children irritated him, because it prompted guilty feelings that he was not as concerned about the effect of a divorce on Mike and Nick. Dwight imagined that if he had an assured annual income of $10,000, he would "leave tomorrow. Is that wrong?" And he had premonitions of a serious writer's block and hatred for his whole life-work relationship. The situation was so stressful, and his self-doubt and guilt so depressing. "I resent everything . . . I resent everybody . . . I hate myself and all else."[32]

The summer optimism about his new writing projects gave way to bleak despair. He acknowledged a lack of motivation and a feeling of sloth. All he could think about the profiles was "how goddamn MUCH labor I'll have to go through." He felt that Gloria misunderstood and simply considered him lazy, distracted by too much social life. He did admit he procrastinated, not by seeing people but by unfocused reading, mostly magazines and detective stories, but also "'good' books which can be converted into escape too." But analysis of how he escaped missed the point. "The point is, WHY or rather WHAT am I escaping from???? What has killed my pleasure in creating, so that I just go on reflexes, and really don't give a damn whether I produce or not and whether people admire or not." Even if he completed the new profiles, he realized he would be on the same old treadmill, and that "like so many, almost all, all things in my life, such a victory would simply pose The Next Problem." He concluded that he could "NEVER catch up . . . nothing I can ever do will make it up, I have failed so terribly in producing."[33]

He felt that his life was slipping by and that he had never established himself as a writer, as an intellectual, or as a man of consequence; he refused to see himself as a distinguished essayist, a first-class intellectual journalist, and to accept such a role for himself. He perceived a decline in his transition from editor of *Politics* to staff writer for the *New Yorker*. He was delighted to write for pay, but he felt that much of his literary achievement lay in the past, when he was more productive and editorially creative at the *Politics* office. Then he had control and direction; now he was dependent on the editorial approval of others and he was, even though he resisted it, tailoring his work for different audiences. His self-analysis in his journals suggests that while he may have been hard on himself for depending on ideas developed nearly a decade earlier, he had a greater respect for the community of *Politics* subscribers than for those who read his prose between the gilded ad columns of the *New Yorker*. He had not made peace with himself, and his uncertainty and anxiety were compounded by his fear that, having already destroyed his marriage and broken up his family, he did not have the emotional or financial wherewithal to begin again. There was great sadness in this self-deprecation, since his refashioning a second career as a major American cultural critic as he was approaching fifty was regarded by at least one observer as an enormous personal achievement marked by great courage and persistence. In the fall of 1953, his self-analysis suggested that he was close to a nervous breakdown.[34]

In October Dwight's long essay review of *The Captive Mind*, a book about totalitarianism by Czeslaw Milosz, appeared in the *New Yorker*. Milosz's views tended to confirm all of Dwight's own convictions. Dwight placed the study in a class with Hannah Arendt's for its subtle and imaginative delineation of the totalitarian mentality. What pleased him most about it was the way it implicitly contrasted the innocence of Americans with the appalling cynicism and despair of Europeans. Dwight took the occasion to comment on the almost virginal purity of the American mind, completely inexperienced in the ways of twentieth-century inhumanity: True, we had troops and money deployed all over the world, but our cities had not been bombed, armies had not destroyed the countryside, and our relatives had not vanished suddenly into death camps and gulags. Nor did Americans "feel obliged to submit in silence to a single political ideology or to inform on their neighbors who resisted." Thus while we might denounce the blunders and perversity of our foreign policy, the fundamental reasons for its failures lay in the discrepancy of the American experience. "Modern history [had] simply passed us by, as though we were some aboriginal tribe placidly living its traditional, idyllic days in an out-of-the-way nook of the globe."[35]

Dwight also dwelt on the role of intellectuals in a totalitarian society as propagandists for the State. As long as they toed the line, for these services they became members of an enslaved privileged class. The only escape was exile, as in Milosz's own case, or suicide, which Milosz documented in his despairing case studies of writers and artists who had sacrificed their imagination and creativity to provide the State with a constant supply of approved trash. Noting Milosz's refusal to conform, Dwight ended his review with the ironic observation that Milosz had been barred from coming to the United States by the McCarran Act because he had once been a functionary of the Polish Communist government. He now lived in France, where, contrary to the United States, "the state was celebrated for its willingness to put up with non-conformists."

Milosz wrote Dwight to thank him for his generous review and for the kind letter that had accompanied a copy of it. He noted, as had Dwight, that the book had not sold well and that some Polish émigré circles denounced it as a defense of Stalinism, making him wonder at just "how far human folly" could go. He added that he found it a pity that Dwight stopped publishing *Politics*, because the magazine had far more influence than Dwight might imagine. A year later, he published his warm appreciation of Dwight and *Politics*.[36]

Hard upon his Milosz review came the appearance of his long, two-installment profile of Alfred Barr. What Dwight liked best about it was its length—for it meant more money from the *New Yorker*. It is in many ways

292 CRITICISM AS A SUBSTITUTE FOR POLITICS

a curious piece, because Dwight managed to curb his own involvement. Throughout there is a detachment, a neutrality not characteristic of most of his journalism. He admired a controlled distance, but in fact what distinguished most of his essays and reviews was the degree to which he made the reader feel he really cared about his subject. Dwight did respect Barr as a man of integrity, principle, and commitment, who was devoted to the cause of modern art and had dedicated his life to promoting it. Dwight recorded Barr's career in meticulous detail as the first director of the museum and later director of collections.

Since the museum stood for modernism and was opposed by reactionary politicians, uneducated philistines both rich and poor, traditionalists and totalitarians of the right and the left, one would expect Dwight to be sympathetic toward Barr and his institution, and for the most part he was. He too appreciated what had become known as the "old fogies of modernism" after the Second World War (Bonnard, Braque, Brancusi, Derain, Dufy, Gris, Kandinsky, Klee, Léger, Maillol, Matisse, Modigliani, Mondrian, Picasso, Rouault, and Utrillo). He also appreciated Joyce, Pound, Eliot, and Yeats. It was the modernism of the late nineteenth century running through the 1920s. There is not much evidence that Dwight had a great interest in contemporary modern art; as a matter of fact, he showed a considerable dislike of the abstract expressionists, particularly the New York School epitomized by Jackson Pollock. Dwight was not above using the derisive label "Drip and Dribble School." This included, in addition to Pollock, Willem de Kooning, William Baziotes, Arshile Gorky, Mark Rothko, Franz Kline, Theodoros Stamos, Bradley Tomlin, Robert Motherwell, and Hans Hofmann. To the degree that Barr and the museum had not only promoted these painters but were in effect responsible for their postwar trendiness, Dwight could hardly have been approving.

Nevertheless, he did not dwell on these matters in his profile. The thrust of the piece was to portray Barr as trying to find a successful way to promote high culture among the Neanderthals. Barr's showmanship, advertising campaigns, promotional stunts were for the most part acceptable to Dwight because Barr, despite what one might think of his judgment and taste, was dedicated to "the conscientious, continuous, resolute distinction of quality from mediocrity." His motives were unimpeachable: He wanted to get the public at large interested in modern art. Dwight could not fault that.

Dwight gleefully exposed the skewed reaction of the mass public to the museum's clever promotion of Whistler's *Arrangement in Grey and Black*. Borrowed from the Louvre, *Whistler's Mother*, a painting denounced and unsold for twenty years, became nationally acclaimed after Barr included it in a tour devoted to American painting and sculpture. Barr then sent it on the road alone, where it was "seen by hordes of mother-venerating Americans." A postage stamp was issued. The Post Office obligingly added a flower vase to the picture, presumably making it more motherly. It was a nice irony, appropriate to modernism's incorporation of the popular in high art, that Dwight's discussion of *Whistler's Mother* was flanked by a sexy ad for Rita Hayworth's portrayal of a prostitute in the film *Miss Sadie Thompson*.

Dwight did not pass up the opportunity to note that Barr's famous van Gogh show was promoted "almost too successfully—to such an extent that a van Gogh print over the mantelpiece became a familiar hallmark of the middle-class American home." Dwight did not explain why such a reception for mass-produced prints of an artistic masterpiece was a bad thing. Perhaps he thought it was because the mass production itself was for profit and that people bought the painting as an act of self-aggrandizement and not because they appreciated the power of *Starry Night*. His vision of hordes of mother-venerating Americans gawking at Whistler's arrangement and masses of middle-class and at best "middlebrow" homeowners pridefully hanging van Gogh tickled his own ingrained elitism. He good-naturedly commented on the egalitarian aspect of the "museum's snobbishness, which had always been democratic after a fashion—the peculiar American fashion in which the aim is not to exclude *hoi polloi* but, on the contrary, to attract them with snob appeal, along the lines of the 'Men of Distinction' whisky ads and the Aqua Velva After Shave Club." He amiably conceded that the promotion "might be justified as a peccadillo in a good cause."[37]

Following his Barr profile, Dwight continued to focus his attention on contemporary culture. His pieces on Milosz and Barr had been positive and approving, but now he gave vent to the scathing negativism which he insisted was an obligatory part of the responsible critic's work. The result was a ruthlessly witty assault on an anthology of contemporary writing, *New Directions 14*, the latest volume in a series published by the leading avant-garde publisher, James Laughlin. Dwight's method of demolition was to compare this collection with a recent anthology of writings from the *Little Review*, a literary magazine that had been edited and published by Margaret Anderson and Jane Heap from 1914 to 1929. Dwight had

guarded his opinions in the Barr piece, undoubtedly because he lacked
confidence in his judgment of contemporary art, but he made up for his
circumspection in this review of contemporary prose, poetry, and essays,
where he felt more at home. He initially wrote his critique as a parody of
the *New Directions* collection's depressing contents:

13 stories

       5 essays

              67 prose poems

                       32 poemy poems

                                425 pages

and pages
and pages
     pullulating   *crepitating*   *ululant with*
words   *words*   *words*
breaking   *breaking*   *breaking*   *on*
cold
grey
stones

GRANDIOSE DIARRHEA OF LANGUAGE.

The bread and butter of 19th century writing, how nourishing compared
to these hors d'oeuvres from the literary deep-freeze!

Those eminent Victorians, long of wind and flat of foot, how terse how
direct compared to All the Sad Young Middle aged Men exacerbating
their insensibilities and taking their time about it.

This poetic lampoon was not acceptable to the editors of the *New
Yorker*. In a marginal note written many years later, Dwight asked himself
whether he had actually tried it on Shawn. He noted that if he had, he had
obviously been unsuccessful and had been forced to translate it under
editorial pressure into *New Yorker* style (or some editor did without losing
a beat—or a point).

The critical imagination and biting satire sharpened by parody were
tamed and neutralized by the *New Yorker* editors. The end of his review
lampooned a *New Directions* essay in which a revolutionary critic asserted
that "the poem in prose is the form of the future. Poems don't rhyme anymore. The saturation point has been reached. One more rhyme and I'll

vomit." Dwight ended his parody: "Indeed and indeed the saturation point has been reached with *New Directions 14* and it's not only rhyming that is sick-making." The editors changed this to read, "Without going into what else besides rhyming is upsetting, one may agree that in 'New Directions 14' a, if not the, saturation point has indeed been reached."

Dwight was determined to carve out a niche for himself as a contemporary critic of American culture. He was making every effort to shun the climate of celebration that seemed to be an inevitable and unfortunate byproduct of cold war tensions. Having given up *Politics* and distanced himself from politics in general, he had decided cultural criticism was to become his vehicle for maintaining an adversarial stance. He would see himself as "standing against the American grain," but he also recognized that his position was not radical in a cultural sense; on the contrary, it was reactionary in its implicit defense of established art, its skepticism and contempt for most young writers and artists. Dwight had little respect for what became known as the "angry young men" in England who railed against the bourgeoisification of English society. Nor did he show any interest in "the Beats," for whom he maintained an acerbic contempt. Since he saw a too easy tendency among a certain segment of the intellectual community to tolerate the shocking for its own sake and not for artistic merit, he felt that preserving standards was a kind of artistic and even moral obligation. Making this stand in the *New Yorker* had its dangers, since the magazine was not highbrow but rather what he would label midcult, and its glittering commercialism was an easy target for those who found its tailored prose in keeping with middle-class caution and circumspection, lacking in spirit and imagination.[38]

In a burst of energy Dwight managed to complete another long *New Yorker* article on the flood of "how to" books, a prizewinning article for the *Reporter* on the governmental misuse of lie detectors, along with book reviews for the *Partisan Review*. This productivity gave him the confidence he needed to finally make the break with Nancy, and it also gave Gloria the assurance she needed that he was on the way to celebrity status and that he would indeed be able to provide for her and her children. She was prepared to ask Robin for her freedom. This was no easy matter for any of the participants, and hard on others as well.

Many of Nancy and Dwight's friends were taken by surprise. Some were aware that both were having affairs, but they could not believe the actual collaboration between the two would end. Some thought of them as the ideal couple, in a very remarkable way. "They were comrades,"

Eileen Simpson recalled. "They had been through the revolution together," Norman Mailer remarked when thinking about his feelings at the time of the break. Simpson remembered actually being at the apartment for dinner when Dwight broke the news. "It was awful . . . a terrible night." Michael Macdonald remembers Nancy awakening him in the morning, holding his hand and telling him about the divorce. It happened the day after he was notified he was being expelled from Exeter for disciplinary reasons. Years later, Nancy could not remember the actual moment when the children were informed, since the marriage had been deteriorating for so long. For Michael the breakup of the family coming on top of his expulsion from Exeter had a devastating impact. He was sixteen, trying desperately to be grown up. He remembers feeling terrible for Nancy and then confronting the thought of Dwight leaving.

The nine-year-old Nicholas found Dwight's departure incomprehensible. He'd had no inkling, although during some of the more rancorous times prior to the actual separation he had broken out with boils on his hands. He could view it only as a personal betrayal. His father had abandoned him. Dwight and Gloria delayed telling their own mothers until the decision was made and a separation agreement signed. Dwight assured Gloria that his mother was not upset; she had "a rather dim view" of Nancy and had given Dwight three cheers. He went on to assure Gloria that his mother thought her charming and "altogether a fit mate for her son."[39]

On June 17, 1954, Dwight and Nancy's legal separation was completed. He had moved out of the apartment and this legal action cleared up all questions concerning the settlement of common property, the custody of the children, and Dwight's obligation for their support and education. Nancy made no claim for alimony, since she had an independent income. Dwight was to be allowed to visit with the boys at the apartment and to have them stay with him every other weekend. He turned over the Wellfleet cottage to Nancy, which meant he was unlikely to make the Cape his summer headquarters in the future. That was another significant break with his past. Gloria had already rented places in Amagansett on Long Island, a community very different from the rustic simplicity of Truro/Wellfleet/Provincetown.

In the same month, Gloria and Robin came to a separation agreement. These agreements were tantamount to divorce, but since divorce proceedings were still so difficult and time consuming in New York State, both Dwight and Robin went to Montgomery, Alabama, to obtain quick legal divorces. Dwight left Monday morning June 21 and was back by Wednesday. Nancy left for Europe with Michael that Friday, June 25. Nick stayed on with Dwight, but after a long discussion with Nancy, who didn't want "Gloria taking over Nick," he was sent off to summer camp. But since the boys were getting along with each other so poorly, Dwight insisted that Mike needed an opportunity to be "the man" in the family without the annoying presence of his little brother.[40]

Dwight and Gloria reassured each other during this period, insisting that the only thing to do was to concentrate on the future. While he was in Montgomery, Gloria wrote Dwight: "No backward looks because they can be destructive and have no positive value. Now is all there is."[41]

Dwight returned to the city after the completion of the divorce proceedings, briefly sharing the apartment on East 10th Street with a summer tenant as well as living out of his *New Yorker* office while Gloria wound up her affairs. Nancy and Mike had left for Europe, and Nick had been sent off to summer camp in Maine. In July Dwight and Gloria drove up to Bigwin, Ontario, above Toronto, where he participated in an adult education seminar sponsored by the Ford Foundation. He and Gloria were not married yet, but they traveled as man and wife; when her divorce came through in early August, they were married before a justice of the peace.

That summer and fall, Dwight and Gloria spent as much time as possible on the Cape and in Gloria's Riverside Drive apartment. After a thorough search they found an apartment in an elegant neighborhood on East 87th Street between Madison and Park. It was an older building, and their north-facing apartment on the first floor was rather dark. In less that a year Gloria managed to arrange a move to a rambling second-floor apartment with a bit more light and larger rooms. So complex was the layout of the hallway passages that it was rumored Gloria at one point literally housed her mother in the apartment for several weeks without Dwight being aware she was there.[42]

13

Innocence Abroad

Dᴜʀɪɴɢ ᴛʜᴀᴛ ʜᴇᴄᴛɪᴄ sᴘʀɪɴɢ ᴀɴᴅ sᴜᴍᴍᴇʀ ᴏғ 1954, Dwight still managed to participate in the battles over McCarthyism and the continuing red scare in America. His award-winning *Reporter* article on the widespread government use of the lie detector was an exposé of rampant police-state tactics used to harass government employees whose political sentiments and affiliations did not meet the standards of government bureaucrats. In the piece he did not hesitate to refer to Senator McCarthy as a "pathological liar."[1]

Disgruntled by the failure of the anti-Communist left to denounce McCarthy, Dwight deliberately sought forums to air his own criticism of the senator. In the spring of 1954 he wrote for *PR* a caustic review of William F. Buckley and Brent Bozell's apology for the coarse demagogue, *McCarthy and His Enemies: The Record and Its Meaning.* He likened their defense of the senator, with its elegant academic style, nice discriminations, and pedantic hairsplitting, to a "brief by Cadwalader, Wickersham & Taft on behalf of a pickpocket arrested in a subway men's room." Dwight compared their excuses for McCarthy's conduct to those of apologists for Stalinism, who always found some other crime elsewhere in the world that made Stalinist atrocities seem mild and understandable. Dwight's most insistent notion was that the McCarthyist period was a bizarre fit of national paranoia. "Like Gogol's Chichikov," he charged,

"McCarthy is a dealer in dead souls. His targets are not actual, living breathing Communists" (Dwight wondered whether there were any left besides Howard Fast). Chichikov had his Gogol, but the senator had yet to find his. Buckley and Bozell served well as "minor comic characters in the mock-heroic epic of McCarthyism," which in Dwight's view was "an interlude in our political history so weird and wonderful that future archaeologists may well assign it to mythology rather than history."[2]

Just how paranoid and weird it actually was came home to Dwight when Nancy told him that FBI agents had grilled her on the phone for details of the divorce, asking her what the grounds were and who Dwight's lawyer was, and calling her lawyer. When they were so brazen as to request an appointment, Nancy shouted "No!" and banged down the receiver.[3]

Dwight's consistent denunciation of "paranoid anti-Communism" and his often joyous invective against it may have been one of the reasons that some in England and Paris were surprised to learn he was being considered for an editorial position on *Encounter*, the Congress for Cultural Freedom's publication in London. Arthur Schlesinger, Jr., had connections in London and had suggested Dwight as a replacement for Irving Kristol, whom his co-editor, Stephen Spender, found difficult to get along with. All of this was done quietly, and Kristol later claimed that he knew nothing about it until Michael Josselson, executive director of the CCF, informed him that he was being replaced.[4]

When Dwight first got wind of this possibility, he was thrilled. Getting out of New York, living abroad and having a secure full-time position on a recognizably highbrow magazine was more than he could have asked for. He wrote Chiaromonte in Rome that Spender was thinking of him as the nonliterary editor, and Dwight relished the idea of going from author to editor again. By April 1955, Dwight was carrying on an extended correspondence with Stephen Spender, who asked him to put down what he thought might be done with *Encounter* if he were to become a co-editor. It became immediately clear that Dwight had in mind creating a London *Politics*. He envisioned a magazine that would compete with the *New Statesman* and *Nation*, as well as with the *Partisan Review*. In addition to its literary content—poems, stories, and criticism—it would also "go in strongly" for political (and cultural) journalism. It would not water down or popularize. As in the past, Dwight wanted to initiate a serious department on "mass or popular culture." Dwight was still negotiating with editor Jason Epstein at Doubleday-Anchor for a book on the subject, and he saw this as an opportunity to combine his own research and writing with steady editorial work.[5]

By the end of April the rumors of the possible appointment of Dwight to the editorial staff of *Encounter* were widespread in both London and New York. Early in May Spender wrote Dwight to say that it had been determined Irving Kristol would leave *Encounter* and do a book on the Milan Conference which the CCF was hosting in September. Michael Josselson was making a trip to New York and would have a talk with Dwight. Spender was sure Josselson would want to determine what Dwight's attitude toward the CCF was and be convinced of his willingness to cooperate with its executive board. Spender assured Dwight that the CCF had never interfered with the editorial direction of *Encounter*. That Spender should take pains to see that Dwight presented an accommodating attitude toward the CCF lent support to the rumors of the organization's editorial control over the publications but also revealed the antagonism between the CCF and the American Committee for Cultural Freedom. Since Dwight had been a member of the American Committee, Spender wanted assurance that Dwight did not hold the unfriendly views toward the CCF that many of the members of the American Committee did.[6]

Dwight sensed the political implications of Spender's letter and dashed off a reply expressing enthusiasm for his appointment and delight that the CCF did not interfere with editorial decisions on the magazine. He noted with approval that the CCF had criticized the American Committee for the very same failing that he had: its reactionary and obsessive anti-Communism. He did tell Spender that the rumors of his selection were circulating within the New York intellectual community and he had been noncommittal, but that recently he had told some people his appointment had been decided except for the matter of salary. He recounted a meeting with Daniel Bell and Sidney Hook at a party in the city. Bell, whom Dwight felt was more sympathetic to his criticisms of the American Committee than Hook was, had told Dwight there was some "bad feeling" on the Committee about his going to *Encounter*. This was partly due to its liking Kristol's politics much better than Dwight's and partly because the decision had been made without consulting the members of the American Committee. This was particularly true of Hook, who was also a member of the executive board of the CCF. Dwight told Spender that he thought Hook should have been informed of the plan to make a change. Hook had denied to Dwight that he was opposed to his selection, claiming only to be annoyed that on a recent trip to Paris his suggestion that *Encounter* should be more political and journalistic was rejected by the editors. Now they were surreptitiously bringing in Dwight presumably to do just that. In typical fashion, while Hook was apparently hiding his anger over the plan to hire Dwight, Dwight spoke directly and honestly to Hook. He said he assumed that Spender and those at the CCF were more in harmony with Dwight's politics than they were with either Hook's or Kristol's. Hook, Dwight recorded, affected not to understand his meaning and dismissed Dwight's suggestion that he and Kristol had sharply disagreed with CCF policy as to the "the proper way to fight communism." Hook flattered Dwight by extolling his editorial talents, but . . . Dwight finished his sentence: "But you think I am undependable politically." Hook laughed and admitted that was true.[7]

Dwight came away from these encounters believing that although there were reservations, his job was all but guaranteed. He assured Spender that if they managed to produce a lively and exciting magazine, his opponents would "grudgingly forget their fears" as to his and Spender's "alleged softness towards the communist issue." He added that he was sure the "veteran front line fighters against communism" were already anachronisms.[8]

Little did Dwight appreciate the intensity of the antagonism toward him felt by Hook and other members of the American Committee who were determined to see that he did not get the job. Dwight had an amiable meeting with Michael Josselson, who came to New York in June. Dwight made it clear that he agreed with the CCF, especially its criticism of the American Committee. As Dwight understood it, the CCF supported cultural freedom against all comers "regardless of whether Communists, Fascists, or mugwumps [were] the parties injured" while the American Committee did not. Dwight asked for $12,000 a year plus expenses and Josselson thought that reasonable. Dwight wrote Spender to report that Josselson had said the deal was practically closed and all was well. Kristol was to leave October 1 and Dwight was to report to London September 1. So confident was Dwight that his position was in the bag that he asked Spender to begin keeping an eye out for a suitable London apartment.[9]

Ten days later the situation looked much less certain. Josselson, who had been out of town, returned to say that the Paris office had yet to approve the deal. There was some question about the salary. But then Josselson admitted there was also concern that Dwight might be too much of a lone wolf, that the letters Dwight had written to Spender about what he would do as editor made it sound as "though he considered *Encounter* as

his own personal journal like *Politics*" and he would resent any suggestions from the CCF. Dwight came away from this meeting with Josselson
mistakenly believing that it was he who feared Dwight's independence.
Josselson then proposed that Dwight come over in September for the
Milan Conference and meet everyone in the Paris office and at *Encounter*.
Then he would return to the States, and if the CCF approved him and he
still wanted the job, he would begin January 1, 1956. Dwight accepted
this as reasonable and felt that it gave him a little more time to try to get a
book on mass culture completed before he left for London.[10]

In July, Nicholas Nabokov, who was also on the CCF board in Paris,
wrote the CCF's official request for Dwight to cover the Milan Conference
for *Encounter*. The Conference was to be a large gathering of intellectuals
from Western Europe and Asia, Africa, and Latin America to discuss
"The Future of Freedom." Michael Polyani, the chief organizer, hoped for
a serious discussion of the "new economic order" that would be neither
socialist or capitalist. It was really a meeting of intellectuals from the
West with their counterparts in the underdeveloped world, one of a number of such gatherings put on by the CCF to counteract the propaganda
meetings of the Eastern bloc. Dwight took Nabokov's solicitation as a
good sign that the Congress was very seriously considering him for the
position of editor.

Dwight and Gloria rented a place in Wellfleet for a part of July and
August 1955. It was on Snow Pond, not far from the Slough Pond cottage.
Dwight spent much time completing a long profile of the Ford Foundation
for the *New Yorker*. He did not really enjoy the writing. His old *Politics*
girl Friday, Bertha Gruner, explained his lack of enthusiasm. It was a profile "not of people or a person but of $$$$."[11] In fact, Dwight's opening
sentence remains a classic in the literature: "The Ford Foundation is a
large body of money completely surrounded by people who want some."
Dwight had interviewed what he called the "philanthropoids," but it had
been an unsatisfying experience. He found it odd that the Foundation,
which spent $60 million in 1954 on many projects that seemed praise-
worthy, cast "a dull patina over the whole business." It was all "unreal,
uninteresting, inhuman." Dwight turned the extended piece into an
analysis of the phenomenon of empirical data gathering and "the language of Philanthropes—"a scholarly jive, with traces of the argot of the
surrounding Madison Avenue account-executive region."[12]

Dwight made a careful study of the Foundation's executives, "middlemen between brains and capital." It was significant that while he was negotiating with the Congress for Cultural Freedom, a creature of presumed Foundation support, he was agreeing with the widespread suspicion that foundations were "quasi-official intelligence agencies working for the State Department under cover of scientific respectability." However, since the Ford Foundation was constantly attacked by such right-wingers as George Sokolsky and Westbrook Pegler and various congressional investigating committees, it must have been doing something right.[13]

This partial exoneration of the Ford Foundation for the enemies it had made suggests the limits of Dwight's critique. He could wittily mock its efforts at mass education and TV (mass culture), its magazine's refusal to criticize the American way of life, its institutionalization of art, and its promotion of bureaucratized social science. He touched on its function in the official American propaganda machine and hinted at the possibility of a covert hand at work. But he never seriously probed its connections—connections which might have suggested that joining *Encounter*'s staff would bring him into collaboration with a similar propaganda network.[14]

Dwight found the research and writing of the Foundation series boring and unsatisfying, and he did not have a very relaxing summer. The divorce and remarriage were hard on seventeen-year-old Michael. To be at the Cape and have his father living with a woman who had long been a member of the community was difficult. Michael tried to live with Dwight, Gloria and Mary Day, Gloria's older daughter. Nancy made it a point not to go to Wellfleet while Gloria and Dwight were there. After a mishap during a beach walk, Michael, who had gotten separated from Dwight, became frightened that something had happened to him. When he returned to the house after his own adventure with a rising tide and found Dwight having a drink and totally unconcerned, he flew into a rage, egged on by Dwight's blasé attitude and high-pitched giggle. Soon he was denouncing Dwight for leaving Nancy and he stormed out of the house. The bickering and tantrums continued until Dwight asked Michael to leave. Dwight wrote assurances to Nancy and to himself that he would be more loving and patient with his troubled son. But the tensions continued and Dwight was increasingly desirous to leave for Paris and Milan in September.[15]

As the day for his departure arrived, the issue of the *Encounter* job remained unresolved. Dwight sailed early in September, spent a few days in Paris and arrived in Milan on September 12, the first day of the

conference. He still thought the job was in the offing and wrote to tell
Gloria how enthusiastic Hannah Arendt was about his imminent return
to serious editing and writing for an important intellectual magazine.
However, he had heard through the grapevine, undoubtedly from Arthur
Schlesinger, that it was not Michael Josselson who opposed his editorship but the American Committee people led by a typically aggressive
Sidney Hook. They were against him for political reasons and for what
they saw as the high-handed intrigue of the CCF in not informing them
of the selection. Hook was so outraged by the maneuver that he threatened to resign from the executive committee of the CCF and "blow the
Congress out of the water." He was infuriated by what he saw as their
backstairs subterfuge in hiring a man of notorious political irresponsibility. In his view, which was quite accurate, Dwight Macdonald was a loose
cannon, unpredictable and not to be trusted to hew to the line demanded
by Hook.

Three days later Dwight got the word that Hook had won out and that
Kristol was to be kept on. Arthur Schlesinger, who was apparently present
and played a role in the negotiations when the final decision was made,
told Dwight that Josselson and Nicholas Nabokov were more vigorous in
their support of Dwight than Spender, who had a reputation for timidity
and peace at any price. Dwight wrote Gloria that he was out as editor. Josselson, obviously embarrassed at the state of affairs, felt obliged after a
negotiating session with Daniel Bell and Schlesinger to make Dwight an
offer to serve as roving correspondent for all of the CCF's European publications. He was to travel around Europe writing for *Encounter, Preuves,*
and *Der Monat* and to serve as an editorial consultant to the three magazines. It was still just a matter of getting the CCF to approve a $10,000-a-
year salary plus expenses. Dwight tried to put the best face on this disappointing turn of events. He wrote Gloria that it might well be an even
better deal than the full-time editorship because it would allow him to put
in more time on his own work. He was still talking about the book on
American mass culture.[16]

As was so often the case, Dwight's initial piece about the Milan Conference for *Encounter* only succeeded in irritating his editorial colleagues. He relished biting the hand that fed him. When Kristol got the
manuscript he wrote a long letter full of complaints representing the
views of the Paris office as well as his and Stephen Spender's criticisms.
They had discussed the article "for more hours than we would like to
count or would prefer to remember." The Paris office and the editors were

so annoyed that they decided to put off publication of Dwight's article and instead run a puff piece by Edward Shils, a prominent sociologist and CCF member, who waxed effusive on the brilliance of the participants ("heads were set into a whirl"). Dwight's article would be presented as a "dissenting opinion" in the next issue in December.

But even though they would allow a dissenting view, they would not permit the piece to be published in its original form. Kristol wrote a paragraph-by-paragraph critique. What offended them most was Dwight's dwelling on the high-roller aspect of the CCF—the posh hotels, the fancy restaurants. It was the critical and mocking tone of the piece that irked Kristol. Dwight had complained about the acoustics, the tediousness of the papers, the "dead formal" quality to the meetings. "There was no fire, no drama, no sparkle": Why hadn't they invited some Russian delegates and perhaps a few fellow-travelers like Sartre to create confrontation and controversy? Dwight observed that the Asian delegates were not interested in the technicalities of Western academic theory; they wanted to know just what Western democracy meant for them in terms of economic development. Despite the editorial harassment, Dwight stuck by his guns on most of the issues. He wrote Spender that if *Encounter* was going to take credit for asking that "well-known independent temperament and dissident, DM, to cover the Conference then . . . the Congress can hardly expect me to take out most of the sting of my piece. They can't have it both ways."[17]

This incident should have alerted Dwight to the role of the CCF in the publications it sponsored. The close monitoring by some of the CCF people in Paris, Kristol's obvious vexation and dismay that Dwight expected a CCF magazine to come out with "what is in effect a complete indictment" indicated that despite all the window dressing, the CCF expected a more harmonious collaboration than Dwight would be comfortable with. Hook, who attended the conference and gave, according to Dwight, a long-winded paper, later wrote that even in 1955 rumors were rampant that the CCF was funded by the Central Intelligence Agency, and he discussed the matter with Julius (Junkie) Fleischmann, a director of the Metropolitan Opera in New York and a Fellow of the Royal Society of the Arts in London, who was the apparent director of the Farfield Foundation, the conduit through which the funds were dispensed. When Dwight and Gloria arrived in London, Gloria, hardly known for a keen sensitivity to political undercurrents, perused issues of *Encounter* lying around the flat and insisted that it was obviously an official publication of some agency of the State Department. As far back as 1952, when Nancy was in Paris she had written that it was generally assumed the CCF and its publications were subsidized by the State Department.

But there is no evidence that Dwight accepted the validity of these rumors in 1955. In any event, he had no fundamental disagreement with the politics of the CCF. His own anti-Stalinism had prompted him to endorse militant political confrontation with the Soviet propaganda apparatus. His only departure was his refusal to celebrate American life and culture as the best of all possible worlds. He did not feel that it was necessary or even tactically sound to apologize for the failures and weaknesses of the United States or its allied associates. For his opponents on the American Committee as well as people like Kristol and subsequently Melvin Lasky, promoting anti-Stalinism meant defending the United States as the leader of the free world. Dwight found that not only a critical failing, but thought it undermined the credibility of the CCF and its publications.[18]

Dwight's career fears seemed to be over. His Ford Foundation pieces were appearing weekly in the *New Yorker* to much praise. His stature as an articulate critic of mass culture was well established; he had been invited to be a visiting professor at Northwestern University and to give a course at the Salzburg Seminar. He was moving rapidly into the role of cultural critic, and he now looked with enthusiasm toward his move to Europe.

Dwight arrived in Paris in mid-January 1956. He spent much of his time at CCF headquarters attending editorial meetings with members of the CCF and editors of its various publications in Europe and Asia. He wrote with pride to Gloria that he had been chosen to enliven the pages of *Encounter*. While in Paris he renewed his friendship with Czeslaw Milosz and set about trying to find a way to finance the work of the refugee poet and writer. Both men were drawn to each other and enjoyed the time they spent together.

Early in February Dwight left for London, where he was to attend editorial sessions of *Encounter*, but he did find time to see the sights, to entertain and be entertained. He took the Kristols to dinner at the Café Royale, the gathering place of Shaw, Max Beerbohm, and Frank Harris in the 1890s. Dwight was delighted with the service as well as the food; there was an attention to detail that was astonishing to "anyone familiar with the rigors of New York." With Spender and Malcolm Muggeridge he dined at the Saville Club. He found it shabbier, smaller, and less impressive than he had expected, but the British were friendly, even talkative, which was not what he had expected. "The famous British reserve and hauteur is a myth—Americans are much ruder and snootier."[19]

Dwight was comfortable in the *Encounter* offices at 25 Haymarket. They were even messier than those of *PR* and *Politics*. Every editorial discussion, whether at the CCF in Paris or at the *Encounter* offices in London, was followed by luncheons or dinners at good restaurants, on the CCF account. Dwight was conscious of these perks and he took both a bemused and at the same time mildly critical attitude toward them, prompted by his innate parsimony and, as he mentioned in his Milan piece, vestigial radicalism. But at this time he did not show any curiosity as to who was picking up the tab for the sumptuous meals and the luxurious printing of the small-circulation, little magazine, which obviously did not pay for itself. Apparently he accepted the story that Junkie Fleischmann was a man inordinately dedicated to the promotion of high culture and independent thought, and Dwight was glad to be a part of that.

He liked most of his colleagues at CCF headquarters in Paris and also at *Encounter*. He got along well with Irving Kristol, although he did not think him a very imaginative editor. He may have had some suspicions of Melvin Lasky, the editor of *Der Monat*, a publication financed by the Ford Foundation in Berlin. Lasky was one of the founders of the CCF and remained very active in establishing its editorial policies. Nancy and Gloria both sneered at the mention of his name and Dwight frequently referred to him as an ambitious self-promoter. But still these people were "Old School Ties," as Dwight frequently referred to alumni of the anti-Stalinist left; they spoke a common language and continued to see the world from a similar perspective, intellectually if not politically.[20]

Dwight found the London intellectual scene in general not dissimilar to the milieu of the New York intellectuals—that is, a man's world. He went with Stephen Spender to a party at *Spectator* magazine's offices. "1,000 men (with very long hair) and .01 women; England, at least, intellectual England, is like the Frontier, a Man's Country." Dwight was taken with the city and repeatedly told Gloria that they would love it together when she joined him in June. Life was better than in New York, more casual, more relaxed, the people more interesting. There were buildings of historical interest on every corner. He tried to hold off sightseeing so as to share the pleasure of discovery with Gloria and the kids. But he did manage to take in St. Paul's ("magnificent") and Westminster Abbey ("disappointing").[21]

In his room at the Athenaeum Court Hotel, in Piccadilly, Dwight worked on his own writing. He found the room idyllic—"ten times more wonderful and livable than any NYC hotel room." He was working on his "Mass Culture" lectures for the Salzburg Seminar and finishing up his critique of William F. Buckley's *National Review* for *Commentary*. Reading Buckley's magazine was painful for Dwight. He would not credit it as conservative, because it showed no respect for law or tradition. It simply was anti-liberal, and Dwight insisted that anti-liberalism was not a political philosophy. In his view, a man like Supreme Court Justice John Marshall Harlan was a genuine conservative because he had respect for the Constitution and insisted on interpreting the 14th Amendment as intended. The *National Review* editors had denounced the recent *Brown vs. Board of Education* decision and used the sophistry of states' rights to defend their reactionary politics. A true conservative like Alfred Jay Nock, Dwight wrote, appealed to law or, if desperate, to tradition but certainly not to the "'hearts of men.'" In this essay Dwight expressed an identification with true conservatism and indicted the *National Review* and its editors as having something in common with the liberals they detested: Both would bend the law to suit their interests; both refused to stand for principle. Dwight conceded that he too "no longer found the left ideology either morally or intellectually satisfying," but it was still "a far more plausible doctrine than the right's crude patchwork of special interests."[22]

After what proved to be a rather unsatisfactory stint at the Salzburg Seminar, Dwight returned to New York in mid-March 1956 and left quickly for his visiting professorship at Northwestern. Gloria was suffering from an unexplainable bleeding of the uterus and she was forced to return to New York to consult her physician. Dwight stayed on alone at Northwestern, teaching two courses, "Mass Culture" and "Literature and Politics since the Thirties." The Mass Culture course was a rerun of his Salzburg lectures. The Literature and Politics course, which later became a course in the political novel, centered around historical and fictional accounts of leftist politics in the thirties and forties dealing with the dominance of Marxism, the split between Stalinists and anti-Stalinists, the rejection of Marxism, and the turn toward anarchism and pacifism—in effect, a course tracing his own political odyssey. A major theme was what Dwight interpreted as the "depoliticization of American intellectuals in the fifties," the "decadence" and ineffectuality of liberal weeklies, left-wing sectarianism, and the rise of an "intellectual right wing." He concluded with an analysis of the role of the intellectual in the fifties as contrasted with the thirties. His own sympathy continued to lie with some form of critical dissent and an obligatory allegiance on principle to an adversary culture. Dwight used his own revised version of "The Root Is Man," recently brought out by a small California publisher, as one of the basic readings for the course.[23]

During this period Dwight clipped endless articles on the entrance of intellectuals into the establishment, their increasing celebration of American life and their repudiation of those whom they considered part of an obsolete adversarial culture. Dwight's marginal notes and markings indicate a suspicion of men like Jacques Barzun and Lionel Trilling at Columbia, both of whom were ridiculing the grumblers in the intellectual community. *Time* in a feature story noted with enthusiasm that "The Man of Protest has to some extent given way to the Man of Affirmation." The United States, crowed Barzun, is "the world power which means the center of world awareness: it was Europe that was provincial." This kind of smug academic self-congratulation infuriated Dwight. Like the celebrators, he too had come out of the Stalinist–anti-Stalinist wars with a chastened view of the world and the diminished possibilities of an effective politics. He too saw the Soviet Union as a threat to world civilization, but he could not stomach the arrogance of the intellectual elite and he did not identify with them. When the *Time* article sought out Trilling, Barzun, Niebuhr, even Irving Kristol to gloat about the entry of the American intellectual into the life of the nation as a supporter rather than a critic, Dwight relished the notion that he could and would continue to be a naysayer, the brusque (even crude, if necessary) outsider.[24]

Dwight remained in Evanston teaching until well into June. He returned to New York briefly before his scheduled departure for London on June 23. While in New York, he made a trip down to Washington to testify in behalf of the Workers Party's case to have their name removed from the attorney general's subversives list. It turned out to be a typical Macdonald episode. Dwight was glad to declare that there was nothing subversive about the Workers Party, but at the same time he insisted on dredging up all his own complaints about their rigid orthodoxy, undemocratic procedures, and refusal to entertain dissent within the organization. At first Joseph Rauh, their lawyer, hesitated to have Dwight testify, but then it occurred to him that Dwight's criticism would lend credibility to the party's case. Dwight did testify, did deny any subversion, and did assault the organization for its rigid, undemocratic practices. On leaving the courtroom, he got into a shouting match with Albert Glotzer, an old

Trotskyist comrade. It was a typical performance, and by that date, despite all the old loyalties, most involved were amused by Dwight's continued truculence and his feisty willingness to testify. Sidney Hook, James T. Farrell, and even C. Wright Mills refused to do so for a variety of reasons. In the end it was Daniel Bell, Norman Thomas, and Dwight who answered the call.[25]

Directly after his testimony, Dwight, Gloria, and her two daughters embarked on the French Line's *Flandre* for France and then on to London to take up his position on *Encounter*. Gloria had had an operation, which appeared to be successful, and she and the two girls were looking forward to the trip. On their arrival in London they found spacious accommodations at 14 Carlyle Square in the Chelsea section, two floors in a fine big house on a block where Whistler, Carlyle, Henry James, and Sir Thomas More had lived. From the start, Dwight was enchanted by England and things English. It became a symbol of resistance to modernity and mass culture. He saw London and England as the antidote to New York and America. He was determined to like the English and their culture. The scale of London was not dehumanizing; one could stroll around and feel like a man rather than an ant as in New York. It was not a grand city like Paris or Rome. Rather, it was a private place in which one was obliged to "poke about to discover the good things." Dwight was particularly taken with the Sir John Soane Museum off Lincoln's Inn Fields, a perfect expression of one man's personal taste. The city was full of these gems. It was great fun to walk around London: there were so many crescents, circles, ovals, and squares, courts and circuses, terraces, gardens, groves, places, roads, alleys, lanes, rows, and mews. Its very lack of plan made every outing an adventure.[26]

Dwight and Gloria spent days walking through the city, lunching in the pubs, reading in the parks. To him, Central Park was a barren heath compared with the great London parks with their vast lawns and flower beds. As for the people, it was a myth that they were "a reserved and phlegmatic race, hostile to strangers and talking in clipped monotones if at all." In fact, their voices rose and fell like Italians'. A London sherry hour "sounded like an aviary compared to the steady surf-like hum of a New York cocktail party." Dwight was astonished by the courtesy of policemen, waiters, taxi drivers, even bus conductors, all of whom, he thought, were notoriously rude in New York.

Dwight remarked on the tension in Manhattan, with its perpetual "atmosphere of potential violence." He had never experienced violence in New York, but he insisted that "the undertone" was there. The unnerving phenomenon of people passing a body sprawled out on the sidewalk in New York, blithely insisting that it was just another drunk, was inconceivable in London, where people felt it their duty to help those in trouble. What accounted for New York's lack of community, this cold inhumanity? After all, it and London were both enormous cities. But London was a collection of small towns. The open plan of New York promoted "en masse living." Even the contemptuous portrayal of the English as a nation of shopkeepers was in error. Dwight noted that, on the contrary, they were haphazard entrepreneurs. They displayed goods poorly and their shops were closed at the strange whim of the proprietors. In New York, the quest for the dollar ensured the availability of goods at all times of the day and night. The English money system was designed to make a financial transaction difficult and time consuming. But the pace of London was far more relaxed and the citizens more at ease.[27]

The social life could not have been more to Dwight's gregarious taste. He and Gloria were absorbed in the literary and intellectual life of the city. He was introduced to and became friendly with Isaiah Berlin, Philip Toynbee, T. S. Eliot, Victor Gollancz, Cyril Connolly, V. S. Pritchett, Francis Haskell, and Richard Wollheim, who later married Gloria's daughter Mary Day. Dwight was invited to join a weekly luncheon club at a restaurant on Charlotte Street in Soho, Bertorelli's. He attended a dinner at the Saville Club to honor Robert Frost, made a London friend of C. Day Lewis, and met E. M. Forster.[28]

Dwight was very well received in London. One Londoner, Margot Walmsley, the office manager at *Encounter*, felt he was viewed as a "great big American cowboy . . . people liked him enormously." They responded to his youthful enthusiasm, his inexhaustible energy, and his receptive appreciation of all things British. They found his self-criticism and skeptical attitude toward his own country refreshing—"so open, so embracing, so generous." Baba Anrep, a delightfully free spirit and daughter of the Russian-born mosaicist and painter Boris Anrep, recalls how Dwight paid attention to her, listened to her ideas, invited her to his parties. She was quickly drawn into his and Gloria's orbit and remembers spirited gatherings at the Macdonald flat as much more lively than the typical English social evening. She loved Dwight because ideas really mattered to him. He had none of the jaded sophistication that was the mark of many a London intellectual.[29]

Irving Kristol recalled that Dwight's literary set was a modern-day

Bloomsbury group with a similar tone of effeminacy, whereas Kristol, whose focus was more on contemporary politics, spent his time with a younger generation of Tory journalists connected with the *London Observer*. Kristol saw Dwight as a naif, a New York provincial with little interest in or understanding of British culture. He noted that Dwight never failed to pronounce the *s* in Grosvenor Square, turning it into a three-syllable word.

Dwight's cultural criticism irritated Kristol because it was, as Norman Podhoretz later pointed out, a substitute for political criticism. Dwight's assaults on American popular culture were a form of anti-Americanism that was anathema to the ideology of the Congress for Cultural Freedom and its loyal editors. Kristol insisted that Dwight's distance from the conservative journalists whom Kristol believed to be at the center of British intellectual life was a form of elitist snobbery. Dwight failed to experience British life with any understanding—he never even spoke to a Cockney, Kristol charged. Actually, Dwight made frequent trips to East End pubs with British friends and enjoyed wandering through the poor as well as the affluent British neighborhoods. As for Dwight's snobbery, Kristol's friend Daniel Bell was amused by Kristol's efforts to be more English than the English. In a skit he wrote for a Paris office party of the CCF he presented Kristol as "faultlessly attired in a black suit, black bowler hat, striped tie (Winchester), black tightly furled umbrella. A thin blond moustache, slightly curled, adorns his lower lip."[30]

On the matter of Dwight's provincialism, Daniel Bell also remembered Dwight's mispronunciation of Grosvenor Square. The error was apparently a standard reference point among Dwight's critics indicative of his alleged provincialism. Bell too insisted that Dwight could not grasp the texture of life in England, particularly the nature of its social classes. He argued that Dwight had no familiarity with the "club land" at the crux of English social life. Academics and intellectuals met in common rooms and clubs such as White's, Boodle's, or the Garrick. Bell had lived at the Garrick on and off for thirty years, and insisted that on this level Dwight was an outsider and had no inkling of that side of English life visible through such old school tie establishments. Upon reflection, Bell conceded that Dwight, not being Jewish and not coming out of an immigrant world, saw this side of English class structure in a dissimilar way. Bell remembered that he, Kristol, and Melvin Lasky, were of eastern European Jewish backgrounds and were overwhelmed by British high culture and by Oxbridge intellectual life. On the other hand, "Dwight coming out of

Yale and such never wholly felt" such awe. Dwight's emphatic Americanism and boyish wonderment at many things British reflected a different
cultural perspective. As a product of the American high cultural establishment, Exeter, and Yale, Dwight tended to poke fun at the pomposity of
the British academy, much as he did the American. Actually, Bell thought
the Yale atmosphere when Dwight went there was probably closer to
Oxbridge than any other place in America, and thus speculated that
Dwight's brusque manner may have been in part a pose as the innocent
abroad.

Bell too seemed to be unaware that Dwight's background in English literary history was formidable. People like T. S. Eliot, Stephen Spender,
Isaiah Berlin, and Malcolm Muggeridge respected Dwight's intellect and
learning; he certainly became a much closer compatriot of these men,
among others, than did his associate editor. They did not find him the
boorish hick his ideological opponents portrayed. Malcolm Muggeridge
liked Dwight because he pursued his essays with the commitment of a
scholar "but a comic one, which is the only honorable kind." Daniel Bell,
on the other hand, was put off by Dwight's constant spoofing and his
refusal to take the pretensions of academia seriously. On balance it would
appear that later memories of Dwight in London by his American associates, so dedicated to the American civilizing mission abroad, reflect their
own career and professional aspirations more than they do Dwight's personal reception among the British.[31]

Despite his anglophilia, Dwight was quick to realize that England was not
immune to the spreading ooze of mass- and mid-cult. Not long after his
arrival he noted that as in America, a poorly written book by an unknown
and muddleheaded amateur could be acclaimed as brilliant by a few
prestigious critics and become an instant success. In the summer of 1956
he wrote another of his famous demolition jobs, about *The Outsider* by
Colin Wilson—for the *New Yorker*, not for *Encounter*, which perhaps did
not want to publish such a sweeping assault on British criticism.
Although Dwight was to meet the young author and find him quite lik-
able, he nevertheless felt that his autodidactic work was humorless,
unbearably pretentious, and superficial.[32]

Colin Wilson, twenty-five years old when his book came out, came
from a working-class background and was for the most part self-educated.
He had attempted to absorb an enormous amount of reading in philosophy, religion, and literature and had come up with a theory concerning

the modern human predicament. It was the kind of book that became part of a trend in the postwar years, when a horrendous chain of events had struck a terrible blow to Western confidence, when theories of the absurd and the tragedy of human existence preoccupied young professors and their graduate students, and when the charting of despair along with Kierkegaardian solutions could be purchased at the checkout counter in supermarkets. Norman Vincent Peale's positive thinking was being replaced by Kierkegaard's *Fear and Trembling* and *Sickness unto Death*. One critic said Wilson was a mixture of both: a kind of instant despair for the handy homemaker. But the sense of the world's irrationality was genuine and deep; hopelessness about the human condition was widespread. Existentialism became a vogue. Students and their professors were devouring Sartre and Camus, *Nausea* and *The Plague* and especially *The Stranger*, with its unforgettable opening, "Mother died today. Or perhaps yesterday, I don't know."

Colin Wilson's book attempted to respond to, even be a part of, this sensibility of desperation in which (perhaps in his own Kierkegaardian way) salvation could be found "on the far side of despair." In a world without meaning, a secular or literary religiosity became attractive. It involved a stock acceptance of absurdity and the tragic human predicament where the inquiring mind of man was, as Sartre asserted, "a useless passion." In effect, the sin of pride became as familiar as Niebuhr's hope for a renewed reception of original sin as a political as well as theological concept.

One might have thought that Dwight, while rejecting the religiosity in Wilson, could nevertheless muster some compassion for the young man's sensibility, or at least an appreciation of his risk-taking and repudiation of mechanistic rationalism, which had once characterized his own youthful intellectual awakenings. Dwight's earlier speculations on the darker side of human existence, his abandonment of Hook and Dewey in the immediate postwar years, suggest a receptivity to the notion of a meaningless world. But it appears that Dwight had long left that stage. He now found it difficult to take these young critics of contemporary civilization very seriously. He was antagonistic toward the "Angry Young Men" of Great Britain, at least insofar as they were represented by *The Outsider* and by John Osborne's play *Look Back in Anger*, which opened the same week *The Outsider* was released. Dwight treated the complaints of its young protagonist, Jimmy Porter, as the petulant ravings of a "psychologically sick person." (This was, incidentally also the response of *Time* editors toward the Beats—Jack Kerouac, Allen Ginsberg, Gregory Corso—a group Dwight had little respect for.) He castigated Osborne for taking his character seriously and treating him as some kind of "red-brick Prometheus, but Jimmy's actual behavior is that of a self-pitying, bad-tempered mediocrity who lacks the strength and talent to live up to his aspirations."

While there was some truth to this criticism, it reveals a lack of both compassion or even interest in the alienation of a younger generation who found the smug complacency of postwar Britain stultifying. As one British critic, Kenneth Alsop, noted, Wilson had taken into account such writers as Sartre, Camus, Samuel Beckett, and Herman Hesse. He had tapped the deep vein of continental nihilism and pessimism. Despite Dwight's charge of "shoddy prose," he might have appreciated Wilson's rejection of a society dedicated to "producing better and better refrigerators, wider and wider cinema screens, and steadily draining men of all sense of a life of the spirit." Alsop argued that Wilson's book would be read long after its publication, because it reflected the time and the feeling of the age. The irony is that this free-floating rebellion against authority, not dissimilar to Dwight's stance for so long, came just as Dwight was taking up the cudgels in defense of community and tradition. In what amounted to his "reactionary mood," he had little patience for the impatience of the Osbornes, the Amises, and the Wilsons, who were a thorn in the side of the custodians of cultural and political orthodoxy.[33]

Nor did Dwight have much interest in the more celebrated literature of despair. He recorded in his London journal that he could not read Camus, Sartre, or Beauvoir. At first he attributed his antipathy to Beauvoir's novel *The Mandarins* as due to its poor translation or simply bad writing. But then as he mused on the whole genre of literature coming out of postwar France he noted that he couldn't read Camus's *The Rebel* or *The Myth of Sisyphus*. He didn't see why he *should* be reading them. Dwight wondered to himself what the attraction was and then declared that he could not read these writers because he didn't share their despair.

> I simply cannot imagine it, it means nothing to me. Why Not? I suppose because I am American. Life seems to me tragic, comic, pleasurable, exciting, boring—but never like death. I am beginning to realize I have been an optimist all my life without ever realizing it. The world of bringing up children seems not to exist in Camus and de Beauvoir.[34]

These private ruminations reveal the postwar generational conflict.
Dwight didn't share the same sense of bewilderment and loss of bearings
that was so often expressed in the reading tastes and declarations of the
younger generation in the angst-ridden bomb-fearing years. He had found
postwar politics perplexing, and was often given to bouts of personal
despair, but he was impatient with youthful ennui and what he felt was a
self-indulgent and, worse, self-pitying tone. This explains in part the
intensity of his assault on Colin Wilson. But there was another aspect to
it. Dwight had commented earlier in his profile of Alfred Barr on the pro-
liferation of classic reproductions in the homes of middle-class people
with no real appreciation of high art but only a sense that to hang a van
Gogh would give them class. In analyzing the whirlwind arrival of Colin
Wilson, he remarked, patronizingly, that the success of *The Outsider* was
just one more indication of "the growth of a public that habitually lives
beyond its cultural means." Wilson's readers were camp followers of the
avant-garde; they wanted to keep up with what was trendy and in the air.
Wilson, a philistine, was one of the barbarians who threatened the world
of high culture with his ersatz philosophizing and religious sermonizing.
Dwight labeled him the Norman Vincent Peale of the new class of poorly,
mass educated, culture vultures with little intellectual background or
equipment.

But what was most appalling to Dwight was how distinguished critics
had fallen for this banal, shoddy, superficial work. Even so prestigious a
critic as Cyril Connolly tolerated the "baggy prose" on the grounds that
Wilson had some important things to say. Philip Toynbee, Elizabeth
Bowen, and Edith Sitwell, among others, had all pronounced the book
significant, full of original insights and acute observations. They seemed
to think what it said was more important than how it said it. Dwight
insisted once again that "an idea doesn't exist apart from the words that
express it. Style is not an envelope enclosing a message; the envelope is
the message." He further claimed that Colin Wilson's prose style was so
debased, it could not convey any serious meaning. This had been
Dwight's ploy when dismissing the Revised Standard Version of the Bible.
But in fact for a great many readers, the message did stand above the
mode of expression. Colin Wilson, a young, ambitious amateur, addressed
himself to one of the major themes of the postwar world: how to live a
rational, fulfilling life in an irrational, absurd world.

Wilson may not have had Dwight's style and the wit but there is much
in *The Outsider* that expresses the same anxieties and hopes as the sec-

ond section of "The Root Is Man." But the mature Macdonald was diffident and apologetic about that piece, and he could see Wilson's treatise
and its success only as another example of the dreadful and corrupting
power of mid-cult.[35]

Not long after he completed his study of the Colin Wilson phenomenon,
Dwight finally took one of the trips that had been the initial purpose of
his role on *Encounter*. In November 1956 he traveled to Egypt to report
on "the Nasser Revolution." Ostensibly the piece was to be a study of the
alleged social reforms instituted by Gamal Abdel Nasser and the nationalistic movement he led. However, in July the United States had reneged
on a promised loan to Egypt for the building of the Aswan Dam. Nasser
responded by seizing the Suez Canal to use its tolls as a means of financing the dam. Dwight reached Cairo at the moment the British and French,
much more affected by the seizure of the canal than the United States,
had come to the end of their patience. Just two days after his arrival, the
Israelis, collaborating with the British and the French, attacked Egyptian
border towns, and four days later British planes began bombing the out-
skirts of Cairo.

One might have thought he would be delighted at being on the spot
with such news breaking. However, his correspondence and writing about
his stay in Cairo and travels in Egypt shows a remarkable lack of reporto-
rial interest. He did not travel to the area of the bombing, did not in fact
leave the city for the war zone, and when he did leave, it was to tour the
ancient tombs and temples at Thebes and El Karnak. Dwight reported his
fear when two "tremendous slamming explosions seemed close by the
hotel" in Cairo and his loneliness during the blacked-out nights. He left
the city and stayed in the suburb of Maadi with an American friend who
had married an Egyptian economist. For several days Dwight was completely out of contact with the *Encounter* office. From his colleagues' perspective he had missed a great reporting opportunity. It was a joke in the
office; Irving Kristol recalled snickering over the vision of Dwight sitting
out the affair in a Cairo suburb, missing it all, knowing nothing, seeing
nothing. The episode confirmed his view that Dwight was not someone to
be taken seriously.[36]

Dwight did make the rounds of government offices and interviewed
several officials, including Anwar el-Sadat. While the focus of Dwight's
interest was not on the immediate Suez crisis, he did talk to some Egyptian intellectuals and recorded their appreciation of Nasser's nationalist

challenge to Western imperialism. He reported that he had not found "one Egyptian who wasn't delighted about the canal seizure; they all regard it as a token of Egypt's at last becoming a real independent country, able to assert her national rights even against the English." Given the nationalist fervor, there was fear of street mobs and potential anti-foreign violence. But Dwight insisted that the people on the street were invariably courteous and friendly, even when they discovered he was an American journalist. He experienced a city well under control and carefully policed. The government was determined to have no "incidents."[37]

Dwight's instinctive need to challenge Western arrogance encouraged him to see the Egyptians in a sympathetic light. He was always irked by the smug superiority of the Western nation-states and their condescension toward such lesser beings as the Arabs. On his return he wrote a letter to the editor of the *Times* of London refuting their accounts of Egyptian mistreatment of enemy aliens and alleged brutality. Although the provocation was great, he had walked about the city with complete freedom; people were friendly and gave him directions. Dwight argued that the good temper of the Egyptians and the policy of their government protected the British citizens in Cairo despite the fact that the British were openly contemptuous of the citizens. Dwight denounced the "foolish and disgraceful" refusal of the Anglo-French troops to leave Port Said, despite repeated admonitions by the United Nations. The *Times* did not print his letter.[38]

Dwight made no bones about his strong opposition to British policy. He wrote Gloria that Prime Minister Anthony Eden was considered "War Criminal #1 in Cairo" and that he agreed with that assessment. In his published report he commented on the ignorance of those who had conceived the Anglo–French intervention. Instead of overthrowing Nasser, they had strengthened his position. While critical of the Nasser regime, Dwight insisted that it wasn't a "Farouked up Egypt." Nasser's Egypt combined democratic reforms (or, he conceded, a skillfully contrived show of them) with extreme nationalism. He quoted critics who saw Nasser as moving toward totalitarianism, but it was Dwight's belief that the regime was popular as much for its extreme nationalism as for its reforms. "For two thousand years the Egyptians have been ruled by foreigners—Persians, Macedonians, Romans, Arabs, Turks, Englishmen. They are just beginning to be masters in their own house and the feeling is sweet." It may have been these sentiments that most irked Irving Kristol and his allies at the *Encounter* office. Their cold warriorism had long since rejected such an analysis. At a party at the Kristols' after he returned, Dwight ridiculed a conservative British intellectual's defense of the British and French intervention. From Kristol's perspective, this supported his belief that Dwight understood nothing of British politics; he remained a naive provincial to the end. But it was Dwight who, almost instinctively, understood what was happening in the Middle East, perhaps more clearly than his conservative critics at *Encounter*.[39]

Dwight was not sorry to leave Egypt. He despised the "flunkydom—the subservience to higher-ups" that he noted among the government officials. It was a hierarchical society where the poor and dispossessed had no dignity and their fate was appalling. The poverty and filthy disorder of the older districts had reached nightmare proportions. He could not fathom how travelers could think this kind of thing picturesque.[40]

14

"Our Best Journalist"

Upon RETURNING TO LONDON FROM CAIRO EARLY IN March 1957, Dwight heard from Nancy that his mother was very ill and that Nancy was bearing a good deal of the responsibility concerning her care. His mother had lost most of her memory and was not really equipped to look after herself. She did not improve, so Dwight flew to New York. Feeling that she had lived a long and good life, he was not terribly upset. Her inability to speak was bothersome, but it made her seem "refined and spiritualized." He took comfort in how she accepted her imminent death without losing her old feistiness. She still managed to lift her chin "in that cocky way." She died March 27; Dwight returned to London shortly after the funeral.[1]

On his return Dwight seemed bent on demonstrating how far he had come from his earlier revolutionary politics by working on a memoir, "Politics Past," which he published in two installments in *Encounter* during the spring of 1957. It was later published in the States as the introductory essay to a collection of his articles that bore the self-mocking title *Memoirs of a Revolutionist*. The tone of self-deprecation, particularly in the first installment, seemed designed to make it clear that Dwight had indeed abandoned politics and was no longer concerned with the kinds of issues that had absorbed him in the late thirties and forties.

By politics, Dwight meant radical politics, which, he asserted, no longer held any attraction for his generation. Describing in detail his deradicalization, his essay served as a political confession. But it was not simply deradicalization; there was more than one reference to his appreciation of conservative thought. Sounding like Albert Jay Nock, who insisted that if it was not necessary to change it was necessary not to change, Dwight argued that "if tradition, privilege, custom and legality restrict, they also preserve." After the Russian experience, he found it hard to respond to the Bakunin slogan of "creative destruction." On the contrary, he was inclined "to endure familiar evils rather than risk unknown and possibly greater ones." Now even private property had a democratic aspect. Leftists ignored the fact that "a citizen with property has that much firmer a base of resistance to the encroachments of the State." He cited not only Jefferson and Madison but Burke, Tocqueville, and others who had been telling radicals this for two hundred years. Dwight insisted that "the revival of a true, principled conservatism . . . would be of the greatest value today. It is a task that would employ the special talents of intellectuals, and if it wouldn't give them the kick that revolutionism did, the hangover wouldn't be as bad either."[2]

Dwight had been for some time insisting on his inherent conservatism and had begun to label himself a "conservative anarchist." The "anarchist" referred to his radical individualism and repudiation of Marxist analysis. It was not long before he was to thank God that his Trotskyist politics had failed. The anarchism also reflected his continued anti-authoritarianism—he still railed at the establishment in politics, education, the unions, while ironically defending his own role as an arbiter of taste in cultural matters. It is thus not surprising that T. S. Eliot, with whom Dwight had a lengthy correspondence and saw often in London, wrote to note just how much Dwight's "form of radicalism [had] in common with [Eliot's] own form of conservatism." Eliot noted how they were both "equally removed from actual politics, either of the left or the right." Even where they did disagree, Eliot said, they were on the same plane of discourse. Eliot insisted that Dwight's lingering respect for socialism was really "theoretical anarchism" rather than socialism. Of course, Eliot's insistence on a religious base for values was in marked contrast to Dwight's individualistic secular morality.[3]

Dwight's account of his past, his amused distance from his earlier passions and commitments, his skepticism toward the working masses—all of this tended to reinforce his departure from the ranks of political radicalism. The second installment had more substance and seriousness;

there was also a hint of nostalgia. Dwight acknowledged that the Trotskyist experience had been "excellent training in political thinking," and he
lamented the prevailing political apathy. Much of their behavior seemed
from the vantage point of 1957 "absurd and even mildly insane. . . . But if
our actions compared grotesquely to our aspirations, if we were intellectually arrogant and morally more than a little smug, we did believe in a
great cause and we did make real sacrifices for it."[4]

Dwight was not so dismissive of his *Politics* years. The journal had
reflected what he felt was his "natural bent toward individualism, empiri-
cism, moralism, estheticism—all cardinal sins, that is deviations, in the
Marxian canon." The magazine's success lay in its openness, its refusal to
have any formula. Dwight was especially proud of having introduced
important writers and intellectuals to his subscribers and of offering a
platform to writers who could not find a place in the "professional specialized world of modern culture." Despite the upbeat memories of editing *Politics*, Dwight, in what appears to be an unpublished conclusion,
reflected the almost smug complacency that characterized the End of Ideology school of Daniel Bell, Edward Shils, Arthur Schlesinger, Jr., and
others. He wrote that in the United States, poverty had become anachro-
nistic, the "common people had the highest living standards in world history," and that what seemed to be dispiriting was the quality of life. These
recipients of material rewards looked "more strained, neurotic, unhappy
than the poverty-struck Italians." All of this had drawn his attention to
what he saw as somehow nonpolitical, or separate from politics, "social-
cultural reportage and analysis."[5]

Dwight's autobiographical reminiscences were well received by his
colleagues at *Encounter*, but there was some criticism of his flippant dismissal of radical politics. Oddly enough, Diana Trilling, in the *New
Leader*, took Dwight to task for denuding life, "robbing everything,
including himself, of importance." She could find nothing that explained
why people of the thirties were drawn into radical politics:

> the suffering and fear of the first years of that decade, the grim idiot face
> which capitalism presented . . . the political developments in Europe
> and the hope with which one looked to revolution in Germany and the
> despair with which one saw the defeat of hope in the first Triumphs of
> Hitlerism; the concentration of all one's imagination of progress and
> enlightenment in one's image of the Soviet Union, and the support given
> these generous fantasies by Roosevelt's rejection of the old attitudes of

conservatism—of all this and so much more that was happening in the politics and culture of that time Mr. Macdonald has literally nothing to say, and less than nothing to suggest. . . . It is a betrayal of history and of its author's contemporaries . . . it is a betrayal of Macdonald too.[6]

Dwight responded by agreeing that the first installment was sketchy and written in an ironic and comic tone, which he insisted was appropriate to an account of sectarian politics in the late 1930s. The second installment on the 1940s was more serious because the times had been more serious. Perhaps unfairly, he compared Trilling's notion of "betrayal" with the old leftist charge of "renegade" against anyone who had second thoughts about the radical position and refused to go along. Trilling did not let this vulgarity pass. The forties were not a more serious decade than the thirties; Dwight was embarrassed by his brief acceptance of Marxism, thus he could only treat it with irony and humor. His frivolity was a "form of self defensiveness." Unfortunately, he had proved that a man can be "both frivolous and a prig." Dwight's unrestrained sense of humor, his readiness to make himself the butt of the joke, his refusal to take himself with solemn seriousness even if the ideas and causes were important invariably irritated his former comrades who were not ready to dismiss their past allegiances with such seeming casualness.[7]

As Dwight wound up his tenure at *Encounter* in the spring of 1957, he wrote a long essay review of three works dealing with the whole issue of mass culture: Bernard Rosenberg and David Manning White's huge anthology *Mass Culture: The Popular Arts in America*; the Englishman Francis Williams's *Dangerous Estate: The Anatomy of Newspapers*, a discussion of the daily press in England; and Richard Hoggart's *The Uses of Literacy*, which dealt with the impact of mass culture on the British working class. Dwight was not pleased with the Rosenberg and White anthology, which contained his own *Diogenes* article. It reflected, he declared, the three bad ways to write about mass culture: the Academic-Cautious, Academic-Baroque, and the Pale Marxist. The first produced tediously documented surveys on who read what, replete with "statistical tables and other academic instruments of torture to produce an innocuous conclusion understood by all before reading the article." The Academic-Baroque was that "pseudo-scholarly style of writing" of pompous pretension. The Pale Marxists were the legion of Cassandras who discerned a terrible conspiracy among the ruling class to corrupt the minds and morals of the masses. He thought *Dangerous Estate* was no better, an ill-written, superficial history.[8]

The one person he accorded some respect to in this omnibus review was the cultural critic Richard Hoggart. What attracted Dwight to Hoggart was his lack of academic pretension, his avoidance of jargon, and his concrete description of just how the British working classes were affected by and responded to mass culture. Dwight's endorsement of Hoggart is interesting because Hoggart had a conception of culture that was a good deal broader than Dwight's. It was based on anthropological foundations and was influenced by Marxist-oriented social historians. Dwight found Hoggart a serious, thoughtful, and conscientious writer. He appreciated Hoggart's ability to reveal the details of working-class mores, drawn from his own boyhood.

Dwight learned something from Hoggart. For while Dwight had much to say about the deplorable taste of the masses, and was contemptuous of their easy accommodation to the consumer society, he really knew nothing of working-class life. He had lost touch with most of his radical comrades from the unions and the sectarian political movements. Hoggart's pungent descriptions as well as compassionate understanding of the artifacts of working-class culture, the gewgaws in the kitchens, the chintz curtains, caught Dwight's attention. Given Dwight's developing theories on the menace of "mid-cult," that is, an ersatz high culture, watered down for the poorly educated middle class, it is not surprising that he was taken with Hoggart's critique of what Dwight now called the "camp followers of the avant-garde." These "half-educated" had "a yearning for culture but neither the intellectual equipment nor the will to be serious about it." In effect, Dwight attempted to enlist Hoggart into the elite who wanted to keep the barbarians outside the gates.

Hoggart's depiction of ill-equipped culture seekers was indeed devastating: "They wander in the immensely crowded and often delusive world of ideas like children in their first fairground House of Thrills—reluctant to leave, anxious to see and understand and respond to all, badly wanting to have a really enjoyable time, but, underneath, frightened." In Hoggart's criticism, however, there is a connection, an understanding and appreciation of the angst and frustrations of the upwardly mobile yet poorly educated citizen seeking entrance to the temple of high culture but at the cost of losing the "healthy untidiness and natural idiosyncrasy" and cluttered homeliness of their working-class lives.[9]

Dwight ignored that side of Hoggart's reflections, possibly because it clashed with his own combative defensiveness of his status. Hoggart wrote autobiographically about the tension between his working-class heritage and his position as a university don. Dwight had experienced an opposite tension: Born into the upper class, he had often attempted to appear alienated from his privileged background as a younger man. But, in the tradition of a Henry Adams, he had come to see his class and its culture as under the gun and in need of a defender.

This essay review on mass culture, his last extended piece for *Encounter* during his tenure as a visiting editor, was provocative enough to draw him further into the mounting mass culture debate. A critique by Edward Shils brought the issue into the political arena, where it was, despite some of the rhetoric, much closer to Dwight's interests and concerns. Shils was an academic sociologist at the University of Chicago, but known by and associated with many of the New York intellectuals. A staunch supporter of the liberal cold war ideology and a devout anti-Communist, he was essentially a yea-sayer who could be counted on to defend not only American foreign policy but also American culture. In the mid-fifties, he was an important figure in the Congress for Cultural Freedom and one of the founding members of what became known as the End of Ideology movement. He was prepared to take on the critics of mass culture as purveying a form of anti-Americanism that did not serve the nation well during the ideological skirmishes of the cold war. Shils also could hold his own in the world of invective and verbal abuse inhabited by the New York intellectuals.

His critique, titled "Daydreams and Nightmares: Reflections on the Criticism of Mass Culture," was aimed at the left-wing foes of mass or popular culture. Dwight must have been pleased to see himself described as the person "who, as editor of *Politics*, did more than any other American writer to bring this interpretation of mass culture to the forefront of the attention of the intellectual public." Shils stressed Dwight's "Trotskyite Communist" past and linked him to the Frankfurt School of German sociologists. Shils claimed that the Frankfurt School and *Politics* constituted the "two fountainheads of a critical interpretation of mass culture that exhibits an image of modern man, of mass society and of man in past ages that has little factual basis." These critics of industrial society had created a nostalgic fantasy covering the primitive brutality of the pre-industrial age. The ex-Marxists' deep disappointment at the failure of the working classes to achieve socialism bred a desire for revenge, which explained their obsessive interest in exposing the low state of popular taste.[10] Looking back on the subject, Alfred Kazin thought "Shils was right" when he wrote "that the ex-Marxists adore the subject of 'mass culture,' because it gives them the chance to show that the people were wrong, not themselves."[11]

By insisting that the culture critics' position was political and ideological, Shils was getting at the heart of the passion and intensity of the arguments, but his belief in the importance of Marxism and the Frankfurt School to Dwight is off base. Dwight was influenced much more by such earlier conservative critics of mass culture as Tocqueville, Ortega y Gasset, and T. S. Eliot, whose approach was both aristocratic and aesthetic in outlook. Nor would it be reasonable to say that Dwight had ever harbored great expectations for the working class or the "masses." It is true that in his initial *Politics* article on a theory of popular culture he rejected the elitist notion of two separate cultures, referring to his own democratic conviction and arguing that "the standard by which to measure Popular Culture is not the old aristocratic High Culture, but rather a potential new *human* culture, in Trotsky's phrase, which for the first time in history has a chance of superseding the *class* cultures of the present and the past." But Dwight had repudiated that sentiment soon afterward, and by 1953, in the *Diogenes* version of his theory, he had edged toward the idea of two cultures, one for the masses and one for the classes. By the time of his review of the Rosenberg and White volume in 1957, he boastfully spoke of his willingness to sneer at the taste of the working classes.[12]

Dwight's tenure at *Encounter* ended in June 1957, when he, Gloria, Mary Day, and Muffin took off on a trip to Italy. Just three years after their marriage, his life with Gloria had settled into a pattern. He had written Nicola Chiaromonte that while his sexual life was "more divine than ever," his day-to-day life, which he had not given much thought to, was difficult and "constantly breaking into painful fragments." It appears that he was often irritated by Gloria's guileless lack of intellectuality. Although he had praised her judgment during the courtship, he often showed little respect for her opinions and was frequently rude and even abusive in his criticisms. Not surprisingly, these irritations led to rows. Gloria was not passive and accommodating. She did have a mind of her own; she had literary, artistic, and, more to the point, political opinions. On an emotional level she was more tolerant of the left, even the Stalinist left, than Dwight.

She was embarrassed by his red-baiting and often rebuked him for it. He felt hers was a knee-jerk lib-lab response to political issues. In one of his written analyses of their relationship, he expressed anger over her apologetics concerning the Russian invasion of Finland and wondered why Gloria's "anti-capitalism always takes a pro-Communist slant and not a Norman Thomas socialist one." He told Mary McCarthy stories about the ridiculous things Gloria would say, and McCarthy would frequently exchange tales of Gloria's absurd comments and conversational gaffes with Chiaromonte. McCarthy claimed that Dwight kept a written record of Gloria stories, because they were so outlandish no one would believe them. However, since Dwight kept everything and no such record is in his papers, it seems unlikely it ever existed.[13]

After a short stay in Siena with the family of his friend Arturo Vivante, they traveled about sixty miles west to the tiny fishing village of Bocca di Magra at the mouth of the Magra River, which empties into the Ligurian Sea. Dwight had learned of the place from Nicola Chiaromonte, who had spent summer vacations there for many years. Mary McCarthy, who was later to spend many summer months there with the Chiaromontes, has left a memorable portrait of the village as an almost utopian community frequented by intellectuals from France, Italy, England, Ireland, and America. Everyone met for aperitifs before dinner and to while away beautiful summer evenings in spirited conversation. There was a secluded beach in a cove with a great marble rock, which could be reached only by small boats piloted by the local fishermen.[14]

Dwight was taken with the place, and he and Gloria were to return on several occasions. For Dwight, Mary McCarthy, and their friends the Chiaromontes, the Tuccis, Isaiah Berlin, and several others, Bocca di Magra was a communal retreat for intellectuals—the old Brook Farm image that Dwight had idealized years before, when he dreamed of turning the Wheeler family farm in Ohio into a colony of like-minded writers and intellectuals. The idyllic spot was threatened by postwar development, however. Mary McCarthy recorded the "tale of encroachment" in her intricate analysis of summer retreats that formed the thesis of her story "The Hounds of Summer." The very tides of humanity, not to say the mob or the masses, were already beginning to lap at the village. But true to McCarthy's perceptive instinct, she found a way to penetrate the relationship between intellectuals and their temporary summer homes. Her story is a critique of that proprietary relationship which in the end is hardly flattering to her friends. It was quite reminiscent of her novella

The Oasis, in which she had caricatured the earnest ineffectiveness of intellectuals.

Bocca di Magra stood in the inevitable path of Dwight's obsession, "the spreading ooze of mass culture." Commercialization, even the mass production of vacations for mass man, was approaching. People were finding their way to the secret place, bringing with them the flotsam and jetsam of nightclubs, cheap hotels, neon-lit seafood restaurants, ad infinitum. They could be seen already sprouting up across the Magra River; it was inevitable that they would soon threaten Bocca di Magra. So bad was the situation that McCarthy, Dwight, and several others organized a kind of improvement association to try to thwart further development.

Gloria found this an absurdity. How could left-leaning intellectuals devote themselves to stifling the aspirations of the hoi polloi? Gloria irked Dwight by pointing out that his and McCarthy's insistence on maintaining the purity of the place was typical of the bourgeois American's notion of proprietorship. It also involved the tacit assumption that one of the main causes of the problem was the breakdown of class barriers, a development they both lamented and exaggerated.

Interestingly enough, as much as Mary McCarthy found Gloria banal, coarse, and vapid, in her own account she did satirize the condescending relationship between the summer people and the poor citizens of "this quaint little fishing village," as they described Bocca di Magra to their hairdressers and liquor store salesmen. Her story was not simply the story of the summer people of Bocca di Magra; it was about summer retreats all over the Western world, where intellectuals and professional people sought to escape their modern urban environment, to even re-create, if necessary, more primitive rural amenities, all in the name of a precious authenticity. Often, McCarthy noted, they used the word *unspoiled* to describe the village, a term they would not use in front of the poor local citizens, "whose interest," she noted, "lay in having it spoiled as fast as possible." There was something suspect, McCarthy noted, "in laudatory remarks that cannot be made in the presence of those being lauded." It was the very poverty of the local peasants that the intellectuals loved so much. Then again, she conceded, how else could one describe the place, "without going into detail, except as a fishing village on the Ligurian coast that had not yet been completely spoiled by tourists." It was a complicated issue, one that a veteran of the Cape, Fire Island, or Amagansett and Southampton could well appreciate. It was the day-to-day encounter with the reality of the formerly abstract concept of mass culture that portended that the inevitable destruction of a truly civilized society was close at hand, unless the remnant of taste and culture embodied by the intellectuals and professionals could maintain inviolate retreats necessary to renew the spirit and revive the energy needed to resist the spreading ooze.[15]

Dwight, Gloria, and the two girls returned to New York early in September. All suffered from culture shock. There was a vast difference between Europe and the United States in the 1950s. The Americanization of the Continent still had a long way to go, but the consumer culture had long since arrived in the United States. A drugstore in America, with its electronic toothbrushes, portable radios, hair dryers, endless gadgets stacked precariously on aisle dividers, was a wondrous display after foraging in London and Italy for the simplest everyday necessities. To Dwight it appeared that the average citizen was drowning in a sea of trivia and gaining no genuine satisfaction. He invariably contrasted bustling New York to Arcadia-like London and wrote friends that he longed to return and expected to find a way by the following year. By November he was increasingly depressed by the "rattrap." He did concede, however, that there was profitable work in New York. He did have friends in the city, the delis stayed open till one A.M. and one could buy whisky for $3.50 a bottle. "Not very GOOD whisky, but still—whisky."[16]

Dwight quickly settled down to work, returning to his mass culture analysis with long essay reviews for the *New Yorker*. The first, on James Agee's *A Death in the Family* was followed by an extended survey of recent books about film. Both pieces carried the message of impending doom. Agee had all the qualifications to become a great writer. He was a master craftsman. He had an old-fashioned interest in good and evil and could write about such things with precision and without sentimentality. There was no condescension in Agee, only a fine sensitivity for ambiguity and nuance. But he was dead at forty-six, having led a wasteful, self-destructive life. Why? In Dwight's view, Agee had failed to find what a writer of his talent had to have: tradition and community. Agee needed definition, limitation, discipline, but without a community of artists and intellectuals to "corset his sprawling talents," with no cultural family to moderate his eccentricities, he was a victim of his own excesses. Dwight was gradually developing a thesis reminiscent of the early Van Wyck Brooks, who had portrayed the poetic and sensitive artist as unable to survive in a mass industrial society.[17]

Before Dwight had left England, he had suggested to Irving Kristol that he might do a New York letter for *Encounter*. In October he reminded Kristol that he was still interested in reporting on life in New York and in the States in general, dealing with the civil rights struggle in Little Rock and discussing what was new in American culture. Kristol seemed receptive and even suggested that he might file the letter monthly. Dwight got quickly to work and produced an acerbic put-down of American society and its culture, especially as contrasted with England, Italy, France— anywhere Dwight had been in his year abroad.[18]

Dwight's piece was deliberately and provocatively anti-American. *Encounter* had accepted Dwight's satirical comments on the Milan Conference with reluctance. He had also caused a ruckus with an adversarial report on the censorship of an anti-Stalinist paper at the International Sociological Association, which had caused Kristol no little pain. This latest belligerent rumination from New York seemed consciously designed to draw their ire. Dwight described Americans as having achieved an equal distribution of wealth far beyond the imagination of Fourier, Proudhon, and Marx, yet they were an "unhappy people without style, without a sense of what is humanly satisfying." Why? Because American values were not securely anchored in the past (tradition), nor in the present (community). The United States, in contrast to England and Italy, was mired in a materialistic ethic; its people had neither manners nor morals; civility was absent, and individualism was translated into "I got mine and screw you, Jack." In effect, Americans did not know how to live together. As a result, one felt a constant sense of gratuitous violence in America—anything might happen. Previewing the sensibility of Saul Bellow's Mr. Sammler, Dwight painted a portrait of a society without limits, order, or decorum. In Europe there was a fundamental appreciation of community, each person differentiated by status and function but each a part of an orderly social structure. In America everybody was *equal* in the sense that nobody respected anybody else unless he had to by force majeure.

This narcissistic, selfish mass not only ignored tradition and community but violated the lovely landscape at every opportunity. The countryside had become a wasteland of hot dog stands, gas stations, motels alongside invading highways "that wind through the country like endless tapeworms"; destroying any structures out of the past, the cities were "vast deserts of the present."[19]

This unflattering portrait of the United States was standard fare during a good deal of the Eisenhower age. Another writer insisted that Americans traveled along in a stupor of fat and the only noise in the land was the "Oink" of greed. But a good deal of this criticism of America in the fifties was promoted by disgruntled Democrats irked by a stand-pat administration that had put the brakes on New Deal and Fair Deal reform, and Dwight's critique hardly came from that direction. He stood solidly in league with an aristocratic, elitist tradition that attacked mass society in the language of classic conservatism.

He also appeared to accept two of the major premises of the End of Ideology school: that the scarcity economy had been overcome and that the fundamental questions were no longer about quantity but about the quality of American life. He noted and seemed to lament the fantastic increase in wealth accompanied by what he saw as a decrease in economic differences. If "socialism be equal sharing of plenty, then we are far along the road. We have more of everything than a human being can conceivably and inconceivably want." Most Americans had far more than the necessities. It was their intellectual, cultural, and moral life that was wanting. People in the poorest quarters of Florence and in the marble factories in the hills above Bocca di Magra were infinitely happier, more a part of their communities and their national traditions, he declared. Of course, it is doubtful he knew much about those communities; there was some truth in Edward Shils's charge of romanticism.

One might add that Dwight knew very little, if anything, about the vast array of ethnic communities in New York. Manhattan's Little Italy had a sense of community and was steeped in tradition, as did the Italian Belmont section of the Bronx, and the Jewish communities on the Grand Concourse and Brownsville in Brooklyn. On the other hand, these communities, with their deep religious commitments and cultural mores, could be seen as representing the ethnic nationalism, parochialism, and narrow-minded provincialism that Dwight denounced and from which many of his intellectual colleagues had sought to escape in the name of a broad-minded, international cosmopolitanism. There were many contradictions in Dwight's impressionistic and idiosyncratic comments. What gave the essay its punch, its power, was its political implications. At a time when Henry Luce's notion of "the American Century" was a constant refrain, when the United States was calling the tune, Dwight's cultural assault was an act of spirited self-criticism that did challenge, at least on the cultural level, the prevailing complacency.

At the same time, he was busy at work on another demolition job, a review of the widely praised novel *By Love Possessed*, by the conservative

writer James Gould Cozzens. *By Love Possessed* launched Cozzens into
instant fame; it outsold the total of his previous eleven novels. He was on
the cover of *Time*, Hollywood paid $100,000 for the movie rights; the film
starred Lana Turner. In "America! America!" Dwight had compared the
enthusiastic raves for this novel of American manners to the Colin Wilson
phenomenon in England, citing both as examples of how the critics and
the mob find a common ground.[20]

The idea for a full-length review of the Cozzens novel was prompted
by Dwight's new friend, the historian John Lukacs, who had been
offended by the novel and found the raves for it absurd. Lukacs was a
student of the Pennsylvania aristocracy, the novel's subject, so he volun-
teered to review it for *Commentary*. The young Norman Podhoretz, soon
to become editor-in-chief, liked Lukacs's review, but his editorial colleagues rejected it. Lukacs then sent it to Dwight for his entertainment.
Dwight thought it was a good job that needed only a little pruning. He
sent it to Spender to see whether *Encounter* might be interested. They
weren't.[21]

In the meantime, Podhoretz had asked Dwight for an essay review of
the book and its wild reception. It was the perfect assignment for him, fitting as it did into his developing mass culture oeuvre. He was influenced
by Lukacs's criticism, but where Lukacs emphasized the intellectual
obscenity, Dwight focused more on the stylistic failings and the message
of bland accommodation to the status quo.

He quoted at length to illustrate Cozzens's failure to write literate
prose. The style was so artificial and unnecessarily complex that it
"approaches the impenetrable—indeed often achieves it." Dwight wondered how the "matrons," whom he patronizingly assumed were the most
susceptible to Book-of-the-Month Club selections and best-seller lists,
managed to cope with it. Cozzens, he charged, thought he was portraying
the trials and tribulations of a decent man when in fact he had produced
an insufferable prig. Dwight took Cozzens to task for having his male
characters speak brutally to women for no adequately explainable reasons. To Whitney Balliett's assertion that the protagonist was "a passionately good man," Dwight responded that he had "the passion of a bowl of
oatmeal." Dwight felt that one of the more dramatic sex passages sounded
partly like a "tongue-tied Dr. Johnson" but even more "like a *Fortune*
description of an industrial process."[22]

But the literary work itself was not the main target; it was the reviewers' enthusiastic reception and inflation of the work that drew Dwight's

initial interest and ire. He found the reviews by critics of established reputation particularly odious. "One can guard against the philistines outside the gates. It is when they get into the ivory tower that they are dangerous." For several paragraphs he quoted from such eminent critics as Brendan Gill in the *New Yorker*, John Fisher in *Harper's*, Malcolm Cowley in the *New York Times Book Review*, Jessamyn West in the *Herald Tribune*, Granville Hicks in the *New Leader*, Whitney Balliett in the *Saturday Review*. The encomiums were unrestrained. "A masterpiece . . . deserves the Nobel Prize . . . the work of one of the country's truly distinguished novelists . . . superbly satisfying, magnificent." Dwight noted that even so sensitive a critic as Richard Ellmann could write, in a more cautious review, that the novel was "a pleasure to read." Dwight responded that every sentence grated on his ear.

How could such rampant misjudgment happen? It was not simply a sign of the lowering of standards. There were two major factors, he wrote. The first was a widespread feeling that Cozzens, the author of eleven previous novels, had been shamefully neglected and deserved a success. But much more important, and in keeping with Dwight's understanding of the dangers of mass culture, the inflated reception of the book was the "latest episode in the Middlebrow Counter-Revolution." The promotion of Cozzens was another example of the MacLeish-Brooks charge that avantgarde literature was out of touch with the real people and the normal life of most citizens.

As was his custom, Dwight sent a copy of his review to Cozzens on the assumption that such an isolated and insulated anti-intellectual would probably miss *Commentary*. Although Cozzens was deeply injured by what he felt was Dwight's "venomous philippic," he wrote a droll and bemused letter feigning astonishment that an educated adult could be so unperceptive, so ignorant of prose style, that shades of meaning and irony were completely lost on him. Dwight's preference for the "stylistic claptrap, crypto-sentimentality and just plain childishness in the work of Hemingway, Lewis, Faulkner and perhaps Tolstoy rather than for the lucid thinking and perfect writing of Somerset Maugham only revealed [Dwight's] literary infirmity." However, Cozzens did give Dwight an A-minus for his mastery of invective. He was sure that "those little-mag people were just loving [Dwight] to death."[23]

They were indeed. Dwight received much mail in response to his "Cozzensiad," as he referred to it. He was quoted as saying he had never had so many letters about an article, and only three out of eighty or ninety

were on Cozzens's side. Isaiah Berlin wrote to say that his Cozzens attack
was an even greater masterpiece than his demolition of Colin Wilson.
Arthur Schlesinger had told Berlin that Cozzens was a good writer. "It
must be a splendid sensation to be able to say that the Emperor has no
clothes on time and time again, even more so when the Emperor thinks he
is clothed and when the courtiers genuinely think so too." Norman Mailer
wrote Dwight that he too had detested the book, but he thought it one of
Dwight's least successful reviews because Dwight's "animus was so complete that even [his] best insights began to be lost in the weight of the
bombardment." Dwight took pleasure in the assertion making the rounds
that his review had knocked Cozzens out of the running for the National
Book Award. He also appreciated the fact that *By Love Possessed* lost out
for the Pulitzer Prize to James Agee's *A Death in the Family*, which he
had described in his *New Yorker* review as a "nearly great book." His criticism had been effective.[24]

Just at the time that the Cozzens affair was making news in the literary
world, the responses began to come in to Dwight's own book, the collection of editorials, essays, and journalistic comment published as *Memoirs of a Revolutionist: Essays in Political Criticism.* Dwight chose the
title as an ironic comment on his earlier political enthusiasms. Coming
when it did, just as he was making a name for himself as a cultural
critic, the book raised the issue of the relationship of his politics to his
new role.[25]

Despite poor sales, the book was reviewed widely and for the most part
favorably.[26] All praised his lively prose style and the freshness it gave old
issues and past politics. Some of the critical notes bothered him, however.
Because the volume contained his wartime criticism and his essay "The
Responsibility of Peoples," the old issue of Dwight's moralism once again
came to the fore. He detected a patronizing, even condescending tone in
this criticism. "The wise old professors" (Daniel Bell, Richard Wollheim)
and the literary philosophers treated him as something of an unruly child.
Bell labeled him a "journalist *cum* intellectual and not a social scientist
or a philosopher. He takes him *self* as the starting point and relates the
world to his own sensibilities. Scientists accept an existing field of knowledge, and seek to map out the unexplored terrain." Richard Wollheim,
the British philosopher and aesthetician, said much the same thing and
labeled Dwight an innocent. Dwight confronted the world directly from
his own sensibilities, his own individual values, but he lacked a disci-

Macdonald family portrait, 1909: Theodore Dwight Macdonald and Alice Hedges Macdonald with Dwight, age three (*right*), and Hedges, age one. (*Courtesy of Nicholas Macdonald.*)

Dwight Macdonald, circa 1920. (*Courtesy of the Macdonald Papers, Yale University Library.*)

DWIGHT MACDONALD, JR.

NEW YORK, N. Y.

"Mac." "Dwight."

Entered Junior Year; Lantern Club (3), Vice-President; G. L. Soule Literary Society, Secretary; Outing Club (2), Vice-President; Chess Club (2), Vice-President; Chess Team; Christian Fraternity (4); Four-Year Club; Class Day Officers, Poet; The Merrill Prize in English Composition (Second); The Johnson-Ennis Prize; Fox Prize in English Composition; The Pitts Duffield Prize; Honor Man, Second Group; *The Exonian* (2), Secretary; *The Phillips Exeter Monthly* (2), Associate Editor, Secretary; College Preference, Yale.

Dwight Macdonald's Phillips Exeter Academy yearbook entry, 1924. (*Courtesy of the Macdonald Papers, Yale University Library.*)

Dwight Macdonald and Dinsmore Wheeler, circa 1935. (*Courtesy of Nancy Macdonald.*)

Dwight Macdonald, Alice Hedges Macdonald (*right*), and Nancy Macdonald, circa 1936. (*Courtesy of Nancy Macdonald.*)

Nancy Macdonald in Majorca, 1936.
(*Courtesy of Nancy Macdonald.*)

Nancy and Dwight Macdonald, circa 1937. (*Courtesy of Nancy Macdonald.*)

Closeup of Dwight Macdonald by Bertrand De Geofroy, 1940s. (*Courtesy of Nancy Macdonald.*)

Dwight Macdonald with Norman and Adele Mailer, Provincetown, Massachusetts, 1960. (*Courtesy of Sabina Lanier.*)

Dwight Macdonald, Robert Lowell, and an unidentified protestor at a Biafran demonstration in New York City in the late 1960s. (*Courtesy of Elizabeth Hardwick.*)

Dwight Macdonald speaking at the Counter Commencement at Columbia University,
June 4, 1968. (*Courtesy of the* New York Times.)

Dwight Macdonald with his son
Nicholas and grandsons Ethan
and Zachary in the early 1970s.
(*Courtesy of Nicholas Macdonald.*)

Dwight Macdonald and his son Michael. (*Courtesy of Nicholas Macdonald.*)

Dwight and Gloria Macdonald in Italy, 1957. (*Courtesy of Sabina Lanier.*)

Dwight and Gloria Macdonald, summer 1974. (*Photo by Ray Lindquist, courtesy of Ray Lindquist.*)

Dwight and Gloria Macdonald and Gloria's daughters, Mary Day Lanier and Sabina Lanier, in the 1970s. (*Courtesy of the Macdonald Papers, Yale University Library.*)

Dwight Macdonald, Robert Lowell, and William Sloane Coffin, Aspen, Colorado, 1964. (*Courtesy of Sabina Lanier.*)

pline; he had no theory, no framework, no developed conception or set of principles, Wollheim wrote.

It was the same old argument that had plagued Dwight when he challenged the Marxists in his essay "The Root Is Man." However, the tone now was one of benevolent head patting: "What a remarkably bright child." Bell, Dwight charged, saw himself as a worldly-wise person "who had to chuckle indulgently" at Dwight's "will of the wisp absolutist-utopian romantic ideas." Bell expressed admiration for Dwight's purity of heart (many praised Dwight's honesty and commitment) that led to such sweeping judgments, while at the same time shaking his head at such youthful folly. Dwight complained that this note was sounded over and over, even in the friendliest reviews. He would not accept the idea that his drastic rejection of "what is" necessarily meant "pure at heart but wooly wit." On the contrary, he insisted to both Bell and Wollheim that he was the one who had seriously attempted to think and to analyze quite objectively and that his "sensible old uncles" were "unwilling to go beyond a certain point in 'facing up to things.'" He charged Wollheim with being a scientized person who thinks "that 'aesthetic' and 'moral' are bad words and that politics is too serious a business to be left to those to whom such adjectives apply, while I think that the only serious aspect of politics is its relation to art and morality."[27]

The same sense of condescension could be found in the view that Dwight had performed an indispensable service as a political commentator but that because he had announced his loss of interest in politics, his book was a voice out of the past. Dwight was sensitive to the suggestion of some that he had sold out by going to the greener pastures of the *New Yorker*, where he was engaged in more genteel criticism. Many lamented his departure from the political scene and expressed the need of his acerbic, protesting negativism in the bland Eisenhower age of "the Nice Guy." Dwight complained to the philosopher of history Hans Meyerhoff that he had made Dwight's post-*Politics* career look like "a resigned acceptance of things."[28]

Dwight had brought this perception upon himself with the facetious title of the volume, which few reviewers failed to comment on. After all, in his Introduction he had referred to his radical politics, at least while in the Trotskyist movement, as "absurd and mildly insane." He had poked fun at the radicals who thought their sectarian battles were at the center of history. He was paying the price for what Diana Trilling had charged was a frivolous dismissal of serious ideological struggles. But Dwight saw this as effective, humorous satire with an important moral point and stressed his and his comrades' serious commitment to an important cause. His point was that he had not been cautious, circumspect, overly protective of career and reputation. On the contrary, he had taken intellectual risks.

In a very flattering review for the *New Republic*, Ian Watt, the British literary critic who had emigrated to the University of California at Berkeley, had written that Dwight in his political and literary criticism was a "kind of Holy Fool" who exhibited a rich combination of critical power, cultural awareness, and cantankerous hilarity. In a paragraph that was cut from his published review, he went on to note:

> Macdonald has genuine analytic power, but he is perhaps best as a satirist; with his quick allergy to cant he in many ways resembles Mencken, and has traces of the latter's tendency to view the spectacle of human inadequacy with disquieting relish. I would guess, though, that the bitterness of Macdonald's scornful and anarchic laughter springs from a foundered idealism, and that his closest analogue is the later Mark Twain. In any case his primary commitment to the Arnoldian role of being the disinterested critic of modern culture is surely what has saved him from collapsing into silence or obscurantism like so many of his politically-oriented contemporaries: his passionate concern for truth, allied to his considerable satiric gift, have armed him to face the accumulating disasters of the century with the pugnacious resilience of a Donald Duck.[29]

Dwight appreciated Watt's recognition that he had not thrown in the towel, that he was still girding for the battles. He rejected the "foundered idealism," however, insisting that he had always been "tough-minded, less open to illusions than Mark." On the question of Dwight's deep moral commitment combined with his satire, he took serious offense at John Lukacs's suggestion that his moral concern was admirably stronger than his unquestionable talent for satire, which made him appear to some sensitive souls as a smart-aleck. Despite the praising tone of Lukacs's judgments, Dwight was quick to reject Lukacs's implication that there was an antithesis between his moral concern and his talent for satire. On the contrary, he insisted, his satire was moral.[30]

Some among his old associates were less receptive, and, in the New York intellectual street fighter tradition, they knew how to dish it out. The

Winter 1958 issue of *Dissent*, Irving Howe's Democratic Socialist quarterly, published a two-barreled blast from the essayist and art critic Harold Rosenberg, who had lived for years in the same building as Dwight and Nancy on East 10th Street, and Dwight's former associate and *Politics* contributor Paul Goodman. These were two old "friends," but in that contentious milieu, criticism was criticism and it often sought the jugular. In a review of the Rosenberg and White anthology on mass culture, Harold Rosenberg led off with a contemptuous appraisal of Dwight as a hackneyed and tiresome critic of kitsch who wallowed in the very culture he decried. There was something supremely annoying "about the mental cast of those who keep handling the goods while denying any appetite for them," Rosenberg wrote, charging that "when Macdonald speaks against kitsch he seems to be speaking from the point of view of art; when he speaks about art it is plain his ideas are kitsch." Kitsch was nature; it would always be with us, Rosenberg declared; one might as well write a daily diatribe against the weather. The thing to do was to promote art. But in fact, Rosenberg charged, modern art revolted Dwight, "tempting him to get to know it about as much as would the gasses in a condemned mine."

Rosenberg was astute enough to recognize that Dwight was a "bloodbrother of Tocqueville, Nietzsche, Ortega" and not a devoted follower of the Marxist Frankfurt School. It may well have been this that most provoked Rosenberg. He noted that Ortega y Gasset "not only disliked mass art, he dislikes the masses themselves. They give him a sense of being crowded. Personally I dislike Ortega more than I do crowds."[31] By implication he surely meant to identify Dwight with the same brand of fear and contempt for the mob. He was on target: one of Dwight's favorite quotations from Ortega was his lament over mobs at places where once only the elite traversed.

In Dwight's response to Rosenberg, he was quick to wonder whether Rosenberg "still had a soft spot in his Marxistical heart for the masses"— an ailment Dwight clearly disassociated himself from. That same spring in an essay titled "Why I Am No Longer a Socialist," Dwight denounced the labor movement and the masses who so readily accommodated themselves to the system. He identified himself as a fellow-traveling pacifist, with no faith in pacifism or politics, and really an individualistic anarchist who saw the main enemy as the State. Despite the aesthetic rhetoric, the political undertones of these controversies were hardly hidden. In Rosenberg's view, the only way to quarantine kitsch was by "being too busy with art." All this denunciation by intellectuals and academics was simply adding to kitsch an intellectual dimension. The only legitimate discussion of popular culture was to expose kitsch as a means of propaganda—that is, as a form of reactionary politics.[32]

Paul Goodman followed up on Rosenberg's critique in *Dissent*'s book section with his review of *Memoirs of a Revolutionist*. Following a dictum of Dwight's, Goodman began his essay with the main point: "The book as a book doesn't add much to the wealth of nations. It is foolishly inconsistent, not often thought provoking, not informative ... and hardly even entertaining." Then in a long, typically rambling essay, Goodman dismissed Dwight as a lightweight. His most devastating epithet was to call Dwight "our best journalist." He then went on to denigrate journalism as a profession that is enamored of the transient and the trivial. Dwight was an expert at this craft: "He thinks with his typewriter about the current material and in the melee of opinions of the readers he hopes to reach."

For Goodman a man who thought with the typewriter "instead of the head, heart and hand failed to rise above the daily headlines and the current mixture of polemical opinions." He noted that Dwight admitted he had been unable to "write a book in cold blood." Goodman was not sure Dwight could write a book at all because he was really just a kibitzer and not a serious thinker "determined to a goal of action or truth." Dwight, Goodman sneered, did not have the "passion of the intellect, of learning and adding to learning, of catching a definition and exhausting the consequences, of affirming a conclusion because it follows." Why? Because he was absorbed in coping with opinions and headline facts, not in the essence of subjects. In a deliberately patronizing and condescending conclusion, Goodman announced that Dwight was a fine writer because he was a "very personal, *sui generis* kind of journalist." Dwight had "animal spirits," was "outgoing, intelligent, scrupulous and unusually free from spite."[33]

This testimony to what Dwight called his "boy scout virtues" hardly made up for the wounding assessment of him as a trivial word monger with little to say about politics, culture, or anything else. Although Dwight insisted that a reviewer had an obligation to ignore friendship and be ruthlessly critical toward work he found wanting, he was personally hurt by both Rosenberg's and Goodman's attacks. He had previously rejected solicitations from Irving Howe and other editors of *Dissent* to contribute to the journal. Now he did send off a lengthy rebuttal.

Because Dwight had always been diffident about his failure to produce a sustained work of literary or political scholarship, Rosenberg and Goodman's sneering superiority hurt all the more. Added to that was Dwight's own respect for and promotion of Goodman during the *Politics* years, when he published Goodman over the objections of many readers who could not wade through his undisciplined prose. Even though Dwight maintained a pose of good-humored irritation about the *Dissent* articles, he was stung by what he considered their "smug and sectarian spirit." He began by responding to the Rosenberg challenge. Contrary to Rosenberg's charges, he liked modern art—Picasso, Braque, Gris, Modigliani, Klee, Matisse, "which is what most people would call modern art." This fell right into Rosenberg's trap: Defending the older moderns while insisting that the current abstract expressionists, the action painters, the "drip and dribble" New York School were "the avant-garde equivalent of rock 'n roll" was exactly what Rosenberg meant by kitsch criticism. Rosenberg wrote in reply to this riposte that Dwight's admiration of old modern art and of artists who were, with one exception, dead was the typical kitsch attitude. It was not kitsch to say that modern art stinks; that was at least straightforward, if conservative or ignorant. Kitsch is never opposed to what is acceptable; it reserves "its contempt for the 'extremists.'"[34]

This was only a surface debate about aesthetics and taste; the issues were fundamentally political. Dwight was insisting that one could not blithely dismiss the menace of mass culture. Ignoring Rosenberg's abstract discussion of art and reality, Dwight declared that "the serious battle today, for all who love literature and art and philosophy and other forms of the culture we have inherited from the past ages, is the battle against the cheapening and debasing of high culture by the lords (and consumers) of kitsch." Dwight was not about to let the masses off by seeing them as victims; they collaborated in their own corruption. He felt that the Marxologists and sociologists who had come stampeding into the field were superfluous. These people, as well as Rosenberg, were simply too abstract. The truth of the situation could only be found in specific dissection. So much for Rosenberg's rumination on art.

Dwight then turned to Paul Goodman, who, despite backhanded and condescending praise, had, Dwight felt, dismissed him as an "ignorant and superficial fellow, a kibitzer . . . a mouther of current slogans, a parasite of the day's headlines," without experience "or passion of intellect. In short a very different sort of chap from Paul," a veritable "one man academy." Dwight put Goodman's pose alongside that of Bell and Wollheim in the category of the wise old professor, the uncle, whose task was to "chide a clever but unsound young pupil." Goodman had questioned Dwight's command of the classics and asked whether he had learned anything at Exeter and Yale. Dwight responded by asking whether Goodman had learned anything at Townsend Harris High School and City College. (These ad hominem darts were ritual exercises in all verbal jousting among the New York intellectuals.)

Dwight then dealt with their different temperaments. What was wrong with thinking with his typewriter? Thought had to begin somewhere. Why was Goodman so irritated with him, why did he feel so smugly superior? Dwight returned to the old battleground of theory versus description. Goodman was a prophet, an abstract thinker, "a wholesaler of ideas," whereas Dwight was drawn to the concrete, the specific. Paul went in for big basic philosophies of life. "He's way up there in the empyrean, and my own earth-bound little typewriter ideas, always limited to some specific aspect of things, must seem intolerably trivial to him. He specializes in answers, while I go in more for questions."[35] Dwight's approach to politics was issue-oriented and moral; it rested on his own individual judgment. Both his old comrades were irritated by his casual mocking of the old left, his easy spoofing of earlier enthusiasms, and his quick assumption that opposing mass culture was the most forward barricade in the fight for artistic and political freedom. While those on *Dissent*, such as Howe, were harsh critics of popular culture, they were inclined to argue from a leftist viewpoint that saw the masses as victims of the rampant exploitation of greedy capitalists—Dwight's Lords of Kitsch. Dwight, on the other hand, insisted that the blame be shared by both the hucksters and their victims, and he seemed to take delight in ridiculing the low tastes of the mob, sneering at the laboring classes, and aligning himself with elite critics like Eliot and Pound to the point of calling for the revival of a classic conservatism in America.

Dwight's expertise at diagnosing the fundamental problems of American society was called into question during the spring of 1958 from another quarter. He had sent his "America! America!" piece to the editors of *Encounter* in late February, and it had been accepted by both Kristol and Spender. Feeling that it was too long and full of unoriginal criticism, Kristol was not as enthusiastic as Spender. He asked Dwight to make some cuts, saying that despite his complaints about it he did like the piece. It was scheduled to appear in the May issue. In April, however, Dwight got a long letter from Spender, who said that Nicholas Nabokov, the secretary general of the CCF, had been very upset by the piece and that he now viewed it differently from the way he had originally. Spender offered several suggestions for giving the piece "more balance." Dwight was irritated by Spender's meandering critique. He didn't care whether Spender thought it balanced or not. He wrote "to Stephenirvingnicholasmike or whoever's around & decides things" that the imbalance was what he liked about the piece: "It says what I want to say, and I didn't write it to express the official, diplomatically sound viewpoint of the Congress pour la etc." The article recorded his personal reactions and they had accepted it for what it was. Now they were backtracking on the basis of criticism from Nicholas Nabokov. He closed with an indirect charge that the London editors were under the thumb of the Paris office and did not have control of editorial policy.[36]

Kristol wrote a crossing letter saying that he had been brooding over the piece ever since Nabokov's comments about it. Kristol could understand Nabokov's complaints, he said, as he too felt the piece was far too "self-lacerating." It was "almost John Osborne-ish." But he agreed that they should give Dwight, as a former editor, as much rope as he wanted. He claimed to have mediated the issue, that Nabokov had "cooled down" and that Mike Josselson supported Kristol's decision to publish the piece. He simply asked that Dwight agree to cut the end where he discussed the breakdown of morale among the military in Korea. In another note he insisted that that parting shot left a bad taste in the reader's mouth.

Dwight felt as if he were back at *Fortune*, or the *New International*, or squabbling with Rahv and Phillips on the *Partisan Review*. He could not let anyone meddle with his copy, particularly on ideological or political grounds. While he was resisting any attempt to edit his piece, he received a long letter from Mike Josselson in the Paris office which put the issue in a more disturbing light. Josselson argued that Dwight was being unfair to Nabokov, to himself, and to America. He pointed out that Spender and Kristol had to eat and that Dwight had to be paid for his work. Why jeopardize the magazine's funding by deliberately irritating the foundations that provided it? Without careful diplomacy and tact, *Encounter* might not exist. He put down Dwight's article as something that might have been written by any naive Fulbright scholar back from his first year in Europe. He suggested Dwight give it to some other publication like the *Reporter*, where it would reach more Americans. While Josselson didn't explicitly say so, the sore point was the anti-American tone, which the Paris office as well as the London editors felt should not be encouraged in England, where Yankee bashing and fears of "Americanization" were already rampant.[37]

Dwight was appalled by these letters. Kristol had also mentioned Nabokov's being upset because the article interfered with the effectiveness of his "fund-raising" activities; Nabokov didn't want to see a CCF magazine "nullify his hard work." Dwight would not accept Kristol's aversion to the downbeat ending. He had wanted to draw attention to a major failing of the U.S. way of life. Dwight remained adamant. He didn't care if it left a "bad taste . . . I have never cared about that." Why must one "speak differently of one's country, or one's party or one's friends or relatives, merely because one has a relation to them?"[38]

Irving Kristol delivered the coup de grace in a letter dated June 12: They had decided to reject the piece. It was a letter of abject apology, but in view of "current negotiations for foundation support, it would be highly inadvisable, and perhaps even fatal to the continued existence of the magazine itself—the Foundations not being much interested in having Europeans and Asians think worse of America than they already do." This was how the situation had been described to him, Kristol said, and he did not see how he could do "anything but bow to its exigencies." The whole pretense of editorial independence, so strenuously maintained by Spender and Kristol, was blasted, in Dwight's opinion. The constant rumors of State Department control were yet to be confirmed. But it was clear that the Paris office of the CCF did have veto power over *Encounter*'s editorial policy, or so Kristol seemed to be admitting. Kristol added that he and Spender would not be living up to their responsibilities toward the magazine if they put it at risk for a subject about which "to be candid—we do not have strong feelings." He closed by saying apologies at this stage would be an impertinence, that they had treated Dwight too shamefully for apologies to act as anything but an irritant.[39]

From Connecticut where he was summering Dwight composed a blistering reply in which he excoriated "those Paris boys": "You can't have sacred cows and double standards in an honest mag." The letter was never sent, however, and apparently a contrite Irving Kristol suggested that Dwight send the piece to the British publication *Twentieth Century*, where it was published in October. When its editor introduced the piece by saying that they would not have published Dwight's "spirited and witty comment on American life were not Mr. Macdonald himself a good American," Dwight immediately dashed off a rebuke. He objected on both factual and logical grounds to being labeled a "good American." How did they know? Patriotism had never been one of his strong points: "I don't know as I would call myself a Good American. I am certainly a Critical American." Dwight charged that the editors had made an error in logic by confusing the source of criticism with its validity. "A Bad American, cynical and traitorous, might still make perfectly sound criticism of his country. . . . It's in that other place to the East that civic virtue is the indispensable passport to print."[40]

The whole issue burst into the open at the end of the year when Norman Birnbaum, an American sociologist teaching in England, saw the piece with Dwight's headnote in *Dissent* and wrote an open letter to the CCF to be published in the English *Universities & Left Review*. Birnbaum denounced the "directive" excluding Dwight's "brilliant article" from *Encounter* as an "unmitigated insolence." This supported rumors that the CCF was "more interested in ideological apologetics than in the substance of the great spiritual issues of our time." He charged the leaders of the CCF with dishonoring the cause they professed to serve—intellectual freedom—and called on them to make a public apology.[41]

This development sent both Kristol and Spender into a rage. Now Kristol completely ignored his previous admission to Dwight that loss of foundation support had been the reason for the rejection and, like Spender, insisted it was a matter of their own editorial judgment. Dwight, nonplussed, declared that both Kristol and Josselson had written him that it was fear of the loss of U.S. foundation support that played a big part in the rejection. Dwight acknowledged that Kristol had never been enthusiastic about the piece, but noted that he had agreed to publish it until the Paris intervention.[42]

This affair was closed until the revelations of the CIA's connection with *Encounter* broke in 1967. Then of course Dwight's dispute with the editors was recalled as evidence of the control exerted by an American government agency over the work of American writers, intellectuals, and scholars. What is interesting is that during the furor over "America! America!" in 1958, Dwight continued to reject the notion that the U.S. government had anything to do with the magazine. He didn't look past the cover of "Foundation" sponsorship, directed his criticism instead toward the CCF's innate timidity and nationalistic fervor, even though Gloria and other friends and acquaintances were deeply suspicious that *Encounter* was obviously an official publication. Perhaps in some perverse way Dwight's low regard for Gloria's political acumen contributed to his blindness in the matter. In fact, he thought she was too easily swayed by leftist clichés. Dwight's insistence on the purity of the CCF and its various literary organs was undoubtedly a reaction against the left-wing attacks. When the story of the CIA connection became official a decade later, however, Dwight's claims of ignorance and of having been taken advantage of seemed a little self-serving.

15

Reactionary Essays

DWIGHT AND GLORIA SPENT MUCH OF JUNE AND ALL OF July 1958 at the home of the professor and architecture critic Sibyl Moholy-Nagy in New Milford, Connecticut. Normal Mailer, who was a nearby neighbor, remembers that summer well, particularly a walk he took with Dwight on July 4 in New Milford. Dwight was doing his usual "patriotism is the last refuge of scoundrels" routine. As they were walking down the street they passed a house with a flag or two covering most of the front porch. Immediately Dwight went into an extemporaneous diatribe about the vulgar display of patriotism by typical bourgeois idiots. Dwight was speaking in his usual high-pitched yet stentorian tone when he noticed a formidable-looking fellow rising out of his chair from behind one of the flags on the porch. Looking down at Dwight, he lectured: "How dare you talk about the flag that way." Hardly intimidated, Dwight delivered a responding lecture: "See here, sir, you don't need to flaunt your patriotism. It is the mark of a rather insecure . . . " At this point Dwight broke into giggles and finished the sentence by suggesting that perhaps the man was making a show of his patriotism to hide his own feelings of ambivalence. The man's jaw dropped and his mouth opened in absolute stupefaction at Dwight's impromptu performance. Mailer recalls being nervous, given the size of the fellow. "He looked bad to me." But Dwight had this fearless, almost disconnected quality, a kind of amiable innocence which simply assumed that the homeowner would understand it was just a discussion and meant absolutely nothing personally. He just stood there staring at Dwight, almost suggesting by his appearance that he thought he had confronted a lunatic whose behavior was totally unpredictable. After completing his thoughts on the subject, Dwight said a courtly "Good day" and they walked on.[1]

Dwight was preoccupied that summer with preparations for a course on mass culture he was going to teach at Bard College in the fall. In addition, he had been invited by the critic R. P. Blackmur to give the prestigious Gauss lectures at Princeton in October and early November. He was worried about these academic assignments and put in considerable time working up his lectures, preparing bibliographies, and lining up guest speakers for the Bard course. Just as he was really getting into the work, however, he got a request from *Esquire* magazine to do a piece on producer and screenwriter Dore Schary's forthcoming movie version of Nathanael West's *Miss Lonelyhearts*. This included a much-desired trip to Hollywood, which would fall during the second week of his class at Bard. Dwight asked James Baldwin, whom he had met through Mailer and whose adverse reaction to American society after nearly ten years abroad he appreciated, to cover the first part of the course for him.

In September Dwight made his first trip to Hollywood. It was every bit as bad as he had expected and obviously hoped, not just Tinseltown but the whole of Los Angeles ("a barbarically provincial non city three thousand miles from our cultural capital and two thousand even from Chicago"). It was a perfect vantage point for making an assault on the most important citadel of mass culture, and Dwight did not let the opportunity pass. He wrote two and a half pages of introduction designed to portray the movie moguls as a crude and illiterate breed who could barely speak English. Jerry Wald, a cigar-chewing producer whose office was full of "Rouault prints and other OK modern art products," had just begun work on Faulkner's *The Sound and the Fury* and was full of cultural optimism. He assured Dwight that there was no longer any such thing as highbrow and lowbrow. There were 25 million college graduates in the country—a market for serious films. He had "just bought *Winesburg*." When Dwight suggested *Ulysses*, he shot back, "I got an option on it." *Sons and Lovers* was next. Wald would have none of Dwight's notion of producing films for a culturally educated audience. The art houses were simply a scrapyard for films that didn't sell.

"An artist tries to reach as many people as possible, don't he."[2]

Dwight then turned to Schary's production of West's *Miss Lonelyhearts* (the film was titled simply *Lonelyhearts*). Schary was considered the most intellectually inclined and tasteful of Hollywood producers, which was why Dwight, now an official custodian of high culture, had been commissioned by *Esquire* to see whether the movie passed muster. It was a perfect assignment for Dwight. He loved West's novel because it was a "miraculously pure expression of our special American sort of agony, the horror of loneliness and of our kind of corruption, that of mass culture." In keeping with what William Dean Howells described as "the smiling aspects of American Life," Schary had agreed to bowdlerize West's bleak work of ultimate tragedy and give the story a happy ending. Schary explained that there was indeed in the West novel "something inescapable about the protagonist's doom [Miss Lonelyhearts is murdered by a cuckolded husband] but I felt that dramatically Miss Lonelyhearts's life did not *have* to have a sad ending." The film's director, working to contract, agreed with Schary that "the Christ figure did not have to be crucified." Dwight added: "The idea of an uncrucified Christ is very American."[3]

Dwight conceded that logically the movie did not require a tragic ending since it had nothing in common with the book except the names of the characters. Schary had achieved the impossible, transforming West's *Miss Lonelyhearts* into *Stover at Yale*. Its pace resembled "the proceedings of the American Iron & Steel Institute"; its photography was "chocolate marshmallow"; its acting was wooden and wanting except for the precision and vitality of Maureen Stapleton. As for the script, it was cornball sentimentality that assimilated the characterizations into "that vast body of clichés that has been slowly built up—as a coral reef is built by the accretion of tiny, identical organisms—in our mass culture."[4]

The real problem was that Dore Schary had sincerely tried to make a serious film, not just an entertainment. He had a reputation as a liberal message filmmaker fighting injustice. But Schary was so prone to sentimentality that "as an artist he is part of the enemy he fights as an idealist." He was intelligent and decent, but he had become so "saturated by mass culture" that he could only see the world through the lenses of corn and cliché.[5]

This initial assault on Hollywood was so Menckenesque, so harshly focused on the foibles of the booboisie, that it made an instant impression, infuriating some readers and even drawing a rejoinder from Arnold

Gingrich, *Esquire*'s publisher. Gingrich charged that Dwight had mistakenly attacked the film for not being the book; Dwight replied that it would be nice if one's publisher read the stuff he printed. If he did he would see Dwight had sneered at the photography, the direction, the acting, the sets, the dialogue, and the "message."

Dwight didn't let it drop. After all, Gingrich was his new publisher, and in accordance with Dwight's long-standing penchant for biting the hand that fed him, challenging his own colleagues, and repudiating his own party, he proceeded to attack Gingrich for duplicitously having called for an "uncompromising exposé of the New Hollywood but at the same time patting the practitioners of this dreadful schmaltz masquerading as art, to show that there were no hard feelings." Dwight insisted there were indeed very hard feelings. The inarticulate corn of *Lonelyhearts* made him angry and, as he had lectured Paul Goodman and Harold Rosenberg, he felt it a critical and moral obligation to fight the growing avalanche of trash. If critics, publishers, and intellectuals insisted on being "positive thinkers, maundering about double standards and peaceful coexistence," then the society "will continue to get the kind of mass culture that Mr. Gingrich seems to enjoy." Dwight was intent on making his film criticism a vehicle for political and cultural activism. He was in the belly of the whale, a mass-circulation mag with mass- and mid-cult trappings, and he had the podium. It would also supplement his income. He was behind in advances from the *New Yorker*.[6]

He returned to take up his weekly class at Bard. He had twelve students and two advisees and he took his job very seriously. But what really absorbed his attention that summer and fall was his preparation for delivering the prestigious Gauss lectures at Princeton. It was probably the influence of Delmore Schwartz and John Berryman that had prompted Richard Blackmur, whom Dwight hardly knew, to extend the invitation. By the time Dwight was scheduled, the seminars had been going on for several years and they had a reputation for being lethal affairs where Blackmur, like a plump old tomcat, played with the speaker and the audience as if they were mice.

Dwight had offered a perhaps not sufficiently esoteric reading list of such classics as Ortega, Leavis, McLuhan, and some articles in the Rosenberg and White *Mass Culture* anthology that included his own *Diogenes* piece. At the first dinner at the Nassau Inn, which Dwight later remembered as one of those typical "genial/suspicious, edgy gatherings,"

Blackmur kept matching Dwight drink for drink and Dwight wrongly assumed it was a sign of "unacademic fellowship." However, during the question-and-answer period, when some "stern-visaged scholar asked one of those whendidyoustopbeatingyourwife? questions" that gave Dwight the "choice between venial error and invincible ignorance," Blackmur intervened "with a rueful smile (but duty is duty) before I could answer (ha! ha!) rephrased the question in a sharper and . . . more unanswerable form, not omitting a sincere sideglance at the speaker (me) that positively radiated hope, against all the evidence of the last two hours, for enlightenment."[7]

That same fall Dwight did the detailed research and writing for a long two-part profile of Eugene Gilbert, the founder and head of a teenage marketing firm. In effect, it was an article designed to tell the middle-aged and older generation of *New Yorker* readers that teenage culture was every bit as banal and menacing as they might have imagined and getting worse every day. The piece was an intricate part of his general thesis of mass culture, illustrating the degree to which the spreading ooze was covering the society from the cradle to the grave. Elvis Presley, rock and roll, the importance of the telephone to teenage life were described from the jaundiced point of view of an adult who had no tolerance for the thing he was describing. As a result, the profile failed to get below the surface phenomenon.[8]

Dwight was indeed distanced from his subject, even though he was the father of two teenagers and the stepfather of two others. Instead of showing an interest in their adolescent enthusiasms, he expected them to share his intellectual interests. And both Michael and Nicholas were keen on pleasing Dwight by displaying bookish tastes. Nicholas was a good student and did well at Dalton. When he transferred to Putney in Vermont he continued to be a successful student who pleased his teachers and was popular with his classmates. Mike, on the other hand, remained difficult. He was now at Harvard but having trouble. He was bright and able but his grades were not good. His erratic temperament and persistent refusal to discipline himself to the routine of school life caused continuous problems. There was in Mike a charm and intelligence that made him appear to be a sophisticated adult. But there was also a streak of almost childish immaturity that often surprised his professors and academic counselors.

To Dwight, Nick remained the easy child, "the admirable Crichton," and Mike the problem child. Nevertheless, Nick sometimes resented

Dwight's distance and his difficulty in expressing affection. He resisted Dwight's aggressive intellectuality, his competitive argumentation, his interest in ideas and issues that seemed more important than the family. It wasn't that Dwight was cold or heartless or lacked feeling—not at all. Mike remembered him as often being a wonderful companion who took him to movies, hockey games, the ballpark, and later on introduced him around at parties. He was inventive and fun and loved to go on outings with his boys. But there was an "emotional remoteness, save in a crunch."[9]

Both sons were scarred by the divorce and unfriendly toward Gloria. As a consequence, Gloria was guarded and defensive when they were around. This was particularly true with Michael, who had several emotional outbursts in which Gloria became his target. The boys also resented the fact that Dwight saw Gloria's children more often than he did them. Actually, Mary Day had left home to attend art school in London, but to the boys it appeared that not only did Dwight have a new wife, but new children who in some ways had replaced them. Dwight did go out of his way to visit his boys in Cambridge and in Vermont. In 1956 Nick visited him in London and they took a trip on the Flying Scotsman to Edinburgh. Dwight had a kilt made in Fort William for Nick from the Macdonald clan plaid and they took the ferry to the Isle of Mull. But even on this trip, what stuck with Nick was the travel experience, not a closer personal relationship with Dwight. Dwight's letters to friends seldom failed to tell of the boys' doings and to express his love for both of them, but he found it difficult to break out of the role of a dominating, larger-than-life, charismatic father figure who seemed to diminish the boys and cast them in a restricting shadow. They found it difficult to be other than Dwight's sons. As such, much was expected of them. Sometimes they felt people were interested in them only on that account. They both grew up to be intelligent and sensitive young men, but being Dwight's offspring hardly made life easier for them.[10]

After another pleasant stay in London in the spring and early summer of 1959, Dwight returned to work on a two-part piece for the *Partisan Review* on mass culture. For several years he had been theoretically writing a book on the subject. Jason Epstein, a brilliant young editor and the originator of the enormously successful paperback line, Anchor Books, was now at Random House and had given Dwight a contract. But as with so many other projects, Dwight never managed to sit down and do the sustained research and writing that such a book demanded. He could talk and write about the phenomenon of mass culture in a descriptive way; he could make one feel its insidious and blighting presence; he understood how it infiltrated the home and the very consciousness of the most resistant citizen, not to mention its appeal as a form of narcotic escape. All of this he could present in clear, precise, analytical, and vivid prose that stimulated self-awareness, forced a recognition of the danger, and called for a response. He became a catalyst for debate and controversy over the phenomenon of mass culture and its implications for nearly every aspect of modern life. This was his forte; he could do it as well (if not better) than any other journalist of his time.[11]

As he was the first to admit, he did not have new ideas about the phenomenon of mass culture, nor was he a profound or theoretical commentator on it. But for intelligent laymen, he expressed their deepest fears concerning the quality of contemporary life. And he did so with flair, style, and a deep moral commitment to recruit a discerning public who would demand a more sustaining cultural environment.

If the task of cultural criticism was a serious one, Dwight leavened it with great humor and wit. Nothing illustrates this more than the *Parodies* anthology he compiled which was brought out by Random House in 1960. Parody, Dwight wrote, was a form of satire more sophisticated than travesty and burlesque, which Dwight also loved and had a wonderful ear for. It was also conservative, for parody deliberately made fun of the new and the strange. It was classical, and its usual target was some form of romanticism. He quoted from George Kitchin's survey of the subject with approval. "Parody was a reaction of centrally minded persons to the vagaries of the modes. . . . Politically it [has been] the watchdog of national interests, socially of respectability and in the world of letters of established forms." Dwight did not fail to note that from the mid-twenties until the fifties the *New Yorker* had produced a school of parodists: Robert Benchley, Peter DeVries, Wolcott Gibbs, S. J. Perelman, Frank Sullivan, James Thurber, and E. B. White. They had "exercised their conservative parodies" at the rise of a "literary avant-garde." He noted that parody had fallen off because the present avant-garde was "too hermetic to be parodied and the world itself so fantastic that satire, of which parody was a subdivision, is discouraged because reality outdistances it." As his editor, Jason Epstein, noted: "How can you parody Beckett?" Dwight's collection was another act of the revival of world-class culture, in its own way a byroad of his "reactionary excursions" in defense of community and tradition. The *New Yorker*, he noted, had been so productive in the use of parody because it combined literary sophistication with a "necessary provincialism—the audience must be homogeneous to get the point." In effect the book was designed for the cultivated remnant.[12]

The volume received an enthusiastic response from the literary community and was reviewed widely in America and in Great Britain, where it was published by Faber. The *New York Times Book Review* carried a front-page review by the British critic Walter Allen, who praised it as a "splendid compilation" and went on to credit Dwight's taste, wit, and discrimination. Charles Poore in the daily *Times* relished it as the most enjoyable book of parodies he had ever encountered. Donald Davie in the British *Guardian* wrote that despite its appropriate lightheartedness, Dwight's work was one of great erudition, discriminating taste, and acute criticism, "the most intelligent treatment of parody that has appeared in English and very probably a standard work for a long time to come." Everyone was taken with Dwight's singular mode of selection. He included material that was not parody but that was funny to him and seemed to fit. He had a section devoted to Self-Parodies, Conscious (Beerbohm, Faulkner) and Unconscious (Johnson, Poe, Wordsworth, and Dickens, among others). Under the general heading "Specialties," he presented wildly comic uses of language, including a parody of Eisenhower alongside one of the president's actual speeches. Dwight ended that section with a "Postscript on Gamaliese [sic] (President Harding)," wherein he explained the roots of Eisenhower's mangling of the language. Perhaps the most imaginative selection was the police blotter record of the mobster Dutch Schultz's bizarrely incoherent dying words presented as a parody of Gertrude Stein.

Parodies was great fun, but it was also clearly the work of a knowledgeable aficionado, a craftsman, and it appealed to a coterie of wordmongers on both sides of the Atlantic. Dwight was trying to reach the educated elite who were versed in the classics from Chaucer to the present and held a common cultural tradition. The left-leaning *New Statesman* in London commented on its conservative and elitist vision and the *Nation* reviewer, while conceding that the work was an "achievement of lighthearted scholarship," made a point of noting that Dwight was "playing a bookish game, best played by literary insiders" and that it often took some "ferreting in the notes to get the point of the joke."[13]

*   *   *

The favorable reception for *Parodies* came at the time that Dwight's extensive two-part article, "Masscult and Midcult," was appearing in *Partisan Review*, in the spring and fall issues of 1960. These two install-ments constituted an essay of over seventy-four pages, which would soon appear in his 1962 collection of mass culture pieces titled *Against the American Grain*. Dwight's long-awaited "book" on the subject carried the subtitle "Essays on the Effects of Mass Culture." Critical essays on Hem-ingway, Joyce, and Twain, a very well received critique of Webster's New International Dictionary, and several other previously published articles rounded out the volume.

Dwight frequently referred to the collection as his "reactionary excur-sions." In an exchange with a critical reader, he conceded his correspon-dent was right, he was a snob because he "put culture before democracy." In "Masscult and Midcult" he made a point of approvingly citing T. S. Eliot for insisting that a defense of high culture had nothing to do with democracy:

> Here are what I believe to be the essential conditions for the growth and for the survival of culture. If they conflict with any passionate faith of the reader—if, for instance, he finds it shocking that culture and equalitari-anism should conflict, if it seems monstrous to him that anyone should have "advantages of birth"—I do not ask him to change his faith. I merely ask him to stop paying lip-service to culture. If the reader says: "The state of affairs which I wish to bring about is *right* (or is *just*, or is *inevitable*) and if this must lead to further deterioration of culture, we must accept that deterioration"—then I can have no quarrel with him. I might even, in some circumstances, feel obliged to support him. The effect of such a wave of honesty would be that the word *culture* would cease to be absurd.[14]

Dwight was hardly as accommodating as Eliot, for he insisted that those who claimed to have an interest in culture and civilization had to resist movements that denigrated it. Alvin Toffler was soon to award Dwight the title "high priest of the culture snobs." Dwight countered by saying Toffler's notion of democracy was to insist that culturally every-body come out on the same level. Dwight maintained that "all great cul-tures had been elitist affairs, centering in small upper class communities which had certain standards in common and which both encouraged cre-ativity by (informed) enthusiasm and disciplined it by (informed) criticism." When Dwight was trying to decide on a title for the collection, he considered calling it *Reactions* but had been dissuaded by the editors, who were sure it would set the "teeth of the children on edge with thoughts of Metternich, Bismarck and Senator Barry Goldwater." He agreed that Goldwater often looked like a chump and a demagogic echo of his two predecessors, but he admitted that in "most non political situations [he] was against progress" and that it was the "reactionary thinkers like Eliot and Beerbohm who seemed closer to the truth than the revolutionaries." Dwight's good friend and admirer John Lukacs thought very highly of him as a cultural critic, noting that when he was at his best he clearly understood that the exposure of untruth was a more worthwhile intellectual task than the exposure of injustice. That Dwight ever intended to maintain such a priority is far from clear, but his cultural essays did encourage such a conservative interpretation.[15]

Dwight led off his essay with a devastating critique of the purveyors of mid-cult, that insidious fifth column inside the temple of high culture that passed off shoddy goods as the real thing. He was explicit and named names: *Harper's*, the *Atlantic*, *Saturday Review*; John P. Marquand, John Steinbeck, Pearl S. Buck, Irwin Shaw, Herman Wouk, John Hersey were all mid-cult novelists. He praised Zane Grey for never confusing his books with literature. It was the sincere pretensions of mid-cult writers that made them so tenacious and insidious. Dwight chose to target Hemingway for his *Old Man and the Sea*, Thornton Wilder for *Our Town*, Archibald MacLeish for *J.B.* ("It is Profound and Soul Searching, it deals with The Agony of Modern Man"), and Stephen Vincent Benét's "orgy of Americana," *John Brown's Body*. All were attacked for clichéd sentimentality, exploitation of emotion, banal homilies, and saccharine middle-brow nostalgia.[16]

Returning to Clement Greenberg's well-known essay on avant-garde and kitsch, Dwight summoned up the glories of the old avant-garde, from Rimbaud to Picasso, lamenting that they had passed on and no one had taken their place except the slick practitioners of mid-cult, who cleverly exploited the discoveries of the avant-garde. Dwight insisted that no new creators equaled the stature of the old avant-garde. Those with creative talent merged into the ranks of mid-cultists. Dwight still had little interest in contemporary modern art and could only describe the work of the action painters as "enormous gloobs and globs"; the school of "drip and dribble" remained nothing more than the painting equivalent of rock and roll. Nor could he tolerate the "beatnik academy of letters." Unlike the old avant-garde, both the artists and writers were cut off from tradition, and had far too much publicity too soon. While Dwight could look back on the spoofs of Dada with amused tolerance, the new attempts to ridicule and shock the bourgeois culture were presented as serious artistic offerings. One might, Dwight mused, call this "the lumpen avant-garde."[17]

There were included in the collection three pieces of literary criticism. The one on Joyce was noticeably thin, perhaps because it was really a review of the magisterial biography by Richard Ellmann. Dwight focused on Joyce's commitment to his craft and refusal to compromise or make any concession to his audience. For Dwight, *Finnegans Wake* was a dead end but a courageous one at that. Under Dwight's lens, Joyce, even though often personally unattractive, became a hero because of his resistance to mass culture and mass society. His other two subjects, Twain and Hemingway, were victims of both because they failed to develop the character and commitment necessary for resistance. Dwight was particularly hard on Twain as a man torn between serious craftsmanship and a desire for celebrity that reduced him into pandering to the mass audience. What gave the story tragic overtones was Twain's awareness of the tawdry backdrop to the "American dream." That sensitivity led to despair and depression during the last decade of his life. As contrasted with Joyce, Twain had found it difficult to speak in his own voice, because he was always tempted to write for a popular mass audience, from which he was inherently alienated, despite his democratic protestations.[18]

Dwight used his mass culture ideas as a Freudian or Marxist critic used their texts to analyze literature in relationship to the culture. It is not surprising that Hemingway did not come off very well. Dwight had found much of Twain's work no longer readable; Hemingway's works, even *A Farewell to Arms*, were dismissed as written in a kind of inspired baby talk. Dwight's theme remained the same: Hemingway, like Twain, was damaged by success. He too could not resist producing mid-cult ooze for a mass audience. Dwight's constant insistence that a true artist was content to write for a small audience almost seemed a kind of psychological projection or explanation for his own experience. He was later to write that it was not possible for a serious writer to sell out and write bad art, but that hardly squared with his analysis of the failures of both Twain and Hemingway. Dwight was personally abusive of Hemingway as an unintelligent philistine who must have had the smallest vocabulary in literary history. Trapped in his initial stylistic innovation, Hemingway never developed, and as he grew older the style became mannered, even absurd. Part of the essay was written as a parody, making Hemingway seem all the more ridiculous.[19]

These essays, perhaps with the exception of a sensitive appreciation of James Agee, did not advance Dwight's reputation as a literary critic, and personal friends of Joyce and Hemingway were not happy with Dwight's cavalier dismissals. George Plimpton and Brendan Gill both wrote to defend Hemingway as a writer and a human being and denied that his public and private personas were one. Janet Flanner, a *New Yorker* staffer living in Paris who was a friend of Hemingway's, found Dwight's piece "caddish," and it drove Nelson Algren into a frenzy. Nevertheless, the essays did find admirers as insightful observations on American culture and the trials of contemporary writers in a mass society.[20]

When *Against the American Grain* appeared in early 1962 it was widely reviewed and it sold better than anything Dwight had previously published. He was pleased with the publicity, which resulted in growing demands for his appearance on talk shows and at conferences on mass culture. He had become one of the arbiters whose views were consulted, and he thoroughly enjoyed the attention.

The capstone of this series of works in defense of culture was his critical attack on *Webster's Third New International Dictionary*. Titled "The String Untuned," it initially appeared in the *New Yorker* in March 1962 and was included in *Against the American Grain*. It was a devastating critique of the direction lexicography had taken in the last thirty years, but it was again, as were so many of these demolition jobs denouncing a decline in taste and standards, at bottom a political treatise. Dwight entered into the "description versus prescription" debate waged by the structural linguists against the "schoolmarms" of grammatical and syntactical tradition. He did so with his usual wit and flair, which amply made up for his ignorance of the technical and scientific tools of modern linguistics. But it was not really the changes in language and how such changes became acceptable that were at the heart of his criticism. What bothered him was that the "permissiveness" of Webster 3 indicated "a plebeian attitude toward language." Respect for the language was more than the appreciation of words and definitions; it marked an "attitude toward life."[21]

He began "The String Untuned" by arguing that Webster 3 reflected the steep decline in culture since the second edition of 1934. Throughout the essay the theme is one of the loss of cultural authority, the erosion of standards, the rise of permissiveness, ignorance of and lack of respect for tradition, for history, for the culture's past. In effect, he saw a connection between the "ignoscenti" (the academic scientists dedicated to destroying the language in the name of scientific description) and advocates of an unlimited, undisciplined vernacular. As though that invasion of barbarism was not enough, those who ought to be the guardians of the culture (the writers, the teachers, and the scholars) had abdicated their responsibility and folded before the pretentious theories of the structural linguists and the pressure of the mob. Dwight expressed his fear of total linguistic chaos and a riot of permissiveness by citing Ulysses's defense of conservatism in *Troilus and Cressida*, which was the source for the title of the essay, "The String Untuned":

> *The heavens themselves, the planets and*
> *this center*
> *Observe degree, priority and place,*
> *Insisture, course, proportion, season, form,*
> *Office and custom in all line of order. . . .*
> *Take but degree away, untune that string*
> *And hark, what discord follows!*[22]

In addition to his Yeatsian fear that the center would not hold was the persistent theme that an inappropriate "scientism" was endangering the language and thus the culture. Language was a matter of values, evaluation, discrimination; it involved an artistic temperament. A scientific approach to its development and expression was simply not relevant. The linguists and their supporters were far more interested in the scientific than the artistic.

Dwight's stand was reminiscent of his argument in the mid-forties concerning the values people live by. To admit change in language too readily, he said, was to divorce a culture from its past, which caused a people to lose their identity, as was the case in the Soviet Union and Red China, where the citizen had become "disoriented, formless, anonymous." Dwight returned to the rhetorical style he had used in the forties when he had imagined the imminent threat of totalitarianism, what the intellectual historian and Macdonald scholar Robert Westbrook has called "a politics of fearful anticipation." In effect, this assault by the linguists and their supporters imperiled the heart of humanistic culture. Linguistic theory threatened the value system in much the same way totalitarian theory did. As if this destructive menace were not bad enough, Dwight connected this threat with the scientific methodology in the Kinsey report on sexual behavior. In Dwight's view, sex, like language, was an area of human activity in which moral or at least value judgments have always been made. Now a quantitative science of measurement was being applied. The scientist, Alfred Kinsey, counted sexual practices in the same way the scientific lexicographer, Philip Gove, editor-in-chief of Webster 3, counted linguistic usage. Dwight insisted that neither linguistic experience or sexual experience could be reduced to quantitative analysis. There was a qualitative factor in both that yielded only to what Gove himself had admitted was the necessity of "subjective analysis, arbitrary decisions and intuitive reasoning."[23]

Dwight saw his piece on the dictionary as taking its place beside his essays on the revised Bible, the Great Books, and the critics' failure in the cases of James Gould Cozzens and Colin Wilson. In every instance, the educated elite who had an obligation to put the brakes on, to subject changes in custom, tradition, culture to severe tests of taste and intelligence, had refused to take a stand or exercised an indiscriminate tolerance. Ironically, in the case of the dictionary it was the mass culture magazines such as *Life, Time, Newsweek, Saturday Review*, along with even the liberal and leftist weeklies the *Nation* and the *New Republic*, that took the side of tradition against the academic structural linguists and their willing tools, the lexicographers: the purveyors of mass-cult and mid-cult were cast in the role of defenders, and the academicians (Dwight seldom missed an opportunity to target them) and scientists represented the new barbarians. Dwight understood the apparent contradiction. "There is nothing an American scholar likes more than a really impressive system with scientific pretensions—while the lay critics, being so to speak on the firing line of actual usage, since they make their living by writing for the public, are more aware of and concerned about the vulgarization of language that is now going on in this country."[24]

Dwight insisted that it was the linguistic scholars allied with the ignoramuses that constituted a mobocracy intent on vulgarization. Dwight was supported by such custodians of the culture as Wilson Follett, America's equivalent of H. W. Fowler, and the consistently elitist Jacques Barzun, both of whom wrote devastating assaults on the new dictionary as a cultural disaster. Follett was nearly apoplectic, calling the dictionary a scandal and a disaster. Its editors were bent on destroying "every obstinate vestige of linguistic punctilio, every surviving influence that makes for the upholding of standards, every criterion for distinguishing between

better usages and worse." Gove and his accomplices had allied themselves with the "school that construes traditions as enslaving, the rudi-
mentary principles of syntax as crippling and taste as irrelevant."[25]

Follett was joined by Jacques Barzun, certainly one of the more erudite
and eloquent of the defenders of the faith. In the Phi Beta Kappa journal,
the *American Scholar*, Barzun got right to the point. He declared Webster's Third "the longest political pamphlet ever put together by a party."
It embodied and preached a dogma that far transcended lexicography.
The editors were unrelenting in their egalitarianism and insistence on
democracy of usage. It followed from the editors' view that "the English
language comprises whatever is intelligible to any group that thinks it is
speaking English, from Puerto Rican children in New York, native
bureaucrats in India or Nigeria, Ozark mountaineers, BBC announcers,
judges of the United States Supreme Court . . . to unfortunate idiots with
cleft palates." Is "the product of the human mind to be treated as a natural object? The answer Yes means that whatever 'the people' utter is a
'linguistic fact' to be recorded, cherished, preferred to any reason or tradition." It is clear that the citadel of culture was under attack and the
elite was striking back.[26]

Although Dwight titled his collection of cultural essays *Against the
American Grain*, he was clearly supported by both the elite literary
establishment as well as the purveyors of mid-cult. When James Sledd, a
scholar of lexicography, put together an anthology dealing with the controversy over Webster's Third, he featured Dwight as the most articulate
and damaging of the dictionary's critics. But Macdonald's review was
"portentously bad," Sledd charged, because it was "disgraced by just the
sort of ignorance and unfairness which cheap journalism in a mass culture would substitute for scholarship." Sledd described Dwight as the
leader of a mob engaged in an "intellectual lynching." Dwight was
undoubtedly pleased to be cast in the role of a leader of the opposition,
but he modestly insisted that he had probably followed the mob rather
than led it. Sledd countered with the lament that friends of his "in four
major universities" had cited Dwight's article with approval. It is also
true that Dwight got more favorable mail in response to his *New Yorker*
piece about the dictionary than anything he had ever written (200 letters, 98 percent of them against Webster's Third, about twice what he
had received after his critique of the Revised Standard Version of the
Bible).[27]

Dwight ended this exchange with Sledd with an eloquent explanation for the impassioned reader response. Language, he insisted, aroused even more passion than religion. Everyone was exposed to language. Even the illiterate had to master it in order to communicate. All people had a proprietary interest in their own language, even if only half consciously. It was their only direct contact with their past. Dwight spoke for the misused people who were discovering that their language was "being eroded under us and the rodents were just those we had assumed would be on the other side, namely the lexicographers." This explained why there had been a "mystical," perhaps more accurately an emotional, reaction.[28]

The collection of cultural literary essays, the *Parodies* book, and the articles and reviews dealing with language had established Dwight as an official arbiter of American culture. He was now in demand as a speaker and visiting professor on campuses throughout the country; he was called on to participate in conferences on language and literature. He was a widely respected literary and cultural celebrity. People recognized him on TV. He had a persona. He was a sought-after guest at New York parties. As one of his old friends put it in a denigrating way, perhaps unfairly, Dwight had traveled from the bohemian literati of the radical Village and its environs to the glitterati of the Upper East Side, of the publishing house world on Madison Avenue. He had definitely not gotten rich, but he was a cultural commentator to be reckoned with. It is not surprising that Arnold Gingrich, publisher of one of the most prominent of mid-cult publications, *Esquire*, in accordance with Dwight's own theories of mid-cult tactics, asked Dwight to take over their regular movie review column. Nor is it surprising that Dwight, with his rising confidence, was ready to take on the challenge.[29]

Dwight's first *Esquire* film column appeared in February 1960 and his last in November 1966. He came to the column with a lifetime interest in film, which he had insisted as a prep school boy was a serious art form. His earliest writings were on film, and he celebrated nights at the movies in his class poem when graduating from Exeter:

> *the gay delight of movie night*
> *the pleasant heat, the stamp of feet*
> *The mighty shout as the lights go out*
> *And the yells at the scene on the flashing*
> *screen.*

By the time Dwight got to Yale he was keeping a notebook about film and reporting to his parents the great things that were happening in the field of cinema.[30]

Given this background as an early "missionary enthusiast," it is hardly surprising that he accepted the offer to be the film critic for *Esquire* in 1960. It all made sense. He was preoccupied with mass culture and its impact on high art. What better place to focus his attention? He had always believed that film might bridge the gap between popular and high culture. The great classic directors were for him creative geniuses; he likened their work and their opportunity to those of the Elizabethan play-wrights, whose work was popular and not considered art at the time. He used the classic films and the techniques of the great directors as stan-dards to judge the current work of the medium. As in his teaching, he took film criticism seriously, insisting that he was a critic and not a reviewer simply touting films for the commercial market.

Because he came to film criticism during an explosion of European avant-garde films, he found his job exciting, and even if he gave little support and encouragement to such experimental filmmakers of the six-ties underground as Jonas Mekas, Jack Smith, and Kenneth Anger, he was wildly enthusiastic about several Italian directors. He thought Fellini's *8½* a masterpiece, in part because he identified with the angst of the protagonist, and he was enthusiastic about several Antonioni films, particularly *L'Avventura*. For the most part, he praised these films while bashing Hollywood productions.[31]

As was to be expected, the bulk of Dwight's screen commentary was negative. Since he was obliged to attend films made in Hollywood (some-times his sons vetted films for him), his film reviews were a continuation of his crusade against the evils of mass culture, Hollywood being the pre-eminent producer of trash. His criticism was carried on with his ever-ready sharp, engaging wit and mocking hilarity: "With the unerring touch of a bad dentist Capote hits and keeps hitting a sensitive nerve of our time"; "These celebrations of Hollywood producers by art museums are as if the A.S.P.C.A. protected the floggers instead of the horses." On the acting of Charlton Heston, Stephen Boyd, and Haya Harareet in *Ben Hur*: "at no point [are they] in danger of lawsuits for impersonating real peo-ple." On some other actors: Robert Mitchum exudes "soggy virility"; Paul Newman "is simply not an actor and possibly not even alive"; Anthony Perkins "becomes more fragile and epicenely contorted with each new film"; Elizabeth Taylor is labeled "Lizpatra with her Bronxville wiles."

Movie *Marty*: "A soggy Bronx bagel." On John Huston playing a cardinal: "He enjoyed every minute of ecclesiastical transvestism." When in the same film (*The Cardinal*) a statue of the Virgin Mary drips red fluid, the Boston Italian peasantry scream and carry on, but Otto Preminger has the cardinal explain it was a rusty pipe—but it was still a miracle because it was God that caused the pipe to leak. Dwight wonders: "But what's the use of being God if you have to worry about plumbing."[32]

Dwight's readers awaited these monthly missiles with anticipation. (Apparently the FBI did also. A review of *Experiment in Terror* claiming that the FBI as portrayed in the film was moronic and the agent even a "greater bungler than the real article" found its way into his file.)[33] Because his column came out a good deal later than the initial reviews in the dailies and the weeklies, his readers could count on a demolition of the most recent Hollywood success and rarely be disappointed. Even readers who could not appreciate Dwight's devastation of films they enjoyed were regular readers judging by the constant mail his column elicited from fans and detractors. As in the past, Dwight turned his column into a public discussion, a dialogue not simply about movies but about the quality of life in America, its culture, its brutality, its frivolity, the sexual immaturity of the American male, the general obsession with sex combined with official prudishness. In his classic assault on Hitchcock's commercially successful *Psycho* as the product of "a most unpleasant mind, a sly, sadistic little mind," Dwight claimed that Americans compensated for their repressive attitudes about sex by allowing the constant depiction of bloody violence. The censors saw nothing wrong in showing in "intimate, suggestive detail a helpless woman being stabbed to death, but had Hitchcock shown one of Janet Leigh's nipples, that would have been a serious offense against morals and decency."[34]

Dwight did not shy away from controversy, of course, nor did he kowtow to his editors. From the very beginning he made a point of reviewing the biblical spectaculars not for their interest as works of art but as reflective of all the worst aspects of mass culture, a menacing, violent mass culture masquerading under the robes of religiosity. It was in his easy razing of *Ben Hur, King of Kings*, and *The Greatest Story Ever Told* that he impishly observed that Hollywood was intent on making the Romans the "fall goys" for the crucifixion of Christ. He cited the Gospels according to Matthew, Mark, Luke, and John to argue that Jewish leaders had urged Pilate to carry out the sentence. It was not reviving this ancient biblical debate that caused the ensuing uproar so much as Dwight's derisive attribution of the motive for this "falsification of the Bible." He noted the spectacular $15 million spent on the film. There were "no ancient Romans around and there are many Jews and $15,000,000 is $15,000,000."[35]

This column, not surprisingly, brought a storm of abuse from Jewish readers. Many charged Dwight with virulent anti-Semitism for raising the old recrimination of the Jews as "Christ-killers." The magazine and its publisher, Arnold Gingrich, were attacked for printing "ugly bigotry" and "literary refuse." Despite the criticism, Dwight would not let the matter drop, returning to the issue in subsequent columns. It got to the point where Gingrich felt obliged to send out a form letter to Macdonald's critics asserting his own and his editors' disagreement with Macdonald's opinions but defending his right to express them.[36]

Dwight did back off from his initial Waspish ethnocentrism and agreed that the argument of biblical scholars suggesting that the Gospels were not disinterested scholarly accounts and that the Romans had certainly played an important role in the crucifixion might have some validity. He defiantly rejected the charge of anti-Semitism, however. His defense prefigured that of opponents to the authoritarian liberalism that demands conformity to a code of "political correctness." Dwight, very much like contemporary critics of PC, argued that people had become entirely too sensitive. The "pendulum of social justice has swung back too far." He felt that "certain racial groups" had become so sacrosanct that they could no longer be criticized like other human beings. He facetiously suggested that if he had written about "what Sitting Bull did to General Custer, some Indian Protective League would jump into action."[37]

It was a typical Macdonald performance. Dwight would not insist that he was not a bigot any more than he would insist he wasn't a liar. "I will even admit that I think of Jews, Negroes, Ugro-Finns, Irishmen and other groups as no better than anyone else, and so feel free to joke, praise or criticize." Furthermore, the liberal tendency to exempt Jews and blacks from criticism was patronizing, he believed. But even in the liberating sixties, racism had not become so removed from American society as Dwight maintained. Many Jewish readers thought his stereotyped innuendo that the filmmakers had exonerated the Jews for the crucifixion to protect their $15 million was an anti-Semitic slur. Since Dwight had railed for years about the commercial zeal of the Hollywood producers and their subservient directors and had also gone out of his way to note their ethnic backgrounds, his critics had a point. As for the charge of racial stereotyping Dwight could tolerate the sentimentality in *Gone with the Wind* and take no notice of its blatant racial stereotyping. As one critic argued in a mostly friendly letter, "Jewish and Negro children still come home from school, weeping over indignities heaped on them because 'the Jews killed Christ' and 'Negroes rape white women.'" Dwight admitted that there had been terrible prejudice and racism in the past, but he insisted that anti-Semitism was not a clear and present danger in the early 1960s and that even though blacks were still shamefully discriminated against, "since 1945 they have been winning . . . the tide is running in their favor." His optimism reflected the heady atmosphere of the early sixties, when enthusiasm for racial justice was widespread, and the belief that it could be achieved in the not too distant future was strong. That enthusiasm was soon followed by disillusionment and frustration. The ethnic and racial nationalism that Dwight had always opposed and that he sensed so vividly in the responses to his column began to reassert itself. By 1966, ethnic and racial nationalism had taken on a new life and the nonviolent, interracial civil rights movement had for all practical purposes come to an end.[38]

His contentiousness, plus his insistence that films should deal with important issues and, when they failed to, his ability to fill in the gaps attracted readers. In 1971 the National Society of Student Film Critics polled its members to select the best critics of the preceding year. By that date Dwight hadn't written regularly on film for over five years. Nevertheless, he outpolled everyone except his most active contemporaries who were writing regular columns, Pauline Kael, Stanley Kauffmann, and John Simon. Dwight had definitely made his mark, not only on his readers but on his colleagues. They respected his honesty, his individualistic verve, and many stood in awe of the clarity and easy precision of his prose.[39]

During the years Dwight was bashing Hollywood as the preeminent example of the menace of mass culture and looking to European films as the avant-garde resistance he so longed for, he was traveling between New York and London with regularity. In England there was still an "educated class" that still "valued the traditions of the language."[40] Dwight's celebrity status as a cultural arbiter, which resulted in article assignments, speaking engagements, and even book contracts, and his regular column in *Esquire* gave him the wherewithal to travel back and forth. Gloria was thrilled. It was exactly what she had anticipated when she first became involved with Dwight. She had seen him as "an influential" who would make his name known in the English-speaking intellectual capitals. Besides, her older daughter Mary Day was still enrolled in the Slade School of Fine Arts in London and Gloria was always eager to make the trip to see her.

But Europe in 1960 was not to be fully satisfying. Gloria left for London in June; busy with writing and speaking engagements, Dwight was to follow in the fall. During the months they were separated, tensions developed over housing. Gloria preferred the bustle of a flat off Bedford Square; Dwight was excited about the possibility of renting John and Vera Russell's place in fashionable, but sedate, St. Johns Wood. Gloria wanted to be near the university and bohemian London; uncharacteristically— but perhaps reflecting his new status—Dwight was drawn toward what he saw as an elegant, posh community.[41] Eventually, Gloria capitulated to Dwight's wishes, which meant dealing with the overbearing Mrs. Russell. Alone in London and feeling neglected by Dwight's friends there, Gloria was further irritated by Dwight's reports of his active social life in New York. Despite the visits to the Dupees, the Lukacses, Gore Vidal, and the dinners with Norman Mailer, Norman Podhoretz, and Jason Epstein, Dwight had his own troubles in New York. There were still tensions with Mike, who had left Harvard and still seemed to Dwight "wildly moody and emotional." As usual, Nick was "the admirable Crichton," and Dwight enjoyed his good humor and his serious questions about art, politics, and letters.[42]

During the spring of 1960, Dwight's momentum had taken him to an Argentine film festival that was aesthetically a failure, but socially a success. In the summer, Dwight visited the Mailers at the Cape in order to write an amusing piece on Mailer's incarceration in Provincetown for "disorderly conduct, public drunkenness, and resisting arrest." When Dwight arrived in London in September he was greeted by the unpleasant news that the Russells' trip to Europe had been canceled and that he and Gloria were to live in only the basement of their elegant house. Difficulties with the Russells encouraged them to look for another place, and they found an apartment in town. Dwight still liked London, but it did not quite have its old charm. He and Gloria gave a celebrity-studded party to celebrate U.S. Election Day in November 1960, but he took little interest in the candidates and urged a no-vote policy in a piece for *Commentary*. He wrote a long piece on the cultural life of London for the *Partisan Review*. And in 1961 he was attracted enough to the militancy of the cam-

paign for nuclear disarmament to participate in the last six miles of the
Aldermaston anti-nuke march and to urge, unsuccessfully, on Shawn a
piece about Bertrand Russell, the leader of the left wing of the movement.
Impressed though he was by the humor and discipline of both marchers
and bobbies, Dwight still wondered whether a little more passion and
civil disobedience weren't called for. A "vegetarian mob" was pleasant,
but in the end ineffective politically and in improving Dwight's increasingly despondent mood.[43]

In this mood, a mood matched by the grim gray London days, he was
delighted to receive an invitation to attend the wedding of Mary McCarthy
and James West, a member of the U.S. diplomatic corps stationed in
Poland. The Macdonalds enjoyed this brief escape to Paris. In that spring
of 1961 he also attended the Cannes Film Festival at *Esquire*'s expense.
He found it yet another rite of mass culture. For the first time, he mani-
fested some sympathy for the Beats, who showed up at the festival seeking "publicity from the philistine world they reject." Lining up outside
the Palais, they chanted at the entering film celebrities: "Give us some
money . . . Give us some money." To Dwight it was a clever spoof of the
festival's commercialism.[44]

Dwight and Gloria spent July and August 1961 in Italy trying to re-
create their 1957 tour. Visits with the Chiaromontes and to "the nest of
singing birds" at Elena Vivante's villa roused Dwight from the depression he had experienced in London during the rainy spring. For the first
time, he looked forward to returning to New York. With his romantic
attachment to London cooled a bit, Dwight now viewed it as "comparatively provincial and airless for all its charm." On the other hand, New
York had an "air of exhilaration and promise (anything goes, anything is
possible) and stimulation and variety." As he embarked for New York in
the fall of 1961 he felt cheerful; the future was full of possibilities.[45]

IV

A RETURN TO POLITICS

16

Warming Up

DWIGHT WAS AN INSTINCTIVE JOURNALIST. HE HAD A NOSE for changes in the cultural climate. Hannah Arendt observed that he could "be so uncannily right in detecting the long run trends."[1] When he wrote Dinsmore Wheeler about the atmosphere of possibility, he already felt more involved and connected; he was anticipating a return to political activism.

In the spring of 1960, the civil rights movement had moved into high gear. The sit-ins in Greensboro, North Carolina, and Nashville, Tennessee, caught the media's attention, and it was not long before the movement of resistance to racial segregation and discrimination exploded all over the South. The Congress of Racial Equality (CORE), with its religious pacifist tradition, quickly lent its services as experienced practitioners of "nonviolent direct action." While respectful of older civil rights organizations such as the NAACP, young blacks wanted to be on the cutting edge of the resistance. Out of their youthful impatience for "Freedom Now!" came the Student Non-Violent Coordinating Committee (SNCC), which quickly developed chapters throughout the South and in the North as well. Students on Northern campuses watched these events with interest. At Berkeley there were several activist student organizations that rallied first around the issue of capital punishment and then that decrepit remnant of McCarthyism, the House Un-American Activities Committee.

They confronted the Committee, which was investigating alleged California Communists, on May 13, 1960, "Black Friday," and were attacked by the police on the steps of the San Francisco City Hall. The mocking repudiation of authority caught the imagination of questioning students throughout the country. At the University of Michigan the first chapter of Students for a Democratic Society was formed. At Harvard, Swarthmore, Oberlin, Columbia, Wisconsin, other organizations were founded and soon were in touch with one another, holding conferences, turning out newsletters, planning protests—making waves. C. Wright Mills, the budding guru of what soon became "the Movement," confidently announced that "The Age of Complacency is ending. . . . We are beginning to move again." In a message to the New Left, Mills launched a devastating assault on the spokesmen for "the End of Ideology," whom he characterized as "complacent, prematurely middle-aged, centered in the . . . rich Western societies." Mills provided a role model of the new "scholar activist" and inspired a growing number of young rebels who had suddenly discovered in their exhilarating confrontations with the institutions of authority, the State and soon the University, that the emperor had no clothes. The once dormant left was ready to move again, only this time it would be led by young people who were indeed ideological but eclectic and skeptical of what they saw as an antiquated and morbid past.[2]

Dwight had the same feeling. He had visited college campuses throughout the 1950s. His own writing had sometimes been influenced by confident announcements from the social science establishment that the end of poverty and racial discrimination in the United States was imminent. He had seen what Daniel Bell applauded as "the exhaustion of political ideas," but instead, he was depressed by the stifling of political debate. The ideological demands of the cold war and the ameliorative gradualism of the social engineers and technocrats who repeatedly insisted that the major problems of Western societies had been solved had created a stultifying consensus; radicalism was considered superfluous.[3]

Dwight had initially come to the academy as a defender of the cultural canon against the onslaught of mass culture. Despite its traditionalist implications, his cultural criticism had always embodied a profound rejection of the status quo. He did not join in his fellow intellectuals' celebration of American society or culture; he explicitly attacked the rampant banality of American life and the passivity of its citizens. Even in the politically lean years, one of his standard talks was on "The Relevance of Anarchism." It was the political counterpart of his attack on mass culture as one of the main ingredients of creeping totalitarianism. Having abandoned all traces of Marxism, he returned to his first love, anarchic individualism and a repudiation of all forms of coercive authority. In the Warfare State of the 1950s, he saw this vision as the only possible escape from the vapid yet menacing stranglehold of centralized power.

In the spring of 1960, Dwight became involved in the defense for the cases of Morton Sobell, who had been tried for espionage with the Rosenbergs, and Junius Scales, who had been sentenced to six years in prison under the Smith Act, which made mere membership in the Communist Party a criminal activity. In both cases, Dwight felt that the government had acted unfairly. So strong were his feelings that he later wrote President Kennedy on behalf of both men, saying that their persecution "was vindictive and destructive of the very democratic principles" that had turned him "against Communism 25 years ago." At a Sobell defense meeting that spring, Conrad Lynn, Dwight's old comrade in the fight against discrimination in the armed forces during the war, noted the broad spectrum of politics represented in Sobell's supporters: they ranged from dyed-in-the-wool Communist sympathizers to such anti-Communist liberals as Reinhold Niebuhr and Norman Thomas. Lynn was sure that there was a new spirit abroad in the land; it could be seen among southern black students as well as in the entire generation coming of age in the 1960s.[4]

Dwight was invited to speak at the closing session of the first national convention of the newly founded Students for a Democratic Society in June 1960 in New York. He was also asked to participate with Murray Kempton and James Wechsler in the opening panel discussion with some students concerned with campus activism. Dwight paternally accepted both assignments. He would be delighted to join in the panel discussion because he was interested to discover "what questions the boys and girls ask."[5] He was soon to discover that many of these young radicals were far more politically sophisticated than he had imagined. It is remarkable how in tune Dwight's thinking was with the thoughts of the early leaders of SDS, Tom Hayden and Robert Haber. "The Root Is Man" became a seminal document in New Left circles, frequently reprinted by young radicals in search of a usable past. The bland, stifling Eisenhower 1950s (the Ike Age) represented everything Dwight had deplored in the mass society. But its chief menace was the increasing centralization of power in the State. There was too much planning from above and not enough problem solving on the communal level, too much sheep-like respect for authority and not enough man-like respect for one's own interests and values, too much herding together in conformist mediocrity and not enough assertion of those individual differences that make life interesting, too many bureaucrats and demagogues and mother-knows-best bully boys in high office ordering or bamboozling us into political behavior that is disastrous to our own interests.[6]

It was a libertarian vision by which, Dwight proclaimed, the individual might once again determine his own destiny and participate in the decisions that so deeply affected his life. Arguing in *Commentary* that a vote in the 1960 election was pointless since there was no fundamental distinction between the candidates, he also called for a revival of an American anarchist tradition, with its emphasis on the individual in defiance of State authority.[7]

He repeatedly insisted that the only solution was to find a way to break up the centralized state into small "communities where individuals could defend their own special interests." There was throughout Dwight's anarchist argument a contrasting mixture of bohemian challenge to authority and a nineteenth-century Jeffersonian liberalism soon to be embraced by Barry Goldwater. Certainly the New Left would have found it hard to swallow Dwight's notion that "the bourgeois free market, insofar as it still exists, is an admirably anarchist device for distributing goods." Probably as a knee-jerk effort to deflect guffaws from the left, he noted that "even the Russians have been forced to edge back toward it lately, the effects of planning and state control being what they are." He declared there was "quite a lot of anarchism going on now in the country." The lunch-counter sit-ins had anarchist features. They were started almost spontaneously by small groups and, much to his delight, they had effectively adopted the Gandhian technique of nonviolent resistance. He praised the Congress of Racial Equality for its organizational structure and tactics: small chapters, nonviolent, grass-roots social action, participatory politics, individuals making decisions, forcing changes in the everyday aspects of their lives. He again criticized Eisenhower's handling of the Little Rock crisis, denouncing the premature and unnecessary use of federalized troops. Presumably he felt that the civil rights movement could withstand the mobilization of racist resistance without federal help. That might have perplexed the young civil rights workers, who were soon demanding federal protection since they could get none from the local police.[8]

But if there were contradictions, inconsistencies, and unanswered questions concerning policy, the entire tone of Dwight's indictment of the status quo, his appreciation of the youthful activism, his insistence on moral values as the very foundation of a serious politics and on the need for active participation in important decision making—all coincided with the thinking of the new generation of student activists and separated him from what Mills described as the prematurely middle-aged. Many of the student activists were deeply skeptical of or uninterested in the old Marxist ideology; they appreciated Dwight's emphasis on individual values. Dwight came to the New Left with a reputation. Theodore Roszak, a serious student of the counterculture, felt that if the New Left drew on the "Old Left" at all, it was a Marxism mediated by the existentialism of Camus. Or, in the American radical tradition, "it is a humanist Socialist text like Dwight Macdonald's eloquent *The Root Is Man* that one discerns behind a document like *The Port Huron Statement*."[9] Staughton Lynd and Todd Gitlin, two leaders in "the Movement," have expressed their indebtedness to Dwight and *Politics*: "There was more good sense and fresh thinking in this one magazine . . . than in all the left journals from that day to this," Lynd declared.[10] Both reread *Politics* to get their bearings during the turmoil of the 1960s.

And the young activists liked Dwight's persona, too. He had developed, even crafted, a distinctive personal logo. He continued to sport his "Leninist" goatee through the years. He often affected a corduroy sports jacket with a chambray work shirt or very loud striped shirts and clashing ties. His lapel was invariably adorned with a button endorsing Emma Goldman or some other old battle-scarred warrior of earlier struggles.

He had a deep and abiding skepticism of the academy and its pretensions and was instinctively opposed to college and university administrators as insensitive to student interests. He continued to nurture his long-held aversion to all forms of pomposity. His age and background were an asset to the youth movement, despite its slogan "Don't trust anyone over thirty." Dwight had connections and access to the media; he knew important people and was respected for his consistent honesty. Perhaps most important, he was ready to listen and to learn. While he toted a good deal of political baggage from the old leftist days, he did not apologize for his investment in past political struggles, but he could acknowledge errors and mistakes in his judgments. He was not afraid to simply abandon old positions. And unlike some of his former associates on the left, he was not about to discourage this potential rebellion—after all, they didn't come along every day.

Dwight's speech on "The Relevance of Anarchism" at the SDS convention in June 1960 had been a political parting shot before a year in London. By the time he returned in September 1961, a genuine student movement was well under way. The election of Kennedy had only heightened the optimism and enthusiasm of many young people, who were certain that the first president born in the twentieth century would cast off the cold war dogma of his predecessors and confront the real issues that concerned the students: racism, poverty, meaningless work, unimaginative, soul-destroying education, and, of course, war and imperialism. There was little in Kennedy's career to support such optimism, but he had run against Nixon, whose career was marked by an extreme anti-Communism and vicious red-baiting undertaken to enhance his political prospects. Nixon made Jack Kennedy look good. There was disappointment over the Bay of Pigs in April 1961, but student activists' spirits were lifted by the Freedom Rides. Inspired by the new possibilities, CORE had begun the rides in May 1961, and if anything, they pointed up the fact that those demanding an end to racial inequality were far out in front of the Kennedy administration.

Dwight observed these events with interest from London. He had shown very little enthusiasm for Kennedy. The American election had simply been an excuse for a party at the Russell house in St. Johns Woods. Only a little more than a week before the Bay of Pigs invasion, Dwight refused to sign a pro-Castro ad to appear in the *Times* on the grounds that Castro's speeches contained incendiary and inaccurate attacks on the United States. Perhaps more to the point, Dwight didn't like Castro's "free-handed use of the death penalty" against "counter revolution" and for things like dope peddling or black market speculation in currency. He cited *New York Times* and *Time* magazine articles indicating that Communist influence in the Castro regime was considerable. Dwight was "skeptical and worried."[11]

The Bay of Pigs invasion, however, was another matter altogether. Castro's frequent charges of an impending invasion proved all too true. Dwight dashed off a quick cable to Arthur Schlesinger, who was now an aide to Kennedy and Dwight's new access to the corridors of power: "My advice on Cuba is whoa!" To his friend Barbara Deming, he wondered how "Kennedy and Schlesinger and all those bright decent young liberals have gotten into THAT mess." He thought they should read Tolstoy's "Stop and Think," which he had reprinted in *Politics* in 1946. His return to Tolstoy's almost mystical document preaching solitude, meditation, and serious questioning about the means and ends of policy reflected his instinctive feeling that the New Frontiersmen were "men of action."

His analysis was reminiscent of that of his "culture hero," Randolph Bourne, who castigated the "pragmatists" of 1916 who were all for getting the job done by going to war without giving much thought to the purposes or consequences. Bourne too targeted the liberal intelligentsia "trained up in the pragmatic dispensation, immensely ready for the executive ordering of events, pitifully unprepared for the intellectual interpretation or the idealistic focusing of ends." Dwight had been rereading Bourne during these years and his name crops up frequently in his correspondence. He wrote Mark Harris in January 1960, "The young people of the USA today need to hear about RB," and in fact Bourne's essays were reprinted and widely used for college courses in the 1960s. It would not be long before anti–Vietnam War intellectuals such as Noam Chomsky would be citing Bourne and Macdonald as seminal critics of American militarism and in particular its liberal apologists.[12]

If a younger generation of dissenters was reading or rereading Dwight's political writings, he was keeping a close eye on their work as well. He was particularly interested in an article by Barbara Deming in the Spring 1960 *Nation* on the militant pacifists known as the Peacemakers. In the 1940s Dwight had been an active member of the group; he wondered now if he should not rejoin them. That he could be so receptive to their stance is revealing, since they were extreme anarchopacifists who believed in a pure form of nonviolent resistance. It was a discipline that, according to Deming, transformed the participants' entire lives, much like a religious conversion. The tone of the Deming piece was reminiscent of the feeling in the last section of *The Root Is Man*, where he called for a totally new moral and ethical way of coping with the world.[13]

That Dwight was so taken with Deming's piece strongly suggests that the cold war pressures were not nearly so intense for him in the early sixties as they had been when he abandoned pacifism only a little more than a decade earlier. The world had changed. Khrushchev was no Stalin, and even if the Soviet Union still had the worst qualities of a mass society, Dwight was not about to be drawn into the same kind of position he had taken during the height of the cold war, explicitly and publicly "choosing the West" and thereby giving support to the State Department and its belligerent policy. As for the immediate situation, he wrote that he would have opposed the Bay of Pigs invasion even if the "US-supported putsch had overthrown Castro, and so would almost all Latin America. Who the hell does Kennedy think he is, the avenging sword of Jehovah?"[14]

When Dwight returned to the city from abroad in the fall of 1961, he was much more interested in all kinds of politics than he had been for more than a decade. He felt the excitement in the air: "The trouble with New York is there are so many things happening—it is indeed lively— but one can hardly respond to all the stimuli." Dwight was convinced that something was stirring. People were waking up.

One of the things that awakened him, as well as the rest of literate America, was a book by Michael Harrington, the former Catholic Worker and now Socialist Party activist who had spoken at the same SDS convention. Dwight knew Harrington from researching his Dorothy Day profile for the *New Yorker* and was impressed with the young man's intelligence and dedication to the socialist cause. In February 1962 the publishers sent him an advance copy of Harrington's book *The Other America: Poverty in the United States.* Dwight was eager to do some serious political work in the *New Yorker* and Harrington's book hit the mark. Dwight thought it was not technically brilliant and its prose was pedestrian, but its message was imperative, since one of the major social problems was that "everybody takes it for granted that there is no large scale poverty in the USA and Harrington shows there is."[15]

Dwight took the Harrington book seriously, perhaps because he too had accepted the notion that poverty in America was a diminishing problem that could be mopped up by a few technocrats. He had admired John Kenneth Galbraith's *The Affluent Society,* but he began his extraordinarily long review of the Harrington book with the up-front statement that Galbraith's assertion that poverty in America was no longer "a massive affliction was inaccurate though unfortunately it was widely believed." Dwight put an extraordinary amount of effort into this review. He read widely in the field, spent a long alcoholic evening with Mike Harrington discussing poverty, and was revising and adding to the piece almost up to the moment of its publication in January 1963.[16]

It was an eloquent summary of Harrington's earnest work. Dwight approached the subject with his dry, acidulous wit, his direct simplicity of language, and a calculated political methodology. After all, here was another of those pieces of serious social criticism that was to be bordered by the exquisite commercial lace of the *New Yorker*'s typical ads for the gimcracks of high capitalism. On one of the more devastating pages of description, one's eye could not help wandering to the left to stare at a travel ad: "We're saving a sunny spot for you right now in the islands of HAWAII."[17]

After presenting the inevitable and necessary statistics, Dwight's summary gave a stout and consistent defense of Harrington's book and his thesis that poverty was widespread in the United States and that though fewer people were poor now than in the 1930s, the chronic nature of poverty had increased because its victims were increasingly invisible and without political representation. Those who lived below the poverty line (roughly $4,000 per year for a family of four in 1963) were far more likely than previous generations to remain poor. "Rags to rags" had replaced "rags to riches." Insofar as Michael Harrington raised and mildly supported the idea of a "culture of poverty," he echoed the famous Fitzgerald-Hemingway exchange. Fitzgerald had insisted: "The very rich are different from you and me." Hemingway is alleged to have replied, "Yes, they have more money." Fitzgerald meant that the rich have a way of seeing the world, of looking at life, a confidence, an invulnerability. Harrington insisted that the poor, too, had a particular way of looking at the world: "Poverty in the United States is a culture, an institution, a way of life." Everything about the poor was different, "from the condition of their teeth to the way in which they love . . . everything is suffused and permeated by the fact of their poverty." This was hard for the middle class to grasp. But Dwight had no difficulty with the concept. He had no experience with the poor, but he did have an appreciation of the effect of community or the lack of it on the development of character and personality. He was sensitive to how the poor might see the world in a different way. At the end of his review, the longest in the history of the magazine, he commented again on how the current generation of poor people seemed to lack the motivation, the hope, yes, the ambition of earlier generations. Dwight seemed to accept Harrington's suggestion that a despairing "culture of poverty" existed and that the newer generation did not seem to have the drive of earlier immigrants, did not appear to value education as much, did not aspire to more than they had, and lacked a vision of potential social mobility.[18]

One major factor, the problem of race, hardly appears in Dwight's review. Harrington devoted an entire chapter to the plight of the blacks, who in the 1960s constituted 11 percent of the population but were nearly a quarter of those listed as below the poverty line. The blacks were hardly invisible, but they were increasingly isolated in central cities, while there was a mass exodus of the middle class to the suburbs. Dwight was conscious of all these factors and they were not neglected in his review, but they were not developed either. Perhaps his recent position during the *Ben Hur* incident that since the war blacks had been winning in their struggle against discrimination explains his reluctance to raise this issue. Or perhaps he simply did not think it wise politically.

He closed his essay review with a spirited demand that the middle class and the wealthy live up to their responsibilities and support a tax program that would provide the poor with enough income to raise them above the poverty level:

> The problem is obvious: the persistence of mass poverty in a prosperous country. The solution is also obvious: to provide out of taxes the kind of subsidies that have always been given to the public schools (not to mention the police and fire departments and the post office)—subsidies that would raise incomes above the poverty level, so that every citizen could feel that he is indeed such. Until our poor can be proud to say "Civis Americanus sum!", until the act of justice that would make this possible has been performed by the three-quarters of Americans who are not poor—until then the shame of the Other America will continue.[19]

Published early in 1962, Harrington's book had quickly found itself in relative obscurity. Harrington assumed it had no future, and he left for Paris to pursue his interest in leftist literary criticism. It was Dwight's *New Yorker* review, appearing almost ten months after the book's publication, that attracted phenomenal attention. Dwight was enthusiastic about the "extraordinary response. . . . The Invisible Poor are rapidly becoming visible. Everybody seems to have been worried and half conscious about the subject—social workers, sociologists and just plain citizens—but the article triggered off their reactions." His catalytic piece was, he boasted, "a brief definition of journalism." This may have been one of the few instances where Dwight seemed to appreciate the dimensions of his talent, his ability to bring to the public marketplace of ideas an important issue, to put it in terms that literate laymen as well as professionals could grasp and deal with. The myth was quickly crafted that Kennedy, in the manner of Teddy Roosevelt being motivated by Upton Sinclair's novel *The Jungle* to get the Pure Food and Drug Act passed, was persuaded by *The Other America* to take action against poverty. More recent scholarship suggests that the president read Dwight's "witty and elegant essay" rather than Harrington's earnest and impassioned but less readable account. TRB in the *New Republic* praised Dwight for the splendid job he had done: "As much as anything it was Dwight Macdonald in the *New Yorker*, summarizing Michael Harrington's book on poverty, that roused President Kennedy." It may well have been the most effective political act of Dwight's career, and he was always proud of the review's direct impact on policy.[20]

Throughout 1962 Dwight continued to view the Kennedy administration with growing skepticism, however. There was an edge to his comments. When Arthur Schlesinger was taken to task for making money by writing as a White House insider and in the same week fell into the swimming pool at Bobby and Ethel Kennedy's house, Dwight sent off a clipping to Mary McCarthy in Italy describing it as "Arthur's catastrophic week." Schlesinger had been pressured by the administration into donating all profits from his writing to charity. While Dwight undoubtedly enjoyed Schlesinger's embarrassment, he felt he had been thrown to the wolves. Still, it might teach the historian "the hard way that Kennedy is no neck-sticker-outer." To his old friend Ping Ferry of the Ford Foundation, who had recently attacked J. Edgar Hoover in the press, Dwight sent congratulations and commiserated with him that Bobby Kennedy "felt obliged to cover up" for Hoover. Dwight thought the administration's toleration of Hoover "sad and shameful" and, apparently unaware of Roosevelt's own role in the FBI boss's ascension to power, compared the "timid" JFK unfavorably with FDR.[21]

The real shock came with the Cuban missile crisis in October 1962. Although in 1960 Dwight had tentatively suggested that a tough policy toward the Soviet Union was the best way to keep the peace, he was appalled by the eyeball-to-eyeball diplomacy that many thought was taking the country to the brink of a nuclear war. At the height of the crisis, Dwight denounced the entire past U.S. policy toward Cuba and pronounced the present blockade as precipitate and reckless. He insisted that the issue of the Russian missiles should have been taken before the UN: "A commission of neutrals should have been established to examine the facts; negotiations should have been undertaken; if after a reasonable delay, the administration felt that it had to act for the safety of the republic, then such an action might have a moral and rational justification that was at present lacking." On October 27, when he read in the press that Kennedy had rejected a deal with Khrushchev that would have called for the removal of U.S. missiles from Turkey and the Russian missiles from Cuba, Dwight dashed off a telegram to the president:

APPALLED BY REJECTION OF UNEXPECTEDLY GENEROUS COMPROMISE PROPOSED BY KHRUSHCHEV STOP HE IS ACTING IN REALISTIC TRADITION OF BISMARCK GLADSTONE AND METTERNICH STOP YOU ARE BEHAVING AS I WOULD HAVE EXPECTED HIM TO STOP IMPLORE YOU TO RECONSIDER.

Dwight's response to this event and his criticism of Kennedy's "reckless provocation" reveals how far he had come from his hard-line support of a much more belligerent position a decade earlier. At one point he even argued that the "overwhelming majority submit to Castro's authority with more enthusiasm than has ever been the case in Franco's Spain." And as for the Russian leader, Dwight had a warm spot in his heart for Khrushchev, repeatedly reminding some of his old comrades that Khrushchev was not Stalin, and thus a far less aggressive and more reasonable and diplomatic foreign policy was called for.[22]

Perhaps Dwight's feeling about the current administration allegedly manned by the "best and the brightest" is most clearly revealed in his review of Schlesinger's collection of essays *The Politics of Hope* that appeared in the first issue of the *New York Review of Books*. This publication had grown out of a strike at the *New York Times* in the early winter of 1963. Dwight had nothing but contempt for the *Times Book Review,* which he was in the process of trashing in *Esquire*. He was one of the enthusiastic supporters of this new enterprise and had been in on its premiere issue, which was promoted with great fanfare. Dwight's piece was titled "Mr. Schlesinger's *Realpolitik*." It was a Bourne-like piece focusing on what he later described as "Arthur's . . . bobby-soxer attitude toward those in power." It also reflected his suspicion of the New Frontiersmen's inordinate pragmatism at the expense of democratic principles. Dwight targeted Schlesinger's "official and reverential" apologetics. During the fifties when the Democratic Party was out of power, Schlesinger had lectured his fellow liberals on the need for "heroic leadership" à la Teddy Roosevelt. Now that the Kennedy brothers were in office, the nation finally had heroic leadership. In his critique of Schlesinger's uncritical portrait of Reinhold Niebuhr, Dwight noted how easily the "intonations of the fashionable preacher blend into those of the ideological con man." Thus the scholar who had written the "historical-mythological underpinning" for past Democratic administrations was back at work again, writing rotund passages of purple prose about the present national leadership, from which Dwight quoted the acolyte's description of the Kennedy administration: "young, vigorous, intelligent, civilized and experimental." The Sons of Liberty had returned, as though "wakened from a trance." Indeed they had; they were in power, and apparently Schlesinger thought they should not abrogate the constitutional rights of citizens "except in the face of war, revolution or economic chaos." Dwight wondered why it was always "the great liberal presidents," Jackson, Lincoln, and Roosevelt, who suspended civil rights. He speculated perhaps it was because they had "good consciences, supplied by intellectuals like Mr. Schlesinger."

Dwight was aware by 1963 that the country was well on its way to a major breakdown of consensus concerning foreign policy and civil rights. He welcomed the renewed debate, and especially the challenge to the State. He was particularly irked by intellectual apologetics for the centralization of power. He noted that when Schlesinger was out of power, he gave lip service to intellectual protest, but now as an insider, Schlesinger was singing a different tune. Dwight wished his witty, clever, sensible, and decent friend had "never gotten involved in high politics."[23]

It was the old argument about purity versus power. Dwight took the classical dissenting position. It was the role, indeed the obligation, of the intellectual to challenge established wisdom, to practice the fine art of negativism, and to remain aloof from corrupting power. But postwar liberal intellectuals had prided themselves on their "realism," their "pragmatism" and practicality, their toughness and willingness to accept the risks and responsibilities of power. This theme pervaded liberal and academic scholarship to the point that even such a genuine scholar and intellectual as Edmund Morgan, the colonial historian at Yale, could turn a brief biographical portrait of John Winthrop into a cold war tract for undergraduates, denouncing intransigent critics of State authority as "simplistic" separatists. "Their forthrightness and courage defied a wicked world. But their defiance was a desertion." Since his initial backhanded praise of Thoreau's civil disobedience as an act of courage and wisdom, if not of intelligence, in *The Age of Jackson,* Schlesinger was forever denouncing those "utopians" whose purity rested on their lack of responsibility: "The pragmatists accept the responsibilities of power, and thereby risk corruption; the utopians refused complicity with power and thereby risked irrelevance."[24]

Schlesinger's conduct reminded Dwight of Bourne's stricture against the "mothlike gyration" of a previous generation of intellectuals who had flocked to Washington from the campuses in 1917 to serve the State, to write apologetics, to espouse the virtues of an antiseptic war. After this brief exchange in 1963, Dwight looked on Schlesinger with a bemused yet jaundiced eye. At one point when Schlesinger titled one of his *Show* magazine columns on old-time comics Harold Lloyd and Buster Keaton "One Vote for Anarchy," Dwight wrote a colleague in exasperation, "What the hell does Arthur know about anarchy." In his view, Schlesinger had become a shill for the State, and Dwight kept a careful file on Schlesinger's role as an insider in the liberal war machine that became his obsession during the remainder of the decade.[25]

Despite Dwight's renewed interest in politics, he was an established cultural figure making radio and occasional TV appearances and university lectures. After listing an endless series of engagements, debates, talks, book contracts, and advances in November 1962, Dwight exclaimed with delight to his old comrade Nick Chiaromonte, "I tell you, Nick, we intellectuals have struck Gold." He was going to be paid $320 for a weekly spot of ten minutes to pontificate on movies for the *Today* show. He quoted a *Variety* headline: DWIGHT MACDONALD THE FILM RIPPER TAPPED FOR TODAY.[26]

It was all very exhilarating. It was also very social as well. He and Gloria were invited to an endless round of cocktail parties, book parties, dinner parties, and receptions, and they were entertaining a great deal in their 87th Street apartment. They both were drinking more. In the spring of 1963, Dwight developed a serious kidney ailment, but decided to try medication and a strict diet to avoid surgery. There was to be no alcohol. He did cut back from a dozen drinks a day, bourbon highballs and martinis to two or three. He kept up this regimen for about a month, and lost 17 pounds in two weeks. His weight had reached 201; his protruding belly bothered his vanity and he often went on crash diets to try to get rid of it.[27]

In August 1963 Dwight and Gloria flew to California for a brief series of lectures at Berkeley. While there he had a frightening and devastatingly painful kidney stone blockage which demanded an operation that took over four and a half hours. The operation was successful, but the painful aftereffects were both physically and psychologically debilitating. The pain was like "nothing in any previous category of my experience. There seemed to be no escape, no exit, only hideous pain," and, he thought at the time with some relief, "very likely death." It was so traumatic that he experienced a helplessness and loss of identity. He also felt that he had "been violated," a feeling that did not pass for weeks. His illness lasted nearly two months, because the wound became infected and had to be kept open. It was not until the middle of October that the ordeal was over.[28]

On his return to New York he found himself being drawn into a wrenching literary-political conflict that drove a deep wedge into whatever remained of the New York intellectual community. Hannah Arendt's five-part *New Yorker* series on Adolf Eichmann had been published in book form as *Eichmann in Jerusalem: A Report on the Banality of Evil*. It elicited an outburst of denunciation and recrimination from the Jewish community, who saw it as accusing Jews of collaborating in their own destruction. Because they were such good friends and he admired her articles and the book very much, Dwight would have come to the defense of Arendt in any case. At the same moment he was being drawn into this dispute, he was in correspondence with Gordon Zahn, who had written an article in *Commonweal* taking issue with some of Dwight's conclusions in his original *Politics* essay "The Responsibility of Peoples." Dwight was impressed and persuaded by Zahn's piece and felt the question of responsibility was important because America was beginning to take a role in world affairs that demanded the conscious and active participation of citizens and a recognition of their responsibility. In effect, the Eichmann controversy and the Zahn exchange were preparing Dwight to shape his own conduct as he began to work his way back into a more activist politics.[29]

Zahn had written his piece as a consequence of the debate surrounding the trial of Eichmann. Eichmann argued that he did not will the destruction of the Jews, he merely carried out orders. "My guilt lies in my obedience," he told the Israeli court trying him for crimes against humanity. In more recent times, Zahn noted, not only had the notion of collective guilt been abandoned, justifiably, but the responsibility of peoples has been turned inside out, and even former architects of Nazi policy had been rehabilitated and held high positions in the NATO military establishment. Russian demands for their arrest and prosecution only added to their claims of respectability. Zahn was not interested in this cold war political thicket. He was concerned with the "ethical and theological principles . . . which permits an individual to rationalize his conformity and even participation in unjust actions or regimes." He was interested in the argument that Auschwitz and Hiroshima offered proof of "a frightful human capacity for 'justified' inhumanity." Zahn was calling for the rejection of the obvious implication of Dwight's argument that the average citizen has no control over the decisions that affect his life, no means of resistance. Zahn was insisting that the individual can and should refuse to participate in actions that contribute to consequences he deems to be morally wrong.[30]

Dwight agreed that the claim of innocence through obedience was not acceptable. He argued once again that it was not the lawbreaker who was to be feared so much as he who obeyed the law. He was moved by Zahn's story of a poor Austrian peasant who was beheaded in 1943 for refusing to serve in the war. As a Catholic, the peasant believed the war unjust, despite the admonishment of his priest, bishop, and a succession of chaplains who sought to quiet his doubts and encourage him to live up to his obligations as a loyal citizen. Dwight felt that such an example "makes one proud to be human, since such people are also human." In his letter to the *Commonweal* editors he wrote: "That Austrian peasant insisting on the primacy of his own conscience against the threats of his State and the sophistries of his Church almost makes one believe that God's creation of mankind—or was it the Devil's?—may not have been such a bad idea after all."[31]

Questions of ethics and integrity were at the core of Dwight's involvement in the "Eichmann Affair." Dwight was enormously impressed with the book, rating it the best thing Arendt had written since *The Origins of Totalitarianism. Origins* had had its critics, but her study of Eichmann prompted a storm of abuse. Mary McCarthy was not far off the mark when, in her defense of her good friend, she wrote of the "Hue and Cry" emanating almost entirely from Jewish intellectual circles. Only a few gentiles— special cases, she noted—had attacked the book. Norman Podhoretz titled his critique "The Perversity of Brilliance." He suggested that Arendt's facile concept that the most heinous murderer may look like the next-door neighbor or an amiable bureaucrat was a mitigation of the horrendous crimes against humanity and led inevitably to a thesis which blamed the victims. She, with aesthetic eccentricity, had given her readers a "banal Nazi" instead of "monstrous Nazi." How could one of the "desk murderers" take the place of the full-fledged monster of classic Nazi depravity? The monocle, the whip, the boot, the gun were essentials in separating these monsters from the rest of humanity. That Arendt's "friendly fascist" was more frightening than the classic stereotype was a crude, intellectual conceit. Worse was the charge that some Jews had collaborated with their oppressors. This was beyond the pale, despite the literature on the subject written by Bruno Bettelheim, David Rousset, Eugene Kogon, and many other survivors.[32]

When William Phillips at the *Partisan Review* solicited a review of Arendt's book from Lionel Abel, Dwight was infuriated. Abel had written a very harsh attack questioning Arendt's philosophical credentials several years earlier in the little Trotskyist magazine *New Politics.* Dwight felt Phillips and Rahv had hired a hit man, a practice they had used in the past. When Dwight called Phillips on this ethically dubious editorial practice, Phillips claimed he had forgotten Abel's earlier assault on Arendt. Dwight's response was that if he and Rahv had actually forgotten the earlier Abel criticism, then they "shouldn't be assigning books for review in a serious magazine." If they had not forgotten, then they obviously wanted Arendt's book roughed up, and Dwight thought they had picked the right person to do it, a "narrow neurotic" with a personal animus toward Arendt. "Lionel is malicious, hot-headed, irresponsible; you can't ever trust him to be fair or even sensible."[33]

Since *PR* let Abel carry on his diatribe against Arendt for nearly twenty pages, Dwight demanded the opportunity to respond to Abel. The thrust of Dwight's counterattack was to accuse Abel, Podhoretz, Bell, and others of propaganda in the name of Jewish patriotism rather than serious criticism. While he found Mary McCarthy's assertion that reviews by gentiles were favorable and those by Jews unfavorable to be unconvincing because there were too many exceptions, he agreed that in most cases and in most personal encounters the reactions did divide along Jewish and non-Jewish lines. He regretted even speaking of "Jewish" friends and "Jewish" critics, claiming he'd thought they "had gotten beyond such labels in serious discussion." Nor could he fathom why they were so incensed at Arendt's critique of the leaders of the Jewish Councils. (She reasoned that if they had been leaderless their organized and efficient destruction would have been a good deal more difficult.) The problem became clear in an exchange with Arendt in *Encounter,* in which the Jewish scholar Gershon Sholem referred to a principle all Jews knew: "In the Jewish tradition, there is a concept, hard to define and yet concrete enough, which we know as *Ahabath Israel:* 'Love of the Jewish people.' . . . In you, dear Hannah, as in so many intellectuals who came from the German Left, I find little trace of this." Dwight noted that even a hardheaded social historian like Daniel Bell, with his "instinct for the center, vital or dead according to one's political temperament," insisted that in a controversy such as this, "one's identity as a Jew . . . is relevant." Dwight ended his defense of Arendt with an accusation of ethnic partisanship on the part of her critics and with a reassertion of his stand in the forties: "I'm old-fashioned enough, my political education going all the way back to the thirties, like Mr. Bell's, to find any exceptional category morally suspect and intellectually confusing. So I take heart in a book like *Eichmann in Jerusalem.*"[34]

With the Eichmann controversy, Dwight had returned to the political-intellectual wars. He had not lost his street-fighter instincts. He insisted on his independent cosmopolitanism, he repudiated any form of nationalism, patriotism, chauvinism. In his dig at Bell he challenged the center, which he found dead rather than vital. Dwight was warming up for the political and intellectual warfare that was already on the horizon. The assassination of John F. Kennedy in November 1963 had, as Malcolm X said, brought the chickens home to roost.

17

Back to the Barricades

UNLIKE MALCOLM X, WHO CLAIMED THAT KENNEDY'S assassination constituted poetic justice for his administration's role in the coup that led to President Ngo Dinh Diem's assassination in Vietnam, Dwight had no belief that history had a rational pattern, certainly not one that could be equated with justice of any kind. He did see two events as highlighting the cruel results of the realpolitik he so distrusted, however. Immediately after Kennedy was killed, Dwight suggested to Mary McCarthy that they form an independent committee to get at the truth of the assassination, which was already engulfed in suspicions of conspiracy and coverup implicating a variety of nations, national organizations, and world leaders, including Kennedy's successor. When McCarthy showed an interest and called her friend Arthur Schlesinger, he replied in effect: "Why don't you amateurs keep your big feet out of history and leave it to us pros?"[1]

Schlesinger's dismissal would only have stimulated Dwight's interest, given his penchant for amateurism and suspicion of experts, particularly those whom he felt served the State. In the first months of 1964 he began making a serious study of the Kennedy assassination and its meaning. While he had always had a skeptical attitude toward Kennedy and particularly the Kennedy mystique of pragmatic politics, he was, like most Americans, shocked by such a brutal act that cut short the life of an attractive and energetic young man. Not long after the assassination, Dwight described feeling worse than he had when Trotsky and Gandhi were murdered. Both those men had lived their lives, but Kennedy had just begun to assume his role on "the great stage." He was handsome, vigorous, with a great deal of panache, and, Dwight added, he was married to a woman who went through the days immediately following his death with a style that approached heroism. The debutante beauty proved to be a "Roman matron."[2]

"The Warren Report is an American-style *Iliad*, i.e., an anti-*Iliad*, full of anti-heroes, retelling great and terrible events in limping prose instead of winged poetry." So began Dwight's detailed dissection of the report. Dwight waded through the entire 912-page document and consulted the twenty-six accompanying volumes of testimony and artifacts. Throughout a good part of 1964 he gathered material, for his long *Esquire* article and for a book on the subject he contracted to do for Harper & Row. Dwight concluded that the Warren Report was a "slovenly mess," a careless and inept investigation. He believed that the overwhelming factual evidence confirmed the report's conclusion that Lee Harvey Oswald shot the president, that he acted alone, and that there was no conspiracy. Although ten years later, after wading through a mass of new material, his confidence that Oswald acted alone was shaken, Dwight never officially changed his mind.[3]

Dwight's analysis in the *Esquire* piece rejected the conspiracy theories of Marxist leftists and liberal ideologues that resulted from their belief that "history is an understandable working-out of the conflict, dialectically progressive, between large, dignified and abstract forces, and is definitely not a chancy game in which small, trivial individuals can absurdly and accidentally affect the outcome." They believed this to be the case because otherwise history would "make no sense." Dwight insisted, as he had when he initially speculated on nations and dictators, that history was indeed a "chancy affair." There was no sense to it. It was perfectly "reasonable" that an incompetent loser like Oswald could change the course of history.[4]

Citing Homer, Plutarch, Poe's "Purloined Letter," a veritable library of literary and historical references, and a survey of all other presidential assassinations in America, Dwight put the tragedy in the historical context of an evolving mass society. He explained the risks taken by the president as due to the need to make contact with the crowd. A nation of 200 million could not really be governed democratically; the decision making was confined to a few men at the top, with everybody else passive observers, and exposure helped bridge the gap between the powerful and the powerless. Pressing the flesh was a "ritual compensation for an imbalance that makes both sides uneasy." The rhetoric of democracy insists that the president is the citizens' equal. They are his "people." But in that open car in Dallas they had suddenly become Hamilton's "Great Beast." The assassination under such circumstances, carried out by such a prototype mass man, fit every theory Dwight had about mass society and the underlying threat of violence and totalitarianism.[5]

In the modern mass media world there was an obsession with image and how one would be counted in History. The irony was that both Kennedy and his killer harbored the same obsession with their images in history. The assassination was also history in the making, and not only Oswald but his wife and mother were soon part of the action. Dwight recorded Marina Oswald's negotiations with her agent over contracts, book and movie rights, and all matters pertaining to publicity, public relations, and advertising. As an expert on cinema, Dwight speculated that *"The Americanization of Marina* might make an interesting movie." Marguerite, or "Mother Oswald," as she was dubbed by the media, "rather like Mother Macbeth," stepped quickly onto the stage. The only problem is that she was flaky and incoherent. Dwight could imagine her doing a duet with Mark Lane, the conspiracy fanatic, before the commission. Dwight held Lane in total contempt; he was either mentally ill or a con man, who blatantly or stupidly or perhaps insanely manipulated "the facts" to fit his conspiracy theses. He was called upon to vet the manuscript of Lane's book *Rush to Judgment* for Viking Press; he wrote a devastating attack on the man and his work and credited himself for their decision not to publish the book.[6]

This bizarre cast of characters drew Dwight into the absurd aspects of history, its gratuitous and unpredictable nature. Individuals may have focused motives, but they have no control over the historical consequences of their acts. He compared Jack Ruby, who shot down Oswald in the inexcusable bedlam of the Dallas police station, with the "anarchist scatterbrain" Van der Lubbe, who torched the Reichstag, and the "ultra-left simpletons" who had recently plotted to blow up the Statue of Liberty. All had "pure" motives of one kind or another, and all expected to be recognized as martyrs in a great cause. In each case these individuals let loose furies totally opposite to their stated goals. They were "pole-axed by History." Dwight had no problem concluding that the Kennedy assassination had been carried out by a demented neurotic. However, given the handling of the case, the stupidity and ineptitude of the police and the FBI, he was left with the question: "Is Dallas America, or is it merely Texas?" Dwight thought it would be comforting to think the latter, but he was obviously not at all sure.[7]

Dwight's piece on the Warren Commission Report was well received by most of his readers. Only the hard-core conspiracy buffs, whom he had directly challenged as psychologically incapable of believing that an isolated oddball killed the president, saw him as offering support to the prosecution.

During his research he had gotten in touch with I. F. Stone, his old "Stalinoid" nemesis, and they became amiable comrades in their united belief in Oswald's sole guilt, and before long in their spirited opposition to American foreign policy. Murray Kempton was another who took the same stance. Norman Mailer nominated Dwight to serve on an independent commission to investigate the assassination. Dwight's article was nominated for inclusion in Crown Publishers' anthology of the best articles of 1965. Sargent Shriver invited him to serve on President Johnson's National Advisory Council dealing with poverty.[8]

Dwight's star continued to rise. He had become a literary, cultural, and political guru. He was awarded an honorary doctorate from Wesleyan University along with Martin Luther King in June 1964. Many doors were opening to him; people sought his views, and paid for them too. When Nat Hentoff wrote to ask him to appear for nothing on a new WINS radio talk show Hentoff was hosting to debate the art of film with Amos Vogel, the head of an art house, Cinema 16, Dwight refused on the grounds that WINS was a commercial station and it could afford to pay him. After all, both noncommercial Channel 13 and Camera Three paid him a small fee. WINS seemed to think "we intellectuals are starved for outlets . . . not true in my case, lots of chances to address an audience and write for largish groups and be well paid also. Things have changed of late and culture is accepted, indeed it's essential to the image of a radio or TV station." Dwight had his doubts about the quality and standards of the "culture explosion," but for the first time in his life he was receiving wide recognition and making money in the process.[9]

Yet there was a tension in Dwight's life at this point. He enjoyed the fast-track role of culture critic, the gregarious life of the celebrity, and at the same time he yearned to return to his old role as a serious critic, not simply of mass culture but of the tawdry and increasingly dangerous political and social life of the nation, its foreign policy, its moral values. When he appeared on college campuses denouncing the thinness and banality of popular culture, he found it difficult to translate the message into a viable dissenting political program. Calling for two cultures, one high and the other low, seemed to many an arrogant cop-out. Some students saw his stance as snobbish and irrelevant.

In the fifties the message did not clash so much with the general atmosphere on campus. But against the civil rights movement, the rise of the New Left, the first budding of a counterculture, rooted in a fervent belief in democracy, even egalitarianism, Dwight's message sounded dissonant. That was no hindrance in itself, since he thrived on just that. The problem was that he himself was in sympathy with much of the student criticism of the status quo. He was not turned off by their style, their contempt for protocol and manners. As a solution to this dilemma, he began to balance his mass culture critique and defense of the canon with his promotion of individualistic anarchism as a relevant political position. He particularly stressed the growing student opposition to authority. It is not surprising, then, that as the student movement began to take on a spirited momentum, Dwight, so sensitive to changes blowing in the wind, took a renewed interest in anarcho-pacifism and especially any opposition to an aggressive global foreign policy.

Like most other Americans in 1963–64, Dwight could not escape witnessing the incredible scenes on television each night as the pictures flickered from police dogs and hoses assaulting demonstrators in Southern cities to the Vietnamese crackdown on Buddhist pagodas and the resulting Buddhist resistance, accompanied by frequent self-immolation for the TV cameras. It was a ghastly display of authoritarian power in both instances and deeply unsettling to a growing number of Americans. American policy in Southeast Asia had hardly been front-page news, but the Kennedy administration had been deeply involved since Kennedy had taken office. In the early years of his presidency, however, the administration played down their keen interest in the area. Even after military personnel were sent first to Laos and then to Vietnam, the American role was kept as quiet as possible. Later one scholar referred to the early years of involvement as "Kennedy's Private War." By 1963 Kennedy had made a solid commitment by dispatching 10,000 "advisers" and letting it be known that America thought this was a place to stand up against Communist aggression.[10]

The early interest and involvement of the Kennedy administration in Vietnam had not prompted Dwight's interest or criticism. He had been no champion of the Communist rebels in Vietnam. In the forties he had rebuked George Padmore, the militant pan-Africanist, in *Politics* for his pro–Ho Chi Minh stand. In July 1963, Dwight signed a statement written by Huynh Sanh Thong, a young Vietnamese lecturer at Yale, to the *Times* denouncing American support of the "oppressive and authoritarian character of the Diem government" but also urging the Kennedy administration, if it was really interested in preserving democracy, to take a far more active role in bringing about Diem's overthrow. An accompanying letter by Thong that was presumably endorsed by Dwight as one of the signers urged Kennedy to act quickly to "help direct" the revolution against Diem into constructive channels. In effect, this letter endorsed the role the Kennedy administration was subsequently to play in bringing down the Diem regime. The statement also insisted that the United States make it clear that it would not tolerate a Communist takeover of South Vietnam. One of the demands was that the American government "send Ambassador Lodge to Saigon right away." Henry Cabot Lodge did go and was intimately involved in the coup.[11]

Julius Lester, a young black activist of the sixties, recalls joining protesters on a picket line in front of the Waldorf-Astoria Hotel in the fall of 1963. The protest was a joint action sponsored by SDS, the Student Peace Union, and the War Resisters League. It was undoubtedly through the WRL and SDS that Dwight was recruited. They were picketing the visit of Madame Ngo Dinh Nhu, the sister-in-law of the South Vietnamese president. Labeled the "dragon lady" after she indelicately referred to a priest's self-immolation as Buddhist "barbecue," she became a notorious political figure in the United States. There were only about twenty pickets, and Lester recognized Dwight. He was the oldest person on the line, and it made an indelible impression on Lester: "Not many people were even aware of U.S. involvement in South Vietnam, and even fewer were protesting about it. But Dwight Macdonald not only knew, he was protesting. That was impressive to a young, unpublished and would-be writer like myself."[12]

Walking a scraggly picket line was often awkward and embarrassing, like marching with a sign announcing the end of the world to stony indifference. But Dwight did his best not to let that feeling dissuade him. Now basking in more renown than he had ever experienced, he meant to use his reputation to serve his ideas. There was more talk among old fans of reviving *Politics*. In June 1964 he reported to his old friend Ping Ferry with enthusiasm that he was "getting all mixed up in dissenting politics again after almost 20 years."[13]

In November Dwight received a personal note from David McReynolds, field secretary of the War Resisters League, asking for his signature on an advertisement against the war in Vietnam. Dwight initially refused to sign on the grounds that it was too emotional in tone and anti-American in sentiment and did not marshal its facts effectively. But only a few hours later he came across another ad sponsored by the same ad hoc committee which backed up its criticism with facts and quotes that convinced him his initial objection to the WRL ad was wrong. He noted in his letter to McReynolds that he had a reputation for changing his mind, but this flip flop set a record even for him. What he found most convincing was the committee's assertion that the war in Vietnam was "a civil war, not an invasion by the North of the South." He agreed to have his name added to the list, but continued to grumble about the emotional tone of the ad.[14]

Later the same month, the WRL wrote requesting Dwight to sign on as a sponsor to a national protest December 19 against U.S. policy in Southeast Asia. The flyer announcing the protest argued that the United States had become involved "in a most shameful way in a war of atrocities by the Government in Saigon against the people of South Vietnam." The signers stated that they were "profoundly ashamed of the role of the American government," and they called on Johnson to declare an "immediate cease-fire," and to convene a conference of all nations involved, including China and the United States. The conference would seek immediate aid to the suffering people of South Vietnam, and would call for free elections and an end of all outside military intervention.[15]

Dwight's marginal note on the flyer from the protest organizers reflects his caution during these first steps toward what was to become an all-out commitment to oppose U.S. policy. He circled the word *we* in the phrase "horrors we now inflict on the people of South Vietnam." "What about 'them,' the invaders?" he wrote. This suggests he still had some difficulty with the argument that the war was a civil war. In any case, Dwight was fast becoming completely opposed to American intervention, but that did not mean he condoned the politics or the actions of the Ho Chi Minh government in Hanoi. He was still very much a part of that old struggle against Stalinist Communism, and he was deeply distrustful of the rhetoric that romanticized Ho as a Vietnamese nationalist and played down his hard-core Communist training, not to mention his earlier willingness to see his Trotskyist opponents slaughtered during his rise to power in the North.

During these early skirmishes Dwight carried on a running argument with McReynolds and A. J. Muste of the War Resisters League about the hard line of the organization, its obvious biases ("after all, Ho Chi Minh and Co. aren't exactly angels!!!"), and its absolute pacifist position, which only drove possible protesters away. But despite his reservations about associates and his consistent criticism of the North Vietnamese, he was convinced that the main issue was to go on record as opposing U.S. policy and not to allow the evils of the North Vietnamese or their apologists in the United States to thwart this essential and necessary criticism.[16]

Dwight was honored and delighted to speak at Muste's eightieth birthday celebration in Chicago. His breezy, off-the-cuff remarks challenging Muste's pacifism angered some of the overserious ideologues in attendance. One person sent Dwight an unsigned card attacking him for his "mental and moral bankruptcy . . . incoherent rudeness . . . intellectual barrenness and political whoring," a preview of the intensity of the political debate as the devastating stakes of the war began to rise.[17]

Dwight was delighted with Lyndon Johnson's landslide victory in the 1964 election, seeing it as a decisive repudiation of the conservative Barry Goldwater. However, with the dramatic escalation of the war the following spring in Operation Rolling Thunder's sustained bombing of the North, Dwight was outraged and felt that those who had supported Johnson against Goldwater had been tricked. They had voted against Goldwater, who had called for bombing of the North, but they had gotten Goldwater's policy. Dwight was still visiting the campuses, and although he had set lectures on poverty, mass culture, and "The Abuse of Language," at every opportunity he offered his "Relevance of Anarchism." He frequently peppered his remarks with caustic attacks on the Johnson administration. In March 1965 he badgered his own senator, Robert Kennedy, to come out with a clear statement in opposition to the war. He pointed out that Senator William Fulbright had done so and it was time for Kennedy to act. He wanted an immediate end to the bombing and the initiation of a peace conference. Dwight charged it was "hypocrisy to pretend we were defending democracy there." He cited Walter Lippmann's advice to "get out as quickly as we can on the best terms we can." Not happy with Kennedy's evasive reply, Dwight fired off another epistle insisting on a direct statement as to his position on the bombing of the North.[18]

Starting in March at the University of Michigan, the teach-in movement swept the country; faculty and students spent endless hours debating the issue. Although Dwight had actually had Schlesinger's name on the list of potential signers of a "Declaration of Conscience" against the war, he couldn't have been very surprised to find him on the team of heavyweight administration defenders who took on the professors on a national hookup out of Washington in May. Schlesinger's insistence that the United States was fighting in Vietnam to preserve the right of the "academic community to debate issues" drew hisses and catcalls. One partisan observer recalls that Schlesinger was visibly shaken by this disrespectful response. The lines were being drawn and the positions were hardening.[19]

Dwight was having his own problems inside the New York intellectual community. When members of the editorial board of the *Partisan Review* invited him to sign a statement opposing American policy, he refused on the grounds that it was too mild. He was impatient with their bland argument calling for some "new thinking." He declared that he had already done his thinking months ago. The present policy was disastrous. As for their criticisms of the anti-war protesters for insufficient anti-Communism and a naive belief "that everything would be fine if only the Yanks would go home," it was an unfair summary of the opposition's stance and it didn't represent Dwight's position. He knew that there were pro-Communists, like Gene Genovese at Rutgers, who had publicly proclaimed their support of the North Vietnamese, but that in no way sanitized the American policy. Speaking as a person who for "the first time in fifteen years had been signing petitions, making speeches and even picketing," Dwight insisted that the protests had "deeper and more reasonable grounds" than the double standard of morality charged by the *PR* editors.[20]

Dwight's own rhetoric was escalating when he charged that the United States in a matter of a year "had become the most feared and hated nation in the world." Citing the statistics of death and destruction inflicted on noncombatants in South Vietnam, he declared that Americans might cite the crude propaganda of the Communists and the gullibility of the "unsophisticated masses," but the facts were that the United States was behaving the way "Castro and Mao want us to." Why, Dwight wanted to know, had the United States adopted "Kiplingesque idealism, shouldering The

White Man's Burden long after more sophisticated imperialisms have let it drop as too weighty an anachronism?" The *PR* editors argued that the protesters had given no thought to what would happen to the South Vietnamese if the United States pulled out. That was hardly the question. What was "happening to them so long as America stay[ed] in?"[21]

In May 1965 Dwight barnstormed the campus of the University of Texas as a visiting fellow. While he was ostensibly to speak on mass culture, he spent much of his time attacking American foreign policy in Vietnam and the Dominican Republic as well as calling attention to the pressing domestic problems of poverty and racism. He praised the activism of the students, and applauded the courage of those who refused to be drafted. Dwight reported to David McReynolds that the students had enthusiastically cheered his criticism. He walked in a picket line protesting discrimination at a local campus bar and grill, and when counterpickets approached carrying Confederate flags, he asked them whether their flag belonged to some foreign country—one of the new African nations, Ghana or Tanzania. He expressed shock when he learned that the university still maintained segregated dormitories. If Truman could integrate the armed forces fifteen years ago, "why in the world cannot a great university be as decent and courageous as Truman?" The student paper carried stories about Dwight all over the front page.[22]

That spring the debate was carried on with increasing rhetorical hyperbole, rancor, and animosity. Even moderates and men of habitual decorum, such as Archibald MacLeish, were speaking out. He publicly declared that Vietnam and the Dominican Republic raised questions as to whether "the nation has become indifferent to the opinions of mankind and outgrown its old idealism." Lewis Mumford, in *A Minority of One*, an anarchist monthly, addressed an open letter to the president in which he denounced his "totalitarian tactics" and "nihilistic strategy," which "shamed Americans" who were revolted by Johnson's "dishonest excuses and pretexts." Mumford did not confine his criticism to the pages of little magazines. In May, in his presidential address to the prestigious American Academy of Arts and Letters, he broke a tradition of disinterested scholarship and devoted his address to a denunciation of Johnson's foreign policy, which he declared was shameful and obliged all those devoted to the arts and humane letters to speak out openly in protest on every occasion when human beings were threatened, whether in Mississippi or in North and South Vietnam, where people were forced to "confront our government's cold-blooded blackmail and calculated violence."[23]

The fast-developing polarity on the campuses, within the intellectual community, and in the country at large was becoming a "we" versus "them." Undoubtedly these intense feelings were at this point confined to a relatively small minority of Americans, but the lack of numbers only contributed to the feelings of frustration and powerlessness, which bred even more anger and suspicion. Given this already poisoned atmosphere, it is not surprising that when the White House announced it was going to present a Festival of the Arts in order to encourage the "cultural renaissance" alleged to be already flowering in many communities, intellectuals smelled a plot. Eric Goldman, the former Princeton historian and at this time "special consultant to the President," was put in charge of the event. He insisted that, contrary to suspicions, the event was not mounted as a means of blunting the rising criticism in intellectual circles of Johnson's dramatic escalation of the war in Vietnam and the intervention in the Dominican Republic. The idea for the festival had been conceived prior to the escalation. He did speculate, however, that after the organizing got under way, it may well have occurred to Johnson that in addition to encouraging the arts it might serve as a "tool to quiet" the mounting opposition to the war.[24]

In times of heated controversy the attribution of devious motives is an inevitable weapon used by all participants in the rancorous infighting. Whatever the initial motives in May, when the festival was announced and the invitations sent out, those who were opposed to the war were convinced that it was an obscene attempt to make it appear as though America's distinguished writers and scholars endorsed the government's foreign policy. When Robert Lowell received his invitation through a personal phone call from Eric Goldman, he accepted, as he put it, "somewhat rapidly and greedily." He thought such an artistic flourish would encourage financial support to the newly established National Council on the Arts. But after consultation with friends, particularly the novelist Philip Roth, and Robert Silvers, the editor of the *New York Review of Books*, Lowell agreed that his acceptance might be construed as an endorsement of Johnson's policy. The very suggestion horrified him, for he had been denouncing the bombing of North Vietnam as an outrageous atrocity. Despite Lowell's emotional instability and frequent bouts with depression, he was an able master of his public personality. He knew that he could have an impact by using the occasion to make a statement. He not only called Goldman to tell him that he had changed his mind and was declining the invitation, but he insisted, despite Goldman's protests, in submitting his rejection to the *New York Times*, which printed it the following day on the front page.[25]

Dwight was not only privy to this development but part of a group that was determined to give Lowell moral support. Led by Bob Silvers and the poet Stanley Kunitz, they organized a telegram to Johnson backing Lowell and expressing a similar dismay at the increasingly belligerent and militaristic stance of the administration. The telegram had twenty signers, six of whom had won Pulitzer Prizes. Hannah Arendt, John Berryman, Lillian Hellman, Alfred Kazin, Philip Roth, and Robert Penn Warren were among the group. Even the beleaguered Eric Goldman admitted that it was "an impressive array of talent," a plaudit he edited from his book chapter on the affair. Significantly, contrary to accounts, the statement by the twenty writers did not urge others not to attend the affair. It merely expressed their respect for Lowell's decision, their opposition to the war, and their hope that people would not "conclude that a White House arts program testifies to approval of Administration policy by the members of the artistic community."[26]

Immediately after the telegram was sent, Dwight received an invitation to the festival as a distinguished film critic. Goldman later described him as "representing criticism of the elitist point of view." What to do? He consulted with his comrades and they concluded that despite his signature on the telegram, he should accept the invitation. If it was revoked on the basis of his signature, they could publicly observe that this festival had a political means test. If the invitation stood, he should attend the festival and write it up for the *New York Review of Books*. This was a tactical decision. Dwight was determined to use the festival to propagandize against Johnson's war. For this purpose he was more than willing to sacrifice "consistency and good taste in the interest of a larger objective."[27]

The Vietnam War posed Bettelheim's question of "behavior in extreme situations." Despite the heat of the rhetoric, the United States was not one huge concentration camp, as Dwight had once described the Soviet Union. But the war provoked a rage in its opponents and also in its defenders that grew in volume and acrimony. One of the things that drove the war's opponents to frenzied distraction was the ability of the society to carry on business as usual. As the critics stood appalled at what they saw as needless slaughter, the great majority of Americans appeared oblivious—either totally unaware or unconcerned. To confront this impenetrable wall of indifference opponents became insistent upon finding ways of "bringing the war home." As Mumford had argued, it was essential to speak out in "breach of etiquette. . . . To go on with our usual routines and rituals . . . is to behave like the crowds who thronged into the arena to witness the games, when the Vandals were hammering on the gates of Augustine's Hippo."

By the summer of 1965 the rhetoric of outrage had grown shrill. Articulate people who normally chose their words with care and precision now threw caution to the winds in a desperate effort to force people to recognize the terrible urgency of the crisis. These assaults only encouraged more grandiosity on the part of those who supported the war as a heroic defense of democracy and freedom. This inflated language was unusual in a society which had for decades been described as consensual and disinterested in ideology. When a distinguished scholar of the culture could compare America's conduct with "totalitarian tactics" and a centrist historian like Schlesinger could see his receptive audience as similar to the crowds at Nuremberg, it is not difficult to understand why Dwight, a thoughtful and caring man, was willing to be, as he put it to friends, "the bad fairy at a Christening."[28]

Thus Dwight went to Washington determined to make trouble at the White House festival, to subvert its ostensible purpose and draw attention to the widening gulf between the American creed and the American deed. He wired his acceptance to Goldman, saying he would attend with the purpose of writing up the festival. Goldman, by this time a harried defender of the president but simultaneously forced to stand for freedom of speech, felt that Dwight was hoping to be disinvited, thereby putting the White House in the position of screening out those who didn't meet their political requirements. He told Dwight in a phone conversation that he was welcome to attend despite his protesting signature and to write anything he pleased.[29]

On the day before flying to Washington, Dwight bumped into Tom Hess, the executive editor of *Art News*. Together they decided to draw up a "Statement for the Press":

> We would like to make it clear that in accepting the President's kind invitation to attend the White House Arts Festival we do not mean either to repudiate the courageous position taken by Robert Lowell nor to endorse the Administration's foreign policy. We quite share Mr. Lowell's dismay at our country's recent actions in Vietnam and the Dominican Republic.

This statement was signed by Dwight and Hess. Dwight saw the circulation of this statement as "another turn of the screw on L.B.J." Throughout the very long thirteen-hour day of the festival, Dwight and Hess approached artists and guests. Dwight solicited signatures from "some forty or fifty" he and Hess knew, but they obtained only seven signatures in addition to their own.[30]

Their critics in and around the White House scorned this feeble endorsement by making it appear the 300-plus people at the festival had been canvassed. As Lady Bird Johnson was later to put it, "I'll take a 397 to 7 majority any time." But in his own account Dwight insisted that of the forty or fifty canvassed, none had refused on the grounds that they favored the president's policy. Some had claimed that as artists they had no knowledge of or interest in politics. Others said that they agreed with the statement but this "wasn't the time or place." Most claimed that they were there as guests and Dwight's action was rude and in bad taste.

It was on this latter score of tastelessness and boorish behavior that he received the greatest condemnation. He obviously did not simply solicit signatures "quietly and privately." From the very moment he reached the White House he began buttonholing people he knew disapproved of the war. He had a brief wrangle with Saul Bellow and a testy argument with Charlton Heston, both of whom he knew held his rude behavior in contempt. All denounced his conduct as disgraceful manners in the "home of the host."

To Dwight, this was a contemptible fawning deference. He did not consider the White House Johnson's private home. It was a symbol of the present administration, "a political not a private, personal place." He also noted that the churlish Johnson, who was in fact already calling the protesters "sonsabitches and fools," did not greet any of his guests, but limited himself to a brief speech and then ducked out "without any handshaking or hostly palaver."

In a response to the Goldman account, Paul Goodman wrote with similar anarchical indignation that Goldman and Dwight's other offended critics didn't understand what a republican society was. The president was simply a public servant and it was the right of any citizen to "call him to his duty." It just showed how far the country had fallen from republican principles that dissenters had to form a cabal and make a demonstration. He felt that the behavior of Dwight and Eartha Kitt, who not long before had breached decorum by speaking out against the war at a White House function, were "more in the original American spirit."[31]

This confrontation between protocol, manners, decorum and Dwight's blunt, aggressive rambunctiousness serves as a micro-model of the event itself. Some guests, like the patrician John Hersey and the gentle and aging Mark Van Doren, were also strong opponents of Johnson's war policy. Hersey had found impoliteness a necessity when he insisted, much to the chagrin of the Johnsons, on prefacing his reading from his *Hiroshima* with a warning that Johnson had embarked on a very dangerous and destructive course. Van Doren, who had initially planned an effusive statement in defense of Lowell, was persuaded to cut it short and find more time to praise his gracious hosts. Johnson would have barred both if the political costs had not been so high.[32]

For Dwight the affair was a scam. Sensitive to the methods of image making and manipulation, he found the show of art and culture emanating from grounds adjacent to the war room of the White House offensive. The pretense of polite society gathering to listen to and view the works of artists and practitioners of belles-lettres while the smiling host was planning further devastation of a land and its peoples was obscene. The paradoxical conflict between the crude, unkempt, boorish counterculture, with its flamboyant lifestyle and foul language, and the warmongers, with their pin-striped suits and Rhodes Scholar backgrounds, seemed to the dissenters an Alice in Wonderland reversal where values and principles were turned on their heads. Using four-letter words, dressing like a hobo or Jesus of Nazareth was obscene and offensive, but ordering the napalming of Vietnam villages, the dropping of weapons designed deliberately to maim but not kill so as to be more costly to the enemy was declared courageous, patriotic, and deserving of praise and support.

Dwight's account of the festival was predictable. He painted it as a midcult affair, a consensual approach to culture, something to suit everyone's taste, a balance between the high and the low. It had been planned to have Lowell leavened by Phyllis McGinley. Dwight claimed to like light verse, but he was sneeringly unappreciative of such efforts as this captivating couplet by McGinley:

And while the pot of culture's bubblesome,
Praise poets, even when they're troublesome.[33]

Dwight's major criticism was that the artists, writers, and musicians were a distinct minority among the guests, most of whom were cultural entrepreneurs, museum directors, presidents and treasurers of symphonies, chairmen of state cultural commissions. There were no directors or composers. Dwight could only conclude from this imbalance that the purpose of the festival was hardly to impress "the actual producers of art or thought with the 'White House's great interest,'" but rather the promoters, money raisers, and of course the expediters of art as a commodity. This should have come as no surprise to Dwight, since he was invited as a movie critic for a mid-cult magazine.[34]

Dwight did admire some of the participants and artistic exhibitions, as well as Duke Ellington's music, which brought the affair to a close. Despite being angered by Bellow's insistence "that citizens have a duty to honor the Presidential office no matter what his views of the president's actions (flimsy) and Vietnam escalation ok because Ike and JFK had started the escalator and LBJ couldn't run back down," he thought Bellow offered "much the best writing we heard that morning." From the distance of twenty-eight years Bellow could only remember Dwight as being frivolous, perhaps a bit ridiculous. He recalled feeling that Dwight "was mixed up that day and pranced into the rose garden like Pan in tennis shoes—a sex symbol on a political mission." He did not think, as some suggested, that Dwight was tight, but rather that "he was simply having a glorious time and was so artfully dressed for the occasion—the uninvited guest in the rose garden wearing sneakers—that he didn't need to have been drinking, the get-up was intoxicating."[35]

But Dwight's general dismissal of most of the artistic contributions had a suggestion of snobbish elitism, despite the sound logic of his political stance. That undertone had been the weakness of his mass culture critiques as substitutes for more conventional political commentary, giving some substance to Goldman's singling out Dwight as an example of a kind of fatuous elitism and cultural snobbery on the part of some intellectuals who adopted an anarchic arrogance toward all things American. Since Johnson had a notorious contempt for the "Eastern intellectual establishment associated with the Kennedys," Goldman took pains to separate himself from that group. In doing so he seemed to suggest that Dwight was a model of that cluster of Ivy League "best and the brightest" who had manned the ramparts of the Kennedy media blitz. On that score he was way off the mark. Dwight's critique of Johnson was not devious and certainly not motivated by intellectual and social snobbery: it was directly tied to his anarcho-pacifist stance that went back to his position during World War II. His alliance with groups like the War Resisters

League, Peacemakers, and conscientious objectors was a world away from the pragmatic crisis managers of the New Frontier. Dwight would have continued to champion Johnson had the president not made the fateful political decision to continue, and escalate, the war in Southeast Asia. But Dwight's portrait of the festival as a mid-cult affair did smack of a cultural contempt that gave some substance to Goldman's criticism.

It is interesting that Saul Bellow and Ralph Ellison, who sharply criticized Dwight's lack of respect for the honor of the occasion, were men of minority backgrounds who in the postwar years had been brought into the church, so to speak. On the other hand, some of Johnson's severest and most distinguished critics, such as Archibald MacLeish, Lewis Mumford, Robert Lowell, Mark Van Doren, John Hersey, Dwight Macdonald, were clearly representative of the Wasp upper class. They could attack the State with a confidence not characteristic of more recent recipients of the nation's respect and honors. Eric Goldman, for his part, was a product of an "impoverished and broken home" who had made his way up from the public schools of Baltimore to a Ph.D. from Johns Hopkins and finally to a prominent position as a popular historian at Princeton and a most readable chronicler of American liberalism. He may well have harbored a distaste for what Irving Howe once described as those "more secure Americans." Surely unlike Dwight, Goldman did "give a damn" about presidential leadership. His portrait of the affair with his insistence that an invitation to the White House normally "was and should be cherished" and repeated references to the special deference that a chief executive is entitled to is hardly a sensitivity that Dwight and his distinguished colleagues had ever developed.[36]

Dwight's performance was savaged by Goldman and Charlton Heston, who described him as a boorish crasher at a dinner party, his face dripping with perspiration brought on by his furious activity, his clothes disheveled, with food stains on his plaid shirt, which hung out along with his underwear, revealing his "round pink belly." This unattractive picture was not confirmed by any other reporters. As political tacticians, Lowell, Hersey, and Macdonald achieved their aims. Lady Bird Johnson and White House aides gathered the next day to discuss "the hammer blows" of the front-page stories in the *New York Times*, the *Herald Tribune*, and the *Washington Post*. Before the month was out detailed accounts had appeared in *Time, Newsweek*, the *New Republic*, and the *Nation*. Dwight was cited in the Soviet Union's *Izvestia* as a leading participant in the "rebellious spirit that prevailed" and was quoted as denouncing American foreign policy "as really abominable. I am ashamed of it."[37]

Eric Goldman, who had done his best at damage control, thought the event had been "an unmitigated disaster." Almost everything that happened after Lowell's letter added bricks to the wall between the president and various groups of intellectuals and "metro-Americans," Goldman's label for a young, well-educated new class determined to make their opinions count. The public reports of the festival had contributed to making that wall "seem as impassable as the barbed concrete between East and West Berlin." It was a genuine challenge to the alleged "consensus" and collaboration of intellectuals with the State that had been so painstakingly developed during the cold war years. Johnson, beleaguered on so many fronts, facing so many vital decisions, understandably thought it a trivial affair, "like a pebble in a shoe," and in the context of those tumultuous times it may well have been. But it was also another step in the early stages of a movement that would contribute to the downfall of the Johnson presidency and lead to the charge that the war in Vietnam was lost at home.[38]

18

The Critic at War

DWIGHT WAS EXCITED BY THE SUCCESS OF THE WHITE House protest. He appreciated the warm response he got from friends who wrote to congratulate him on his conduct at the festival. The affair had galvanized the intellectual and artistic communities, and Dwight was quick to give his support and name to an Artist Protest Committee, which published a large ad in the *New York Times* with over 630 signatures. The ad called on all Americans to end their silence and speak out against the war. Dwight's name could now be found among the likes of the screenwriter Alvah Bessie, the actor Howard Da Silva, Howard Fast, and Corliss Lamont, all of whom he had seen as contemptible apologists for Stalin a little over a decade earlier. The ad was signed by liberals and anti-Stalinist radicals of all stripes. Within the liberal-left community, there was a growing consensus that regardless of past affiliation or even present politics, opposing the war was all-important. This ad charged that Johnson's stated concern for peace and democracy was "mocked by eleven years' maintenance there of brutal police regimes assisted by American money, American guns and finally . . . American blood."[1]

Dwight's letters were full of enthusiasm for his return to activist politics. His son Nicholas, a junior at Harvard, had recently married Elspeth Woodcock of Cambridge, Massachusetts, in Paris, where Mary McCarthy hosted their wedding reception. Dwight was persuaded to make a phone call from

New York to congratulate the couple and wish them well. After briefly talking to the young people, Dwight breathlessly urged Mary to return to America because it was "becoming just like the thirties." The excitement of the meetings, the planning, the camaraderie had shattered the anonymity and lack of community Dwight always complained about. People were getting together once again to talk about important things and to do something about them.[2]

On August 12, 1965, Dwight testified before a congressional hearing on the war conducted by New York Representative Robert Fitz Ryan. Dwight pictured Johnson as attempting to emulate the Russian czars, whose armies served as the "gendarmes of Europe." Now the United States was the "self-proclaimed policeman of the world." Dwight dwelt on the "absurd anti-communist crusade" as a new form of "global McCarthyism." Like its predecessor, it was based on ignorance and did no harm to Communists but injured innocent bystanders. It was splitting the country, as McCarthyism had, into a disapproving minority—the liberal intelligentsia of teachers, students, writers, and other members of the educated classes—and an actively or passively approving majority.

Dwight's personal distaste for Johnson was not veiled. He spoke of the president's "neurotic craving for consensus" and his "moral bullying." Dwight insisted that he was on the American side but that the administration's policy only served the nation's enemies. He now wondered whether a vote for Goldwater might not have been a shrewder tactic. The Democrats would have forced him to go slow, but they offered no challenge of any consequence to their own party leader. Dwight railed at the smug satisfaction of so many Americans and closed by quoting from Mark Twain of December 30, 1900:

> I bring you the stately nation called Christendom returning bedraggled and dishonored from pirate raids in Kiao Chow, Manchuria, South Africa and the Philippines with her soul full of meanness, her pocket full of boodle and her mouth full of hypocrisies. Give her soap and a towel but hide the looking glass.[3]

Dwight's bitter recriminations against the Johnson administration did not wipe away his annoyance at the growing support, particularly among the young New Left, of the North Vietnamese and the Viet Cong. He was even more distraught when members of an alleged pacifist organization endorsed a Viet Cong victory. He wrote Dorothy Day complaining that this was hardly pacifism. Nor could he accept those with a penchant toward anarchism lining up with Communist state power. It wasn't simply the contradictions in principle that bothered him. He thought it was stupid politics. It was hard enough, he wrote, "to get Americans to see our point of view, namely that this disgusting war in Vietnam should be ended at once and we should stop bombing the people of Vietnam, without having to take on the burden of being tagged as adherents of the other side."[4]

Through his long years of opposition to the Vietnam War, Dwight never abandoned that position. He overcame the hurdle of red-baiting isolation and was willing to march, to sign, to speak on platforms with people whose ideology he held in contempt. But he insisted on arguing against texts and positions that offered support for the Communists in North and South Vietnam. As the frustration mounted and Johnson continued to escalate the war, some opponents began expressing their hope for a Communist victory if only to end the conflict and save Vietnamese and American lives. But Dwight refused that temptation. He was in favor, by 1966, of immediate U.S. withdrawal, unilateral if need be, but he never showed any enthusiasm for a Communist victory and always felt such a stance was not only morally wrong but politically indefensible. In October 1965 he refused to be listed as a sponsor of a Fifth Avenue Peace Parade, writing in the margin of McReynolds's appeal: "No—against united front on such issues with groups who favor victory of anti-USA side." Dwight often attended such affairs in order to do his duty and swell the crowd, but he refused a blanket endorsement of what he saw as deceitful Communist ideologues.[5]

Despite Dwight's growing absorption in the activities of what was now being called "the movement," he was still involved in the intellectual and cultural battles that were part of the New York intelligentsia's natural environment. He had already compared himself to an aging gunfighter, and in the mid-sixties there was no question that an extremely talented, shoot-from-the-hip young gunslinger had arrived in town and was determined to make a reputation for himself. He was Tom Wolfe, a Southern graduate of Yale with a Ph.D. in American studies, a crime Dwight never let him live down. Wolfe had worked as a newspaper reporter in Springfield, Massachusetts, and on the *Washington Post*. His first real splash was a series of articles on teenage culture, specifically fashion, and the customizing of cars, demolition derbies, and other rituals of mass culture.

He had actually submitted his notes for the first article to *Esquire* rather than a traditionally edited text. Those notes were an explosion of ono-matopoeic sounds ("Bonk, Zonk!!!!!!!!!!!! Floonk . . . ."), manufac-tured words that captured the frenetic enthusiasm of the youthful subcul-ture and its style-conscious rituals. Wolfe took great liberties not only with syntax but facts as well, producing a glitzy, fast-moving, sensational-istic piece of innovative journalistic sociology. If Wolfe had a thesis, it was that much of the modern fashion in clothes, in cars, in music came from the bottom up rather than the reverse. Teenage culture was the wave of the future. The "arteriosclerotic," "infarcted" mummies like Dwight did not appreciate Wolfe's irreverent style.

Immediately sensing their disdain, Wolfe took on the sacrosanct *New Yorker* in *New York*, the Sunday magazine of the *Herald Tribune*. This was an audacious mugging of a national icon, made all the worse by his bully-ing put-down of the *New Yorker*'s quiet, gentlemanly editor, the exquis-itely diffident William Shawn, as "the museum curator, the mummifier . . . the smiling embalmer" dedicated to the preservation of Harold Ross's antique journal. Wolfe later claimed that he had no idea his spoof of the magazine would elicit such howls of rage. After seeing an advance copy, the usually taciturn Shawn had lost all composure and had phoned and written Jock Whitney, the owner of the *Tribune*, urging him to stop publi-cation of Wolfe's "false, murderous attack." Shawn insisted that Whitney would do himself a favor by keeping the *Tribune* out of the sewers of "gut-ter journalism."

In retrospect, one might argue that the reaction of Shawn and his *New Yorker* tribe launched Wolfe's career. His two-part article, titled "Tiny Mummies! The True Story of the Ruler of 43rd Street's Land of the Walk-ing Dead!", was a talented piece of malicious buffoonery that could only have been squelched by silence. Wolfe's piece did not come from the tra-dition of avant-garde cultural modernism mixed with the remnants of Marxism as had earlier critiques of the magazine. Wolfe was what would now be seen as a postmodern "destructivist." He had contempt for the old-fogy standards of Dwight and his comrades. Whereas Dwight had written of the teenage culture with chagrin and disdain, Wolfe really pen-etrated that world and captured its flavor, its enthusiasms as well as its bizarre absurdities. He had an ear for sounds, for intonations, and of course his work was often celebratory. True, the barbarians were at the gates, but it was time for the walking mummies to wake up to their own death and lie down.[6]

In behind-the-scenes consultation, Shawn and others chose Dwight to take on this vulgar upstart in the appropriately highbrow *New York Review of Books*, in contrast to the tawdry vehicle used by Wolfe. Dwight first targeted Wolfe's recently published collection of essays, *The Kandy-Kolored Tangerine-Flake Streamline Baby*, which he labeled "parajournalism—a bastard form . . . exploiting the factual authority of journalism and the atmospheric license of fiction." Dwight's professional judgment told him that Wolfe was an acute observer and had a good ear for the city's style; in fact, he would do well "as a writer of light pieces for, say, the *New Yorker*." But that was his last effort at jocular irony. He turned to a much heavier sarcasm and ridicule of Wolfe's stylistic excesses, adoption of wild and nonexistent words, and attention-grabbing gimmicks that only served as an assault on language. One would have expected Dwight to take on this novice with a sharper stiletto, poking holes in Wolfe's vulgarity. Instead, he came on as a stuffy professional whose occupation was being demeaned by a trendy new form of kitsch he thought would pass. On that score he was definitely wrong. His piece was defensive, heavy-handed, and lacking in his customary dismissive derision. In his second installment, devoted to Wolfe's attack on the *New Yorker*, Dwight responded like the other wounded *New Yorker* writers who were appalled that their sweet-tempered editor had been mugged by a street thug. Dwight spoke of Wolfe's cruel lampooning of Shawn's shy demeanor, his distortion of his character, and his wildly false suggestion that Leopold and Loeb had first chosen their classmate William Chon (Shawn) as their murder victim. This was Wolfe's far-fetched explanation of Shawn's pathetically withdrawn demeanor. The *New Yorker* and the *New York Review of Books* went to astonishing lengths to find errors in Wolfe's piece, sending Renata Adler to Chicago to wade through the Leopold–Loeb trial testimony looking for references to William Chon (Shawn). *Review* editor Bob Silvers called Nathan Leopold in Puerto Rico to verify or deny Wolfe's obviously spoofing reference.[7]

Why Dwight approached this task with such solemnity is not clear. His defense of Shawn's character and reputation was sincere. He had had a good relationship with the editor over the years and was grateful for the way he treated him. At this time he was writing regularly for *Esquire* out of his *New Yorker* office and had not written a word for the magazine since the Harrington review three years earlier. Loyalty and gratitude toward the editor and magazine surely played a part. But there was more to it than his being irked by Wolfe's disrespect for Shawn and the *New Yorker*. Wolfe was treading on Dwight's turf. Dwight had dissected teenage culture in the fifties. His *New Yorker* study was contemptuous of the growing influence of the teenage market. It was a part of his assault on the menace of mass culture. By contrast, Wolfe, the ultimate vulgarian, celebrated that culture. In fact, Dwight could find only one value in Wolfe's work: "Old he bad, new he good." Wolfe's success was his identification and sympathy toward his youthful subjects. Dwight had been distant and disdainful, and that made him resent all the more the constant references to the "arteriosclerotic old boys trying to hold on to the whole pot with their arms of seersucker."[8]

There was irony in this confrontation, because Dwight in his frequent trips to college campuses was seeing much more of America's youth and identifying with their critique of the older, that is, his generation. Dwight was attracted to the young people's new appreciation for political activism. Wolfe had only a sneering cynicism toward any form of activism. But Wolfe was sensitive to this phenomenon of old fogydom soliciting the friendship of the young. The subject of one of his articles on teenage culture had been Murray the K, a rock-and-roll disc jockey, who had used his entrée with the kids to urge them to resist the temptation to drop out of school. On a radio talk show Wolfe had asserted, "I don't think Vietnam is as important as Murray the K. I really don't." As for the anti-war activism of such antiquarians as Norman Mailer, Nat Hentoff, and Dwight Macdonald, Wolfe thought they were "going through intellectual menopause." They believed the only role for a thinking man was to be a young radical. "Well," he added, "a man can be radical as long as he lives, but he can't keep on being young." He thought these "aged bisonheads" were desperately "doing the fountain of youth thing. . . . This is a way of hanging on to the nineteenth-century idea that if you can attach yourself to a cause, that somehow you can keep the vital juices flowing."[9]

Wolfe's critique of the *New Yorker* was a double-barreled shot. It displayed a shrewd sensitivity to the complicated dimensions of mass culture and it ridiculed Dwight's and his colleagues' political commitments while dismissing the magazine as "the most successful suburban woman's magazine in the country."

The American Studies Ph.D. was obviously keeping up with the literature and anticipating some of it, frequently describing the opponents of the war, if young, as rebelling against their fathers or, if older, desperately trying to preserve their youth. Americans often have trouble accepting the very idea of political or social commitment that actually affects conduct. Wolfe was to make his reputation as a skillful deflater of those dedicated to any cause other than the pursuit of style and pleasure. To Dwight, this prince of vulgarians, in his tailored white suits, parasols, and broad-brimmed Stetson hats, was nothing but "a dandified poseur," even worse, a "poor man's Max Beerbohm." For some reason Dwight had a penchant for dismissing a writer, any writer, as not the equal of Beerbohm. He had done the same with Colin Wilson, who was totally perplexed by the comparison. Wolfe took no notice of it, but one suspects he thought it an archaic eccentricity of Dwight, who, he wrote to the editors of the *New York Review of Books*, had become his Boswell, devotedly annotating his laundry slips. Dwight could only respond with a sputtering rage at the unfeeling practitioner of "hit-and-run journalism."[10]

There was another side to the issue. Under the seemingly cautious William Shawn, the *New Yorker* had been for some time turning out, between those notorious ads, some of the most trenchant social and political criticism to be found in America, such as Dwight's piece on poverty, Hannah Arendt's study of Eichmann, James Baldwin's warning in *The Fire Next Time* (Wolfe dismissed this as a hair shirt for liberals), and most recently Robert Shaplen's eye-opening reports from Saigon, not to mention the pointed editorial attacks on the war that were appearing with increasing regularity in the previously light-hearted "Talk of the Town" section and were having an effect. Shawn was changing the direction of the magazine. It was not long before the *New Yorker* became, along with the *New York Review of Books*, a platform for the most articulate and acerbic critics of the war in Vietnam. Wolfe's mannered disdain for serious discussion was another example of that impenetrable wall of indifference that so riled the likes of Dwight and his comrades. However, in this skirmish, they may have helped promote the banality they detested.

The absurd madness of the escalating war continued to absorb most of Dwight's attention and fed a growing contempt for the intellectuals he felt were dragging their feet in opposition. In January 1966, the choreographer and dancer Shirley Broughton held one of her "Theatre of Ideas" in a second-story Chelsea loft. Well over sixty members of the New York intellectual community turned out to debate the war, led by a panel of "experts" who, after introductory remarks, were questioned from the floor by lively participants. The panel was composed of Arthur Schlesinger, Jr., as the insider historian; Michael Waltzer, a political scientist from Princeton and an editor of *Dissent*; Staughton Lynd, a Yale historian and eloquent critic of the cold war establishment; and Irving Kristol, who stoutly represented the defense. Schlesinger was on both sides of the question at various stages of the debate. Since it was still a Democratic administration and Kennedy men were still involved in managing the war, Schlesinger, while critical of American policy, expressed an inordinate compassion for the complexity of the decisions that had to be made.[11]

Dwight was irritated by Schlesinger's waffling. He opened his comment and question from the floor with the suggestion that perhaps Schlesinger, despite his pulling rank as both a historian and a former government insider and therefore more knowledgeable than his critics, was now "more of a public figure than a historian." But Dwight's real point was to insist that President Johnson only had two alternatives: to get out or escalate. He had chosen the latter in order to "recoup a political failure by military means." Dwight then proceeded to explain the "lessons of history" to the historian. Political problems, which best described the situation between North and South Vietnam, could not be solved by military means. We might be winning the war militarily, as the administration and the joint chiefs insisted, but we were losing it politically. There had been a political vacuum in Vietnam since 1964, Dwight noted; did the historian-insider know whether the United States had any viable alternative to the Viet Cong?[12]

As was his wont, Schlesinger agreed with "nearly everything" Dwight had said but reminded him that when LBJ decided on escalation he had good reason to believe success was possible. Things had not worked out and now his decision "look[ed] ridiculous in retrospect." At the time of the escalation the Viet Cong seemed to be winning and it was thought that the only way the United States "could move them toward negotiations was to persuade them that our commitment to the South was so strong that they could not win the war." It was a perfectly rational argument.[13]

Since Schlesinger, the consummate pragmatist, would not address Dwight's real question, the political aims of the war, Irving Howe made a stab at it by dismissing Schlesinger's assertion that the war managers in Washington were "people just like us." They had, rather, "a whole different set of ideological preconceptions that distorted our whole Vietnam policy over the past ten years." The war managers simply had no conception of or interest in the "indigenous roots of the social revolution going on in Vietnam." Howe agreed that Dwight's phrase "to attempt to recoup by military means a political defeat" was the "key to the whole thing." The only solution was to negotiate a political settlement with the Viet Cong.[14]

Given the tenor of this crowd Schlesinger drew boos and hisses when he insisted that only the men in Congress and the administration had any effect on policy. The demonstrations only embarrassed those in the Senate who were fighting the good fight, quoting one anonymous senator to the effect that more demonstrations would silence the more effective senatorial opposition. The only use of demonstrations was to offer emotional release. As the cool and consummate insider, Schlesinger believed that the politically responsible person acted in a way that had the most political impact; he did not seek "emotional orgasm" for himself. It was a typically myopic insider's view, which denied any relationship between elected officials and what was going on in the streets. Nor did Schlesinger entertain the notion that some of the congressional opposition to the war might well find support in the demonstrations.[15]

Dwight remained silent during these exchanges with Schlesinger, but in an exchange of letters with Nicola Chiaromonte he indicated his irritation. Chiaromonte replied that he found Schlesinger "insupportably stuffy and pretentious and equivocal as well." He also reported that he had heard from Lionel Abel that Isaiah Berlin had whispered "while S was speaking that 'he [Schlesinger] thinks history was made up of blameless acts.'" Chiaromonte charged that Schlesinger's self-importance and smugness went as far as saying "'I, who am an historian' (while you are not) or 'I who was in Government.'" He opined, "Now that's just intolerable."[16]

In February 1966 Dwight left for another visiting professorship at the University of Texas. His main obligation was to teach a course on film history and film criticism. It did not take him long to become a colorful, provocative, and seemingly eccentric figure on the Austin campus. He was not reluctant to use every opportunity to express his political views, and the local papers were full of his acerbic commentary. The campus newspaper quoted him as saying that Johnson's policies had made America "the successor, in the eyes of the world, to Nazi Germany and Stalin's Russia as a disturber of the peace." Dwight followed this with a historical summary of Johnson's subterfuge of gaining election in 1964 as the peace candidate when in effect he was going to borrow Goldwater's bombing policy. This outburst was headlined with a quoted statement: "As an American I am ashamed of my government's action in Vietnam and the Dominican Republic."[17]

The piece brought an irate response from a geographer on the faculty who questioned Dwight's scholarly credentials and the lack of balance in his polemical accusations. Dwight, delighted, was quick to assert his political expertise. He also took the opportunity to dismiss his critic for his typically anemic "departmentalized thinking," a major weakness of the academic mind. He did not think it immodest to match his experience "as a veteran Communist-fighter with that of such more recent recruits as Mr. Rusk or Mr. McNamara or Mr. Johnson." As for special credentials in politics, he felt certain that "politics was too important to be left to the specialists, who in the Vietnam and Dominican imbroglios have made as big a mess as any amateurs could." The distinguishing mark of a democracy, Dwight lectured, was the citizen's right, indeed his obligation, to speak out when his values and interests were threatened by the policies of his government. It was not only a right but "also a duty to kick up a row about it in public." As a parting shot he invited his geographer expert to explicate the "geographic inwardness of America's massively increasing commitments to 'save from Communism' a small nation of no strategic significance" whose people did not seem particularly interested, "at least not at the price of the daily destruction of their villages and their lives by our liberating bombs. Genocide, we used to call it in a more innocent age."[18]

Dwight was at home. This was just the sort of thing he liked—the give-and-take of political controversy. He relished the opportunity to speak his mind as a "Distinguished Visiting Professor." When Mary McCarthy wrote from Paris asking what he was doing about Vietnam, he was proud to be able to send her a raft of clippings on his anti-war activities. He had been the only faculty speaker at a campus rally sponsored by the local branch of the Students for a Democratic Society. He had opened his talk with an ironic salutation to his "fellow Americans." But in recounting the event, Dwight wrote that despite the obvious ironic inflection, he really did think of himself as an American and was "horrified by what Johnson was doing to the country"—turning it into something he didn't like, didn't recognize, didn't feel was his own: "This is becoming our Peloponnesian War, I'm afraid, and it shows, as with Athens, that despite all the good things about our internal political-social-cultural life, we have become an imperialist power, and one that, partly because of these domestic virtues, is an inept one." In a brooding tone, he mused that Vietnam might mark the end of the United States, "not only as a world power, but as a civilized nation internally." He found the lack of conscience on the part of the government frightening and wondered whether the small band of opponents,

Walter Lippmann, Fulbright, a few other senators, plus George Kennan and the intellectual-academic community, was enough to stop the war machine. He was appalled by what he saw as the "back tracking of the *N.Y. Times.*"[19]

Despite these grim forebodings and the feeling of frustration as to ways of making a difference, Dwight was having a good time as the itinerant professor. He found the students lively, and he and Gloria also enjoyed Austin. They liked the neat efficiency of their suburban motel-type flat with a pond in back where they frequently went canoeing. They liked the charming Southern courtesy.[20]

When they returned to New York, Dwight was greeted by the death of his old friend, Delmore Schwartz. Dwight was the executor of Schwartz's "estate," such as it was, and given the poet's notorious bouts of paranoia, alcoholism, drug abuse, and the general shambles of his life, Dwight had been unable to avoid becoming involved in Schwartz's constant disputes with ex-wives and friends. In fact, he was often the target of Schwartz's rage. But even though Schwartz often abused Dwight and accused him of conspiring with his enemies, Dwight had a genuine love and compassion for the tormented man. He identified with Schwartz and felt that in some ways they were alike. Because Dwight was often given to depression and regret over not having lived up to his talents, he could sympathize with Schwartz's despair. In a moving tribute delivered at the memorial service July 18, 1966, and later published in the *New York Review of Books*, Dwight recalled that they had been taken with each other when they first met in 1937: "We were alike: New Yorkers by birth and upbringing, restless, impatient, fond of argument, pushing ideas as far as they would go . . . we could say almost anything to each other without hurt feelings or bloody noses."

They found each other exotic. Dwight recalled Schwartz's ironic perception of Dwight's Wasp "Yale-gentile background, which struck him as picturesque but slightly primitive." Dwight never could fathom what he considered Schwartz's "obsession" with his Jewish childhood. It was the banter, the ceaseless intellectual "dozens" they played, that Dwight recalled with relish. It was certainly something that he sorely missed.

But there was something else about Schwartz that interested Dwight: his victimization. Schwartz's precocity was astonishing. He had achieved his best work while in his twenties and thirties, and then it had been all downhill, a tormented descent of bitterness, unfulfilled ambitions, bizarre accusations of conspiracy and the careful placing of blame. Schwartz's life confirmed for Dwight his belief that those of inordinate insight and sensitivity were vulnerable in American society. "Poetry is a dangerous occupation in this country," Dwight declared. He put Schwartz in a class with James Agee. "The gap between what they might have done and what they actually realized in their work [was] heartbreaking." This was Dwight's constant inner fear about himself. He quoted a long passage from Baudelaire on Poe, claiming that it could just as well have been written about Schwartz. As a disciple of Poe, Baudelaire felt that Poe would have found comfort in Paris or Germany, but in America "he had to fight for his bread. . . . He went through life [in America] as if through a Sahara desert and changed his residences like an Arab. . . . For Poe America was nothing more than a vast prison which he traversed with the feverish agitation of a being made to breathe a sweeter air."[21]

To Dwight, Schwartz was a victim of that mass society that destroyed what it could not absorb and make into a commodity. But he recognized that Schwartz, like Agee and Poe, also had a "genius for self-destruction." Perhaps, as Philip Rahv suggested in his own appreciation of the poet, Schwartz was more depressed by life than fearful of death. The manner of his lonely departure in a tawdry Times Square fleabag shocked Dwight. Schwartz's body lay unclaimed for several days before a *New York Times* reporter recognized the name and a belated obituary appeared in the paper. On seeing the announcement, Dwight called the hotel and was put in touch with Schwartz's aunt, who asked him to pick up his "literary remains" from the room. Dwight found the room full of "crumpled girlie magazines," cheap paperbacks, and a half-dozen indecipherable note-books. A "solitary lair," it reeked of loneliness and despair. "He died, like Poe, alone and unknown."[22]

Schwartz's death, combined with the unpleasantness of New York in the summer (the city even then stank of sweat and urine by the dog days of August), drove Dwight and Gloria out of the city and up to Maine for two weeks. Since their return from Texas, Dwight had suffered from severe culture shock. The city seemed so seedy. Early mornings he was awakened by the "mad whining, clanging, banging, screaming-geared" cacophony of garbage collection. During the night there were the "shrieking, earsplitting" fire engines. Dwight wondered how London survived the flames with their "tinkling little apologetic bells."[23]

When they returned from Maine, where they had visited with Cal Lowell, Elizabeth Hardwick, and some other friends, Dwight was thrown back into his work. He had decided after much thought to abandon the movie column. At a dinner meeting at the Plaza, he and Harold Hayes, the managing editor at *Esquire*, agreed that Dwight would undertake a column on contemporary politics. John Lukacs and Nicola Chiaromonte were delighted; neither thought film was worthy of Dwight's talents. Lukacs would have preferred him to return to his cultural essays and begin work on a serious memoir. Chiaromonte had no objections to serious analytical journalism. Both were worried that the column might simply be a monthly diatribe against the war and not the serious kind of work Dwight was capable of and had wanted to do.[24]

Dwight too was worried about the column, simply because he was having more and more difficulty writing. That was one of the reasons he sought out teaching jobs. He found preparing lectures so much easier than writing finished articles. There was none of the pressure. During the period since he achieved celebrity as a cultural commentator, he had sought and accepted contracts for books, chapters in anthologies, and introductions, but most of them never came to fruition. He had promised an entry on mass culture for *The International Encyclopedia of the Social Sciences*, but he never even began the essay. In a letter to the disgruntled editor, he expressed his anguish and his belief that he had simply been writing too long. On a visit with Chiaromonte, who was a visiting professor at Princeton and giving the Gauss lectures, Dwight complained of his writing block. Chiaromonte insisted there really was no such thing. Dwight responded that years ago when he was under pressure at *Politics*, he seemed to have so much to say that the writing came relatively easy. Actually, he had struggled then with the deadlines. But he was younger, and if he drank as much, he could hold it better. Writing had always been demanding, and he had toiled over it, producing several drafts for even short pieces.[25]

In a letter to Mary McCarthy, Chiaromonte expressed his impatience with Dwight. Like Lukacs, he felt Dwight was not using his talents for serious work. His "block" was simply due to the triviality of his subject matter. Nor was he happy with the proofs of Dwight's first "Politics" column for *Esquire*. Chiaromonte found the prose appalling, incoherent, and Dwight nervous about appearing to be radical or committing himself to some extreme position. Chiaromonte considered the piece so poor as to be beneath serious criticism.[26]

This was a harsh verdict. The piece made political sense. *Esquire*'s readership was much younger, and Dwight was known to them as the entertaining curmudgeon movie critic with tough standards. He was now going to use *Esquire* as a platform for effective political propaganda. It was a mid-cult magazine with a mass circulation, and he meant to make the most of that. His first piece was designed to establish his credentials as a man of evolving common sense. He said all the right things to reassure his readers that he was no longer a Marxist, or even an idealist. On the contrary, he sought the practical, the possible, and, yes, the lesser evil. As expected, his ridicule of the administration was delightfully acerbic. He referred to Secretary of State Dean Rusk as a living fossil "more at home in Australia with the wombats and platypuses" than in the modern civilized world. Dwight argued that what had awakened him to the lunacy of Johnson's vision was the "crisis he had created, not inherited, in the Dominican Republic." He asked rhetorically that even if one accepted their unfounded charges of a potential Communist victory, what right did the United States have to intervene? Why couldn't the Dominicans institute a Communist government if they so desired? In making his case, Dwight invoked all the charges of the New Left critique of an obsolete imperialism. He established his mainstream sanity by stating that he voted for Roosevelt in 1932 and 1936. At another point, when discussing his criticism of Kennedy and the Bay of Pigs as well as the missile crisis, Dwight spoke of himself as being disturbed, "not as a radical but as a moderate, a conservative, you might say, with a leaning toward anarchism." After all, writing for *Esquire* was an entirely different matter than preaching to the choir at *Politics* or *Encounter*.[27]

Dwight was pleased with the column. It was exactly what he wanted to do. It was not scholarship, nor investigative reportage, just an intelligent critic's commonsense view of things, which he hoped would engage his readers and provide them with arguments and information buttressing an anti-war stance. It was conversational, it did tend to ramble, and he was to constantly struggle against the deadlines—but he was always motivated by that mass audience and he felt that through this work he could make a difference.

Toward the end of the year, when his social calendar was jammed with benefits, cocktail parties, dinner engagements at the Epsteins', the Podhoretzes', the Lowells', Hannah Arendt's, he managed to turn out a provocatively favorable review of Barbara Garson's *MacBird*, an irreverent satire of the political establishment by way of a burlesque of Shakespeare's *Macbeth*. Dwight was delighted with what he readily conceded was a "tasteless, crude, wholly destructive satire." But it was just what the subject called for, "precisely the approach most congruent to the atmosphere of Washington under the presidency of Lyndon Johnson." What Dwight liked most was Garson's harsh assault on Johnson's total insensitivity toward all opposition, as epitomized in this scene, which Dwight quoted with glee:

MESSENGER: Beatniks burning draft cards.
MACBIRD: Jail 'em.
MESSENGER: Negroes starting sit-ins.
MACBIRD: Gas 'em.
MESSENGER: Asian peasants arming.
MACBIRD: Bomb 'em.
MESSENGER: Congressmen complaining.
MACBIRD: Fuck 'em. Flush out this filthy scum; destroy dissent. It's treason to defy your President. (His followers start to move doubtfully.) You heard me! Go on, get your ass in gear. Get rid of all this protest stuff, y'hear.

Garson, a Berkeley radical, had deftly captured the "reek of Johnsonian politics, satirizing a reality so grotesque that it sometimes defied even Ms. Garson's luscious exaggerations." What lifted the play into "the realm of impeccably bad taste" and made it so amusing to Dwight was that no person or group escaped her poisonous frivolity. Johnson was an easy target, for he was indeed as vulgar and tasteless as his satirist portrayed him to be, but the Kennedys and their brightest and best acolytes did not come off any better. Every statistically significant group or tendency—the old left, New Left, black nationalists—all received their lumps. "Only an anarchist like me," Dwight concluded, "could find much comfort in the play and cold comfort at that."[28]

Dwight did his best to promote the play's publication and to encourage its off-Broadway production, which turned out to be a great success. He also helped solicit support for the Garsons' small independent publishing company, Grassy Knoll Press. To those infuriated by the play, the company's name was beyond the pale. Dwight did not take seriously the play's actual plot, which implicated Johnson in the Kennedy assassination. He simply took it for granted, and the author agreed in a letter to him, that the scenario simply resulted from using *Macbeth* as the vehicle for her burlesque. It is a mark of the times that the insinuation drove her critics to ill-humored hysteria. Lionel Abel, in the *Partisan Review*, was indignant that Dwight, Robert Lowell, and Robert Brustein, among others, had enthusiastically praised the spoof. (Lowell had proclaimed it a "work of certain genius"; Brustein saw it as a companion piece to Jean-Claude van Itallie's *America Hurrah*.) All these critics were united in their revulsion for Johnsonland. Abel poignantly asked: "Is our country so horrible? Is our President so evil?" The Johnson character's collaboration in the assassination was crucial to the play, and to try to dismiss it, as Dwight had, was an act not only of bad taste but of bad faith. The *Times* of London, in endorsing Abel's solemn and humorless criticism, said that Dwight was eager to seize the radical moment, that he had hurriedly assembled a literary-critical mask to disguise the broad, delinquent grin beneath. In denying the seriousness of the assassination implications, Garson and Dwight were simply trying to have it both ways; they wanted to come across as revolutionists while neutralizing the rebelliousness they were so anxious to be praised for. In *Fortune*, Max Ways used Dwight's praise of "a sophomoric play" as a typical example of "New Left nihilism and revolutionary fanaticism."[29]

No one was really talking art or criticism, they were talking politics. And it was on that level that the play became a cause célèbre and a most revealing document illustrating the degree of frustration, anger, and alienation to be found in certain quarters of the intellectual and political community. The play was an anti-American diatribe, and it was that element that marked the polarity in perception. There were those who were against the war because it represented everything that America was against. And there were those who were against the war because it represented everything America had become. Personally torn with anguish, Dwight was often on both sides of this divide.

Johnson had indeed taken on a kind of personal villainy unusual even in American politics. The protesters' constant chant "Hey, hey, LBJ, how many kids didja kill today?" captured the bitter animus abroad in the land. Dwight felt this same personal hatred for Johnson. When Nicola Chiaromonte wrote to say that he had come to hate Johnson, Dwight replied "Bravo, Bravissimo dearest Nicola! For months now I have been coming reluctantly, shamefacedly, but irresistibly to the same conclusion," that Johnson was a "catastrophic presence on the world scene." Dwight worried about this, because to express such "subjective" feelings when looking at history went against his Marxist and anarchist training. But he admitted that he could not help "hating, detesting Johnson, quite personally" as he had Stalin and Hitler. Johnson was the worst president the country had ever had, "the one who's done dirt more than any other on the best of our political tradition, all the more effectively because of his very abilities." Johnson was no fool, Dwight observed; he was endowed with a "vigorous and terrifying will, like Hitler and Stalin." Dwight was convinced that neither Eisenhower nor Kennedy would have tried to bull it through in Vietnam as Johnson had. They were normal, calculating politicians, concerned with profit and loss, but Johnson was bent on catastrophe for the United States and for the Vietnamese.[30]

It is not clear what brought on such a personal and somewhat distorted perception of Johnson, who was as calculating and thoughtful about political profit and loss as any other man in American politics. Johnson had concluded that a loss in Vietnam would consign the Democrats to the wilderness for another twenty years, surely a rational perspective given recent history. It may have been simply the need for personal animosity to sustain the constant demands of opposition when even as late as the fall of 1967 Johnson and the war stood up in the polls. Some, like the British journalist Henry Fairlie, felt that the vindictiveness directed at Johnson came from the Ivy League camp, with their smug sense of superiority and contempt for the grass-roots vulgarity of Johnson—in effect, a matter of style.[31] But Dwight, like Barbara Garson, was even-handed in his contempt. He scorned the courtiers of Camelot as well as the boorish rancher from the Pedernales. For they too were deeply involved in the intrigue, the lying and deceit that drove the country on in its relentless pursuit of a futile war.

19

Bringing the War Home

On THE WAY HOME FROM A BRIEF FORD FOUNDATION junket to West Germany in January 1967, Dwight stopped in London and visited the *Encounter* offices. They were in turmoil due to the public outcry over charges that the Congress for Cultural Freedom and its publications, including *Encounter*, had been subsidized covertly by the Central Intelligence Agency. These accusations were hardly new; Dwight had been hearing them since he first went to work for the magazine in 1957. Starting in the early sixties, Dwight had queried Stephen Spender about the constant allegations, but Spender always gave him evasive answers and official explanations of foundation support.[1]

This seething cauldron of gossip and intrigue first bubbled over in a series of *New York Times* articles in April 1966. Such well-known participants in CCF affairs as Arthur Schlesinger and John Kenneth Galbraith denied the charges of covert funding, insisting that *Encounter* had maintained absolute editorial freedom. A year later, in March 1967, the California New Left magazine *Ramparts* published a detailed account of CIA funding of the National Student Association as a means of countering the propaganda of the Soviet peace offensive at the Warsaw Bloc student conventions. In the *Ramparts* story the CIA's support of the Congress for Cultural Freedom was also mentioned. When a letter to the *New York Times* was being composed, Daniel Bell had asked Dwight to sign it, but he had declined because the denial was so evasive and simply refused to address the issue of secret financing.[2]

The *Ramparts* exposé was soon followed by the proud confession of Thomas Wardell Braden, a confident liberal cold warrior who had established the International Organization Division of the CIA, which dealt with the Congress for Cultural Freedom. Braden described the CIA's early role in funding the non-Communist left organizations in their propaganda war against the Communists. What caused real embarrassment was his blunt assertion that the CIA had not only funded *Encounter* but had placed an agent in the CCF and another as an editor at *Encounter*. Speculations remain to this day as to whether the agent was Irving Kristol or Melvin Lasky. Braden's statement that an "agent *became* an editor" would seem to implicate Lasky. Both men have denied any formal CIA connection.[3]

During his stopover at the *Encounter* offices, Dwight had talked with Lasky, who portrayed himself as an innocent and hapless victim of the "duplicity of The World." Dwight had always had misgivings about Lasky, whom he compared privately to LBJ. "The moral sense seems to be lacking in their make up." He confided to Stephen Spender that he would put nothing past Lasky, since "power and success were the only criteria for determining what was right and what was wrong." By the time Dwight reached New York, the issue was a cause célèbre. Outraged and hurt, *Encounter* editors Frank Kermode, Stephen Spender, and John Gross resigned, insisting they had no knowledge of the CIA funding. This only highlighted the deceit of the Congress for Cultural Freedom. Dwight wrote Michael Josselson, whom he respected, advising him to make a straightforward statement apologizing for participating in the deception and explaining the reasons for doing so and the good that came of it. Dwight insisted that an open break with the past was essential if the CCF and the publishing operations were to have any integrity; continued evasion and stonewalling would only make matters worse. Dwight said he would like to sue the CCF for putting him in "a position in which [his] personal and professional reputation [was] not exactly strengthened," but he couldn't because he couldn't *prove* that he wasn't in on it. Dwight felt he had been betrayed, "played for a sucker."[4]

In another letter to Malcolm Muggeridge, one of "the big bonzes of the Congress," who he thought might be sympathetic, he wrote that he found it easy to believe that Lasky was in on the deal for several years. Lasky conceded that he had learned of the CIA support in 1963 and had gone to

work to get legitimate private foundation funding. Lasky admitted that he
and CCF officials were guilty of "insufficient frankness." Of course, there
were many who claimed to have been well aware of the financial arrangements. William F. Buckley, who had worked for the CIA in the 1950s,
claimed he knew of the funding operations as early as 1954. Another
unnamed conservative made the reasonable claim that anyone close to
the operation who did not figure out who was paying the bills was a "dope
or a dupe," or "a hypocrite or an agent." There were literally legions on
the left who had made the charges of covert government agency funding
but without the hard evidence. Paul Goodman claimed to have heard it
from a boozy horse's mouth at a party in 1962, and he made reference to it
in a *Dissent* article that year. Goodman declared that "when Cultural
Freedom and the *Encounter* of ideas are instruments of the CIA, nothing
could be more treasonable to cultural freedom and the encounter of
ideas." Goodman at this time tried to get Dwight to dig up confirmation
for these charges. But Dwight, simply because it was such a commonplace rumor on the left, continued to dismiss it.[5]

When Irving Kristol got wind that Dwight planned to publish a piece
on the affair in *Esquire*, he wrote to him to state bluntly that he had never
known of any CIA funding and would sue anybody who claimed he did. In
Kristol's view, the entire issue had been blown out of proportion and had
become a "witch-hunt" as part of the anti-Americanism campaign so virulent among leftist critics of U.S. foreign policy, life, and culture. He
accepted the official version that *Encounter* was financed by the Farfield
Foundation through the good offices of Julius (Junkie) Fleischmann, who
warmly supported such cultural activities. Kristol was irritated by
Dwight's priggishness and his complaint that he had been played for a
sucker. He did not think it was justified. Kristol knew what he was doing
and had edited the magazine the way he wanted to without undue interference from any outside agency, and Dwight had written just what he
wanted to. Kristol seemed to have forgotten his own letter of apology for
being influenced by the Paris office in rejecting Dwight's "America!
America!" article. Kristol claimed that if he had been aware of secret
funding he would not have gone along with the deception. But *Encounter*
was and remained a fine magazine and he would not partake in any
lynching party.[6]

Lasky also wrote a long account (four pages, single-spaced). He too
denied any knowledge of the secret funding and asserted his interest in
pursuing truth and beauty, urging Dwight not to publish anything that

would hurt the future of *Encounter*, which he would continue to edit. The last to write Dwight before he printed his own account was Stephen Spender. In his letter, which was thick with marginal additions, he wanted to make his own innocence absolutely clear. Most people familiar with the affair agree that Spender probably never grasped what had been going on.[7]

This avalanche of self-serving explanations did not satisfy Dwight. An open confession was inevitable, and he published his statement in his column in the June 1967 issue of *Esquire*. It began:

> I confess that from August 1956 to June 1957 I was on the payroll of the C.I.A., unwittingly and, as they used to say in the National Student Association, unwittily. . . . The C.I.A. paid some and perhaps all of my salary as a special advisory editor of *Encounter*.

He went on to describe in detail the activities of the Congress for Cultural Freedom and its publications. He conceded that he had heard the rumors of U.S. government money from some of his "more radical and less temperate friends, to be tautologous," and that he had pooh-poohed their warnings because they were based on "nothing more substantial than the fact that the Congress was openly anti-communist, as for that matter, was and am I." He went on to say that his resistance to the charges could be laid to his "petty bourgeois prejudice in favor of hard non-ideological evidence." He recalled Paul Goodman urging him to do an exposé but that he had refused because he felt that Goodman, "like the others, was arguing from logical extrapolation, rather than factual knowledge." Dwight also recalled encountering at a party at William Phillips's apartment the executive secretary of a small foundation which Dwight took to be the chief underwriters of the CCF and asking him point-blank whether government money was involved, and the answer had been far from reassuring. But even this was not "proof in the petty bourgeois sense." Given Dwight's penchant in the past for shooting from the hip when his instinct moved him, his scrupulous insistence on a smoking gun in this case seems a bit ingenuous.

Dwight told the story of his rejected "America! America!" piece and how it had been turned down because the Paris office felt his spirited anti-Americanism would jeopardize foundation support. He now thought he should have been more curious. He concluded that the three main explanatory justifications—(1) the CCF and its publication did good work; (2) the publications were independent; (3) due to the climate of McCarthyism, non-Communist–left activities in Europe had to be funded surreptitiously—did not address the fundamental issue:

> Secrecy in such matters is corrupting in itself, regardless of the practical success of the operation, because it means there can be no control or criticism of the unknown source of money so that the recipient is responsible for policies that may be shaped by forces of which he is not even aware.

This was a classic means-and-ends issue for Dwight; it was at the bedrock of his political morality. If a man supported the policies of the CIA in its effort to encourage liberal-left opposition to Soviet Communism, that was a perfectly legitimate political decision. It might well be appropriate for such a person to knowingly do work that was covertly funded. But to be used for such work and not be informed was an immoral and corrupting deception. Dwight could see no possible justification for such conduct in an organization allegedly seeking to encourage cultural freedom. He *had* been played for a sucker and had been used as a stooge. As for Kristol's assertion that he was not ashamed, Dwight did not name him, but he wrote in *Esquire* that if he knew, he should be ashamed not for accepting the funds from the CIA but for doing it on the sly and "deceiving practically everybody."[8]

In retrospect, there is more than a little reason to accept Daniel Bell's suggestion that at the time, government funding to help the non-Communist left's fight against Stalinism was readily accepted and even endorsed by many in the intellectual community. There was what was known as "Counterpart Funds" from the Marshall Plan, money directed at the cultural–propaganda side of the battle. That the funding was covert was an obvious political necessity. The issue of independence was meaningless, since those who participated were anti-Stalinist liberals and radicals, who were glad to support American policy, especially one that was aiding liberal to leftist organizations in Europe. This would have been true of Dwight in the late forties and early fifties. It was a question of political pragmatism. Arthur Schlesinger, Irving Kristol, Mel Lasky were in no way ashamed of their work and made no excuses. Dwight in a letter to Spender conceded that *Encounter* was "one of the few real successes in the CIA's history." When questioned about the issue in the 1980s, Bell and Lasky remained incredulous at the naïveté of the innocents who made such a fuss. But in the late 1960s, the liberal-left consensus was in a shambles, the notion of the Soviet Union as the great menace had declined, and the reputation of the CIA as a rogue agency responsible for unjustified violence and chaos in Latin America and Asia was widespread.[9]

For Dwight the entire affair was another example of the immoral conduct of an irresponsible State that would employ any methods to achieve its ends. The collaboration of intellectuals willing to use their friends and colleagues and deliberately deceive them was as appalling as it was depressing. When Lasky was kept on at *Encounter* and dismissed Spender and Kermode's criticism as no more than a tempest in a teapot, Dwight wrote an angry letter to the *New York Times*. It was not published, but he reproduced it in his *Esquire* column. He again insisted that Lasky should resign or the trustees should fire him, because if the magazine was to have any integrity it had to have a new start and that meant new editors. Spender, he wrote, had resigned with sobriety and a sense of responsibility. Lasky had shown neither in his comments or his behavior.[10]

Dwight's anger at having been used and his public confession did not endear him to some of his old comrades. When he signed a statement in the *Partisan Review* condemning secret subsidies, Nicola Chiaromonte was not at all pleased. As an editor of *Tempo Presente*, the Italian publication of the CCF, Chiaromonte was getting a good deal of criticism from the left. He and his co-editor Ignazio Silone had been singularly independent of the Paris office. In fact, they had published Dwight's "America! America!" after *Encounter* rejected it. Chiaromonte was also irritated by the editorial tone of the *New York Review of Books*, to which Dwight was increasingly devoted. He thought a piece by Noam Chomsky that cited Dwight as an inspiration was "not without a certain bad faith & falsification of historical perspective—was being an anti-Stalinist tantamount to being a conformist?" When Chiaromonte saw Dwight's name on the *PR* statement denouncing secret funding and expressing a lack of confidence in magazines that had received it, he dashed off an angry letter asking Dwight whether he meant that he didn't trust the *Tempo Presente* editors. Dwight quickly reassured him and promised to make that explicitly clear in a subsequent commentary in his *Esquire* column. In his apology for hurting his friend's feelings he said he was driven by the notion that the entire affair was on its way to being swept under the rug by the CCF and *Encounter*. He was appalled that Lasky was staying on as the editor, with no expression of outrage by British intellectuals. He wanted "a public squawk."[11]

Chiaromonte felt Dwight should have informed him of his plan to sign the *Partisan Review* statement. Dwight considered that unfair. "Come off it," he wrote, "we're friends, not rival UNESCO bureaucrats." Dwight believed that Chiaromonte, who was in many ways his intellectual and moral mentor, profoundly disagreed with him over principle. Chiaromonte deemed secret funding permissible if the magazine could maintain its editorial independence, while Dwight insisted that secret funding was immoral and unethical in itself. Dwight would not let it drop. "Either one thinks the CIA secret funds were justified by the ends: a liberal indepen-dent leftist voice against Communist totalitarianism . . . or else one doesn't, in which case one says so now, after the secret has been revealed, openly and bluntly." What irritated Dwight was the "smoothing it over"; this silence seemed to him "not the thing at all."[12]

Chiaromonte's grievances got to the essence of the dispute. Dwight was increasingly tolerant of his younger New Left comrades; he was less receptive to the old anti-Stalinist circumspection. He frequently declared that it was not an issue anymore. Chiaromonte, on the other hand, was not at all happy with what he described as the New Left's bad-faith assump-tion that there was no such thing as Stalin and anti-Stalinism, only McCarthyism and anti-Communism. He was appalled by the young revo-lutionaries' attachment to Mao, Castro, the late Che Guevara. It was "false and unserious." Chiaromonte certainly had not joined forces with the yea-saying conservatives who were relatively tolerant of the war in Southeast Asia, but as a dedicated veteran of the leftist opposition to Stal-inist totalitarianism in Europe, he was irked by the indifference and naïveté of the younger generation. While Dwight hardly supported the New Left's romanticism toward third world dictators, he was intent on focusing his left politics on the anti-war movement in America. That was the most important political issue of all.[13]

Dwight often sounded like a Dutch uncle as he scolded such New Left leaders as Tom Hayden and Ivanhoe Donaldson for their intemperate and ill-informed rhetoric, but he heartily agreed with their attack on American foreign policy and disrespect for establishment authority. In a *Commentary* symposium that asked where he stood with respect to anti-Communism, he replied that it "*tout court* had become a dead horse use-ful only for ritual flogging by our less sophisticated and more cynical politicians." He also conceded that the obsessive, black-and-white anti-Communism of the 1950s had helped "to create a favorable climate of opinion for Johnson's war." Given this, he was at a loss to know what the alternative position could have been, other than acting as an accomplice of the opposition. But he continued to insist that it was "no longer a central issue" for him after LBJ's crusade in Vietnam.[14]

As 1967 wore on, the sense of despair and frustration increased among the liberal-left intellectual opposition. The administration seemed isolated and impervious to criticism, a beleaguered band of "misled and misleading men." In October, Dwight was one of the sponsors behind "A Call to Resist Legitimate Authority," a leaflet that declared there was now a legal right and moral duty to exert every effort to end the war and to give support to draft resisters and deserters.[15]

Dwight responded quickly to Noam Chomsky's idea of withholding a portion of his income tax as a confrontational way of challenging the government's policy. After consultation with Mary McCarthy and others, however, he did not actually withhold taxes. His entire income was as a freelancer, and the financial hardship would simply be too difficult. He wrote McCarthy that he envisioned a "declaration of civil disobedience to be signed by at least 100 Americans—intellectuals, artists and writers, musicians, movie-makers, scholars," advising young people of draft age to refuse to be conscripted because the war was illegal, immoral, and a shame and disgrace to every American citizen. As Dwight worked on this mobilization of the intelligentsia, he also joined with David McReynolds in an "Impeach Johnson" campaign. In August a group of war resisters including McReynolds, Ashley Montagu, Richard Falk, Robert Silvers, and Jason and Barbara Epstein met at Dwight and Gloria's apartment to draw up a statement calling for Johnson's impeachment. Dwight was all for the propaganda value of an impeachment campaign, provided they could work up a genuine legal brief based on the Constitution and Johnson's usurpation of power by waging an illegal and undeclared war.[16]

This flurry of activity in the despair-ridden days of late 1967—the calls for resistance, tax evasion, draft evasion, and impeachment proceedings—led naturally to the great March on the Pentagon in the fall of 1967. That dramatic and ultimately violent episode has been searingly captured in Norman Mailer's vivid account, *The Armies of the Night*. Mailer pointedly subtitled the work "History as a Novel, the Novel as History" to suggest the illusiveness of contemporary history and the different ways of trying to get at the essence of the event's significance for

Americans at the time and for future generations. Dwight proclaimed it the best piece of writing in Mailer's career.[17]

Dwight's judgment was seconded by Robert Lowell. What astonished them both was how Mailer, often drunk, enraged, exhausted, totally absorbed in the events, and finally arrested, could, without taking any notes, remember and record the factual reality with such astonishing accuracy. Mailer, for his part, had respect for Dwight, and his judgment of Dwight's contribution to American intellectual life was wonderfully direct:

> Of all the younger American writers Mailer was the one who had probably been most influenced by Macdonald. Not so much from the content of Macdonald's ideas, which were always going in and out of phase with Mailer's, but rather by the style of Macdonald's attack. Macdonald was forever referring the act of writing to his sense of personal standards which demanded craft, care, devotion, lack of humbug and simple *a fortiori* honesty of sentiment. . . . Macdonald had given him an essential clue which was: look to the feel of the phenomenon. If it feels bad, it *is* bad. . . . Macdonald's method had worked like Zen for him.[18]

If Mailer had an "unquestionable" fondness for Dwight, the feeling was reciprocated. Even when the beast in Mailer's bosom burst forth as it had in his drunken tirade at the Ambassador Theater on the Thursday evening preceding the more organized events of the weekend, Dwight was loyally defensive. Clark Loomis Herbert, an English professor at Northwestern University and closet poet, sent Dwight an acerbic denunciation of that tawdry and vulgar performance in verse. It began

That slob up there
     gasses
hot air escapes
     his bloated belly
          billows pigeonwise
this inflated author
     swells,
a pregnant spinnaker,
     violated by a summer squall

Blowhard

Herbert's trashing ends,

> *O come, let us adore him,*
> *Christ!*
> *but he*
>> *profanes these solemnities*
>> *desecrates the temple*
>> *betrays the faith!*
>>> *not intoning the beatitudes*
>>> *but bawling out obscenities*
>> *Holy apostle, he gives peace the kiss of Judas!*
>
> *Look how this bullyboy*
>> *planks his overweening jackboot heels*
>>> *as to dragoon these gentle mutineers*
>>> *into indulgence*
>>> *and keep his keel against the rock and roll*
>>> *of his big head awash in seas of alcohol.*
> *Dissent does not present its protest here;*
> *this word-merchandiser only belches.*[19]

Dwight could appreciate this ridicule, for he had been upstaged and dismissed, even threatened, by Mailer that evening. His friend's performance, he conceded, had been "ghastly." "But," he wrote, "that's not all of him. I've known him pretty well for many years and I really love the guy, he's not only shrewd, humorous, decent (yes), he's generous—also touched w. genius, but also one of the least vain or overbearing or show-off (yes) writers I know." Dwight insisted that despite that awful first evening, during the next two days, the Friday afternoon at the Justice Department draft resistance rally and Saturday at the March on the Pentagon, where the ever resourceful Mailer outwitted the guards and managed to get arrested, he had redeemed himself. Dwight maintained that Mailer had been sober, modest, and concerned.

Dwight saw the two sides of Mailer: the ever present subterranean, lurking beast, described so vividly by Mailer in his extraordinary account, as contrasted with the dignified, intelligent, thoughtful participant and phenomenally acute observer. It was, of course, this latter achievement that meant so much to Dwight the critic and Mailer's mentor on those three brisk October days of discontent. As a craftsman, Mailer had captured not simply the facts of the event, but its substance, its in-depth meaning. That Jewish American on the make had the inherent distance and disinterest to size up the "Ivy League intimacy" of Dwight, Lowell, and the cadre of assorted academic intellectuals who allied themselves with the youth brigades, the guitar armies, the pot-smoking, acid-tripping rebels against the madness of America as epitomized by the war.[20]

Through Mailer's benevolently egotistical lens, one could follow the march of "America's best poet, best novelist and best critic," dedicated to offering their services to this adversarial crusade. Mailer gave voice to Dwight's conception of himself as the "conservative anarchist." Mailer declared that he was a "left conservative" who tried to "think in the style of Marx in order to attain certain values suggested by Edmund Burke." Lowell, the New England aristocrat, "his features at once virile and patrician," marching alongside gave the image validity. Here they were, the novelist, the poet, the critic, marching to defend the "root values" trampled on by the best and the brightest of "technology land." A useless, purposeless life of consumption was Mailer's portrait, which simply echoed and coincided with Dwight's own analyses. Here culture and politics met to do battle. And if the new younger generation was absorbed in the technological authority of the corporate economy, they had a hatred of the authority. That's what appealed to these middle-aged custodians of the culture. Instead of being repelled by the extremes of the younger generation's denial of deference, its repudiation of and contempt for authority, they shared it.

There was no respect or deference because the authorities

> had lied, through corporate executives and Cabinet officials and police officers . . . newspaper editors and advertising agencies and . . . mass magazines, where the subtlest apologies for the disasters of the authority . . . were grafted in the best possible style into the ever open mind of the walking American lobotomy: the corporation office worker and his high school son.[21]

Mailer had captured and given voice to the anger, the alienation, the contempt that had brought what Lowell described as these "absurdly locked" protesters to that pretentious monument to world power, the National Mall. Now the critic, the novelist, and the poet were bearing witness to the violation of that democracy. Across the Arlington Memorial bridge they moved at a snail's pace, and only after endless rhetoric in the shadow of the Washington obelisk. Then more tedious speeches, and finally on to the exorcism of the Pentagon. There General Abbie Hoffman and other gurus were chanting "Out, demons, out. End the fire, end the war, end the plague of death," and from the background the long sustained *ommmmmm*. It was indeed street theater, but always in the background was the fear of a wounding confrontation with the defenders of that humorless arsenal.[22]

And then, taken by surprise, Mailer was gone, an end run around the helmeted guard and into a no-no zone, followed by a filmed arrest. His two less skillful comrades, the critic and the poet, failed to entice the guards. As Dwight later recalled, Mailer, "the man of action, had taken the men of letters, as well as the enemy, completely by surprise." For Dwight it was embarrassing and unquestionably disappointing. He was a spokesman for resistance, disobedience, confrontation, he wanted and needed to bear personal witness, and yet in the end, even after a shoving match with a guard who didn't care who he was and had never read his *Esquire* columns, the probing of the Pentagon no longer seemed worth the effort. The battle ended with a whimper, and Mailer's fearless comrade, "one of the oldest anti-Communists in America," along with the aristocratic poet, retired from the field and made their way back to New York.[23]

If Dwight was disappointed by his failure to get arrested, he may have also experienced some pangs of envy at the astonishing literary and journalistic achievement of Mailer's account of the event. But in a delightfully good-humored parody of Mailer's introduction to the work, he reviewed *Armies of the Night* in his *Esquire* column, declaring that Mailer had carried "journalism into literature." It was "an astonishing literary performance . . . [of] Jamesian control . . . a literary gem." In a second column dealing with the march and Mailer's account, Dwight quibbled in a revealing way with a few of Mailer's perceptions. While Dwight, Mailer, and Lowell agreed that the rhetoric of the endless speechmaking was boring, Dwight was "charmed by the exorcist rites." The "out, demons, out" chants paled after a while, but Dwight did not accept Mailer's charge that "Dwight hated meaninglessness even more than the war in Vietnam." On the contrary, if the Fugs, a scatological folk-rock group, could chant America out of the war, he would stick with them as long as it took. Mailer underestimated Dwight's skepticism toward the computerized linear thinking of the war managers. The spoofing mockery of the exorcists, the mayhem of Abbie Hoffman's Yippie brigades appealed to him. Their rollicking nihilism was anarchic, and he liked that.[24]

It was this penchant for seeming irresponsibility that irritated Dwight's old left and intellectual associates of earlier days. His apparent acceptance of much of the New Left agenda appalled them. Dwight's encounter with this younger generation of naysayers was far more comradely than that of Irving Howe, Michael Harrington, William Phillips, Daniel Bell, and other veterans of the ideological wars of the thirties and forties. This comradeship was facilitated by his renewed enthusiasm for anarcho-pacifism, which had become a viable means of organized resistance and civil disobedience. While Dwight continued to express impatience with the anti-intellectual strain that ran through "the movement" and often reproached the young radicals for being ill-informed about the radical tradition, he was not as incensed by their refusal to take Stalinism seriously as his generation of battle-scarred activists were. Dwight continued to repudiate any romanticization of the Vietnamese Communists in the North or the South, but he was increasingly scornful of anyone preoccupied with anti-Communism. To Diana Trilling, Dwight was one of the more extreme examples of an intellectual with no real connection to reality, with no stable beliefs, simply caught up in the currents of catastrophe without a clue. Dwight readily accepted the United States' global interventions as a far greater menace than contemporary international Communism. When Martin Luther King charged that the United States was "the greatest purveyor of violence in the world," Dwight wrote the civil rights leader agreeing that "our country has become the main center of violence in the world today." He enclosed a check for $100 as a token of his support for King's "new radical dissent from Johnson's genocidal war." In a 1967 interview concerning his current reading tastes, he remarked that "Communism is no longer a central issue with me after LBJ's crusade in Vietnam." Rather than rail against international Communism, he was more interested in exposing the darker side of our technical productivity and mass industrial society as "a sinister extension of the American way of life."[25]

Accompanying what his former liberal-left colleagues now saw as an inordinate and unbalanced anti-Americanism was his apparent attraction not only to New Left politics but to the lifestyle of the youthful counter-culture. In contrast to his distaste for the crisis managers of the cold war, he saw many of the young cultural critics as imaginative, innovative, and interesting. He was soon to become friendly with Abbie Hoffman, whose antics he had observed with some pleasure during the March on the Pentagon. In Dwight's view, Hoffman combined creativity, wit, and shrewdness in an uncommon way. Abbie, he wrote in an introduction to the published transcript of the Chicago conspiracy trial, "had mastered his peculiar style so thoroughly" that he could "play around . . . with it like a frisky dolphin." Soon the Yippie leader was a frequent guest at Dwight's apartment.[26]

Around the time of the Pentagon march in 1967, Dwight received a particularly ethereal counterculture manifesto by a young sociology professor, Frank Lindenfeld, that suggested making love could end making war:

> We want a world of peace and love and beauty; a world of brotherhood and flowers . . . in which work is play and in which each produces according to his abilities and takes what he needs; a world in which there will be free bread, free wine, free love . . . a world in which men can be honest and open with each other, in which lovers can lie naked and fuck under the trees . . . We want a communist world, that is capitalism without the profit motive, socialism without the state. We want a decentralized world built more to human scale.

It is hardly surprising that Lindenfeld had been greatly influenced by "The Root Is Man" and had enclosed his own peroration in a request to reprint the last section of Dwight's essay in an anthology. Granting his permission, Dwight expressed his kinship with the author and was amused to think how his old comrades Irving Howe and Philip Rahv "would snort in contempt . . . infantile leftism, petty bourgeois romanticism . . . but can't you see, Dwight, it's not POLITICAL." Dwight added, "And they'd be right," for how could we "practice 'self defense' while continuing to love everybody and fuck everybody in the sun, everybody that's willing of course, and not be a part of the system of violence and death." But still, Dwight "lik[ed] the spirit" and went along with every word "philosophically, you understand. I am all for utopia, an experiment noble in purpose, as Herbert Hoover said of prohibition. But what's a nice libertarian like you doing in a sociology department?"[27]

Dwight's toleration of the romantic excesses of the counterculture and his apparent endorsement of the New Left's unrelieved anti-Americanism caused consternation and contempt among the yea-sayers and old anti-Communist stalwarts. Their condemnation began to take on an inevitably chauvinistic edge. As in the fifties, when Dwight's cultural criticism was attacked by Edward Shils and Irving Kristol, it was seen as an aesthetic substitute for political anti-Americanism. Norman Podhoretz recalled it was the attitudes toward Communism on the one side and Americanism on the other that determined one's allegiance. Many of Dwight's critics were upwardly mobile intellectuals with immigrant backgrounds who had now found a home in the cultural life of the nation, and they took pride in their achievement. The carping, naysaying critics who had the temerity to suggest that America was a corrupt, greedy, and violent society offended these new members of the intellectual and cultural elite. They had chosen the West, a decision that Dwight was never really comfortable with.[28] He was returning to his old negativism and alienation. On that score alone, the youthful generation's challenge to authority, their skepticism and even contempt for the platitudes of democracy and equality and higher learning while the country waged a monstrous war, only increased his admiration for them.

Nor was Dwight scandalized by the penchant of the young for pot. Always open to new experiences, Dwight and Alistair Reid, a friend at the *New Yorker*, took psilocybin in the 1950s at the apartment of the pioneer drug experimenter David Solomon, the editor of anthologies on marijuana and LSD. For some strange and inexplicable reason, Solomon decided to accompany the drug trip with German martial music. Dwight was extremely moved, even to tears. When he came down, he was disturbed at how much the drug had affected him and how he had lost control of his faculties. It was an experience that he had wanted to have but not one that he desired to repeat. He had also, on several occasions, tried pot, mostly with Norman Mailer at the Cape. His reaction was typical of the habitual drinker. The drug didn't seem to have much effect, or at least he didn't think so. By the time drug use was widespread on college campuses in the sixties, Dwight felt the illegality of drugs was unwarranted. He thought it "absurd that pot was illegal while it was thought normal and respectable to drink oneself purple and commit vehicular and other homicides as a result." People became far more aggressive when drunk than when stoned. That may have been the underlying reason pot was banned, he theorized; after all, we might become a "nation of peaceful pantywaists and where would our next LBJ come from?" Dwight simply did not see these changes in social mores as a threat to civilized life; he did not view the youthful defiance of the drug laws a challenge to any essential authority. His instinctual libertarianism made him far more receptive than many of his generation. Oddly, Norman Mailer became far more threatened and worried by youthful drug experimentation than Dwight did.[29]

But there were developments on the left that did worry Dwight. As frustration grew, the tendency of some elements in the anti-war movement to engage in ever more violent rhetoric and conduct increased. In September 1967 Dwight covered the Conference for a New Politics in Chicago for *Esquire*. It was a depressing experience for him because the guilt-ridden and frustrated white left had caved in to the swaggering black nationalists, who were by this time apparently bent on destroying the nonviolent, integrated civil rights movement. In his column he conceded that he had always had some reservations concerning the moral and intellectual aspects of the New Left but that this conference surpassed his gloomiest suspicions. But even this deplorable debacle did not push Dwight into a rejection of the cadres of the New Left; he continued to support the young radicals of SDS and Vietnam Summer, the campus rebels, the draft resisters, the community organizers of the poor and the blacks, whose alienation from America went far beyond electoral politics into civil disobedience.[30]

While he proclaimed he would support all alternatives to Johnson, by the end of 1967 Dwight had come to the conclusion that confrontation was the only possible form of effective protest against the war. His own and his comrades' rhetoric of disdain and alienation seemed to demand the personal commitment of a nonviolent resistance. Dwight ended the year by making a series of speeches defending the need for civil disobedience. He debated William F. Buckley on *Firing Line* and engaged in argumentation over the legitimacy and necessity of civil disobedience in the *New York Post* and the *Village Voice*. For Dwight it was a moral imperative. One could not continue to allow the war to go on without overt and illegal disobedience. He publicly encouraged young men to defy the draft.[31]

This notion of "from politics to resistance" fomented by Dwight, David McReynolds, David Dellinger, Noam Chomsky, and others was criticized not simply by upwardly mobile intellectuals rapidly on their way to a neo-conservative defense of American life and culture, but also by Irving Howe, Michael Harrington, Lewis Coser, and Bayard Rustin—people with genuine radical credentials. They felt civil disobedience, which disrupted the life and routines of American citizens in order to "bring the war home," would encourage a backlash and thus retard efforts to end the war. Howe called it irresponsible "adventurism," and Harrington argued that it only answered the psychological needs of the despairing. Such "middle-class tantrums" alienated people from the anti-war movement with no viable political justification. Having a sound political strategy was also a moral imperative, Harrington argued. While these opponents of resistance prefaced every statement with sincere declarations of their opposition to the war, their alternatives appeared to the more militant as delay and compromise at the high cost of continued destruction of life in Vietnam and mounting body bags returning to the United States. McReynolds wondered in frustration what it would take to get Mike Harrington and others on the old left back on the picket lines again.[32]

But Dwight knew that the politics of resistance had its risks. It could and did in all too many instances lead to irresponsible conduct, frivolous and self-indulgent street theater, racist backlash, and sporadic violence. By the end of the year, after the race riots across the nation, the continued escalation of the war, the total repudiation of the integrated civil rights movement by the radicals pushing Black Power (as a slogan Dwight felt it outdid "Lucky Strike has gone to War" for meaninglessness), and other forms of black nationalism and New Left sectarianism, Dwight had little confidence in the opposition movement. In a brilliantly acerbic commentary attacking icons on the left and right, he described the chaos that gripped the nation and the world. All reports suggested that the world was going to hell, he wrote; the young were as prejudiced as the old, the left as "alarmingly abstract" in their thinking as the right. "Black power demagogues were giving white power demagogues a run for their honky money." For Dwight, things had been so much clearer in the days of the old radicalism, when principles were principles and revolutions were revolutions, "not random self-destructive outbursts of looting and arson and rioting by what Marx would have called the lumpenproletariat." In the old days there were no "anti-cultural 'cultural revolutions,'" in which "hysterical mobs of teen-agers [were] mobilized by an elderly mystagogue to tear apart a great nation for his own private ends." Dwight described Mao's *Red Book* as a compilation of platitudes called thoughts. In the old days one had to weigh when to choose the lesser evil, a "problem involving calculation of ethics, rationality and other old-fashioned factors." Now it was simply a matter of greater evils. The choices, he thought, were invented by somebody named Hobson.[33]

Dwight then launched into a marvelously prophetic satirical account of how democracy registered the views of its citizens. Ideas were handled the way products were hawked on the TV screen. It was almost as though he was watching the tube in the 1990s:

Mao or Ky? Nasser or Ben Gurion? Tshombe or Bourmediene? Governor Wallace or H. Rap Brown? Johnson or Nixon? Hanging or shooting? Check the ballot for the so-and-so of your choice. And what is *your* view, sir, on Vietnam, should we escalate or negotiate? pull out or push in? use The Bomb or make a survey of land tenure? hope for the best? fear the worst? is the President handling it badly, well, fairly well, fairly badly, don't know? And, dear sir or madam, the riots, what of them? Yes? No? Don't know? And what do they portend, would you say in confidence, the dawn of hope (Andrew Kopkind in the *New York Review of Books*), the twilight of disaster (Dwight Macdonald in *Esquire*) or Don't Know (Governor Romney in the Michigan statehouse)? The Gallop Poll [*sic*] will protect your anonymity to death—and, sir or madam, what might be your views on that perennially pressing problem: afterlife? no afterlife? heaven and no hell? hell and no heaven? both? neither? *nada*? maybe? don't know? Check one please. Indeed, the Gallop Poll will not only respect your anonymity as one blank unit in our great multi-blank democracy, but will absolutely insist on it. So what is your free, democratic, anonymous opinion about the late disturbances a) sociologically b) ecologically c) logically? Check the box of your choice, like what turns you on, what do you dig, are you a burn-baby-burn cat or do you favor urban renewal and the Great Society?

Such was the atmosphere while waging war in the mass society—a cacophony of ill-informed consumers of contradictory ideas, all treated as being equal. What appalled Dwight was the lack of any discrimination. But as a political commentator he confessed that as he watched the society going down the drain, he was overstimulated by the shotgun bombardment of events like a switchboard with too many calls coming in, or an overprogrammed computer loaded with too many contradictory questions. In Dos Passosian headlines he reproduced the unintelligible scramble of meaningless mayhem taking place in various parts of the world, including American cities, while pious platitudes pronouncing the march of civilization accompanied the slaughter. "PRESIDENT JOHNSON HEARS SERMON BACKING VIETNAM AIMS." The Archbishop of San Antonio invoked the late Pope Pius XII in support of his thesis that "apathetic neutrality is harmful to world peace." The intervention in Vietnam was not merely allowed and lawful, it was "a sad and heavy obligation imposed by the mandate of love." "Oh," sighed Dwight, "if only we could have a little more apathetic neutrality and a little less love."[34]

Then Dwight got down to specifics. He had given strong support to Israel in the recent Six-Day War, but he remained appalled at their treatment of Arab refugees. Their callousness was all the more distressing in a people who had suffered throughout history as victims of oppression. Their handling of the refugees was "disgraceful and disgusting, and let my anti-anti Semitic Jewish friends make the most of it. I am once more an anti-anti-anti Semite."[35]

Moving on from that provocation, Dwight took on the Student Non-Violent Coordinating Committee for their racist, anti-Semitic attacks on Israel, but more for their adoption of black nationalism and violence. He wondered why they retained their original name: "cynicism? laziness? absent-mindedness?" He conceded that criticizing SNCC at this time was like taking candy from a baby, but even the baby needs a spanking when it acts as "these callow *luftmenschen*—you'll excuse the Yiddish—who think they're *ubermenschen.*" He denounced their newsletter as the product of an ignorant and ill-informed malice. "God help us if they ever get hold of the Protocols of those Zionist Elders! But they probably won't— they're not strong on library work." He concluded his blast with the observation that "the deterioration of SNCC has been appallingly rapid even for these speedy times: racial hatred, a neurotic delight in violence, corny melodrama, ignorant fanaticism—how did all that dedication and ideals sour so rankly in two or three years?" Dwight recalled that he had once sent them $25: "This honky wants a refund, blackey."[36]

This outburst against the racial nationalism of SNCC had an interesting denouement. A few years later, Dwight was a visiting professor at the University of Massachusetts in Amherst. Sitting with Norman Birnbaum and an editor of the *Massachusetts Review*, the group was approached by Mike Thelwell, a black writer and member of the university's African-American Studies department, and Julius Lester, another militant activist who was interviewing for a job there. Thelwell turned to Dwight and said, "Here's your SNCC refund," handing him a check for $25 and exiting before Dwight knew his head had been handed to him. Perhaps by that point Julius Lester was entertaining doubts about the racial nationalism that had gripped so many of the young black rebels as well as their tolerance of anti-Semitism only a few years before. He recalled feeling uneasy about Thelwell's dramatic gesture and said he had great respect for Dwight.[37]

Although his critics ignored them, Dwight wrote some of the most scathing criticisms of New Left excesses. That he had cast his lot with the resistance did not mean he had abandoned his skepticism and his ever-ready ability to expose the weaknesses in his own camp. Despite his unease about the infantile leftism he saw developing in certain New Left quarters, he ended 1967 determined to up the ante in his own opposition to the war. He was determined to pursue civil disobedience to the point of risking a jail term, which at the time seemed highly possible.

20

Kulturkampf

Typical of Dwight's divided consciousness during these days of turmoil was his ability to be absorbed in plotting civil disobedience and then dash off to give a lecture on "The Necessity for an Elite Culture." Though viewed by many as an aging fomenter of the new forces of barbarism among the younger generation, he frequently visited campuses to preach the need to maintain cultural standards. He wanted a "revolution," but one that appreciated high culture.[1]

On January 5, 1968, a grand jury in Boston indicted five of the ringleaders of the group known as Resist for "conspiracy to counsel, aid and abet young men to violate the draft laws." Dwight had been in on the operation since its conception and was a principle author of its "Call to Resist Legitimate Authority." He signed several recruitment letters soliciting friends and associates to join the movement. Prior to and after the indictment, he spoke with Dr. Benjamin Spock and others at press conferences deliberately designed to provoke government prosecution. Now with the actual indictment of Spock, Yale chaplain and former CIA operative the Reverend William Sloane Coffin, the writer Mitchell Goodman, Marcus Raskin, a scholar at the leftist Institute of Policy Studies in Washington, and Michael Ferber, an activist graduate student at Harvard, Dwight stepped up his activities in support of the Boston five.

Dwight had very mixed feelings. He was disappointed not to have been arrested, for all the obvious reasons. He wanted the recognition that an arrest would bring as one who acted on his principles. This was important to him. On the other hand, he had a genuine fear of imprisonment. Marcus Raskin worried that some of the people involved might not realize the possible penalty of spending the next decade in jail. Dwight responded that he was sure that at worst they would not do more than five years, but that "you and I and the other ringleaders of the Resist group started the whole action in order to challenge the Government to arrest us so there would be a 'Show Trial,' with maximum publicity . . . which would challenge the legality and would impugn the morality as a side effect of the Vietnam war." He added that a secondary object had been "to share the legal guilt of the young draft-resisters." So if they had not realized the possible penalties, "we damn well should have—else what was the point of the whole business." In effect, Dwight felt that arrest was an essential part of the tactic.[2]

The indictment of the leaders of Resist was merely the first of a series of astonishing events that made 1968 the "fulcrum year" politically in the course of the war in Vietnam. First had come the Tet offensive, which exposed the lie of administration propaganda concerning the effectiveness of the war. While it may not have been a military defeat, it was a devastating political defeat. It made the discussion of withdrawal legitimate, and it brought the *Time*, *Newsweek*, CBS mainstream into the discussion. Tet was followed by the astounding success of Eugene McCarthy's challenge to the president in the Democratic primary in New Hampshire. Dwight had endorsed the Minnesota senator, whose 42 percent of the vote encouraged opponents of the war to feel they could indeed oust an incumbent president. McCarthy's showing finally induced the circumspect Robert Kennedy to agree to launch his own candidacy. Despite the hurt and anger of the McCarthy supporters, this was a real shot in the arm for the anti-war forces, since Kennedy had been denouncing the war for nearly six months. The speed of these hammer blows made it difficult for even the most experienced of political analysts to determine their meaning or their consequences. But one thing was certain, Bob Dylan's movement anthem, "Blowin' in the Wind," had reached the inner sanctum of the White House and even into the highest councils of State and the war room as well.

On March 31, after consultation with a group of Washington "wise men," Johnson announced on nationwide television that he would de-escalate the war, cease the bombing of the North, and make every effort to negotiate an end to the war. At the close of this speech he startled the nation and his advisers by announcing that in the interest of peace at home, as well as abroad, he would not seek reelection. He would leave office in order to prevent further acrimony and polarization. But his withdrawal came too late to recapture the consensus he had prized so much. On April 4 Martin Luther King was gunned down in Memphis, which was followed by riots in several major cities across the country. Shortly after King's death, Dwight's attention centered on a battle almost in his own backyard.

Columbia University stands majestically between Broadway and Amsterdam Avenue at 116th Street on the Upper West Side of New York. When one passes through the formidable wrought-iron gate at 116th Street and enters the campus, one sees to the left the imposing Low Library, set on an Olympian pedestal overlooking the campus. Directly opposite, on the other side of the vast Low Plaza, College Walk, and the manicured south lawns, stands Butler Library. Across its impressive gray expanse are the names of Thucydides, Socrates, Herodotus, Demosthenes, and other classical luminaries. Columbia University looks like a hallowed sanctuary designed to protect and nurture a community of scholars engaged in the life of the mind. One of the oldest and most respected institutions of higher learning in the country, it boasts a distinguished faculty with more than its share of Nobel Prize winners. Situated in the cultural capital of the nation, the university exudes an atmosphere not only of learning but of power. Columbia maintained a close and fraternal relationship with the national newspaper of record, the *New York Times*, whose board chairman was also on the Columbia University board of trustees. The paper recorded with care every relevant intellectual and social event that occurred on the campus.

Columbia University is the heart of the Upper West Side intellectual community. The neighborhood's shops, restaurants, bars, and cafés are the haunt of distinguished scholars, who mix with freelance intellectuals, writers, students, assorted university hangers-on, as well as hustlers, drug dealers, peddlers, street people, and homeless vagabonds. It is exactly what an urban university should be, an integral part of a great city, enmeshed in its pace, diversity, aspiration, culture, history, and ambition, which make for an exciting and sometimes dangerous place to live.

In the professional world of higher education, a tenured position at Columbia was a much-sought-after plum. Faculty salaries there were lower than at a number of competing institutions, because the city itself was such a drawing card that scholars would take less to be in the Big Apple. Many of the faculty were forced to seek supplementary income. Consulting, freelance journalism, textbook writing were for many a necessity added to the normal demands of scholarship. This may in part explain the faculty's reputation for an often distanced and indifferent treatment of both graduate and undergraduate students. Columbia was believed to be, more than any other university in America, designed for the faculty, and that meant a place where one could work in peace, without interruption by the administration or the students. It was only after the students took over President Grayson Kirk's office that Fred Dupee, who had been teaching on the campus for a quarter of a century, learned where the president's office was.

Thus the image of great power and learning was accompanied by a reputation for indifference to the human scale of life, indifference toward all but exceptional students. Values were passionately studied and embraced in the abstract but seemed unrelated to the political and social realities of this often romanticized "community of scholars." Some of the Columbia faculty had a special and close relationship to the New York intellectual community. Daniel Bell, Lionel Trilling, C. Wright Mills, Meyer Schapiro, Richard Hofstadter, and Fred Dupee, just to name a few in Dwight's circle, were all on the faculty at one time or another. And of course many of the city's freelance intellectuals were alumni of the college or the graduate schools.[3]

Dwight was conscious of Columbia all his life. He knew many of the faculty, and after his rise as a cultural critic he was frequently asked to lecture to various groups on the campus. As recently as 1964 he had briefly taught there as a visiting professor. He had a slight run-in with Jacques Barzun, a Columbia professor of history, for not helping a young woman who worked at the *New Yorker* get into the graduate program. He felt she had been ignored, and summoned up Kafka's "Castle" as a metaphor for the bureaucratic indifference of the professoriate. There was something about the studied elegance of Barzun and Trilling that irritated Dwight, as it had C. Wright Mills. When a Londoner questioned him about their preeminence, Dwight wrote that his opinion of Barzun was not very high. He appreciated Barzun's scathing indictments of mass culture, but he was so "damned civilized, urbane and genteel."[4]

On April 23 this bastion of learning exploded. Tensions had been mounting for at least two years. Radical student organizations complained

of the university's complicity with the war machine. Its Institute of
Defense Analysis was involved in weapons research. At the same time,
relations with the surrounding Morningside Heights community had
soured over the years as the university expanded and evicted the largely
poor and minority population in an effort to stabilize the deteriorating
neighborhood. This conflict was symbolized by Columbia's decision to
build a gymnasium in Morningside Park on city-owned land. The park
was thought so dangerous as to be unusable to anyone except drug dealers and other criminal types, but symbolically it looked as though the
university was grabbing land from the Harlem citizenry. Columbia
intended the gym to be shared by the local citizens and planned an
entrance from the east at the bottom of the steep incline in the park, but
this sensible plan was seen as having master–slave overtones. There is
reason to believe, as the critics of the radical students maintained, that
these issues were not the real reasons for the rebellion. The war in Vietnam and the obvious complicity of such an establishment institution
made the university an obvious target. After all, the war was the work of
"the best and the brightest," and Columbia was staffed by them and
served as a training ground for them.[5]

In a perceptive, if sometimes unpleasant, portrait of Lionel Trilling,
Alfred Kazin observed that the Trillings, Lionel and Diana, preeminent
citizens of the Columbia community, "lived on the edge of the abyss." He
meant that as Jewish intellectuals in the immediate post-Holocaust years,
they were keenly sensitive to the horrendous possibilities in the modern
world. But there was that other abyss that lay down below Morningside
Heights. One could see it stretching out to the east toward Queens and
north toward the Bronx: Harlem. It was largely ignored by the Trillings and
other faculty who lived over toward the Hudson on stately Claremont
Avenue, on Riverside Drive, or out in the suburbs. Kazin noted that on
once meeting Trilling on Broadway, he suggested a cup of coffee in the
Bickford Cafeteria. He got the impression that Trilling hardly knew of its
existence and like so many other ex-radicals who had become a part of the
establishment, he did not have much time for the life as it was lived even
in the streets of his own neighborhood. He merely tolerated his environment, or on occasion could be amused by it. In Trilling's America, Kazin
charged, there were "no workers, nobody suffering from a lack of cash, no
capitalists, no corporations, no Indians, no blacks." Trilling was the preeminent master of deradicalization. While having a keen understanding
that the modernist literature he relished and taught with such sensitivity

was deeply subversive and part of an adversary culture that held the prevailing bourgeois society in contempt, he could only anguish at a new
younger generation who saw him as a defender of that bourgeois culture
and insufficiently appreciative of their insistence on engagement. If the
student radicals romanticized the downtrodden, the good respectable citizens of the academy romanticized the operations of the "knowledge society" as exemplified by Columbia and the officers who ran it.[6]

Dwight supported with enthusiasm the defense of the canon and Western civilization, but even a suggestion of genteel pomposity annoyed him.
Despite his vitriolic attacks on mass culture, he was a genuine egalitarian
in social and political relationships. He hated hierarchies, the pecking
order, and authority. Still, even with his instinctive distrust of the pretentiousness of the upper reaches of university life, he had no intention of
going up to Columbia when the troubles began. He was impatient with the
anti-intellectualism of "the movement" and found the overheated radical
rhetoric tiresome. The first accounts in the *Times*, his main and hardly
objective source of information, did not entice him. Oddly enough, it was
Gloria who first encouraged him to go. Her younger daughter, Sabina, had
attended Barnard, and through her and her friends there Gloria was hearing sympathetic reports of the student challenge to the administration.

The clincher was a phone call Dwight made to Fred Dupee to inquire
about the situation. Dupee responded with enthusiasm: "You must come
up right away, Dwight. It's a revolution! You may never get another
chance to see one." Since Dwight had castigated himself over the years
for failing to leave his desk due to deadlines and missing great occasions,
he was not going to risk another such instance, especially when the revolution was right across the park. He went up to the campus on Friday
April 26, the day the math building was occupied and H. Rap Brown and
Stokely Carmichael entered the campus and spoke briefly to the black
students occupying Hamilton Hall.[7]

When Dwight arrived on the campus, four days prior to the great bust
during the predawn hours of April 30, he was immediately entranced by
the atmosphere of "exhilaration, excitement—pleasant, friendly, almost
joyous excitement." He found a campus where everyone was talking to
everyone else. "Hyde Parks suddenly materialized" and then broke up
into small intimate groups—"even the jocks were arguing" about something other than the point spreads.[8]

Dwight loved it. It was as though a "Victorian father had been removed
from his family's bosom (or neck)." Later when he saw Grayson Kirk on

television complaining of student anarchy and barbarism, he felt his image was on target. Always treated as children, the students were suddenly free, even obligated, to figure things out for themselves. Dwight was thrown back into his own past, recalling his own revolutionary fervor. He was not the only one; the Trillings, Daniel Bell, Fred Dupee, and others were remembering their youthful radicalism and comparing it to the present situation. Those opposed found the students wanting, crude, vulgar, even malicious, while those in favor found them inventive, full of humor and high spirits. Reminded again of Stendhal's description of Napoleon's liberation of Milan, he compared the students' jubilation with that of the Milanese. "The Columbians had realized with a start how dull and mediocre their existence had been under the Kirk administration."[9]

In the view of many of his former leftist friends and others of his generation, Dwight's participation in the Columbia revolt was beyond the pale—an act of total irresponsibility. It was charged that Dwight was captured by the youthful radicals and really had no notion of what they were about. The chutzpah of his encouragement of criminal activity, and of his revolutionary analogies that lent legitimacy to a band of malicious vandals, shocked his critics. They believed he made a ridiculous spectacle as one of the "literary fellow-travelers of the revolutionary young," an "aging intellectual camp follower of the Youth Cult." He was so enthralled by the emotional tantrums of a student generation brought up in an "overly permissive society" that he had lost all respect for the university. Some, like Daniel Bell prior to the uprising, felt that the Columbia administration was attentive to the students, and had addressed their concerns after the Free Speech student rebellion at Berkeley in 1964. The opponents of the student radicals dismissed as irrelevant their demands concerning the Institute of Defense Analysis, the building of the gym on public land in Morningside Park, and the non-negotiable demand of amnesty for the ringleaders of SDS, insisting that these issues were cynically designed to agitate the student population. The real issue was maintaining the university community through the "rule of law." The radical ringleaders had violated the code of the university community and left the administration with no choice but to call the police to restore order and protect property.[10]

Dwight, received as a sympathetic celebrity, was escorted across windowsills into occupied buildings, where he offered support and encouragement to the students. Why? Was he, as his critics charged, simply an aging romantic trying to recapture his youth? That may have some validity, but it doesn't make his conduct any less valid. He had for some time been an untenured, unconnected, itinerant circuit rider on the visiting professor trail, where he had met and dealt with numerous administrators and self-important faculty members. He often said that the "tenured gentry" espoused a high-minded, abstract morality, but they seldom acted on their principles. His feelings toward the Columbia administration were instinctively negative. The place, he said, had become completely "fossilized." When Diana Trilling, her patience tried past all endurance, rebuked him for his tolerance, asking, "Do you know they were peeing out of the president's windows?" he replied: "If anybody's windows can be peed out of, it's his windows. He was a big stuffed shirt and of course they got rid of him, thank God."[11]

What really attracted Dwight was the communal organization in the student-occupied buildings. In the math building, he visited the most militant of the "communes," led by the "professional revolutionary" Tom Hayden, no hero of Dwight's. But Dwight, invoking landmarks of the Russian Revolution, described the math building as "the Smolny Institute of the revolution, the ultraleft SDS stronghold." Fayerweather Hall, which he saw the following Monday, became "the Menshevik center," led by students willing to compromise with the administration if some of their restructuring reforms were entertained.[12]

Where the students' opponents concentrated only on the incidents of rudeness, disrespect, and violence, Dwight and others were impressed with the "calm, resolute, serious" attitude of the students. He conceded that there had been scuffling and some fistfights, inexcusable vulgarity in the taunting of the police, and bottle-throwing when the police arrived. But it was his judgment that the cops had monopolized the violence and used it freely against the offending protesters and indiscriminately against bystanders. As for the activist students, he saw them commit no vandalism; on the contrary, they made serious efforts to keep the buildings clean. He attended meetings in occupied buildings and was impressed to see participatory democracy in practice, and the "resourcefulness" with which the students met the problems of their mass occupation—food, bedding, communications, security. He was relieved to discover that "intellectuals can be practical when they have to be."[13]

Dwight was also impressed with the sheer educational value of the undertaking. Thousands of students milled around the campus arguing tirelessly the questions raised in the first place by those SDS zealots: the relationship of the university to the State; its position on class, race, minorities, women, and the surrounding community. Looking back on his experience, he wrote his old Exeter teacher Wells Kerr: "They're the best generation I have known in this country, the cleverest and the most serious and decent." He wished they read a little more, and he regretted their penchant for obscenity. But Dwight easily agreed with Paul Goodman that the obscenities of "the best and the brightest" in the government and on the faculties were far more lethal, and that's what they should be concentrating on. Still, the obscenities did bother Dwight; they suggested weak vocabularies. He wanted a revolution with high standards, thus his persistent lecture on the "Necessity for an Elite Culture."[14]

Given the political atmosphere in the country in the spring of 1968, there was a strange distance from reality on the part of many of the faculty and administration at Columbia. According to James Shenton of the history department, many of the faculty knew that change had to come, but they hated to give up that atmosphere of quiet dignity and security that had made the academy such a pleasant place to be. It was Shenton's belief that many of the senior faculty were so distanced and aloof from the world of the students that they did not have the faintest notion what was going on. Lionel Trilling recalled that he had had no idea there was student unrest, no inkling that his students had an interest in anything political. He conceded at the time that he had been "walking around in a state of complete innocence and vagueness about this." Even when the students took a dean hostage, Trilling considered it a student prank. It wasn't till the students occupied the president's office that he felt that things were getting serious. But Trilling was certain that "the [Institute of Defense Analysis] issue, when examined, was more symbolic than substantive."[15] Since few denied that the entire atmosphere was poisoned by the war in Vietnam and since it was agreed that the great research universities of the country were dependent on and contracted to the war-making state, there was something extremely literary in Trilling's assertion. Significantly, the anthropologist Marvin Harris took a completely opposite stance. He credited the radical students for alerting the oblivious faculty to the involvement of the university with an institute devoted to developing some of the most lethal weapons being used in Vietnam.[16]

Two worlds were colliding here and perhaps a "distorting intimacy" can be found on both sides of the issues.[17] It was not simply a generational struggle, as some insisted, for if it were, how could one explain the enthusiastic support of the student protesters from professors like Fred Dupee, Marvin Harris, Alex Erlich, Eric Bentley, Erich Fromm,

Immanuel Wallerstein, except by ad hominem accusations of aging romanticism? These were not nomads like Dwight Macdonald with no deep roots in the academy. They were distinguished members of their disciplines and profession. The struggle had generational overtones, but it was basically political. The defenders of the university during this crisis lamented what they called the politicization of the campus, but the students did not politicize the campus; they only presented a counterpolitics to existing politics. Surely Lionel Trilling's *The Liberal Imagination* was one of the most profoundly political books written in the postwar years. Daniel Bell's collection of essays *The End of Ideology* was an undeniably political and ideological work. One could say the same of Seymour Martin Lipset's *Political Man*, and the writings of Edward Shils on Western intellectuals. The idea that the scholarship of the New Left was "presentist" and bent on "politicizing" the universities at the expense of disinterested, objective, scholarly inquiry is a sincerely held view, but it reveals a profoundly partisan and political understanding of what happened during these years and very little self-examination.[18]

As for romanticization, there is no question that Dwight and others romanticized the student radicals, drew analogies to the revolutionary history of the Western world, and on occasion entertained the hope that the events of the late sixties would mark the beginning of much-needed change. They were suffering from the wishful thinking characteristic of radicals and reformers and for many a necessary ingredient of activist commitment. But the defenders of the university also romanticized the "community of scholars." On the one hand they applauded the rise of intellectuals to positions of authority and prestige, but they tended to ignore the complicitous relationship between the universities and government, in both domestic and foreign affairs. The university was a place of business and politics as well as learning. Surely Columbia University under the presidency of Grayson Kirk, the enthusiastic promoter of the Strickman cigarette filter, and a board of trustees made up almost entirely of corporate executives could not persuasively deny that they represented and indeed were an intricate part of the prevailing establishment.[19]

If Dwight's visit to the campus, presumably as a sympathetic journalist intellectual, angered his old leftist friends, his letter of May 20 soliciting funds for SDS infuriated them. It looked and read much like his communications with the readers of *Politics*. He assumed a tone of comradely discourse. SDS needed money because it was about to be evicted from its headquarters. It had an opportunity to obtain a place in a "friendly cooperative building." He conceded he had mixed feelings about the New Left and in particular SDS. Their political line, which was "anarchistically porous," was sometimes "alienated to the point of nihilism," and their methods had been deplorable from a libertarian point of view. The only justification for their ideology and tactics would be the existence of a revolutionary situation in the country; this was obviously not the case in general, Dwight wrote. Then came the inevitable *but*, which was the crux of what motivated Dwight: "But on two particular and major issues today, Vietnam and race-*cum*-poverty," there was a revolutionary situation. "The follies of the Establishment, in these two cases, are so extreme . . . as to make necessary the use of extralegal pressures." He then condoned the students' seizing of buildings at Columbia "to which the students had a moral right, from concrete use and interest . . . against the abstract ownership of the trustees." Dwight insisted that "the other condition, also met by the Columbia sit-ins, for revolutionary, extralegal tactics" was the fact that the majority of the Columbia undergraduates had recognized that while such actions were "unlawful and even, at first, statistically undemocratic," they were the only tactics "adequate to the historical situation." He compared the Columbia students to those of the Sorbonne. SDS had adopted the kind of "outrageous defiance of the Establishment which can shove it off its dead-center stasis toward basic reform." So on balance, Dwight declared his support of SDS because he thought "the Establishment needs its shoving" and he hoped the SDS would survive and "keep shoving."[20]

This solicitation was sent out probably indiscriminately by SDS to Columbia faculty members, while Dwight mailed it to those he thought would be sympathetic. It was published June 20 in the *New York Review of Books*. In some quarters it was met with outraged astonishment. Diana Trilling began her response "I think you must be out of your mind to ask Lionel and me to contribute funds to the SDS!" She went on to portray Columbia as the victim of anarchy and destructiveness. The university's participation in "defense programs" did not have to be shared by the faculty. The university was not racist. The proposed gym was a useful community project that had been turned into a racist symbol by Harlem and SDS agitators. The students had no rights of ownership. The ownership of a private university was a "delicate balance between property rights vested in a private corporation and . . . its spiritual ownership, which was shared by all those dedicated to its proper use." As for reform, SDS was hardly to be supported, since it was interested "only in destroying the institution." She closed by saying neither she nor Lionel would ever again be able to think of Dwight as "a politically serious man."[21]

Other letters of condemnation came from Columbia faculty members Fritz Stern and Ivan Morris; the writer Shirley Hazzard; the critic Bernard Haggin; and many others. Daniel Bell, Edward Shils, Seymour Martin Lipset, and a host of subsequent neo-conservatives published their critical responses. Some simply scrawled epithets across Dwight's letter and returned it to him: "You are a nut!"; "How mis-directed can a person be?" He took the opportunity to answer this clamor in an exchange with Ivan Morris, a Columbia professor of Japanese literature, in the *New York Review of Books*, and personally responded to several other critics during the summer. Dwight continued to question SDS's rhetoric and some of its tactics while insisting that it deserved credit for initiating a beneficial subversion of the status quo. He again denied the charge that the students had engaged in much violence. Most of the violent acts had been committed by the police called onto the campus by the administration, he said.

Dwight also denied the accusation that most inflamed the students' critics, that the SDS leaders were responsible for the burning of Professor Orest Ranum's research notes as an act of malicious vengeance because he had been one of their more outspoken opponents. Ranum was a campus character, known as "Batman" because he frequently wore his academic robes to class. Shortly after the students took over the president's office, he was admitted, attired in his robes. He proceeded to sit down at Kirk's desk and accepted one of the president's cigars, convinced that he could negotiate a withdrawal. As Fred Dupee noted with some delight, it was the "brave conscientious professor that did the departing." Dwight deplored the destruction of Ranum's property as a base and disgusting act and said he would have nothing to do with an organization that used or tolerated such tactics. But all the evidence made police agents provocateurs (the campus had been crawling with plainclothesmen) far more plausible culprits. In fact, the fires had broken out in Hamilton Hall *after* the building had been cleared. There was much dispute about this incident. One faculty member (who does not wish to be identified simply because it would make life difficult) claims he went with Ranum to his office after the fire and that Ranum told him some graduate course notes were burned. But soon the press was quoting Ranum as charging that ten years of research notes were destroyed. The executive committee of Columbia referred to "the irreplaceable notes of two years." In any case, the striking students adamantly denied the charges. SDS leader Mark Rudd claimed it was the work of plainclothes police.[22]

Irving Howe and many others accused Dwight of betraying his own principles, of forgetting what he had always preached: The means had to be compatible with the ends. Dwight conceded that initially he had opposed such direct action on principle when it interfered with others' access to buildings. But that was a general principle. As had happened so often in his life, "the general yielded to the pressure of the particular." Ultimately he concluded that despite his frequent criticism of SDS, and the entire student movement, on balance the students, even SDS, represented "more of a life-enhancing force than the authorities."[23]

He held a fund-raising party on May 24 at his apartment for SDS attended by Frances FitzGerald, George Plimpton, Art D'Lugoff, Bob Silvers, plus Norman Mailer and his usual "eclectic entourage." Mark Rudd made the pitch, only to be kibitzed by an inebriated Mailer.[24]

Dwight gave his support to the very end. The members of the more moderate, reformist Students for a Restructured University planned a countercommencement to take place at the same time as the official ceremony. They asked the distinguished Columbia economist Alex Erlich to give the welcoming address and invited Dwight, Harold Taylor, and Erich Fromm also to address the students. The audience of graduates would be those seniors and graduate students willing to walk out of the official commencement, which for security reasons was held in the Cathedral of St. John the Divine. Replacing Grayson Kirk in an effort to head off an angry demonstration, Richard Hofstadter, the immensely respected historian, was to address the students. As soon as Hofstadter rose to speak, 271 graduates also rose, quietly left the cathedral, and made their way up Amsterdam Avenue to Low Plaza on the campus.

It was a balanced and highly appropriate ending to the tumultuous semester. Hofstadter gave a speech marked by quiet dignity and genuine compassion in which he lamented the recent disruptions as a disaster that had done unreckoned damage to the university. He argued with persuasive elegance that to "imagine that the best way to change a social order is to start by assaulting its most accessible centers of thought and study and criticism" was to "show a complete disregard for the intrinsic character of the university" and also to develop a curiously self-destructive strategy for social change. In his closing remarks he conceded that the students were understandably troubled by "two facts of the most fundamental consequences for all of us—the persistence at home of poverty and racial injustice, and abroad of the war in Vietnam." Hofstadter recognized the same two revolutionary conditions that Dwight had argued justified the student rebellion. He agreed that the poverty, racism, and particularly the "cruel and misconceived venture in Vietnam" had done more than any other thing to inflame the students and undermine their belief in the political processes. He closed with the hope that now that the crisis was over, Columbia was on the road to much needed reform. It was a reconciliatory, thoughtful speech with a fine appreciation of the significance of events and compassion for the participants. Richard Hofstadter, like so many of his generation, had made a long political journey from thirties leftism to a pragmatic Trilling-like modulation, but through it all he had maintained his deep-rooted humanism.[25]

Over on the Low Plaza the countercommencement got under way. There were from 2,000 to 4,000 in attendance, depending on whose estimate one chose. It was a beautiful sunny day and the spirit of the assembly was festive, with about 300 graduates and some faculty and their guests, wives, children with balloons, parents, and curious spectators. Young women cooled their bare legs in the fountain basins and one small tyke wandered around naked. In an impassioned welcoming speech, Alex Erlich said it was a tragedy that there were two commencements but it would have been a greater tragedy if they had not had the countercommencement.

Dwight was sartorially splendid for the occasion in a black-and-white checked suit with a purple-and-white striped shirt and a lavender silk tie. On his lapel a McCarthy for President button replaced his usual Rosa Luxemburg pin. After comparing the students' rebellion to the Boston Tea Party, he looked around and, seeing six red flags, asked why there were no black ones to suit "my anarchistic tastes." Then in a most Dwight-like fashion he went out of his way to draw a chorus of cheerful booing and hissing for his pointed criticism of the radicals' misplaced revolutionary rhetoric and in particular Tom Hayden's recent call for "two, three, many Columbias," à la Che Guevara, who had urged "Create two, three, many Vietnams in Latin America." While Dwight found the strike productive, he said, he did not believe they could use the same tactics of building blockage indefinitely without destroying the university. It would be a tragedy if Columbia should come to resemble those Latin American universities where education was impossible because of chronic strikes and disruptions. In a tone not dissimilar to Hofstadter's, Dwight warned that the objective must not be to destroy the university, for "they would have nothing to replace it" and would only "stimulate the already oppressive standards in the nation."[26]

The *Times*'s education columnist, Fred Hechinger, wrote a scathing attack on the countercommencement, singling out Dwight and Harold Taylor as camp followers of the youth movement. He interpreted the hisses that had greeted Dwight's criticism of the Hayden tactics as a warning that even his fawning deference to the young radicals would not be accepted unless he was totally acquiescent to their revolutionary line. This was a singular misunderstanding of Dwight Macdonald. In a spirited response signed by Taylor, Erich Fromm, and Dwight, they denied they had any dispute with Hofstadter's description of what a university should be. In an open university, why should students not pick their commencement speakers? As for being camp followers of the young, they declared that the views they had held for years were now shared by a growing number of young people, including the view that the universities, here and abroad, had failed in their mission "as centers of humane learning." Evidence could be found in the fact that Columbia, among other universities, had "not been humane enough, imaginative enough, nor democratic enough to gain the respect of their students or to find ways of dealing with them without the help of the police." If Hofstadter had spoken eloquently on the ideal of the university, Dwight, along with Fromm and Taylor, spoke even more eloquently of the reality of Columbia and the poverty of its imagination when faced with crisis.[27]

21

The Cultural Cold Wars

In KEEPING WITH DWIGHT'S SPLIT PERSONALITY, THE agitating activist vs. the incisive cultural critic, he received, in the midst of the troubles at Columbia, a note from the Guggenheim Foundation asking him whether he was ever going to use the fellowship they had awarded him in 1963 for a book on Edgar Allan Poe. They informed him that he had to make a decision. Since it was "take it or leave it," he took it, with the promise that he would get down to the book before the end of the year.[1] He and Gloria were planning to go to England for the summer; he would take his Poe material with him.

They had been spending time in the Hamptons on Long Island, renting and visiting. Gloria was always on the lookout for a place of their own. In July 1968 Dwight received the $4,000 Guggenheim stipend. Just four days before they were to sail for England, Gloria chanced to walk down shady Spring Close Lane in East Hampton; seeing some people in a yard, she asked whether any of the houses were for sale and was told that the village plumber, Mr. Bly, and his wife might be interested. At the end of the lane she discovered a secluded summer cottage tucked away behind a wall of trees and shrubbery. She simply went up to the house and when the tradesman appeared, she told him she would like to buy his house. In practically no time the arrangement was made and they called off the England trip and made preparations to move in.[2]

It was quite modest; Gore Vidal called it a shack. Dwight was initially skeptical and did not show much enthusiasm, but Gloria, with the moral support of their old friend Joel Carmichael, persuaded him that it was a good deal. Presumably Dwight contributed his Guggenheim stipend toward the $17,000 purchase price. Gloria's mother gave them the balance. By August he and Gloria had moved in. Before long he was delighted with the place; it was "so quiet and sensible," offering the rustic privacy he had loved so much on Slough Pond. Somehow it helped distance him, at least psychologically, from the glitz and glamour that was so much a part of the East Hampton atmosphere. He had a charming hexagonal study surrounded by thick shrubbery built not far from the house. It was a secluded little hideaway with no phone. He adorned the walls with photographs, clippings, and other memorabilia. Here, despite his writing block, he spent a good deal of time, with his newly acquired dog, Queeny, as his constant companion.[3]

Although Dwight often stayed out in the Hamptons for weeks at a time, he was still very much involved in anti-war activities. He spoke frequently for Eugene McCarthy, although he found him an irritatingly ineffective candidate. "He called us to Armageddon and then located it in Scarsdale, New York—or perhaps Pasadena, California." The senator was "wellbred to the point of boredom." Dwight longed for the aggressiveness of that other McCarthy. But he resolved to stick with the candidate because he saw him as morally and intellectually superior to Bobby Kennedy. But Dwight did not have the animus toward Kennedy that those who saw his decision to enter the presidential race as an act of supreme political opportunism did. He told Rust Hills at *Esquire* that he would be glad to go along with Bobby as "a lesser evil compared to LBJ, Humphrey or any Republican except Lindsay." He saw little chance of either Nixon or Humphrey "dropping the white man's burden."[4]

Actually, the man Dwight most admired in higher political circles was the anguished Secretary of Defense, Robert McNamara. He had planned a long profile for *Esquire* on McNamara as a man of intelligence and compassion who had come to see the dimensions of the tragic mistake he and his administration had made in Southeast Asia. Dwight's article was scrapped when McNamara was pushed out of the Johnson administration. Journalistically, McNamara had instantly changed from an "*is* (journalism) to a *had been* (history)." He had become "a non-person, a back number."[5] The real tragedy was that Robert Kennedy also became a "nonperson" with his senseless assassination on June 5.

The riots and disruptions at the Chicago convention only confirmed Dwight's contempt for the Democratic Party. As in other instances, Dwight regretted he had not attended. Of course he was riveted to his TV set watching the debacle. Appalled by Humphrey's support of the war, he characterized him as nothing but a ventriloquist's dummy, an ersatz Johnson. Dwight toyed with voting for Dick Gregory, but he thought his running mate, Mark Lane, was a neurotic demagogue. Benjamin Spock and Eldridge Cleaver would have been acceptable, but they were unable to get on the ballot in New York. A write-in so upset the woman manning the voting booth that Dwight ended up casting his ballot for Paul O'Dwyer and other "lesser good guys." Dwight could see no substantial distinction between Nixon and Humphrey, even if that position meant helping Nixon gain the presidency. At the time, Dwight thought that might even be okay, since he assumed no Democrat could get the country out of the war. He regretted his decision when Nixon proved to be "as nutty an idealist as LBJ when it came to the necessity of admitting the country was in the wrong and acting on it." Despite Dwight's intense antagonism toward black nationalism and the wild rhetoric of black militants, he declared that a "1968 vote for Gregory or Cleaver was qualitatively significant" since "each represented a combination of New Left and Black Power challenges to the way things are, while Nixon and Humphrey stood for an extension of the status quo with not enough difference between them to justify the choice."[6]

Written in the early days of 1969, that provocative remark reveals the anger and frustration he was feeling politically and the degree to which he identified with the militant anti-establishment position. He wrote the column in the midst of a bitter fight over control of the public schools in New York that came to a violent and disrupting climax during the fall of 1968 and that ran well into the following year.

He had become involved in a rather careless way. Michael Harrington, as a leader in the Socialist Party and an officer of the League for Industrial Democracy, sent out a flyer supporting the American Federation of Teachers, who were contesting what they considered a breach of due process committed by the new experimental community school board in the Ocean Hill–Brownsville section of Brooklyn. It was an extremely complicated dispute involving traditional union issues of job security and seniority in a direct confrontation with what was thought to be the progressive move toward decentralization of the bureaucratic public education system in New York City and the establishment of local community control of the schools.

In September, after the local board had failed to reinstate ten teachers it had transferred out of the district, the United Federation of Teachers went out on strike. This prompted Harrington's statement urging support of the union, with a cover letter asking the recipients to sign the statement. Harrington argued that "decentralization was not the issue" and insisted that the United Federation of Teachers had "pledged its full cooperation" to make the decentralizing reorganization successful. The real issue, he said, was job security: the "right not to be fired arbitrarily by your employer because he doesn't like the color of your skin, or the way you wear your hair, or the political opinion you hold." He went on to defend the rights of union contracts which were being broken, the right to be free of harassment. These rights had been secure until the local governing board in the Ocean Hill–Brownsville District "fired" (that is, transferred) the ten teachers. After a judicial hearing, the teachers were found to have been dismissed without just cause, but when they returned to the district they were subjected to abuse, shouting, and shoving by "community representatives" with, Harrington charged, the support of the local board.

Harrington argued that these protesters were not community representatives but a core group of fifty to a hundred individuals bent on unfairly harassing and intimidating teachers. The union should not yield on this issue, because the future of teacher unionism was at stake. If teachers in Ocean Hill–Brownsville could be "fired without cause, teachers' rights and union membership will be eroded in one decentralized school district after another." It would mean the liquidation of "the most effective organized force for improved quality education." He warned that "genuine parent participation and teacher unions" would both collapse if the freedom to teach and the freedom to learn are snuffed out by mob rule. He closed by picturing a situation where "small groups of demagogues and self-appointed vigilantes can make and break people, institutions and laws . . . with the politically calculated acquiescence of the city administration."[7]

Dwight was one of the recipients of Harrington's cover letter, and as was often his habit, he sent off his signature with little thought or apparently much knowledge of the complicated events that surrounded this situation. When the statement appeared September 20 in the *New York Times*, Dwight was contacted by friends and associates acquainted with the issues who told him he had made a mistake and that Harrington's statement was an inaccurate, partisan account of the situation and the forces confronting one another in one of the bleakest and most despairing quarters of Brooklyn. It is also interesting that Gloria, from what Dwight saw as her "deep wells of feminine intuition, savvy and prejudice, was appalled" when she learned that he had signed the "Right to Teach" manifesto.[8] Although Dwight frequently charged that Gloria was given to a mindless infantile leftism, she had been suspicious of *Encounter* long before he was. She had urged him to go up to Columbia, and she was quick to see that the politics of the UFT was not the sort of stance that Dwight at that moment was prepared to take.

After talking to Nat Hentoff and reading his attack on Michael Harrington's Ad Hoc Committee to Defend the Right to Teach and realizing that the "Albert Shankerized" UFT was determined to resort to "chronic paralysis of the whole school system," Dwight called and wrote Harrington that he was resigning and wanted his name removed from the committee's literature. He immediately joined Donald Harrington's opposing group, the Emergency City Committee to Save School Decentralization and Community Control.[9]

Dwight made it his business to learn the details of the political warfare that had been going on for well over a year, during which time the UFT had sunk considerable money and energy into watering down and trying to thwart the move toward decentralization. Dwight was incensed that he had been so blind to the realities of the situation. He was further irked when Michael Harrington, now wearing the hat of national chairman, Socialist Party USA, invited him to participate in a panel on "Morality and Radical Political Action." Appropriately, Dwight turned to the *New York Review of Books*, which was already on record as supporting the local board, and wrote "An Open Letter to Michael Harrington." He began by turning down the invitation to the panel discussion. Since he felt Harrington had deceived him, he had "no stomach for talking about political morality, and certainly not the radical kind" under Harrington's auspices.[10]

This affair rapidly became a gang war among factions of the New York intellectual community, and Dwight entered into it with all his usual aggressiveness. He spoke as one who had allied himself with the New Left while Michael Harrington was still maintaining his ties to the Social Democrats, who had become the sharpest critics of the younger generation of radicals. They had also been much less militant in their stand on the war in Vietnam and had been reluctant to join in the demonstrations.

Dwight acknowledged that he often had come to regret signing thousands of cause statements, but that this was the first time he was ashamed of his quick response. He had been had. Organizers like David McReynolds appreciated Dwight's willingness to respond promptly to calls for action without carping about the details. But this political thicket, with so many barbs, booby traps, racial and ethnic land mines, demanded caution and circumspection. Dwight admitted it was his own "ignorance and more painful . . . sloth" that in part explained his original acceptance of the statement. But it was also his "confidence" in Harrington that made him assume the prima facie validity of the statement. He had been wrong.

He then launched into an attack on Albert Shanker and the union, which deliberately "increased the fear and hatred dividing Negro and Jew in the city" by circulating "nutty racist tracts by minuscule . . . and even nonexistent covens of black racists in Jewish neighborhoods." How could Harrington refer to the union as a progressive force when its own leaders "cynically circulated racist filth"? Harrington had to know the context of the situation, that in fact the union had lobbied against decentralization legislation. Harrington had to know that of the teachers Rhody McCoy hired, 70 percent were white and 50 percent were Jewish. Why did Harrington ignore these facts? Not only had Harrington's statement supported the socially retrogressive party in the dispute, but the statement contained drastic misrepresentations of fact. Instead of being for community control, as the statement claimed, all the evidence pointed to the opposite. Secondly, the assertion, repeated three times, that the local board "fired" the teachers was patently false. They "had been involuntarily transferred" back to the central board for reassignment. This was a routine process. It was true that McCoy chose to do this in a flamboyant and provocative manner and he might well have handled it better. But using the term "fired" gave the charge of threatened "job security" substance. There was some truth in Harrington's insistence that there was right on both sides and also in the censure of Dwight's ethics by the longtime union activist Brendon Sexton for disclosing a private confidence.[11]

There *was* right on both sides of this complicated issue. And apart from the specific dispute between the union and the community board or between centralization and decentralization, this was a conflict between the old left and the new. Just after Dwight had decided to withdraw from Michael Harrington's committee and join up with the opposition, he got a letter from his old friend Bertha Gruner, who had long ago worked for him on *Politics* and who was now employed by the Board of Education and was a member of the union. She wrote Dwight expressing astonishment at seeing his name on Harrington's statement:

> Ever since I left the Trotskyists I find myself in constant disagreement with most of my ex-"comrades," who have broken away from the movement but who got stuck with a Marxist ideology and a anti-Stalinist fobia [*sic*], which stifles their free reaction to newly developing situations.

She went on to discuss how this wing of the old sectarian left, still under the influence of Max Shachtman, had been so slow to support withdrawal from Vietnam. In fact, Shachtman and Bayard Rustin had remained hawks to the bitter end, while Irving Howe and Harrington had surrounded their anti-war position with such carping criticism of the New Left that they had alienated the younger generation. She then launched into a detailed defense of the "scabbing" teachers who were working for the community board and describing how she, a lifetime unionist, was proud to cross the UFT picket lines. The entire tone of the letter supported Dwight's tendency to oppose Shanker, Bayard Rustin, A. Philip Randolph, Michael Harrington, and others of this "esoteric old left sect." Dwight praised her analysis of the situation as "masterly," both as to "argumentation and . . . as to new information" concerning the funds Shanker vowed to spend to defeat decentralization and the 600 teachers who went to Albany to lobby against it.[12]

When Dwight was criticized for not knowing some of the intricate details of bureaucratic intrigue in the Ocean Hill–Brownsville battle, he went out to the district, as he had to Columbia, to see for himself. He published another letter, this time in the *New York Times*, giving an account of his impressions that contradicted much of what he saw as their biased reporting.[13]

After a day spent exploring four of the schools in the district, he had a similar response to that of Alfred Kazin, who had also come away with a sympathetic appreciation for what the community leadership was trying to do. Dwight thought that something remarkable was happening out there for a change. It was a "deeply imaginative experiment that may have lessons for all ghetto schools." He found the atmosphere "friendly and relaxed." He was free to wander around without a guide. He observed no fighting or violent horseplay in the halls, which he understood was unusual in ghetto schools. One minor altercation was dealt with quickly with "patient firmness."[14]

Dwight was particularly impressed with the young, mostly white, sophisticated teachers and their easy relationship with their black pupils. Discipline did not seem to be a problem. It was in fact irrelevant, since "some kind of dialogue and sympathy had been miraculously established between the teachers and the students." He wrote that it would be tragic if Shanker and his UFT were "able to destroy this promising experiment in community control." In a letter soliciting support from Mayor Lindsay, Dwight insisted that at long last "there were some very promising beginnings of real, as against professional, teaching" in Ocean Hill. Dwight made a big point of his devotion to amateurism in defense of some of the appointed principals who did not have the normal union seniority required for the jobs. They had not been corrupted by the bureaucratic malaise.

There was more than a little irony in the angry response of Dwight's opponents. They saw him as an ignorant dupe who had seen nothing but Potemkin villages in the vast wasteland of this Brooklyn ghetto. Anti-Stalinists had always hated fellow-travelers more than the genuine article, just as radicals held liberals in more contempt than reactionaries. Dwight had shown the same contempt for those liberal Stalinoids who had swallowed the line. Now he was at the receiving end of similar accusations from the old left union supporters who saw him as a hoodwinked naïf. Some simply wrote him off as irresponsible and mentally over the hill. Others pictured him as a delightful old fool reminiscent of Dostoevsky's ridiculous idealist Verkhovensky in *The Possessed*, who romanticizes the desperately dangerous radicalism of his fanatical nihilist son and his associates. But whether ridiculed as a fool or the object of rancorous contempt, Dwight was not deterred by adverse reaction to his political positions. On the contrary, it may have given him some sustenance. Contrary to his critics' view, Dwight often found it difficult to make the activist scene, and he assured his severe critic Nicola Chiaromonte that he had to feel strongly about the issues, feel the urgency and a "sureness about the crisis" before he could be "flogged . . . into action." Certainly one of the motivations was his critics' opposition. Murray Kempton, who supported community control of the schools, opposed the war, and had been one of the signers on the Cleaver defense committee, saw Dwight as being in "a wild stage" of his life and attributed his growing anarchic tendency to having been burned in the CIA/*Encounter* affair.[15]

Dwight had once again gone out on a limb. It was not difficult for him to oppose stodgy unionism, since he had long been a critic of the co-opted union movement. His skepticism about black nationalism might well have kept him away from this issue, but his anarchic interest in small community control attracted him. It was his belief that only by breaking up the massive bureaucracies could any kind of innovative and imaginative education and even life begin. What he had seen on his brief tour of the schools in Ocean Hill convinced him that the experiment was worth the gamble.[16]

The dispute ostensibly involved black ghetto communities contending for power, but this brought broader issues to a head, which led to more extensive warfare: the Vietnam War, the State of Israel, charges of black and New Left anti-Semitism and third world radicalism, with which militant blacks were identified (an identification they proudly accepted), and anti-Stalinism/Communism, since the Soviet Union supported the wars of liberation in the third world and were hostile to Jews and Israel. This complicated mix all came together in a political confrontation between two generations of American leftists and between acrimonious wings of the New York intellectual community.

This conflict was perhaps best illustrated by the warfare between *Commentary* and the *New York Review of Books*. A *Commentary* article by Nathan Glazer titled "Blacks, Jews and the Intellectuals" published in April 1969, charged that many New York intellectuals, Norman Mailer and Edgar Z. Friedenberg in particular, along with other Jewish members of the group, had given legitimacy to black extremists by signing an ad that had run in the *Black Panther* magazine. Glazer did not mention Dwight, Murray Kempton, or Fred Dupee, all of whom had signed the ad. Intent on denouncing Jews who "taught violence, justified violence, rationalized violence," Glazer confined his identification to those of the 140 signers who were Jewish. But he did write Dwight a curt note denouncing his signature on the ad. Kempton also received an irate letter from Glazer. It is true that Dwight did contribute to the Cleaver defense fund and also to the defense of the Black Panthers a year later.[17]

Dennis Wrong, a *Dissent* editor writing in *Commentary*, traced the New Left involvement of the *New York Review of Books*. Its tone, he charged, had become "extravagant, self-righteous anti-Americanism," and he named as examples some of Dwight's associates, Noam Chomsky, Mary McCarthy, and I. F. Stone, who, he noted, had been a "Stalinist fellow-traveler (although it was considered dirty pool to recall such things today.)" But that, of course, was exactly what the battle was all about. These people challenging the Board of Education establishment, the government establishment, the university establishment were not sufficiently anti-Stalinist. They had bought the vague eclecticism of the New Left and abandoned traditional old left positions. Wrong argued that Dwight and other contributors to the *New York Review of Books* stood firmly with New Left sentiments on trade unions, black power, public bureaucracies, official liberalism, and, implicitly, the concerns of the Jews, whereas Max Shachtman and those allied with him had thrown in their lot with the AFL-CIO outlook, meaning any progressive movement had to be based in the working class and also meaning support for the likes of George Meany and Hubert Humphrey.[18]

Dwight's experience in Ocean Hill–Brownsville may explain his interest in the incredible events at Yale in the spring of 1970. Bobby Seale, a high official of the Black Panthers, had been indicted for ordering the murder of another Panther accused of betraying the organization. He was extradited from California to New Haven, where the alleged crime had taken place. The Panthers, described by J. Edgar Hoover as the greatest threat to the internal security of the country, were indeed under the gun; their headquarters had been raided by the FBI in several cities, and in Chicago two of their leaders had recently been murdered in their beds. The New Haven Panthers initiated plans for a huge May Day rally in defense of Seale. Some of the local black militants contacted student radicals, who immediately wanted to cooperate by bringing Yale into the rally and using the occasion to agitate for better relations between Yale and the surrounding community: to the north of the campus was one of the most despairing black ghettos in New England.[19]

The ersatz Gothic architecture of Yale fit the image of a medieval walled city that would be surrounded by peasant revolutionaries demanding justice. One of the younger black faculty members saw Yale as symbolizing "the contradiction that plagues our society today: the elite and the oppressed." Here were the offspring of the rich living in "an enclave of privilege, surrounded by those who are suffering." He referred to Yale as "a jewel in a swamp." Yale, like Columbia, had a reputation for land grabbing and indifference to the plight of the evicted tenants. There were always promises of services to the community that were not carried out. Resentment toward the university was widespread.[20]

In the combustible situation brewing that spring, Yale administrators, especially the fifty-year-old president, Kingman Brewster, had no intention of responding in the manner of Columbia's Grayson Kirk or the inflexible administrators at Berkeley six years earlier. At the outset of preparations for the weekend of rallies and demonstrations on behalf of the Panthers and particularly Bobby Seale, Brewster astonished his own faculty and outraged assorted politicians and conservative citizens by expressing doubts as to whether the Panthers could get a fair trial in American courts anywhere in the country at this particularly volatile time. He was given support by the ubiquitous William Sloane Coffin, who was Yale's chaplain when he was not in court defending himself against charges of promoting draft resistance. Coffin argued that it might be legally right but morally wrong for the trial to go forward, because it would do nothing to assure domestic tranquillity and was punitive rather than curative.[21]

Dwight felt a need and desire to attend the affair and lend support to the Yale students who were mobilizing the campus. No longer nurturing hostility toward his alma mater and not having the same antagonism toward the Yale administration as he had toward Columbia's, he took a great interest in the university. He had recently been made a Fellow of Ezra Stiles College, recommended by two distinguished faculty members, Paul Weiss and John Hollander. But what may have guaranteed his attendance was an incident that had occurred April 15, just two weeks prior to the scheduled rally. Dwight spoke at an anti-war rally in Hammarskjöld Plaza next to the United Nations, and on his way back to his *New Yorker* office he stopped in Bryant Park, behind the public library on 42nd Street, where Mayor Lindsay and William Kunstler, the radical lawyer, were to speak. He milled around with his friend Nora Sayre, whom he described as the "stormy petrel of such demonstrations," because of her vivid and sympathetic articles covering the student and counterculture movement. However, he found it trying enough to give a speech, let alone listen to them, so he left and went back to his office. Soon the phone rang; it was Gloria, who had seen on television that the Panthers had taken over the speakers' stand at Bryant Park and confusion and fighting had ensued. Dwight had seen nothing. In fact, Sayre had said to him shortly before he left that there "was no anger . . . no tension" as there had been at the Pentagon and at some other recent demonstrations. He was sure that if he returned he would discover that he had missed "a karate attack on Kunstler by a lady from the women's liberation front or an ultra black militant, the kind that regards the panthers as uncle toms bec. they have decided to work w. white radicals."[22]

On Thursday April 30 he made the trip to New Haven to show solidarity with the liberal-left elements on the campus. His sons, Mike and Nick, and Nick's wife Elspeth also went to the campus. Tom Hayden, who was also to journey to Yale, recalled that on the eve of the May Day rally the potential for violence was "more frightening" than it had been in Chicago in 1968. The tension had been building for nearly a month; at least 10,000 demonstrators were expected to make the trip to New Haven. There were sensational rumors of violence, of bombings, of vigilante groups of both right and left descending on the city, of the Panthers flying in 36,000 of their members from as far away as California, and of SDS busily making bombs. On the Wednesday before the event President Nixon announced increased military aid for the invasion of Cambodia, which only heightened the anger and widened the appeal of the demonstration.[23]

Store windows were boarded up throughout New Haven, and a great many people left town. So did many of the students not interested and involved in the protest. Even Hayden agreed that Brewster and his administration at Yale had openly separated themselves from the Nixon administration. Vice President Spiro Agnew denounced Brewster for his skepticism about the justice of Seale's trial and called for his dismissal by the board of trustees. A member of the Yale Corporation and a banker stoutly defended Brewster as one of the "most distinguished University Presidents in the United States" and questioned Agnew's qualification to judge Brewster's actions. Coffin claimed that he prayed each night for the safety of President Nixon.[24]

Looking back on the event, the lack of bloodshed, given the situation, was miraculous; the majority of the students, the administration, the faculty, allied with the Panthers and the New Haven police, deserve the credit. Except for the actions of deliberate provocateurs spreading rumors on Saturday night, which led to a confrontation with the National Guard on the Green and the bombing of Ingalls Rink at midnight, in which two people were hurt by flying glass, there was no significant violence. Only thirty-seven people were arrested, mostly from out of town. Radicals looking for a violent confrontation were disappointed, and charged that the shrewd Brewster had co-opted the situation. Indeed he had. He had made his university a vast hostel housing visiting dignitaries, providing college auditoriums for open forums and students as marshals who worked with the Panthers to quiet volatile situations. It was only two days later that the National Guard was shooting and killing students at Kent State.[25]

Dwight, Mike, Nick, and Elspeth spent the time there attending the speeches and seminars in the evenings and the rallies on the Green during the day. They were impressed with the calm, pragmatic radicalism of Rennie Davis and Tom Hayden. Both played down violence and disruption and encouraged study of the imperialistic system that had brought the nation to such an impasse. Jerry Rubin gave a less temperate speech, punctuated by constant expletives, denouncing the older generation and suggesting that the place for the revolution to start was in the family by children shooting their parents. Mike took notes the first evening and following day, but Dwight seems to have spent the time milling around and meeting friends, although he spoke of writing a piece for *Esquire* (it never appeared). He did attend a forum on gay liberation and for the rest of the weekend sported a pink Gay Rights button along with a button bearing the Cleaver quotation "If you're not part of the solution you're part of the problem."[26]

He came away feeling good about the entire affair. Defending his activities and his alliance with the New Left to the harshly critical Nicola Chiaromonte, Dwight stoutly defended his support of the Panthers and said Yale had been a "lovely scene—radical and sensible and serious, how those Yale student marshals—abetted by 12 masters of the colleges—worked to tamp down the potential violence, and w. complete success." He insisted that the whole "New Left youthful thing changed drastically, a 180 degree turn" during the five days between the invasion of Cambodia and the Kent State killings. He insisted that the young were getting back on the track. They had begun when their campaign for McCarthy in the 1968 primary "gave LBJ the shove heard around the world." They were acting in a responsible political manner, putting massive pressure on legislators and on parents of draft-age students that was not only nonviolent but "also respectably inside the system." Also, faculty members and even presidents of universities were sympathetic. Among the 50,000 demonstrators who gathered in Washington a week after the May Day rally at Yale there were a thousand Yale students and faculty led by their president, Kingman Brewster, lobbying their congressional representatives. Dwight was for the moment hopeful, even though he conceded that things could change just as rapidly. But at the end of

May he felt that even Nixon and Attorney General John Mitchell would have to pull in their horns and let up on the protesters—"after all, US industrial-capitalist society can't work without the products of the great (and lesser) universities, and if the campuses continue to be recalcitrant (profs as well as students) something will have to give, like defending Indo-China against the Communists at the expense of American society—and American anti-Communist imperialism too for that matter."[27]

This may well have been the high point of Dwight's optimism concerning the resistance to the war and the challenge to American foreign policy. It's clear that he accepted the notion that a serious reformist movement could be mounted on the basis of student rebellion and middle-class and professional backing, apparently without working-class or labor support. Since he had such enthusiasm for the youth movement and had made an alliance with the activist professoriate, it is not surprising that he would now turn toward the university as a way of making his living. His writer's block at this time was severe, undoubtedly made worse by his constant anti-war activity. He knew from his previous teaching experience that he could get both pleasure and intellectual satisfaction out of the classroom, and if it didn't cure the writer's block it was a way around it. He had finally arrived at that point in his life where a secure slot in the academy was most appealing.

22

Dwight's Last Tapes

"I'M TIRED OF WRITING AND MIGHT ENJOY TEACHING" was a typical opening in a series of letters Dwight wrote in 1969–70 seeking visiting professorships. To closer friends he was more candid in conceding that he wanted to teach because his once inexhaustible flow of written words was "dammed to a trickle." To others he complained openly about his writer's block and cited teaching as a necessary source of income and something he enjoyed. In the spring of 1969 he began a full-time teaching career, landing visiting professorships of varied duration at the University of California Santa Cruz, the University of Wisconsin at Milwaukee, Hofstra University, the State University of New York at Buffalo, where he taught for several semesters, and John Jay College of Criminal Justice of the City University of New York.

He liked to teach on commuter campuses, where he found the students more motivated. They came because they wanted an education and not because "papa sent them." Dwight noted that his best student in a little seminar on parody was a "switchman on the Milwaukee railroad, w. five years union seniority, 25 or 26 years old, done his war service as have many others—average age much higher than at . . . ivy league colleges—all to the good, they have come back because they want to learn more."[1]

Dwight was indeed a teacher, and his students were astonished and grateful for the time and interest he put into their work. One student wrote, "You are the only professor I have met . . . who deserves to be called a teacher." His sessions were so full of rollicking amusement that students went to class just to have fun. But he took his teaching very seriously. His courses had genuine content. One graduate student, who became his assistant, recalled they were "tough" and "exacting." The film classes were not "gut" courses. The students were required to read a good deal of film criticism and to write weekly papers, which Dwight graded with painstaking detail, writing marginal comments all over them, often with addenda longer than the papers themselves.

Dwight usually offered seminars on Poe, or political fiction from Hawthorne to Mailer, and sometimes parody, in which students were expected to read a novel, a work of criticism, or its equivalent a week, be prepared to discuss it in class with specific references, and also to turn in written responses to the reading. For undergraduates he discouraged term papers, so as to avoid the inevitable scissors-and-paste jobs that only unnerved and depressed him. He wanted the students to think for themselves, and to give their own personal responses. All he asked was that they make a good case.[2]

At the end of each seminar class he would give his students an inscribed volume of his writing, usually one of his remaindered titles. For a seminar on political fiction he gave each student an original copy of *Politics* with personal inscriptions. Many of these students were adult working people who had never had anyone pay that much attention to them, not even parents, let alone teachers. Myrna Greenfield, a student of his at John Jay, called Dwight the "democrat of democrats; aware of class distinctions, he paid no attention to them." He had none of the snobbery so common among the disgruntled professoriat, disappointed they weren't at Harvard, Yale, or some other prestigious institution. His open friendliness and ability to treat everyone as an equal can be seen in his relationship with the staff. Secretaries, assistants to the chairmen, screen projectionists always remember Dwight with enthusiasm. A phone call thirty years after his appearance on campus elicits an immediate "Oh, he was wonderful!" Why was he wonderful? "Because he was delightful, expansive, full of fun. He treated everyone as an equal, talked seriously about politics, literature, any subject that came into his head." He assumed that they were all part of the same intellectual enterprise. In some wonderful, almost magic, way, Dwight had touched their lives and they had immense respect and affection for him, for his sensibility, his integrity, and the inordinate attention he gave to their work. As Dwight wrote in a little piece for the *Antioch Review*, he did not know how to sell out. His teaching, like his cultural criticism and his politics, rested on a bedrock of personal moral principle, and his young associates always recognized that.[3]

Early in the 1970s Daniel Okrent, Dwight's young editor at Knopf and later at Viking, visited him in East Hampton in connection with an edition of Alexander Herzen's autobiography that Dwight had edited. Dwight was anguishing over an introduction he was obliged to write for it. As he showed Okrent his book-lined hexagonal study, he remarked, "This is where I don't write." He was spending considerable time in the study staring at blank paper, looking through notes and clippings, trying to settle down to his own "intellectual autobiography." Periodically, John Lukacs and others would plead with him to write a full-scale autobiography. Publishers had shown a genuine interest, and Dwight did keep growing files, which he referred to as "Dwight's last tapes." They mostly consisted of one-page notes to himself, comments, scraps of paper. There was no outline, no organization, no coherent plan of attack.[4]

The year 1972 was a bad one for Dwight. He wasn't writing, he had no regular teaching appointment, and Nicola Chiaromonte died suddenly. Dwight was shaken by this "most depriving death" in his life since his father's sudden heart attack nearly fifty years before. Dwight literally had spasms of grief every time he realized he would never talk to Chiaromonte again. He was the one person Dwight felt closest to, morally, intellectually, and also personally. He, like Dwight's father, had given him the feeling of being valued for himself individually, "not for my brains (though also for them too) or achievements, just for myself—a personal level of love and respect—as I felt for him too." Chiaromonte was the one person Dwight had confided in through much of his adult life.[5]

With no teaching assignment, Dwight spent much of the spring and summer in East Hampton feeling he was wasting his time. John Lukacs kept up a steady stream of letters urging him to settle down, stop the occasional reviewing and other trivial occupations, and turn to his memoirs. He called them Dwight's "auto-history," as distinct from autobiography, in that it would cover a span of American intellectual history as seen, felt, experienced by Dwight. Not objective history, and not subjective history, but "participant history."[6]

Lukacs believed Dwight had been at the center of American intellectual life for fifty years, knew the protagonists, and had extraordinary powers of observation and insight. Dwight was "so good at describing things

and people and ideas that he ought not have bothered with the task of trying to define them," Lukacs insisted, and Dwight agreed. This case for
Dwight's obligation to work on his "auto-history" marked Lukacs's correspondence with him for well over twenty years. Lukacs had an eye and an
ear for Dwight's talent and a deep affection for him, but while he loved
Dwight's company, took infinite pleasure in his enthusiasms, he wanted
Dwight to be something he wasn't: a dedicated, detached, and disciplined
writer.[7]

Lukacs advised Dwight to think of the auto-history as a book and not a
collection of articles, but to "write it in the form of articles in the way the
spirit moves you." Lukacs was leading him by the hand. He did write
three pieces for publication, the preface to the Alexander Herzen book, a
long front-page review for the *New York Times Book Review* of W. A.
Swanberg's biography of Henry Luce, and a hilariously bitter dismissal of
Norman Cousins and his new mid-cult publishing venture *World*. All were
written in a way that followed Lukacs's advice. They were Dwight looking
at men, ideas, and movements through the prism of his own intellectual
participation in the cultural events under scrutiny. Ever sensitive to what
Dwight was up to, Lukacs quickly noticed this and continued to encourage him, perhaps hoping that the auto-history might be written by way of
this rather oblique approach.[8]

Discriminations, the collection of Dwight's later essays, was as close as
he ever got to writing his "auto-history." The footnotes, forewords, and
afterwords to the essays allowed him to enter into the account, to chastise
himself for poor judgment, to bring his evaluations up to date, to comment
on the relevance of past history to the present. He was again thinking
with his typewriter, as he was quick to concede to his students and colleagues. "It's in the actual process of composition that I discover, gradually, by trial and error, hit or miss, what I really think about the subject."
His running commentary covers the war in Vietnam, the Nixon presidency ("the most subversive in our history"), and his own personal controversies, the Eichmann and *Encounter* affairs. Through all of this, he is
quick to point out when he was wrong, how blind he was about this, how
stupid about that.

Like Dwight's other collections, *Discriminations* was widely reviewed,
and the response was varied. Those who loved his prose and his wit were
effusive in their praise, while those who were irked by his provocations
pounced on the endless contradictions, changes of position, bizarre political misjudgments. Jose Yglesias, an old leftist warrior himself, wrote a

warm appreciation of Dwight's steadfast dedication to acting upon his values. Yglesias saw *Discriminations* as both an enlightening and entertaining account of the madnesses current in the awful transition period of the mid–twentieth century—the very best kind of intellectual history. Many readers were grateful for the service he performed in detecting fraud, cant, and pretentiousness and warning his fellow citizens.[9]

In the spring of 1974, while teaching at Buffalo, Dwight suddenly realized that the one space essential to his professional identity was about to be lost: his *New Yorker* office. His voluminous papers, "Dwight's Last Tapes," were mostly housed there. After considerable negotiation Ruth Gay, who had formerly been married to Nathan Glazer and had known and admired Dwight for years, persuaded him to sell his papers to Yale, where she worked acquiring significant collections. In her plea she told Dwight that the "whole golden epoch when you were publishing *Politics* was probably as much defined by *Politics* as by any other force." At this point in his life Dwight needed the praise and, more important, the money. He collected as much unemployment money as he was entitled to and at one point was picking up welfare checks. He was delighted to get $15,000 for nearly seventy-five boxes of papers. He had not expected his "anal-retentive . . . pack ratism" to pay off so well.[10]

The librarians from Yale ventured into Dwight's "Augean" office and looked over the accumulation of manuscripts, letters, and books—in effect, his life's savings; as Dwight always said, he saved everything. He had made carbons of most letters he had sent, even his love letters, much to the discomfiture of Gloria. The fateful moving day was to be during the spring vacation from SUNY Buffalo in 1974.

Dwight persuaded his son Michael to help him make the final move out of the office. Most of the material was going directly to East Hampton, where Dwight was to look it over and set some material aside; the rest would be picked up by the Yale librarians. They took dolly loads down to 43rd Street and packed them into a borrowed station wagon. Both Dwight and Michael were depressed. It seemed such a forbidding milestone. Dwight had been ensconced in his *New Yorker* office for over twenty years. He was seen as a permanent fixture and felt so himself. The move seemed to seal the end of Dwight's career as a "writer." Even though he had not written anything of significance for the *New Yorker* since his piece on Harrington's book in 1963, he was still a staffer on the magazine. As long as he had the office, with its inimitable atmosphere, he was "Dwight

Macdonald, the journalist-intellectual." William Shawn, who was very fond of Dwight, must have been aware of how important the office was to him and thus had allowed him to keep it long after he stopped being an active writer on the magazine. Somehow moving the stuff out to East Hampton marked the end.

The staffers at the magazine were equally saddened to see Dwight go. Many had made a point of stopping in to see him. Hunched over his dirty cluttered desk, he would peer up at his visitor behind a wall of stained coffee cups. Immediately his face would light up. "How good to see you. What did you think of Nixon's latest caper? That man is world-class, I tell you," followed by bemused giggles, shrill cackles, rollings of eyes, grimaces of despair. A discussion was launched. "He was a great presence," Alistair Reid wistfully recalled. Dwight's reputation at the *New Yorker* was similar to his reputation on college campuses: outgoing, warm, open, endlessly interesting, entertaining, and caring. He was sorely missed.[11]

In Beckett's play *Krapp's Last Tape*, the elderly protagonist is fascinated by who he was, who he is, and the changes in persona that take place in a lifetime. Dwight knew the play and was fascinated by his own past and present and the changes he had experienced. He wanted so much to get a handle on his life, yet he seemed to feel lost. *Discriminations* was his last writing with the exception of a few occasional pieces and a spirited appreciation of Buster Keaton in the *New York Review of Books* in 1980. After the spurt of August and September 1972, Dwight resumed his nonwriting ways. Commitments went by the boards. In 1973 he wrote Elizabeth Phillips, a Poe scholar, that he could not write anything more than a letter and not many of them either, by way of excusing himself for not writing a promised preface to her forthcoming volume on the poet. Dwight's own long-planned critical study of Poe never came to fruition. He was at Buffalo at this time, teaching a full schedule and very much taken up with his course work and his life there. He explained he was teaching his political fiction course and that his reason for not writing was his "Bartleby neurosis . . . the non-reasons of Bartleby the Scrivener." It was all "very scary," but he thought Poe might have understood "the imp of the perverse."

Dwight's only explanation was that he was sixty-seven years old, and had been writing (copiously until the last decade) since he was fifteen. He claimed to be simply tired of expressing himself "in written words (spoken words, as in teaching, or arguing or just chitchat gossip, okay) and my boredom/fatigue increases in direct ratio to how easy the job is—

by now, I know I can do a lot of critical-journalistic kinds of writing and it doesn't interest me." Dwight hoped that these ruminations would help Phillips understand his "peculiar stasis." Despite the literary references to visions of hopelessness and the perverse, he affected a jocular, offhand tone. But there was also a lurking fear that he had nothing more to say, or at least nothing more to say that people wanted to hear.[12]

Teaching and being absorbed in campus activities was an enormous help in keeping Dwight intellectually alert and interested in life and all its myriad conflicts. But still, as painful as he had always found it, Dwight saw himself as at bottom a writer. Among his body of notes and snippets for the memoirs is a page he wrote in the late 1970s observing that he had made his money for nearly ten years from teaching more than writing. He had become a teaching bum ("like a tennis bum"). He admitted that he enjoyed teaching because of the students—"the young are the last, best hope of this self-destructing society." But for him, teaching was "more output than input." It was like "mining, an extractive industry," not a creative one like farming or manufacturing. Dwight found that what he did was to "mine his past experience, rather than, as in writing, cultivate it for new harvests or using it as raw material to make new products." He thought he might feel differently if he were not "a writer." He could see creative rewards for others in teaching. "But I am a writer and I must keep in contact with my mother earth or like Antaeus I begin to die. If character is destiny, MY character is a monochrome = 100% writing." Writing had been his pleasure and, along with his reading, his only hobby. For fifty years it had been his profession. Looking back, he was impressed at how confined to one vocation his life had been, like "one of those weirdly overspecialized parasites." Then he voiced a recurrent theme of his later years by alluding to how little living he had done away from his writing desk. He concluded this ruminating note by remarking that if some graduate student (they were now constantly plaguing him for interviews about others of his circle) "unwisely attempts 'The Life and Letters of Dwight Macdonald,' he'll find damned little life to flesh out the letters." There was Bartleby again—in his cubicle facing a blank wall.[13]

At the same time, he maintained the hope that he would get down to the memoirs himself, as so many were urging him to do. He must have felt that his life of intellectual activity was important and worth recording. In a note in the late seventies, after a particularly bad asthma attack for which he was hospitalized, he wrote that he was for the first time concerned with death, because he did not want to die until he had finished his memoirs, "which I can see as the climax and meaning of my whole literary career—a possibly major book in US letters."[14]

Though he recognized that he had lived at the center of the nation's intellectual turmoil, other snippets of late self-analysis designed for his memoirs reveal his nagging feeling that his life had been a "succession of failures." He still recalled the visit to Felix Frankfurter in Cambridge in 1936, when he felt he had not measured up, that he had been out of his depth, and that Frankfurter had wondered: "Is this all these clever journalists know?" Analyzing this recurring memory, he wrote that maybe Frankfurter had been too tough and not receptive to Dwight's journalistic flair. But still it haunted him.

His life a "succession of failures"? John Lukacs perceptively wrote that the tragic side of Dwight's character was "diffidence." Dwight had a deep and, for the most part, well-hidden "self-distrust." He felt that he had taken short cuts. He once told Lukacs that he had never had "an original idea," as though they were everywhere apparent in the works of successful writers. He did not elaborate. Did he mean he had not been a Marx, an Einstein, a theoretical social scientist? At other times he acknowledged that his masscult–midcult writings were a contribution to an understanding of mass culture, but that was about it. During these years he was constantly fighting off self-deprecation, which inevitably led to depression. He seemed unable to make peace with himself, to accept himself for what he was, what he was good at, what he *had* achieved. He was recognized as one of the finest prose stylists of his time, a first-rate intellectual journalist with an ability to describe and make clear complex ideas, make them come alive to intelligent laymen.

It may have been a simple problem of maturity. When questioned, Daniel Bell agreed that Dwight's quixotic exuberance was not naïveté, as some thought, but that he was "boyish in his enthusiasms." Perhaps the price for that boyishness was his inability to come to terms with his life. Not long before he died he talked to his doctor, Stephen Smiles, who, though he had been seeing Dwight for ten years, really didn't know much about his background and literary prominence. In a quiet, diffident manner, Dwight remarked that he had been an important person in the intellectual life of New York City and, by implication, the country. Smiles, somewhat skeptical since Dwight often arrived at his office rather disheveled, looked it up. He was astonished at Dwight's easily verified reputation. When Dwight's half-page obituary appeared in the *Times*,

Smiles was impressed. "Many people die," he thought, "but not many get a half-page in the *New York Times*."[15]

Dwight wanted more renown; he had wanted his books to sell more; he wanted people to recognize him. One of his most pleasurable incidents at Buffalo was being approached on a country lane by a total stranger who asked: "Aren't you Dwight Macdonald?" The man, Ray Lindquist, a local clergyman, poet, and intellectual, was an avid reader of Dwight's work and had seen pictures of him. Dwight was thrilled. He seldom experienced that kind of recognition. Usually the case was quite different. At a birthday party for his daughter-in-law, Elspeth, Dwight had a good time talking to a number of the younger people. One young woman was fascinated by his conversation and quick wit. She finally turned to him and asked: "Mr. Macdonald, what do you do?" Dwight was taken aback and began to stammer, "Well, I, I, I *was* a writer, an editor of *Partisan Review* and *Politics*, wro . . . wrote for the *New Yorker*," and his voice simply trailed off. He was perplexed, but worse, hurt by the fact that the younger generation didn't know him. It only confirmed his fears that he had not really made his mark.[16]

During this period Dwight was teaching at John Jay and very much involved in activities on the campus. When he had negotiated for the job, Theodore Gill, the head of the art, music, and philosophy department, had thought he wanted a Distinguished Professorship, which would mean a high salary and very little teaching. Dwight said he was interested in the money but he wanted to teach, he wanted access to the classroom and the students. The last thing he wanted was to get a teaching job and have it not provide the escape from the writing pressure. To Gill and the president of John Jay, Don Riddle, it was love at first sight. They were delighted and proud to have Dwight on the staff and he was most pleased to be there. For the two years he taught at John Jay he carried a full teaching load, lectured to many campus groups, and took part in several campus activities, including a stellar performance in the college production of Jean Anouilh's *Thieves' Carnival*. Dwight played the role of the crotchety but shrewd Lord Edgar. A picture and story appeared in the *Times*: "Critic Takes a Role In 'Thieves' Carnival.'" Dwight took it seriously, and at the same time had a wonderful time joshing with the students, fellow actors, and directors, hamming up his role, improvising wry comments, enlivening the production.[17]

When Dwight was active and occupied, he had inexhaustible energy. When the college became embroiled in a budget crisis he was there nearly every day and took part in parades and protest marches. It was only when the long summer months came, if he was not lecturing, not doing a mini-course here or attending a conference there, the depression waiting in the wings returned. Dwight was so gregarious that he had to be around people, had to go to the office. He was never a solitary writer. He had written in his apartment on 10th Street with his family all about him, in his crowded offices at *Politics* and the *New Yorker*, in a cabin in the Wellfleet woods with children, dogs, friends coming and going. That was one of the attractions of college teaching: he had a place to go to and there were the office staff and other writers to see.

When his position at John Jay was terminated in 1976 because he had reached the age of seventy, he hoped for future assignments as a visiting professor, but budget restrictions made that difficult. Dwight did teach at the University of California in San Diego for six weeks in the spring of 1977 and at Yale twice a week in 1978 and again in the spring of 1980. He also continued to give talks on campuses through the *New York Review of Books* lecture agency. But after the full-time position at John Jay ended, he was more and more cut off from his circuit-rider existence. The gypsy scholar life demanded a little more energy than he was now willing to expend, and Gloria was also tired of the packing and traveling.

Without the *New Yorker* office and the regular schedule at the college, he felt increasingly isolated and cut off. During these years Dwight could be difficult to live with. Although he and Gloria had genuine love and affection for each other, he was capable of harsh verbal abuse. He was impatient with what he considered her fey, scatterbrained demeanor. He would ridicule her opinions on literary or intellectual matters. At times he drove her to despair. She complained of being victimized by him and said that her friends wept at his verbal abuse of her. On one occasion during a drive in Buffalo they stopped for a hamburger. As they were walking across the parking lot they were talking about Joseph Conrad, and Gloria remarked that Conrad bored her; she found his novels tedious and difficult to wade through. Dwight became so agitated at her presumption to make such a judgment that he began shoving her, pushing her hard with his open hand. Some people saw this and called the police. They came into the restaurant where Gloria and Dwight were arguing and asked Gloria whether she wished to press charges. She was aghast. "Of course not; he's my husband." It was just his way. Sometimes things would become so unpleasant that Gloria would leave the apartment or the cottage in East Hampton to visit friends or even leave the city to stay with one of her daughters. At these times Dwight seemed to have an uncontrollable inner rage. He may have regarded Gloria with contempt, but there is much to suggest he was more contemptuous of himself, of what he considered his failure as a writer, his waning powers of expression.[18]

On the other hand, there were times when he spoke admiringly of Gloria. He had credited her good taste and judgment in persuading him to change the title of *Memoirs of a Revolutionist* to *Politics Past*. In a discussion with Erik Wensberg about Agee's being influenced by Blake, Wensberg mentioned S. Foster Damon's exhaustive critical study. Dwight called into the kitchen, "Gloria, what do you think of Damon's study of Blake?" Gloria responded immediately that it was a marvelous work that had told her everything she wanted to know about the poet. The picture of Gloria as a frivolous airhead is not a fair one. She had an affected manner, a fluttery tone to her voice, a flirtatious eye, and she was manipulative and defensive around some of Dwight's intellectual friends, but his relationship with her was a good deal more than physical. He dedicated *Discriminations* to "dearest Gloria." With her, even late in his life, he was given to public displays of affection. But he was becoming increasingly erratic and unpredictable. Gloria sometimes found him irrational in his spontaneous rages and depressions.

His increasingly frequent outbursts of ridicule and contempt as well as joy and love became a predictable part of his personality. At one point when he arrived at a party at the Lowells', Elizabeth Hardwick shepherded him through the room with a quiet plea: "Now Dwight, don't raise a row, don't cause a ruckus, don't be rude." He would go to a play, a film, an opening, and make his judgments known in a boisterous manner. Even quiet, scholarly symposiums and seminars did not intimidate him. On one occasion John Lukacs invited him to a round-table discussion of Lukacs's recent book, *Outgrowing Democracy*, at the Lehrman Institute. Although Dwight accepted the invitation, he had an instinctive antagonism toward such affairs, driven by a contempt for the academic milieu and also by his own defensiveness about not being prepared in a scholarly way. When he arrived he settled into his chair with "an air of monumental gloom." As the chair of the discussion explained the rules, Dwight interrupted to say that this was all too long-winded and boring. During the next two hours, he fidgeted, looked up at the ceiling "with a dramatic expression of dismay," talked with his neighbor about irrelevant matters in a loud voice.[19]

On other occasions, at a play that William Arrowsmith had translated for production at La Mama in the Village, and at a showing of avant-garde films out in the Hamptons, Dwight rose up from the audience, making loud, howling protests. In the first instance he was irked by what he felt was gratuitous and unnecessary obscenity and he shouted at the actors before leaving. In the second, he felt the films were arty trash and insisted on making his views known while they were being shown.

Dwight's erratic and unpredictable behavior in these last years was frightening and disturbing to Gloria and some of his friends. John Lukacs thought that Dwight's poor health had affected his mind. Tom Dardis, a Keaton biographer and also the author of *The Thirsty Muse*, a study of drinking writers, was convinced that alcoholism explained Dwight's outbursts. To some it did appear that Dwight was coming apart, losing his way, and that it was a tragedy because the man had so much energy, so much talent, so much potential still within him.

There were others, however, who only remember Dwight as charming, with an infectious good humor that enlivened the party. While recognizing that Dwight drank a lot, these friends did not think he was impaired by it in any way and they did not see his drinking as the cause of his writer's block. If anything, it was a consequence of that block, a way of escaping from the despair over finding nothing to say when he went out to his study. Mary McCarthy claimed that Dwight often began drinking in the morning in his office, particularly when he was writing something. But Alistair Reid, one of his cronies at the *New Yorker*, said he often drank with Dwight in nearby restaurants and bars but he never remembered Dwight drinking in his office. Later when Dwight shared an office with Bob Montgomery at John Jay, they were known to have a drink now and then in the office. It appears not to have been a regular practice, however, since Dwight saw many students, and none recalled ever seeing him under the influence of alcohol in classes or in his office. He would occasionally stop and have a drink with one of his students on the way home, but nothing more than that.

Dwight had no doubts about his drinking. Once when Gloria asked in exasperation why he drank so much, he replied impatiently, "I'm an alcoholic, goddamn it!" The term is so imprecise, so subject to debate, so freighted with moral implications that one hesitates to employ it. There is no question, though, that Dwight was prone to alcohol abuse and it certainly injured his health and goes some of the way toward explaining the dark side of his last years.[20]

Some people attributed Dwight's occasional insensitive provocation to his drinking. On those occasions when he engaged in bitter arguments concerning Israel and the treatment of Palestinians, he had usually been drinking. At a large party in London he encountered a woman who was a staunch defender of Israel and also a concentration camp survivor; her camp number was tattooed on her wrist. In the heat of the argument over the Israelis' treatment of the Palestinians, Dwight dismissed her concentration camp experience and "loudly and repeatedly laughed off Auschwitz." There is little doubt he was tight, but drunk or sober, he was always irked when he thought that Jews were using the Holocaust as license to defend the policies of Israel. In the Hamptons, where many Jewish people summered, he continued to take a critical stance concerning the treatment of the Palestinian refugees and, as in the past, would show impatience at what he felt was their using the tragic history of the Jews to explain away the reactionary nature of the present Israeli government. Some guests at a party given by Helen Rattray, editor of the *East Hampton Star*, became angry at Dwight for what they felt was an apology for Nazism. On another occasion during a party of his own, he had a testy argument with Helen Weinberg, his good friend and East Hampton neighbor, on the same issue. It is quite clear that Dwight was deliberately provocative on this question, as he had always been. As in the past, he felt his Jewish friends were too nationalistic and uncritical toward Israel and that they interpreted any challenge to Israeli policy as anti-Semitism.[21]

Dwight had become a colorful, eccentric curmudgeon. Many people loved being with him, loved his generally good humor. Others found him garrulous and irritating. Some people he had known well and worked with in the sixties, such as Jason Epstein, Norman Podhoretz, and others in the publishing and editing world, became impatient with him and avoided him when possible, ignored him at parties, and were not averse to labeling him an old fool and a has-been. In that competitive intellectual world, Dwight had never nurtured his career, had never been cautious and circumspect. He had never prepared for his retirement, never looked for a sinecure. There were many who owed him debts, for he had been so outgoing, so helpful to aspiring writers and editors throughout his life. He never called in any debts. To some it seemed that Dwight was, at a crucial moment in his life, being marginalized by a younger generation of intellectuals who were obsessed by "making it." John Lukacs sees him as a tragic victim of the "bureaucratization of intellectual life," which produced a new breed of intellectuals whose "unceasing concern with public success or academic careers" was pursued "at the expense of authenticity and originality and often personal probity itself."[22]

It was usually when Dwight was inactive, not engaged in teaching, writing, or lecturing and simply hanging around the cottage in East Hampton or the 87th Street apartment, in a kind of lonely isolation, that he most often became depressed and irritable. And when Dwight was less active he drank even more than normally. Leslie Fiedler claimed that Dwight, like himself, had a dangerous talent: he could drink a good deal and still manage to get his act together in the morning. In fact, it was Dwight's morning cheerfulness combined with apparent loss of memory concerning his conduct on the previous evening that irritated Gloria. He might have insulted the guests and abused her before the night was over, but in the morning he would appear as though nothing had happened. Even if he did remember, he did not take his conduct seriously, had no anger or resentment, and wondered why Gloria was sulking and making so much of it. In the last years of his life Dwight was psychologically dependent upon his daily ration, which meant at least a couple of size-able martinis at cocktail time and a bourbon or two after dinner. Gloria claimed that he often drank a pint bottle of whiskey on the drive out to East Hampton from the city.[23]

On a visit to the States in the late 1970s Annabelle and Igor Anrep, the brother and sister-in-law of Dwight's friend Baba Anrep, took a bizarre trip from the city to East Hampton to pick up Dwight's bomb of a car. From there they took the ferry from Montauk to the mainland and drove up to the Slough Pond cottage in Wellfleet. On the drive Dwight sipped from a whiskey bottle he produced from under the seat. Since he was dri-ving rather erratically and had had several martinis at lunch, Igor offered to take the wheel, even though it meant getting used to driving on the "wrong" side of the road. Dwight was delighted to accept because it meant he was able to "have a decent drink." Igor quickly discovered that the car swerved sharply to the left when the brake was applied, that the horn did not work, and, when it became dark, that the lights didn't either. Dwight pooh-poohed the danger, reassuring them that they would soon come to their turnoff onto the sandy road leading through scrub pines to the cottage. It was pitch dark by that time. Dwight managed to extricate himself from the back seat and got behind the wheel to navigate the wind-ing lanes, assuring them that he knew the way, although it had been some time since he had been there. They puttered along in the maze of sand roads, passing silent, darkened cabins in the woods but not coming to the cottage. Dwight would leap out of the car at the various forks in the road and light matches to see the name signs nailed up to trees, always to no avail. Finally Annabelle and Igor persuaded Dwight to stop at a house on stilts at the top of a hill. He agreed, and as he did so a light appeared. A man with a great Doberman on a leash came down the steps. Dwight stepped forward and announced: "I'm Dwight Macdonald." The man approached with open arms and there was much shouting and laughter, exclamations and handshaking. He turned out to be a film director to whom Dwight had given good notices, although they had never met before. The director immediately pointed them to the cottage.[24]

In East Hampton Dwight and Gloria had taken a very active part in the life of the village in their first years there. He lectured and participated in symposia at the Southhampton Community College. He conducted dune walks as he had in Wellfleet. He was thrilled to go out of Montauk harbor to watch the Tall Ships pass by in 1976, when many of his jaded neighbors showed little interest. When the motor went out in their boat and they were adrift in fairly rolling seas, Dwight showed no anxiety. He had brought a bottle along, managed to make drinks and regaled his friends with stories of his own sailing days on Long Island Sound until they were towed in.

Dwight participated in some of the Theatre of Ideas meetings that were held out in the Hamptons. On one occasion, he, Harold Rosenberg, and Joel Carmichael were the principal discussants on Trotsky and Trotskyism. This revolutionary reminiscence took place poolside on the estate of Lillian and Michael Braude, wealthy patrons of artistic and intellectual affairs, giving a nice ironic twist to the meeting that the combatants relished. Because of Dwight's writer's block, two of his neighbors, Helen Weinberg and Galen Williams, who was the head of the Poets and Writers series in New York, invented for him the Spring Close Highway Literary Seminar. Dwight was the guide and mentor. He would select the books, invite anyone he wanted, and they would sit around and discuss the works. Dwight was reading contemporary fiction at the time and they began with *Gravity's Rainbow*, which Dwight liked enormously, and *Portnoy's Complaint*. After these two Gloria said, "Let's read *great* literature." She had a passion for the nineteenth-century English novel. So they read *Bleak House* and *Middlemarch*. Then Dwight said they must read *The Possessed*, perhaps because he had heard it rumored that his critics of late compared him to the father of the revolutionary. He declared the novel was particularly relevant to recent times. He maintained the same spirited enthusiasm for and dedication to the enterprise as he did with his college classes; he loved the discussion, the give-and-take. He listened to the others and always appeared to be learning something new from their responses to the works. Dwight's complete lack of pretentiousness, his treating everyone as though they had the same background and training that he had, his respect for their individual experiences deeply affected those who participated in these seminars. There were levels of personal experience, particularly with members of his own family, where Dwight found it difficult to communicate, difficult to express love and affection, even though he felt it. But when it came to matters of the mind, when it came to the exchange of ideas, he had a marvelous ability to be inclusive, to draw people out, and to make them feel that what they were saying was important. It was a gift. He had achieved it in his writing, a tone of voice that broke through all formality and got to the crux of the issues. He was able to do the same in conversations at parties, small groups, book clubs, wherever people came together to exchange ideas.[25]

Typically looking for trouble, Dwight came to the defense of an East Hampton librarian dismissed by the staid library board. It was a conflict of generations. Mrs. Ford, the forty-year-old librarian, was simply too close to the counterculture. She had put up posters that offended the conservative members of the board. One was a rather mild quotation from Lincoln to the effect that "Silence is complicity." It was interpreted by a board member as advocating dissent. Dwight wrote that he assumed that was part of a library's function. Mrs. Ford also refused to post a notice denying entrance to barefoot readers. Again Dwight argued that it seemed reasonable that "children, hippies and adult bathers who drop in from the beach shouldn't be discouraged; in sum that use should precede decorum in library priorities." Dwight got his comeuppance from none other than Jean Stafford, who had earlier ridiculed his support of the student rebels at Columbia. She quickly labeled Dwight and his associates as "protest people" who jump on juggernaut bandwagons. They had initiated this "noisy brawl" because they would do anything to raise a ruckus. Dwight treated all this with great good humor, but underneath he was concerned about the librarian, a young woman with two small children and an unemployed husband in a "not so fat job market." Yes, he had ideas about libraries; yes, he liked people who liked books; yes, he was defending culture in the heartland of well-heeled mid-cultism. But he was also defending a young, powerless person's job. Not surprisingly, the librarian did not win her case, so there was another notch in Dwight's belt of lost causes.[26]

Another activity kept him on the fringe of belles-lettres and gave him some pleasure. In 1970 he was selected for membership in the National Institute of Arts and Letters. For the first few years he didn't do much more than attend the annual awards ceremonies in the spring. However, as his teaching drew to a close, he accepted assignments on various awards and membership committees. For the best novels of the last five years he voted in 1974 for *Gravity's Rainbow, Armies of the Night,* and *Portnoy's Complaint.* He championed Jane Jacobs, claiming that her *The Death and Life of Great Cities* was the book that "all by itself changed the thinking and practice of city planning and urban renewal." He fought hard for the broadening of categories to include film directors. The Institute, he argued, "should put its imprimatur on the open secret that cinema is art. Institutions (and Institutes) need such periodical self-shakings, like a dog emerging from water."[27]

Working as a judge of writing for the Institute kept him reading current literature, meeting with people who loved books, arguing, and debating matters of culture. It helped him feel he was still a part of the intellectual community and also that he was recognized as an important contributor to the culture. He needed that reassurance. But it too was not writing. In one of his earlier bouts with writer's block, Dwight had written a friend that he loved editing and would like to do more of it. He was a world-class editor, but no one was offering him that kind of work, and there is no evidence that he made an effort to seek any out. However, in 1974, while he was still teaching at John Jay, he received a note from a young Harvard graduate, James Atlas, asking to speak to him about a biography of Delmore Schwartz he was writing. After an amused skepticism, Dwight did become cooperative and then actually agreed to edit the work in progress.

Why did he take it on? In addition to his love for Schwartz and a desire to see him done justice, he undoubtedly hoped it would start the engine. If he couldn't write, perhaps editing someone else would do the trick. That was a part of the motivation. The job he did for Atlas went far beyond the call of duty. He read the manuscript with a probing, wary eye, catching every cliché, every overworked phrase, every unnecessary adjective, every weaseling evasion, every qualifying adverb. The pages were scrawled with marginal notes on all four sides. Most important, Dwight took the academic tics out of the book, the pedantry, verbosity,

pretentiousness, all a part of a "stylistic disease." He kept insisting that
Atlas not let the research, the data push him around. He must push the
data around, leave out the stuff that didn't interest him. He warned Atlas
not to allow the data to overwhelm his personal pleasure in writing—"and
good writing never comes but from such pleasure":

> Your general approach should be that you're the boss, not the Facts, and
> that you have no obligation to trudge through D's life month by month
> and record all the details you've so admirably dug up, that in general if a
> factual item doesn't interest you (i.e., if you can't think of much to say
> about it from YOUR point of view), you should omit it . . . you should
> only record what gives you pleasure, your focus should be narrowed to
> your own whims and crotchets (within reason). . . . Be a literary man, not
> a research mouse. Enjoy yourself as a stylist, simplify and generalize
> according to your taste and pleasure and fun, kick those Facts around,
> ignore them if they don't strike you as important to YOUR view of D.

If Dwight was usually critical, even harsh, as though he were, as he put it,
"a sundial that only measures shade," he could and did offer praise and
encouragement. He assured Atlas that he had the ability to produce a
"superb and definitive biography of Delmore." Atlas was astonished. He
actually did not think of himself as an academic and felt some confidence
in his prose style. He wrote Dwight after receiving the edited manuscript:
"I have never before been *shown* by anyone just what it means to write
more forthright prose."[28]

For a young writer, such masterly advice, such caring interest, such
generosity with time and expertise were a priceless gift. Once again, as he
had so often during his life, Dwight had brilliantly translated his ideas,
beliefs, and values into conduct. He wanted a good, caring biography of
Schwartz and he was willing to give all that he could to that end.

An essay on Buster Keaton was Dwight's last piece of published writing. It appeared in December 1980, just two years before Dwight's death.
Like the introduction he wrote for Vladimir Nizhny's *Lessons with
Einstein*, this too took an extraordinary effort on his part to complete. He
missed his deadline, causing much consternation at the *New York Review
of Books*. It was republished as the introduction to the Da Capo edition of
Keaton's autobiography, *My Wonderful World of Slapstick*. Despite the
angst involved in its production, the piece had much of Dwight's old verve
and critical acumen. As with all his later pieces, following perhaps

unconsciously the advice of Lukacs, he returned to his past and brought himself into the history of film. He recalled movie outings with his family, walking up Broadway after a dinner at the "spaghetti palace" to the "movie palace (the Rialto, long since reduced to porn flicks, is the only survivor)" to see some of the great silent classics. Dwight was a survivor too, and this piece was a part of his attempt to revive his waning literary powers.[29]

Dwight died at 12:35 A.M. December 19, 1982, at the Metropolitan Hospital in New York City. Gloria as well as his two sons, and Elspeth and Zachary had visited him earlier in the evening. When Gloria left at about nine she had no notion that he was near death. The official cause was congestive heart failure complicated by alcohol abuse, atrial fibrillation, bronchial asthma, and gastrointestinal hemorrhage. In short, Dwight was in very poor shape. He had been waning rapidly since 1977, in and out of hospitals in East Hampton, Hyannis on the Cape, and in the city. Shortly before his death he had made a trip to Wellfleet and a local remembers seeing him sitting on the curb in town; at first she took him for a derelict, so disheveled was his appearance. On that occasion he had to be driven by Nick to the hospital in Hyannis twice for asthma attacks. Intermixed with these episodes, Dwight was admitted to the hospital on more than one occasion to dry out. He was diagnosed as having prostate cancer and had an operation to remove the inner core of the prostate gland. In his last year Dwight was hospitalized for chest pains and found to have an irregular heartbeat. In December 1982 he was transferred from the Southampton hospital to the Metropolitan Hospital.[30]

Throughout these late years in the 1970s, Dwight was often full of energy and hardly a man of moderation. Despite a serious chronic asthmatic condition that he had had since childhood, he was a heavy smoker and now affected a long filtered cigarette holder, which gave him a kind of jaunty air à la FDR and made for an aggressive pointer when engaged in serious conversation. He smoked a lot, drank a lot, and ate a lot of almost anything with no care about his diet. At cocktail parties he would gorge himself on the canapés, clams, oysters, lobsters with drawn butter, rich creamy desserts, after-dinner cordials, and late evening highballs. Whatever heed he had ever given before, weight-watching never penetrated Dwight's consciousness any longer. Nothing fazed him. He developed a plump belly that he wore above his belt line; often his shirt tails would be out and unbuttoned and this round pink tummy seemed to be attempting an escape. Since he was so tall he resembled a gangling kangaroo, or some kind of great pregnant bird with a nose like an eagle and flashing eyes. Yet he remained formidable. Helen Rattray recalls a beach party on Gardiners Bay in East Hampton, where she had set candles afloat on the bay to help light the picnic area. She still sees Dwight "wending his way through the little points of light like a great Father Neptune."[31]

Intensely engaged in the present, Dwight had taken little care of his health. To the very end, even when he was developing more serious health complications, he continued to smoke, drink, and break his doctors' diets. He was reckless with the medications he took, often experimenting and mixing barbiturates with alcohol. He had also obtained the paraphernalia to inject himself with adrenaline, which eased his asthma and gave him the necessary energy to enjoy an evening out. On one occasion he openly gave himself an injection at a party, to his host's astonishment and his host's children's horror. Nothing could be more of a metaphor for Dwight's involvement. He was dying, he knew he was dying, but he would not put up with it. He would do anything to go on with the next day. He died the way he lived. At one point in one of his endless memos to himself he described a hair-raising emergency attack of asthma that took him into the Southampton Medical Center. It was the worst attack of his life, with hours of vomiting and suffocating, but a new regimen and new prescriptions and a motorized self-operating breathing machine had revived him and he felt rejuvenated. But he was now more concerned with death because he wanted to finish his memoirs before he died.[32]

Though the last years were filled with the sadness and confusion of a man who had, on some levels, lost his way, his endurance, his interest, his enthusiasms, while at times diminished, still sparkled. The year he died he was meeting a class once a week at the Yale Drama School, where he had a remarkable taped conversation with two young Yale scholars, Robert Westbrook and Robert Cummings. Stimulated by their probing questions, Dwight responded with alacrity about events ranging over the past forty years. His mind was clear, and while he was ready to concede the mistakes he had made, the times he had been wrong, he remained adamant that a person had to have a set of values that came from within, from their own experiences, to which they gave loyalty and which determined their conduct. Unquestionably, he wasted some of his talent. If he had been more disciplined, more focused, he might well have produced that book "in cold blood" he had always imagined. But his achievements were hardly insignificant. His collected writings, some still in print, others likely to be reprinted, were remarkable for their descriptive clarity, their precision, their amiable self-criticism, their humor. They remain uniformly readable and cogent descriptions of the times in which he lived. He knew, as John Lukacs observed, that "every word was not only an aesthetic but a moral choice." Dwight took himself seriously, but was never pompous. He was a man deeply involved in the life of his times; the issues and the people he engaged with were important to him.

And he was important to those who knew and worked with him. A great many turned out for the funeral service on that cold gray December day at which John Simon, Danny Rosenblatt, and Nicholas and his wife Elspeth made statements. Many recalled their warm remembrances of the delight Dwight brought to their lives. For many he had been an inspiration. Few had been better able than he to dramatize with such clarity the central problems of the modern era, the dehumanization, the nurturing of callousness, the combination of routine bureaucracy and organized terror. Whether writing about the bomb, fascism, Stalinism, the Holocaust, or the necessity of rebellion, Dwight got to the root. The most spirited and most idealistic of a younger generation of intellectuals and activists have returned to his essays in their ongoing attempt to revive a countertradition that will continue to remind us that the root is man. For Dwight, like Randolph Bourne, coming out of a tradition of radical humanism, invariably opened spaces for opposition during times when the political choices seemed severely limited. It was that tradition of humanistic radicalism that he so eloquently defended. It is encouraging that younger rebels can still find instruction in the conclusion of "The Root Is Man":

We must emphasize the emotions, the imagination, the moral feelings, the primacy of the individual human being once more, must restore the balance that has been broken by the hypertrophy of science in the last two centuries. The root is man, here and not there, now and not then.[33]

Notes

Unless otherwise indicated, all Dwight Macdonald correspondence, notes, and manuscript material are contained in the Macdonald Papers in the Archives and Manuscripts Division of the Sterling Library at Yale University. There is a comprehensive index to the papers. Only where there may be some confusion have I included box and folder numbers. I have referred to Dwight Macdonald as DM throughout the notes.

PREFACE

1. On the connection between Dwight Macdonald and Camus and particularly Vaclav Havel, I am indebted to the work of Jeffrey Isaac, whose *Arendt, Camus, and Modern Rebellion* is a brilliant discussion of this tradition, and Casey Nelson Blake's *Beloved Community: The Cultural Criticism of Randolph Bourne, Van Wyck Brooks, Waldo Frank, and Lewis Mumford*, which also deals with the historical development of this intellectual tradition.

CHAPTER 1: SCHOOLBOY AESTHETE

1. DM, "Letters to Dinsmore," box 57, folder 1367, p. 6. In the 1970s when DM was making an attempt to work on his memoirs, he went through his correspondence with Dinsmore Wheeler and interspersed quotations from the letters with his reactions.
2. Much of this biographical material on DM's parents comes from Biographi-

cal Notes in the Macdonald Papers as well as the published Yale Class of 1897 Quarterly and Half Century Records in the Sterling Library at Yale. Nancy Macdonald provided me with a copy of the Macdonald family tree as copied from Alice Hedges Macdonald's family Bible.

3. DM to Mrs. Bostrom, an unidentified correspondent, Aug. 5, 1965, box 8, folder 171; see also Biographical Notes in the Macdonald Papers.

4. The records of his grades at the Barnard School may be found in the Macdonald Papers, box 158, folders 3 and 4. The school, which is still in existence under the name Horace Mann/Barnard Elementary School, has copies of the yearbook, the *Barnard Brick*.

5. DM to parents, n.d. (1920–24), diary letter, box 3, folder 1.

6. DM to Alice Macdonald, n.d., 1924; see also DM, "Huc Venite and All That," in Henry Darcey Curwin, ed., *Exeter Remembered* (Exeter, N.H.: Phillips Exeter Academy, 1965), pp. 18–24.

7. Curwin, *Exeter Remembered*, pp. 18–24.

8. Letterhead of the Hedonists in Macdonald family letters, Dec. 7, 1923. "Huc Verite," p. 20 note. See also Diana Trilling, "Interview: Dwight Macdonald," *Partisan Review*, anniversary edition, 51, 1984, p. 804. In this interview he refers to himself as a rather "priggish conservative."

9. DM to parents, Feb. 2, 1923, box 1, folder 23; Dinsmore Wheeler to DM, March 1, 1926. An undated 1924 letter in Family files discusses an argument with Hedges about which schoolmates are gentlemen and which are not.

10. DM, "My Notebook, July 13, 1925–August 24, 1926, Containing Various Passages Relating to Persons, Books and Ideas of Interest Chiefly to Myself," pp. 75, 117.

11. DM to parents, March 16, 1924.

12. DM to parents, April 14, 27, 1924.

13. DM, "To the Second Rapids," in "My Notebook," pp. 14–15.

14. DM, "The Wall," *Phillips Exeter Monthly*, April 1924. James Agee to DM, June 16, 1927. DM to Alfred Barson, June 18, 1971.

15. DM to Dinsmore Wheeler, January 1925.

16. DM, "Letters to Dinsmore," his later selection of passages from this correspondence—there are several versions of these typescripts with editorial comment. All his life, DM read earlier correspondence and commented on his reactions.

17. For the Agee-Greenberg exchange on fears of homosexuality, see Laurence Bergreen, *James Agee: A Life* (New York: Dutton, 1984), p. 284. DM to Wheeler, March 1, 1926. Material on DM's perjorative references to homosexuals from interview with Danny Rosenblatt. For DM's concern with and fear of homosexuality, see his marginal notes on Correspondence and his Dream notes in restricted materials, Macdonald Papers.

18. DM to parents, n. d. (pre-1924); to Wheeler, May 24, 1928, May 1, 1928; to parents, Feb. 2, 1924.

19. DM to parents, n.d. (1924), box 1, folder 3.

20. DM, "My Notebook," pp. 38–43; DM to family, undated but summer of 1926 about Miriam Isaacs, box 1, folder 7.

21. DM to Wheeler, Feb. 20, 1929; Wheeler to DM, Feb. 25, 1925; DM to Wheeler, Feb. 27, 1929; DM to parents, February n.d., 1926.

22. Interviews with Lionel Abel, Joel Carmichael, William Phillips, Irving Howe, Daniel Bell, Daniel Aaron. Sidney Hook, "The Radical Comedians," *American Scholar* 54, no. 1 (Winter 1984–85): 48–49.

23. DM to Wheeler, fall 1925.

24. DM, "The Private Papers of Dwight Macdonald: Private and Personal," p. 12. DM to Wheeler, October 1924, box 57, folder 1367.

25. DM to parents, n.d. 1925, box 1, folder 70; Dec. 8, 1924. For an extended essay on DM's intellectual and literary career at Yale, see Robert Cummings, "The Education of Dwight Macdonald, 1906–1928: A Biographical Study" (Ph.D. diss., Stanford University, March 1988). I am indebted to Robert Cummings for countless insights and his help and cooperation in our mutual effort to research the life of DM.

26. DM to Wheeler, February 1925; for further literary criticism, see DM to Wheeler, May 14, 1925.

27. DM, "My Notebook," pp. 63–64.

28. DM to Wheeler, March 2, 1931, box 58, folder 1379.

29. DM to James Rowland Angell, president of Yale University, Dec. 1, 1925. DM wrote in a marginal note on a copy of his letter to Angell, "Dean Jones gave me a roasting for paper and general sloppiness of the letter." Angell did say that he would pass along DM's sentiments to the Chapel Selection Committee.

30. DM, "My Notebook," p. 95.

31. The handbill appears in *Yale Record* writings of DM. DM wrote an account of his exchange with the dean on the back of a copy of the handbill. See also Diana Trilling, "Interview: Dwight Macdonald," p. 805.

32. DM to his mother, February 1926 and mid-March 1926; to parents, June 3, 1926. Dean Frederick Jones to President J. R. Angell, June 3, 1926; Angell to Jones, June 5, 1926. Throughout the correspondence, his mother frequently pleads with DM to cultivate the right people and to act with more diplomacy.

33. Alice Macdonald to DM, spring 1927.

34. DM wrote a series of columns for the *Yale Daily News* comparing the college unfavorably with Harvard and other institutions. In the "Inquisitor" column of the *Yale Record*, March 9, 1927, he denounced the "Tap Day ceremonies whereby the Societies selected their membership." DM, "My Father," in "The Private Papers," pp. 46–47.

35. DM to Wheeler, fall 1926, shortly after DM's father's death.

36. DM to Wheeler, late 1926. DM, "The Private Papers," pp. 26–27; "My

Father," pp. 46–47. On missing his father at breakfast, Feb. 2, 1952, in private journal covering the 1950s in restricted materials.

37. This account of his working odd jobs at Yale in 1927–28 is from a dream analysis written in preparation to see his analyst, Max Gruenthal, dated May 1, 1951, box 159, folder 15.

38. DM to parents, fall 1924. In his sophomore year he organized a campus protest against the mandatory wearing of beanies; pleading for support, he wrote in his column: "Are there no Reds, no anarchists, no radicals, no thrower of bombs among us?" ("Hats," *Yale Daily News*, April 22, 1926).

39. DM to Wheeler, winter 1926; April 2, 1929; May 30, 1939. All these letters discuss the notion that they might turn the Wheeler family farm into such a retreat for like-minded literary souls. Wheeler to DM, Dec. 28, 1927, proposes maintaining the idea of a retreat to the farm as a "consoling and inspiring thought to keep in the background that whatever the immediate future may bring we can eventually come here and live to the best of our abilities and capacities."

40. DM to Wheeler, April 12, 1928.

41. Ibid.

42. DM, "Sherwood Anderson," *Yale Literary Magazine*, July 1928, pp. 209–43. Anderson wrote saying DM was one of his most insightful critics; see Howard M. Jones and Walter B. Rideout, *The Letters of Sherwood Anderson* (Boston: Little, Brown, 1953), pp. 44, 188. Sherwood Anderson to DM, n.d. 1932; DM, "The Romantic Lecturers of Yale," *Yale Record*, May 4, 1927; DM, "The Teaching of English at Yale," *Yale Literary Magazine*, June 1928, pp. 171–80. George Wilson Pierson, *Yale University 1921–1937* (1955), p. 305.

43. DM to Dinsmore Wheeler, June 26, 1928. For relationship with Dot Wheeler, see the "Homer Page Chronicle" (Homer Page was the name of the Wheeler family farm); see also "From the Long Chronicle About Dot and Me," 1927, box 64, folders 25, 27; DM to Dinsmore Wheeler, September 1928, written from on the job at Macy's.

CHAPTER 2: FROM LUCE TO LENIN

1. DM to Dinsmore Wheeler, October 1928.

2. Ibid.

3. DM to Wheeler, Jan. 8, 1929.

4. DM to Wheeler, March 12, April 28, 1929. DM, "Introduction: Politics Past," in *Memoirs of a Revolutionist* (New York: Farrar, Straus and Cudahy, 1957), p. 8. DM to Wheeler, March 1, 1929.

5. DM to Wheeler, Dec. 12, 1929.

6. DM, "Sherwood Anderson," *Yale Literary Magazine*, July 1928, pp.

209–43; see p. 241 for comparison with Proust and Joyce. DM to Wheeler, April 2, May 30, 1929.

7. DM, *Against the American Grain* (New York: Random House, 1962), p. 33, footnote. See DM "Journal 1928–43," originally in box 54 but subsequently moved to "additional material."

8. DM to Wheeler, Nov. 6, 1928, a thirteen-page account of politics at election time in New York; Dec. 10, 1928, on voting for Smith.

9. DM, "Journal 1928–43," p. 26.

10. DM to Wheeler, Jan. 14, 1930. See also "Journal 1928–43" for an account of his amorous interests in the early 1930s.

11. Much of this material on his experiences with women may be found in DM, "Journal 1928–43." See pp. 28, 33, 34. DM to Wheeler, July 23, 1929.

12. DM, "Journal 1928–43," p. 33.

13. DM to Wheeler, late 1932; May 30, 1929; Sept. 21, 1930; Feb. 2, Aug. 6, 1932.

14. DM, "James Joyce," in *Against the American Grain*, pp. 125–26. In this review of Richard Ellmann's monumental biography, he offered a hilarious account of his encounter with Joyce.

15. DM to Wheeler, Nov. 15, 1931. Interview with Eunice Clark Jessup, March 19, 1987.

16. Wheeler to DM, June 9, 1931, on wretched conditions during Depression. DM to Wheeler, Sept. 21, 1930, on party life in Manhattan.

17. DM to Wheeler, Aug. 1, 1929.

18. Wheeler to DM, Jan. 24, 1930. DM to Wheeler, Feb. 2, 1930.

19. Edmund Wilson, "The Literary Consequences of the Crash," in his *The Shores of Light: A Literary Chronicle of the Twenties and Thirties* (New York: Farrar, Straus and Young, 1952), pp. 498–99.

20. DM to Wheeler, Oct. 9, 1932.

21. Much of this information on Nancy Rodman is based on her own unpublished memoirs, in her possession; on the records in the Vassar Library; and countless conversations with her. See also Norman Louis Levey, "The Radicalization of Dwight Macdonald" (M.A. thesis, University of Wisconsin, 1966), pp. 50–51. DM to Wheeler, Dec. 14, 1933; Jan. 8, 1934.

22. DM to Nancy Rodman, July 9, 1934; Nancy to DM, July 12, 1934.

23. DM to Nancy, July 1, 1934.

24. DM to Nancy, July 16, 1934.

25. Nancy to DM, July 18, 1934.

26. DM to Nancy, July 20, 1934.

27. DM, "U.S. Communist Party," *Fortune*, September 1934, pp. 69ff. This article was unsigned, as were most *Fortune* articles in those years. The article was also condensed in *Reader's Digest*, May 1935, pp. 36–40.

28. DM to Nancy, Aug. 31, 1934.

29. Wheeler to DM, Feb. 11, 1935; DM to Wheeler, Feb. 14, 1935.

30. DM to Wheeler, Feb. 14, 1935. Wheeler to DM, Feb. 15, 1935.

31. DM to Wheeler, March 8, 1935. The page in the notebook with Trotsky's and Mellon's pictures is undated, but it would appear that the comments were written in 1927 or 1928.

32. Nancy Macdonald, unpublished memoir in her possession. DM, "Letters to Dot," in Dot Wheeler correspondence. This has commentary that DM added to the edited selection of letters. DM to Dot Wheeler, Oct. 19, 1934, professes a continued love for her after telling her he planned to marry Nancy. DM, "Journal 1928–43," p. 43. *New York Times*, Oct. 17, Nov. 17, 1934. Alice Macdonald's warning to Nancy in Nancy Macdonald memoir.

33. Nancy Macdonald, unpublished memoir. DM to Wheeler, Nov. 26, 1934.

34. Interview with Nancy Macdonald, Nov. 6, 1988. Nancy Macdonald, unpublished memoir.

35. DM to Dinsmore Wheeler, March 20, 1935.

36. DM to Wheeler, July 11, 1935. DM to Esther Dette, a friend from New Haven spring 1935; DM, "Notebooks on European Trip," box 107, folder 510.

37. DM to Wheeler, June 6, 1935. Nancy Macdonald to Esther Dette, June 8, 1935.

38. DM, "Section Two, Technique (Of Dictators)," box 107, folder 510. I am indebted to Robert Cummings, who painstakingly transcribed most of these Dictator Notebooks, attempted to put them in some kind of chronological order, and wrote interpretive commentary.

39. The entire discussion of dictators may be found in a series of notebooks and loose pages in box 107, folder 509, dated 1935–36. This material was written on DM's trip to Europe with Nancy and shortly after his return.

40. Granville Hicks, *Where We Came Out* (New York: Viking, 1954), pp. 36–37; this is quoted with some elaboration in my biography of Oswald Garrison Villard, p. 216.

41. DM, Dictator Notebook, March 1935, pp. 21 ff. on the Klan. See also Robert S. Lynd and Helen Merrell Lynd, *Middletown* (New York: Harcourt Brace Jovanovich, 1959), pp. 484–85.

42. DM to Wheeler, Jan. 8, 1929. This letter was written when DM was six months out of Yale. Despite its ill-informed confidence it expressed a sensibility that was to grow to significant proportions in the postwar years. In a schoolboyish way it is not entirely removed from the ruminations of Roquentin in Jean Paul Sartre's *Nausea*. Given Dwight's later attack on Colin Wilson's *Outsider*, he might have had more sympathy for that young man if he had reviewed his own notebooks and correspondence written when he was a young man.

43. DM to Esther Dette, July 20, 1935. Interview with Nancy Macdonald. Nancy Macdonald, *Homage to the Spanish Exiles: Voices from the Spanish Civil War* (New York: Insight, 1987), p. 33. Nancy remembers their return as due to a call from Luce.

44. DM to Esther Dette, Aug. 28, 1935.
45. DM to Cathy Hellman, Oct. 5, 1977. This was written to Hellman's wife at the time of his death, wherein DM went back over their long friendship since college days. Interview with Mary McCarthy, Nov. 13, 1987. See Mary McCarthy, *Intellectual Memoirs: New York 1936–1938* (New York: Harcourt Brace Jovanovich, 1992), p. 19.
46. DM to Nancy, Oct. 4, 1935. DM, review of the Dennis book, *Common Sense*, February 1936, pp. 26–27. DM reviewed regularly for *Common Sense* up until 1937, when he began writing for the Trotskyist publications.
47. DM to Nancy, Nov. 26, 1935.
48. DM to Nancy, Nov. 28, Dec. 2, 1935.
49. "'Luce and His Empire': A Radical Critique of a Liberalistic Biography of a Reactionary Tycoon," in DM, *Discriminations: Essays and Afterthoughts* (New York: Grossman, 1974), pp. 273–74. This was a reprint, with added footnote commentary, of his review of W. A. Swanberg's biography *Luce and His Empire*, which appeared on the front page of the *New York Times Book Review*, Oct. 1, 1972. It contains a good account of DM's struggle with *Fortune* editors over his steel articles. DM, "Republic Steel," *Fortune*, December 1935, pp. 77ff. DM to Wheeler, Nov. 6, Dec. 20, 1935; DM, "Fortune Magazine," *Nation*, May 8, 1937, p. 529.
50. DM to Wheeler, Nov. 6, 1935.
51. Wheeler to DM, Oct. 26, 1935. DM to Wheeler, Nov. 6, 1935.
52. Harry Roskolenko, *When I Was Last on Cherry Street* (New York: Stein and Day, 1965), p. 58.
53. DM to Wheeler, March 16, 1936.
54. Ibid. (p. 4 of a seven-page letter).
55. DM, "U.S. Steel IV," Mscpt. April 14, 1936; DM's unedited manuscripts are in the *Fortune* file Box 16, folder 417, Macdonald Papers; "Fortune Magazine," *Nation*, May 8, 1937, p. 530.
56. DM to Wheeler, June 10, 1936.

CHAPTER 3: A DISTINGUISHED GOY
AMONG THE PARTISANSKIES

1. DM to Dinsmore Wheeler, June 10, 1936.
2. Ibid. Nancy Macdonald to Wheeler, Oct. 30, 1935. Interview with Nancy Macdonald, Nov. 6, 1988.
3. Geoffrey Hellman to DM, March 24, 1936.
4. DM to Hellman, March 24, 1936.
5. Charles Biederman to the author, Jan. 16, 1989. DM was an admirer of Biederman's early abstract paintings and wrote a thoughtful unpublished appreciation of his work predicting fame for him. Biederman later became

what he called a "structuralist," mounting abstract designs on wood and canvas. He has a reputation in England, and in April 1989 had an extensive showing of his work at the Borgenicht Gallery in Manhattan.

6. DM to Dinsmore Wheeler, Oct. 24, 1936.

7. Ibid. DM, notes on the meeting recorded much later in life, March 9, 1972. Felix Frankfurter to DM, Nov. 18, 1936. At a lunch with Richard Goodman, a Kennedy aide, in the 1960s, DM told the story of the interview with Frankfurter as an embarrassing mishap.

8. Fred Dupee to DM, Oct. 21, 1936. This reference to DM's vote is the only evidence I actually have that he voted for Browder. I have never seen his letter declaring such an intention. He never mentioned a vote for Browder later in life; on the contrary, he wrote that he voted for "either Roosevelt or Thomas in 1936." See DM, "The Candidates and I," *Commentary*, April 1960, p. 288. This is the only instance I have discovered of a possible Macdonald prevarication, since it is extremely unlikely that he would have voted for Roosevelt and equally unlikely that he would have forgotten his vote. DM's later intense anti-Stalinism may have made it impossible for him to admit how far he had gone in his fellow-traveling in 1936. Although I received DM's FBI file through the Freedom of Information Act, I did not get the material containing this evidence. I am indebted to Natalie Robins, who gathered the material for her book on the FBI's surveillance of American writers, *Alien Ink: The FBI's War on Freedom of Expression* (New York: Morrow, 1992).

9. DM to Dinsmore Wheeler, Dec. 6, 1936. DM, "Pittsburgh: What a City Shouldn't Be," *Forum*, August 1938.

10. DM to Wheeler, Dec. 6, 1936.

11. DM, undated self-analysis in box 170 of Macdonald Papers. Reference to not writing a book "in cold blood," DM, *Memoirs of a Revolutionist* (New York: Farrar, Straus and Cudahy, 1957) p. 9. DM's worry over not writing a book is also evident in a letter to editor in response to critical review by Paul Goodman, *Dissent* 5 (Winter and Autumn 1958), 82–86, 397–99. Throughout his life, Dwight castigated himself for being eclectic, "slack," and lacking a writing discipline.

12. The best account of this initial meeting with Rahv and Phillips may be found in Daniel Aaron, "The Thirties: Now and Then," *American Scholar* 35 (Summer 1966): 510. See also William Phillips, *A Partisan View: Five Decades of the Literary Life* (New York: Stein and Day, 1983), pp. 47–48;

13. Interview with Mary McCarthy, Nov. 14, 1987. Mary McCarthy, *Intellectual Memoirs: New York, 1936–1938* (New York: Harcourt Brace Jovanovich, 1992), pp. 83ff; Carol Brightman, *Writing Dangerously: Mary McCarthy and Her World* (New York: Clarkson Potter, 1992), pp. 144–50.

14. David Bazelon, *Nothing but a Fine Tooth Comb: Essays in Social Criticism, 1944–1969* (New York: Simon and Schuster, 1969), pp. 19–20. Reading

her memoirs, DM discovered that Mary McCarthy had a Jewish grandmother, which made him feel even more of an outsider.

15. DM to Dinsmore Wheeler, May 9, 1937.

16. Selden Rodman thought DM was scornful of *Common Sense* because he
    wanted to take the most extreme position and hence believed there was
    something soft and naive about radicals who simply wanted to reform the
    New Deal. Interview, May 7, 1987.

17. Mary McCarthy, "Portrait of the Intellectual as a Yale Man," in *The Com-
    pany She Keeps* (New York: Farrar, Straus and Cudahy, 1942). For good
    account of the impact of the Moscow trials in America, see Alan B. Spitzer,
    "John Dewey, the 'Trial' of Leon Trotsky and the Search for Historical
    Truth," *History and Theory* 29, no. 1: 16–37.

18. DM, review of John Strachey's *Theory and Practice of Socialism*, in *Com-
    mon Sense*, January 1937, p. 26. DM, introduction to Mary McCarthy at the
    YWHA, Nov. 10, 1963. DM, *Memoirs of a Revolutionist*, p. 10.

19. Malcolm Cowley, "The Record of a Trial," *New Republic*, April 7, 1937, pp.
    267–70. DM, "Trotsky and the Russian Trials," letter to the editor, *New
    Republic*, May 19, 1937, pp. 49–50.

20. Talk with Joel Carmichael, a friend of DM since the 1930s, Nov. 10, 1986.
    Interview with Joel Carmichael, May 14, 1987. DM, "The American Writ-
    ers Congress," letter to the editor, *Nation*, June 19, 1937. Henry Hart, "The
    American Writers Congress," *Nation*, June 26, 1937. DM to George Abbott
    White, Nov. 13, 1972, responding to a biography of F. O. Mathiessen with
    details on the dissidence at the Writers Congress. See also Aaron, "The
    Thirties."

21. Gold's *Daily Worker* article cited in *Partisan Review*, December 1937, p. 3.

22. DM, "Laugh and Lie Down," ibid., pp. 44–53.

23. Leon Trotsky to DM, Jan. 20, 1938, Trotsky Papers, Houghton Library,
    Harvard University.

24. DM to Trotsky, July 7, 1937; *Memoirs of a Revolutionist*, p. 15.

25. DM to Trotsky, Aug. 23, 1937.

26. Trotsky to DM, Jan. 20, 1938.

27. Philip Rahv to Trotsky, Feb. 21, 1938.

28. Trotsky to Rahv, March 21, 1938. Fred Dupee to DM, Sept. 18, 1937.

29. William Phillips, *A Partisan View*, p. 44.

CHAPTER 4: A TROTSKYIST EDUCATION

1. DM to Dinsmore Wheeler, Aug. 4, 1938.

2. DM to Wheeler, Aug. 24, 1937.

3. DM to Wheeler, May 30, 1938.

4. DM, "They, the People," *New International*, August 1938.

5. Rose Stein to DM, July 6, 1938.

6. DM to Rose Stein, July 10, 1938.

7. Carlos Hudson, editor of the *Northwest Organizer*, to DM, April 5, 1939. Ironically, Dunne felt that if DM wrote articles for the *Northwest Organizer* in the tone of sarcasm and irony of his pieces in the *New International*, his column might have more appeal.

8. DM, "Notes on the First Convention of the CIO, Pittsburgh, Nov. 14–18," box 115, folder 573.

9. DM, *Fascism and the American Scene* (New York: Pioneer Publishers, 1938).

10. DM to John Chamberlain, May 3, 1939.

11. DM, "Sparks in the News," *Socialist Appeal*, Aug. 8, 1939. His column changed names several times. Initially it was "Off the Record," then became "Sparks in the News," then "The War Deal." It was always directed at the liberal-left press and academy, whom he considered to be selling out to the capitalist war-makers.

12. Interview with B. J. Widick, Nov. 19, 1988. Harry Roskolenko, *When I Was Last on Cherry Street* (New York: Stein and Day, 1965), pp. 157–58. B. J. Widick, "Trotskyism, Factions and Functionaries," a chapter of his forthcoming memoirs, goes over this period in great detail.

13. Roskolenko, *Cherry Street*, ibid. DM, notes for *Memoirs of a Revolutionist*, box 72, folder 70. He describes the Trotskyist meetings with great humor.

14. DM, "Once More: Kronstadt," *New International*, July 1938, pp. 211–13. His column, titled "They the People," first appeared in this issue.

15. Ibid.

16. Interview with Felix Morrow, March 12, 1987. In a reply to a letter from Stephen Whitfield, author of a short study of DM's politics who had asked DM about his courage in attacking Trotsky, DM said: "I'm sure I never told Leon Trotsky how to make a revolution, if only bec. my chutzpa (w. I agree exists, thank God) is always critical, negative, destructive, never positive how to do it. . . . My first contribution . . . was so characteristic of my whole political writing career, always refractory, esp. to my comrades, taking away with one hand what I give with the other, joining the emperor's procession while pointing out that he was naked, my arrogant advice to Trotsky was not how to make a revolution but—my specialty—how not to make one." Some have thought that Dwight and Nancy's adoption of party names (he was Joyce, she was Elsie Dinsmore) was frivolous; see Nancy Macdonald, *Homage to the Spanish Exiles* (New York: Insight, 1987), p. 59; Stanley Kunitz, ed., *American Authors* (New York: H. W. Wilson, 1977), pp. 272–73; Jane Brown, "The Saga of Elsie Dinsmore: A Study in Nineteenth-Century Sensibility," *University of Buffalo Studies*, 17, no. 3 (July 1945). I am indebted to Ann Pierce for digging out this informative article for me, and to my copyeditor, Anne Montague, who suggests in a more plausible explanation than any other that "Elsie Dinsmore" might have been an homage to Dinsmore Wheeler.

17. Leon Trotsky, "Hitler and Stalin," *Liberty*, Jan. 27, 1940, pp. 6–9. DM,

"Shamefaced Defensism: Some Notes on Comrade Trotsky's *Liberty* Article," Internal Bulletin, February 1940. Joseph Hansen, "From Science to Slander: Some Notes on a Student in the Burnham School of Science," Internal Bulletin, Feb. 20, 1940, box 150, folder 41. DM, "Guilty as Charged, Your Honor," Internal Bulletin, Feb. 19, 1940. Trotsky, "Back to the Party," Internal Bulletin, Feb. 21, 1940. Trotsky's reply to Macdonald (he spelled the name wrong all through his piece) is in Trotsky, *In Defense of Marxism* (London: New Park Publications, 1971), p. 153. It is also in the Macdonald Papers in its Internal Bulletin form.

18. DM, "Back to the Argument," unfinished and unpublished, dated spring 1940, in the Trotsky files of the Macdonald Papers.

19. Alan M. Wald, *The New York Intellectuals: The Rise and Decline of the Anti-Stalinist Left from the 1930s to the 1980s* (Chapel Hill: University of North Carolina Press, 1987), pp. 203–4. See also DM interview with Alan Wald, Nov. 5, 1973, box 55, folder 133.

20. DM, "National Defense: The Case for Socialism," *Partisan Review*, July–August 1940, pp. 251ff.

21. Ibid., pp. 253, 258. See John Patrick Diggins on Bruno Rizzi in his *Up from Communism: Conservative Odysseys in American Intellectual History* (New York: Harper & Row, 1975), pp. 186–87; Alan Wald, *The New York Intellectuals*, p. 181.

22. DM, "National Defense," p. 266.

23. Ibid.

24. Ibid.

25. Interview with B. J. Widick, Sept. 18, 1986; DM, interview with Alan Wald, Nov. 5, 1973. See also Paul Buhle, *The Artist as Revolutionary* (London: Verso, 1989). "Liberty, Equality, Fraternity: 1789–1940, An Editorial," *New International*, August 1940, pp. 131ff.

26. DM, "Fraternity," Internal Bulletin #5. "Statement of the Political Committee (On Comrade Macdonald's Letter to the Political Committee on 'Fraternity')," Internal Bulletin #5, p. 8.

27. *The Fourth International*, I, October 1940, a magazine collection of the letters and articles written by Trotsky at the end of his life, edited with an introduction by Joseph Hansen, pp. 126, 135ff.

28. For accounts of the Trotsky charge of stupidity, see: James Burnham, "Politics for the Nursery Set," *Partisan Review*, February 1945; Harry Roskolenko, *Cherry Street*, p. 158; Stephen J. Whitfield, *A Critical American: The Politics of Dwight Macdonald* (Hamden, Conn.: Archon Books, 1984), p. 20, citing Roskolenko; Richard Pells, *The Liberal Mind in a Conservative Age: American Intellectuals in the 1940s and the 1950s* (New York: Harper & Row, 1985), p. 81, citing Burnham citing Trotsky; Sidney Hook, *Out of Step: An Unquiet Life in the 20th Century* (New York: Harper & Row, 1987), p. 518, citing no one. Joel Carmichael recalls having been present when DM read the Trotsky attack and says initially his face blanched but

that he quickly recovered his poise and made light of it. Hook says DM felt hurt and that when he was solicited for funds to upgrade the security at Trotsky's Mexican villa he replied "Fuck Trotsky." Since DM later made a great effort to get Trotsky's papers to Harvard and since he almost never used four-letter words, Hook's story does not have the ring of truth.

29. *New York Post,* Aug. 30, 1940. *Socialist Appeal,* Sept. 9, 1940.

30. Michael Macdonald to author, Oct. 29, 1988. DM's reading of Trotsky (*Literature and Revolution, The History of the Russian Revolution,* and *The Revolution Betrayed*) is cited in the extensive notes he took in Majorca in 1935, his "Dictator Notes."

31. DM, "Trotsky Is Dead: An Attempt at an Appreciation," *Partisan Review,* September–October 1940, pp. 340, 344–45.

32. Ibid., pp. 349–50.

33. Flyers and pamphlets for Shachtman's campaign are in box 147, folders 7–10. Some of the titles of DM's pamphlets were "Why the Negroes of the Bronx Should Send Max Shachtman to Congress," written for the *Listener* (1940); "For a Socialist Defense Against Hitlerism: Ten-Point Platform of the Workers Party"; a draft: "Towards the *American* Revolution"; "Marxism: Religious Dogma or Scientific Discipline?"

34. DM to George Warren, chief of Visa Division, U.S. Department of State, Sept. 30, 1940.

35. Nancy Macdonald to Victor Serge, June 1, October 9, 19, 1940. Nancy Macdonald, *Homage to the Spanish Exiles,* chapter 3, and particularly pp. 59–60. Alan Wald, "Victor Serge and the New York Anti-Stalinist Left," in his *The Responsibility of Intellectuals: Selected Essays on Marxist Traditions in Cultural Commitment* (Atlantic Highlands, N.J.: Humanities Press, 1992). The Victor Serge file in the Macdonald Papers shows how important a role Nancy played in all these affairs and how close she was to many of the refugees, particularly Serge, whom she referred to as "Cousin."

36. Max Shachtman, "Again on the Question of Party Responsibility and Discussion," Jan. 3, 1941, an internal party letter. All the material on DM's case can be found in box 150, folder 46 of the Trotsky files.

37. Shachtman to DM, March 8, 1941. DM to Shachtman, March 10, 1941. DM, "Letter of Dwight Macdonald to the Plenum," Internal Bulletin #9, March 22, 1941. "The Statement of the Political Committee on Behalf of the Plenum in Reply to Dwight Macdonald: The Plenum Rejects Comrade Macdonald's Ultimatum," May 5, 1941. DM to Dear Comrades (resignation letter), July 1, 1941; it was accepted unanimously by the political committee.

CHAPTER 5: FROM TROTSKY TO BOURNE: WAR IS THE HEALTH OF THE STATE

1. DM, "War and the Intellectuals: Act Two," *Partisan Review,* 1 Spring 1939, pp. 3–20. During the Vietnam War, Noam Chomsky referred to Bourne as

one of his models of the proper conduct of intellectuals, in *American Power and the New Mandarins*, Introduction, p. 5. In Chomsky's essay "The Responsibility of the Intellectuals" (also in *American Power*), he refers to DM in the same vein.

2. Randolph S. Bourne, *War and the Intellectuals: Collected Essays, 1915–1919*, edited with an introduction by Carl Resek, p. 41 and "The Twilight of Idols," pp. 53–64. When a student at Antioch College compared passages from DM with those of Bourne, DM was astonished and pleased at how similar they were (Tom Curtis, "Skeptics in a Time of Faith: The Anti-War Writings of Randolph Bourne and Dwight Macdonald," for Professor Lawrence Grauman's course Literature of American Social Criticism, Aug. 10, 1968, in the Grauman file, box 19, folder 458).

3. DM, "War and the Intellectuals," pp. 8–11 passim.

4. Ibid., pp. 5, 9. Whereas Bourne's "War is the health of the State" was a bitter pronouncement, a young professor at Harvard, Arthur Schlesinger, Jr., in making a defense of a large military budget, observed cheerfully: "Whatever else is said about a 'permanent war economy,' at least wages are high, employment is full, and the economy is relatively stable and productive." (Arthur Schlesinger, Jr., "The Future of Socialism, *Partisan Review*, May–June 1947, p. 241.)

5. Ibid., p. 15.

6. Ibid., DM, "Trotsky Is Dead," *Partisan Review*, September–October 1940, pp. 339–40.

7. Stephen Spender, "September Journal," *Partisan Review*, March–April 1940, pp. 92–93.

8. DM, "Notes on a Strange War," *Partisan Review*, May–June 1940, pp. 170–75.

9. Stephen Spender, manuscript letter to the editor, Aug. 16, 1940, published as "The Defense of Britain, A Controversy," *Partisan Review*, September–October 1940, pp. 405–406.

10. DM, editorial reply to Stephen Spender, Ibid., pp. 407–408, 410. On Churchill's attitude toward both Hitler and Mussolini in the 1920s, see Virginia Cowles, *Winston Churchill: The Era and the Man* (New York: Harper, 1953), p. 271.

11. DM, editorial reply to Spender.

12. DM to G. L. K. Morris, Sept. 2, 1940. See also DM, "Proposal for a New Magazine," undated but clearly September 1940, box 74, folder 92.

13. DM to Morris, Sept. 9, 1940. DM sent these ideas to Harold Rosenberg at the same time.

14. DM, "Proposal."

15. Morris to DM, Sept. 7, 1940. Harold Rosenberg to DM, Sept. 8, 1940.

16. DM to Morris, Sept. 9, 1940.

17. Name change announcement, *Partisan Review*, September–October 1940,

inside cover. Reader response, Ibid., November–December 1940, back cover. For summary of reader comment, see "Letters," Ibid., January–February 1941, pp. 77ff.

18. Nancy Macdonald to Victor Serge, June 6, 1941. See Elizabeth Pollet, ed., *Portrait of Delmore: Journals and Notes of Delmore Schwartz* (New York: Farrar, Straus and Giroux, 1986), pp. 119, 120. See Delmore Schwartz, "New Year's Eve," in *In Dreams Begin Responsibilities and Other Stories* (New York: New Directions, 1978), pp. 94, 101, 107; the character Grant Landis is obviously modeled in part on DM. See also James Atlas, *Delmore Schwartz: The Life of an American Poet* (New York: Farrar, Straus and Giroux, 1977), p. 102.

19. Interview with Nancy Macdonald, Nov. 6, 1988. Emily Hahn, *Romantic Rebels: An Informal History of Bohemianism in America* (Boston: Houghton Mifflin, 1968), p. 285; Joseph Mitchell, *Joe Gould's Secret* (New York: Viking, 1965). On one occasion when Joe Gould was in Bellevue Hospital drying out, he listed DM as next of kin.

20. DM to Victor Serge, June 15, 1941. Nancy to Serge, July 23, Aug. 11, 1941.

21. DM to Serge, June 15, 1941. Nancy to Serge, July 1, 1941. DM, "Project for a Book on Nazi Germany," in box 111, folder 539. There are also notes and outlines for this project. DM's papers abound with these initial outlines for books that never came to fruition.

22. James Burnham, "The Theory of the Managerial Revolution," *Partisan Review*, May–June 1941, pp. 181–97. DM, "The End of Capitalism in Germany," ibid., pp. 198–220. Letters to the editor, *PR*, September–October 1941, pp. 440–42.

23. C. Wright Mills, review of Franz Neumann's *Behemoth: The Structure and Practice of National Socialism, Partisan Review*, September–October 1942, pp. 432–37. DM, "Political Notes," *PR*, November–December 1942, pp. 479–82. DM to Victor Serge, June 15, 1941.

24. DM, "The Burnhamian Revolution," *Partisan Review*, January–February 1942. DM to the young American historian John Patrick Diggins, July 1, 1972, not sent, letter in files. Nancy to Serge, Aug. 11, 1941.

25. DM to Serge, Feb. 10, 1942. DM to Diggins, July 1, 1972, not sent. DM refers to a lunch with Burnham and Burnham's loss of heart and faith in a very short time. See also James Burkhart Gilbert, *Writers and Partisans: A History of Literary Radicalism in America* (New York: Wiley, 1968), p. 241.

26. DM, "What Is Fascism: The Discussion Continued," *Partisan Review*, September–October 1941, pp. 418–30.

27. Clement Greenberg and DM, "10 Propositions on the War," *Partisan Review* July–August 1941, pp. 271ff. Footnote on Quixote, p. 272.

28. Ibid. George Orwell, "London Letter," *PR*, July–August 1941, p. 317.

29. Greenberg and DM, "10 Propositions," p. 275.

30. Ibid., p. 278.
31. Philip Rahv, "10 Propositions and 8 Errors," *PR,* November–December 1941, pp. 499–506.
32. Clement Greenberg and DM, "Reply," ibid., pp. 506–509.
33. Lewis Coser, a relatively recent immigrant from Europe who later wrote for *Politics* and *Dissent,* came to DM and Greenberg's defense in a letter to the *Partisan Review,* January–February 1942.
34. Nancy to Victor Serge, Aug. 11, Nov. 9, 1941.
35. DM to Serge, Nov. 25, 1941.
36. Ibid. DM, "A New Dimension," *Common Sense,* January 1942, p. 19.
37. DM to Serge, Nov. 25, Dec. 15, 1941.

CHAPTER 6: A MAJORITY OF ONE

1. DM, "Kulturbolshewismus & Mr. Van Wyck Brooks," *Partisan Review,* November–December, 1941, reprinted in DM, *Memoirs of a Revolutionist* (New York: Farrar, Straus and Cudahy, 1957), pp. 203–14.
2. Ibid. S. A. Longstaff, in *"Partisan Review* and the Second World War," *Salmagundi,* Winter 1979, pp. 115–16 (see footnote 13), points out that as the *PR* editors abandoned their Marxism, they made their support of modernism substitute as a radical political stance. Longstaff's scholarship on the New York intellectuals is perceptive and full of original insights.
3. "On the "Brooks-MacLeish Thesis," *Partisan Review,* January–February 1942, p. 38. This was a collection of responses to DM's article.
4. DM, "Kulturbolshewismus," p. 214. For other accounts of the issue of the role of literature and politics, see Terry A. Cooney, *The Rise of the New York Intellectuals: Partisan Review and Its Circle* (Madison: University of Wisconsin Press, 1986), p. 206; Alexander Bloom, *Prodigal Sons: The New York Intellectuals and Their World* (New York: Oxford, 1986), pp. 116–17.
5. Irving Howe, "The Dilemma of Partisan Review," *New International,* February 1942.
6. DM to "Dear Ex-Comrades", "The Partisan Review Controversy," *New International,* April 1942, pp. 90–92. This issue also contains the editors' response to DM.
7. DM, "The British Genius" (review of Orwell's *The Lion and the Unicorn: Socialism and the English Genius*), *Partisan Review,* March–April 1942, pp. 166–69.
8. James Agee, "Films," *Nation,* May 22, 1943, p. 743. DM, letter to the editor, *Nation,* June 19, 1943, p. 875.
9. Paul Goodman, "Better Judgement and Public Conscience: A Communication," *Partisan Review,* July–August 1942, pp. 348–50. DM to G. L. K. Morris, June 25, 1942.

10. DM, "The American People's Century," *PR*, July–August 1942, pp. 294–310.

11. D. S. Savage, "Pacifism and the War," *PR*, September–October 1942. DM to G. L. K. Morris, June 25, 1942.

12. DM to Morris, June 25, 1942. See also S. A. Longstaff, *"Partisan Review and the Second World War,"* p. 117. Longstaff quotes from an interview he had with Fred Dupee, who discussed "the boys'" reluctance to take risks.

13. For the manipulative conduct of "the boys," see William Barrett, *The Truants* (New York: Doubleday, 1982), pp. 40–41ff; Sidney Hook, *Out of Step: An Unquiet Life in the 20th Century* (New York: Harper & Row, 1987), pp. 512–15; Mary McCarthy, "The Oasis," *Horizon*, February 1949, for a devastating portrait of Rahv ("concealment was second nature to him"); Irving Howe, *A Margin of Hope* (New York: Harcourt Brace Jovanovich, 1982), p. 119.

14. "Under Forty: A Symposium on American Literature and the Younger Generation of Jews," *Contemporary Jewish Record* 7, pp. 3–36. Daniel Bell, "A Parable of Alienation," *Jewish Frontier*, December 1946, p. 15 and passim; "The Intelligentsia in American Society," in *The Winding Passage* (Cambridge, Mass.: Abt Books, 1980), p. 134 and passim. For a thorough discussion of the relationship between Jews and gentiles in the New York intellectual community, see S. A. Longstaff, "Ivy League Gentiles and Inner-City Jews: Class and Ethnicity Around *Partisan Review* in the Thirties and Forties," *American Jewish History*, Spring 1991, pp. 325–43. As Delmore Schwartz was to remark in a later explanation of the celebration of American life and culture during the 1950s at the peak of the cold war: "Civilization's very existence depends upon America, upon the actuality of American life. . . . To criticize the actuality upon which all hope depends thus becomes a criticism of hope itself," James Atlas, *Delmore Schwartz* (New York: Farrar, Straus and Giroux, 1977), p. 343, citing Delmore Schwartz, "The Present State of Poetry," in Dike and Zucker, eds., *Selected Essays of Delmore Schwartz* (Chicago: University of Chicago Press, 1970).

15. Daniel Bell, "The Intelligentsia," pp. 132ff. Dinsmore Wheeler to DM, April 28, 1944.

16. DM to Victor Serge, Aug. 11, 1942.

17. Interviews with Nancy Macdonald and Michael Macdonald. Letter to the author from Albert and Roberta Wohlstetter, May 21, 1991.

18. DM, "Political Notes," *Partisan Review*, November–December 1942, pp. 476–77. DM to Dinsmore Wheeler, Nov. 29, 1943.

19. Delmore Schwartz to DM, Dec. 19, 1942.

20. DM to Schwartz, Dec. 22, 1943.

21. DM, with research by Nancy Macdonald, "The War's Greatest Scandal: The Story of Jim Crow in Uniform," published by the March on Washington Movement, 1943. DM, "The Novel Case of Winfred Lynn," *Nation*, Feb.

20, 1943, pp. 268–70. This case concerned a black man who refused conscription into a segregated service.

22. DM, "Outline for a Pamphlet, 'What Do We Stand For?'", for the MOWM. "Memo on the All-Negro Policy of the MOWM," n.d. 1943, box 119, folder 606. Text of rally speech, n.d., box 118, folder 599.

23. Ibid.

24. Sidney Hook, "The New Failure of Nerve," *Partisan Review*, January–February 1943, pp. 21–23.

25. Sidney Hook, "The Radical Comedians: Inside *Partisan Review*," *American Scholar*, Winter 1984–85, pp. 48–53, reprinted in his *Out of Step*, pp. 512–18.

26. Sidney Hook, "The Failure of the Left," *Partisan Review*, March–April 1943, p. 159 passim.

27. Meyer Schapiro (he used the pseudonym David Merian), "The Nerve of Sidney Hook: Socialism and the Failure of Nerve: An Exchange," *Partisan Review*, May–June 1943, pp. 248–57. Hook, "The Politics of Wonderland," *PR*, July–August 1943, pp. 258–62. Schapiro (David Merian), "Socialism and the Failure of Nerve—The Controversy Continued," letter to the editor, *PR*, September–October 1943, pp. 473–76. Hook response, "Faith, Hope, and Dialectic: Merian in Wonderland," ibid., pp. 476–81. For a more extended discussion of the details of this exchange, see Cooney, *The Rise of the New York Intellectuals*, pp. 193ff.

28. See Irwin Edman, *Candle in the Dark: A Postscript to Despair* (New York: Viking, 1939), pp. 9–10, for a most moving expression of this tragic disillusionment.

29. Job Leonard Dittberner, "The End of Ideology and American Social Thought: 1930–1960," (Ph.D. diss., Columbia University, 1974), Appendix III, interview with Daniel Bell, p. 430. This dissertation has been published. Morton White, *Social Thought in America: The Revolt Against Formalism* (Boston: Beacon, 1957), p. 3. Daniel Bell, *The End of Ideology* (New York: Free Press, 1961), pp. 300ff.

30. DM, Dictator Notebook and MSS, Macdonald Papers.

31. DM, book notes on Niebuhr, box 154, folder 43.

32. DM, "The Future of Democratic Values," *Partisan Review*, July–August 1943, pp. 321–22.

33. Ibid., p. 324.

34. Ibid., pp. 343–44.

35. Ibid., p. 343.

36. DM to Delmore Schwartz, Dec. 22, 1943.

37. Dinsmore Wheeler to DM, April 28, 1944.

38. DM to Philip Rahv, June 1943. Rahv to DM, n.d. (late June 1943). DM to Rahv, July 3, 1943. DM to Victor Serge, June 25, 1943.

39. G. L. K. Morris to DM, May 12, 1943.

40. Rahv to DM, Monday, n.d. [June 28?], 1943. DM to Rahv, n.d. Monday [July 6], 1943. For a slightly different account of these matters, especially the identification of the initial angel, see William Phillips, *A Partisan View: Five Decades of the Literary Life* (New York: Stein and Day, 1983), pp. 138–39; also Terry Cooney, *The Rise of the New York Intellectuals*, pp. 190–91.

41. DM, undated copies of draft of letter of resignation to his colleagues, June–July 1943, box 42, folder 1030.

42. DM, letter of resignation, *Partisan Review*, July–August 1943, p. 382.

43. Ibid., pp. 382–83.

44. Philip Rahv to DM, July 28, 1943.

45. G. L. K. Morris to DM, July 11, July 22, Aug. 10, 1943.

46. Nancy Macdonald, interviews with the author and her book, *Homage to the Spanish Exiles* (New York: Insight, 1987), pp. 61–62.

47. Murray Kempton, "Dishonoring *PR*," in Ben Sonnenberg, ed., *A Grand Street Reader* (New York: Summit Books, 1986), pp. 415–16. For DM's opinion of Rahv and their relationship, see S. A. Longstaff, *"Partisan Review* and the Second World War," p. 126.

48. Nancy Macdonald to Victor Serge, Aug. 16, 1944.

CHAPTER 7: *POLITICS* AND THE SEARCH FOR RESPONSIBILITY

1. I am indebted to Michael Macdonald for remembering Greeley's tenancy; see also J. Muller's *New York As It Is* (New York: J. Miller, 1866, reprinted by Schocken Paperbacks in 1975 as *The 1866 Guide to New York City*) for a good description and brief historical sketch of the building.

2. DM, "Why *Politics*," *Politics*, February 1944, p. 6; see also DM, "Publisher's Preface" in *Politics, Vol. 1–6* (Westport, Conn.: Greenwood, 1969), a reprint of *Politics*'s 1944 editions. *Politics* was initially a monthly, but later on there were gaps in its appearance until in the end it was a quarterly. It was priced at 25 cents a copy and $2.50 for a year's subscription. It contained no advertising, only announcements concerning events and meetings of interest to readers as well as book promotions. It was always a shoestring operation and contributors were paid the minimum rate.

3. Gertrude Buckman, a former wife of Delmore Schwartz, recalls one session at which DM became peeved at those who offered jocular suggestions, letter to the author, Nov. 30, 1986; interview, Oct. 27, 1989. DM, "Publisher's Preface," Greenwood reprint of *Politics*. DM to Alfred Kahn, Nov. 17, 1943, on the press and the left wing. C. Wright Mills to DM, Oct. 25, 1943, rejecting *Gulliver*. DM to Nat Hentoff, April 21, 1960.

4. DM to Dinsmore Wheeler, Nov. 29, 1943. DM, "Why *Politics*," pp. 6ff.

5. DM, "Why *Politics*," pp. 6ff.

6. Interview with Lionel Abel, Feb. 16, 1985. See also Lionel Abel, "Reconsideration: Nicola Chiaromonte: Innocence and the Intellectual," *New Republic*, March 24, 1986, p. 41.

7. DM, "Afterword" to "The Mills Method," in DM, *Discriminations* (New York: Grossman, 1974), pp. 299–300.

8. Daniel Bell, "A Parable of Alienation," *Jewish Frontier*, December 1946, p. 12; "The Coming Tragedy of American Labor," *Politics*, March 1944, pp. 37–42.

9. DM to Daniel Bell, Nov. 26, 1946, accusing him of making a "snug career" at the University of Chicago and no longer being interested in the world DM inhabited. Interview with Daniel Bell, March 9, 1990; he attributed the "floating kidney" label to Herbert Solow.

10. Interview with Lewis and Rose Coser, July 29, 1988. See also Lewis Coser's "The WP, *Politics* Magazine, and Dwight Macdonald" in *The Legacy of the Workers Party, 1940–1949: Recollections and Reflections,* transcript of Tamiment Library Oral History of the American Left Conference, May 7, 1983, pp. 36–39. (Transcript in the Tamiment Library, New York University.)

11. Nancy Macdonald to George Demetrian, a *Politics* reader, July 30, 1946. DM to William Hesseltine, Nov. 18, Dec. 8, 1943. Melvin Lasky to DM, Nov. 13, 1944; this correspondence began as early as 1940, when Lasky as a student at Michigan was interested in Workers Party politics.

12. Andrea Caffi, *A Critique of Violence*, with an introduction by Nicola Chiaromonte (Indianapolis: Bobbs Merrill, 1970). Chiaromonte's introduction has a good biographical sketch of Caffi.

13. DM, "Publisher's Preface," Greenwood reprint of *Politics*, on Nancy's role. DM, "Introduction: Politics Past," *Memoirs of a Revolutionist* (New York: Farrar, Straus and Cudahy, 1957), pp. 25–26. Interview with Niccolo Tucci, Sept. 13, 1988.

14. DM, "Why *Politics*," pp. 6–7.

15. Quoted in DM, "Comment: The End of Europe . . . ," *Politics*, February 1944, p. 33.

16. Daniel Bell, "Politics in the Forties," in his *The End of Ideology: On the Exhaustion of Political Ideas in the Fifties*, rev. ed. (New York: Free Press, 1962), pp. 305–307; this is a reprint of Bell's review in the *New Leader* of DM's *Memoirs of a Revolutionist*. DM, "The Psychology of Killing," *Politics*, September 1944, p. 239. DM, "Randolph Bourne," *Politics*, March 1944, p. 35. Bourne ad in *Politics*, Summer 1948, p. 188. DM, "Comment: The Shape of Things to Come," *Politics*, June 1944, pp. 129–31.

17. Walter Oakes, "Toward a Permanent War Economy," *Politics*, February 1944, pp. 11–15. Daniel Bell, "The Coming Tragedy of American Labor," *Politics*, March 1944, pp. 37–42. DM, "Wallace and the Labor Draft," *Politics*, February 1945, pp. 35–37.

18. DM, "The Intelligence Office," *Politics*, March 1944, p. 62. "Bombardier" (pseudonym for a black airman), "The Story of the 477 Bombardment Group," *Politics*, June 1944, pp. 141–42. DM, "Free and Equal," *Politics*, May 1945, pp. 150–51. DM, "A Theory of Popular Culture," *Politics*, February 1944, p. 23. Unsigned review of Georgene H. Seward, "Sex Roles in Postwar Planning" (an article published in the *Journal of Social Psychology*), *Politics*, February 1944, pp. 152–53. DM (as Terrence Donaghue), "American Woman's Place," *Politics*, April 1944, p. 95 (Donaghue and Dryden are described as partners in a Staten Island ferret farm in the list of contributors, *Politics*, August 1946, p. 256). Michael Macdonald believes, and it seems reasonable, that DM adopted the two names with the initials T. D. after his father, Theodore Dwight Macdonald.

19. Robert Duncan, "The Homosexual in Society," *Politics*, August 1944, pp. 209–11. See also Ekbert Faas, *Young Robert Duncan: Portrait of the Poet as Homosexual in Society* (Santa Barbara, Calif.: Black Sparrow Press, 1983), pp. 156–60. A sensitive scholar of American history and the history of homosexuality in America, Martin B. Duberman, feels that Duncan's non-self-hating essay written in 1944 was "indeed a very special piece" that was approximately two decades ahead of its time; phone conversation with Duberman, June 19, 1989.

20. Maurice Zolotow, "On Highbrow Writing," *Politics*, August 1944, pp. 217–18. DM, "On Lowbrow Thinking," *Politics*, August 1944, pp. 219–20. DM, who seldom resorted to vulgar colloquialisms, could not resist quoting Engels's assault on bourgeois platitudes.

21. DM, Introduction to Greenwood edition of *Politics*, p. 4. DM to Victor Serge, Jan. 20, 1944.

22. C. Wright Mills to DM, Feb. 12, 1944. George P. Elliot, "'Where Are You Going?' Said Reader to Writer," *Politics*, September 1944, p. 245. William Palmer Taylor, letter to the editor, "The Tribune's Little Helper," *Politics*, April 1944, pp. 95–96.

23. DM, "'Here Lies Our Road!' Said Writer to Reader," *Politics*, September 1944, p. 248.

24. Ibid.

25. Ibid., pp. 250–51. For an eloquent defense of DM's intransigent positions, see Robert Westbrook, "Horrors Theirs and Ours: The *Politics* Circle and the Good War," *Radical History*, September 1986, p. 14. DM later conceded that his anti-war position was a mistake, but Westbrook persuasively argues that it was a "creative mistake" because it "freed him to openly criticize all sides in the conflict." For DM's sarcastic assault on the Trotskyists, see "The Only Real Moral People," *Politics*, May 1944, pp. 109–10. David Bazelon, a *Politics* contributor, to DM, July 12, 1944, criticizing DM for his attack on leftist "revolutionaries." James Cannon to Rose Karsner, Cannon's wife, Letter #56, June 7, 1944, #57, June 12, 1944, in James

Cannon, *Letters from Prison* (New York: Pathfinder Press, 1972), pp. 89–93.

26. DM, "'Here Lies Our Road,'" p. 251.

27. DM to Dinsmore Wheeler, Nov. 29, 1943. DM, "The World's Biggest Union," *Common Sense*, November 1943, pp. 411–14. Notes on "The New CIO Ideology," box 116, folder 578.

28. DM, "'Union Security': Two-Edged Sword," *Socialist Review: Monthly Discussion Supplement of The Call*, April 1944, pp. 1, 4; Part 2, July 1944, pp. 2, 4.

29. Nancy Macdonald to Victor Serge, March 28, 1942.

30. DM, Introduction to "Modern Texts: Chapter 37 of Melville's *Redburn*," *Politics*, March 1944, p. 56.

31. European (Andrea Caffi), "The Automatization of European People," *Politics*, November 1945, p. 337.

32. DM, "Some Questions to a Democratic Committee," *Politics*, June 1944, pp. 132–33.

33. Paul Tillich with introductory remarks by DM, "Some Answers of a Committee," *Politics*, July 1944, pp. 190–91.

34. DM, "Comment," ibid., p. 162.

35. Bruno Bettelheim, "Behavior in Extreme Situations," *Politics*, August 1944, pp. 199–209; Bettelheim's article first appeared in a more extensive version as "Individual and Mass Behavior in Extreme Situations" in the *Journal of Abnormal and Social Psychology*, October 1943, pp. 417–52. The notion of the practice of unspeakable crimes as a matter of daily routine was one of the more shocking revelations coming out of the Nazi genocide and was part of the thrust of Raul Hilberg's monumental study, *The Destruction of the European Jews*, as well as Hannah Arendt's speculations on the character of Adolf Eichmann.

36. DM to Dinsmore Wheeler, May 11, 1944.

37. Interviews with and extensive notes and reminiscences provided by Michael Macdonald.

38. DM to Al Goldman, Aug. 18, 1944.

39. Al Goldman to DM, Aug. 3, 1944.

40. DM, "The Jews, 'The New Leader,' and Old Judge Hull," *Politics*, January 1945, pp. 23–24.

41. DM, "The Responsibility of Peoples," *Politics*, March 1945, pp. 83–84, reprinted in *Memoirs of a Revolutionist* (New York: Farrar, Straus and Cudahy, 1957).

42. Ibid., p. 86; DM quoted a Swedish source to the effect that the ghastly murders were carried out in secret by the SS and the vast majority of German people did not know what was going on.

43. Ibid., pp. 86–87.

44. Ibid., pp. 89–92.

45. DM, "The Responsibility of Peoples: An Essay on War Guilt." This *Politics* pamphlet edition had two new sections, one on "The Führer Principle, Here and Now," which was a discussion of leadership cults, and a final section, "The Community of Those Who Endure." Neither of these appeared in the initial *Politics* article of March 1945, nor in the Cunningham Press edition of 1953 or *Memoirs of a Revolutionist*.

46. DM cited Hannah Arendt, "Organized Guilt and Human Responsibility," *Jewish Frontier*, January 1945, pp. 19–23, reprinted in Roger W. Smith, ed., *Guilt: Man and Society* (Garden City, N.Y.: Anchor, 1971). See also Hannah Arendt, "The Concentration Camps," *Partisan Review*, July 1948, pp. 747–48.

47. Solomon Bloom and Gordon Clough, letters to the editor, *Politics*, July 1945, pp. 203–204.

48. Jim Cork, ibid., p. 207. Reinhold Niebuhr, ibid., May 1945, p. 160.

49. DM, exchange with Guenter Reiman, ibid., p. 155. DM, "The Responsibility of Peoples," p. 88. For this discussion of some of the inconsistencies, confusions, and contradictions I am indebted to the insightful discussion in Robert Westbrook's article "The Responsibility of Peoples: Dwight Macdonald and the Holocaust," in Sanford Pinsker and Jack Fischel, eds., *America and the Holocaust: Holocaust Studies Annual*, vol. 1 (Greenwood, Fla.: Penkewell Publishing, 1984) and also in correspondence with him.

50. Jean Malaquais, letter to the editor, *Politics*, September 1945, pp. 283–84.

51. Nicola Chiaromonte, "Rome Letter," *Politics*, May–June 1947, p. 118.

CHAPTER 8: IN A TERRIFYING WORLD
THE ROOT IS MAN

1. Interviews with and extensive notes and chronology compiled by Michael Macdonald.

2. Virginia Chamberlain to author, July 20, 1989. Interview with Chamberlain, Aug. 13, 1987. Interviews with Nancy and Michael Macdonald.

3. DM, untitled comment on cover, *Politics*, August 1945, p. 225.

4. Interviews with Mary McCarthy, Nov. 13, 1987, Aug. 9, 1988.

5. Niccolo Tucci, "Victory," *Politics*, May 1945, p. 129, cover; Tucci's comments had initially appeared in the November 1944 issue, p. 304.

6. DM, "The Bomb," *Politics*, September 1945, p. 258.

7. Ibid.

8. Ibid. Virgil Vogel, "The Bomb (2): Birthplace," about the Hanford Engineer Works where the bomb was manufactured; Harold Orlansky, "The Bomb (3): Observations from an Asylum," an analysis of the mass insanity that produces such a thing as the atom bomb, written by a conscientious objector who was working in a mental hospital at the time, *Politics*, September 1945, pp. 261–63.

9. DM, "The Bomb," p. 260.

10. "Whither Politics," an exchange between DM and his critics in the letters section of *Politics*, May 1946, p. 142.

11. DM to *Politics* reader Art Wiser, a conscientious objector and pacifist, Jan. 16, 1946. Herbert Orloff, "Atomic Fission and Revolution," letter to the editor, *Politics*, October 1945, pp. 317–18.

12. Simone Weil, "*The Iliad*, or The Poem of Force," translated by Mary McCarthy, *Politics*, November 1945, pp. 321–31. For an interesting view of Weil's influence in America and her legacy to the New Left, see Staughton Lynd, "Marxism–Leninism and the Language of *Politics* Magazine: The First New Left . . . and the Third" in George Abbott White, ed., *Simone Weil: Interpretations of a Life* (Amherst: University of Massachusetts Press, 1981).

13. DM to Art Wiser, Jan. 16, 1947. The Hans Gerth statement is part of a collection of responses to the "New Roads" series of articles challenging Marxist analysis of contemporary affairs, undated in February or March 1946, box 155, folder 47.

14. DM, "The Questionnaire, Preliminary Report," *Politics*, May–June 1946, p. 124.

15. DM to Ed Seldon, Dec. 12, 1945. For DM's account of this and other meetings, see DM, "Politicking," *Politics*, January 1946, p. 30.

16. Calder Willingham to DM, Dec. 1, 1945. On Lewis Coser's talk, see DM, "Politicking," *Politics*, January 1946, p. 30. An ad for the series of discussion sessions appears in *Politics*, November 1945, p. 352. The other speakers were Lionel Abel, Paul Goodman, and Frank Fisher.

17. Seymour Martin Lipset to DM, Oct. 4, 1945.

18. Ibid. Andrea Caffi ("European"), letter to the editor, *Politics*, November 1945, p. 336. DM, editorial response to a letter from Dachine Rainer, an anarchist activist who often hung around the *Politics* offices with her philosophical radical companion Holley Cantine. They edited their own journal, *Retort*, in the 1940s out of Bearsville, New York. Rainer was a pure, absolutist anarchist who often took DM to task for backsliding.

19. DM, "A Report on Food Packages," *Politics*, December 1945, pp. 383–84. Phil Heller, letter to the editor, *Politics*, February 1946, p. 63.

20. Interview with Danny Rosenblatt, Dec. 13, 1985. I am much indebted to him for his recollections of both DM and Nancy and the atmosphere of the *Politics* circle.

21. DM, "Food Packages," pp. 362–63.

22. DM on the "monstrous" developments in Russia to Kurt Glasser, Office of the United States Military Government in Germany, Oct. 10, 1946. Anonymous U.S. Army sergeant, "500 Red Army Men," *Politics*, October 1945, pp. 304–305.

23. Ibid., p. 305.

24. Balticus, letter to the editor, *Politics*, December 1945, p. 383. Edouard Roditi to DM, Nov. 5, 1945.

25. I am indebted to a correspondence with the late Irving Howe in July 1989 for helping me to grasp the historical context in which DM shaped this editorial stance. Howe, of course, is in no way responsible for the conclusions reached.

26. Irving Howe, after reading this and other related material, termed it not anti-Semitism, "but it's a kind of bleached narrowness of spirit." I have anguished over these passages and have discussed the issue with Jews and gentiles. The latter balk at any suggestion of anti-Semitism. Jews for the most part also do not consider Dwight's conduct in the category of anti-Semitism. Coming as I do, from a Wasp and Jewish parentage, I may be overly sensitive having lived in a Protestant background where anti-Semitic remarks during my youth and young adulthood were very common. All I can say is that when I read through this material I was unsettled.

27. Nicola Chiaromonte, "Proudhon: An Uncomfortable Thinker," *Politics*, January 1946, pp. 27–29. Salwyn Schapiro, "P. J. Proudhon, Harbinger of Fascism," *American Historical Review*, July 1945, pp. 714–37. Schapiro offered a good deal more evidence than the stereotype of Jewish money-lenders to support his charge of Proudhon's anti-Semitism, which Chiaromonte did not discuss in his article.

28. DM to Arthur Steig, Jan. 31, 1946.

29. Arthur Steig to DM, Jan. 31, 1946. In this connection, years earlier when Dwight had written on Hollywood producers he seldom failed to point out their Jewish ancestries, particularly if they had changed their names. The critic and novelist John Gregory Dunne referred to this practice as an "ethnic code, cryptological anti-Semitism." He did not wish to imply that Dwight was an anti-Semite but that "the tactic in the hands of those who were, had an ugly whiff of nativism." Dunne was referring to the collection of DM's writing *On Movies*, pp. 96, 100. John Gregory Dunne, "Goldwynism," a review of several books on Hollywood producers and directors in the *New York Review of Books*, May 18, 1989, notes 11, 30. See also DM, "Notes on Hollywood Directors," *Symposium*, July 1933, p. 193.

30. Will Herberg, "Personalism Against Totalitarianism"; European (Andrea Caffi), "Towards a Socialist Program"; Paul Goodman, "Revolution, Sociolatry and War," *Politics*, December 1945, pp. 363–80.

31. Constant Reader, letter to the editor, and DM's reply, *Politics*, January 1945, p. 32. Calder Willingham to DM, Nov. 13, 1946.

32. Nicola Chiaromonte, "On the Kind of Socialism Called 'Scientific,'" *Politics*, February 1946, pp. 33–44. DM to Victor Serge, Nov. 25, 1944.

33. DM, "Whither Politics," *Politics*, May 1946, p. 141.

34. James Farrell, "New Roads Discussion," *Politics*, March 1946, pp. 89–93. Virgil Vogel, ibid., February 1946, pp. 46–48. For lengthy critiques of the

egocentricity of *Politics*, see letters columns throughout this period.

35. DM, "The Root Is Man," *Politics*, April 1946, p. 99.

36. DM, editorial rejoinder to James Farrell criticism, Ibid., March 1946, p. 92.

37. *Ibid.*, pp. 99–100.

38. *Ibid.*

39. *Ibid.*, p. 101.

40. *Ibid.*, p. 102

41. *Ibid.*, pp. 104–108.

42. *Ibid.*, p. 102.

43. Ibid., pp. 105–106ff. DM, "Politicking," *Politics*, May 1946, p. 176.

44. DM to Mary McCarthy, Aug. 24, 1946.

45. Interviews with Paul Magriel, Feb. 22, 1990. Interviews with Michael Macdonald, Danny Rosenblatt, and Norman Mailer.

46. Interview with Michael Macdonald, Nov. 3, 1988.

47. Interview with Mary McCarthy and Andy Dupee, Fred Dupee's wife, Nov. 13, 1987. Taped interview with Nancy Macdonald in Peter Manso, *Mailer: His Life and Times* (New York: Simon and Schuster, 1985), pp. 176–77. Adele Mailer in Manso, *Mailer* p. 295. Ann Birstein, talk with the author, Feb. 1, 1989. On the nudity at cocktail parties there is some difference of opinion. Many close friends do recall nudity on the beach but not at parties in the cottage.

48. Saul Bellow, *Humboldt's Gift* (New York: Avon, 1975), pp. 311–13. Bellow's spoof in its totality was a patronizingly unfriendly satire.

49. Interview with Mary McCarthy, Nov. 13, 1987. Interview with Virginia Chamberlain, Aug. 13, 1987. Interviews with Michael Macdonald previously cited. DM to Mary McCarthy, July 30, 1948.

50. Interview with Eileen Simpson, April 1, 1986.

51. Michael Macdonald to the author, July 16, 1989.

52. DM to Nancy Macdonald, April 1, 1948.

53. Eileen Simpson, *Poets in Their Youth* (New York: Random House, 1982), p. 180. Interview with Eileen Simpson, April 1, 1986. John Haffenden, *The Life of John Berryman* (Boston: Routledge and Kegan Paul, 1982), p. 211. Paul Mariani, *Dream Song: The Life of John Berryman* (New York: Morrow, 1990), p. 223.

54. The anecdote about the vegetable garden was told to me by Gloria Watts, a longtime Wellfleet resident.

55. Interview with Daphne Hellman, a harpist and wife of Geoffrey Hellman, DM's Yale classmate.

56. Nancy Macdonald to Andrea Delacourt, June 13, 1946.

57. DM, "The Root Is Man: Part II," *Politics*, July 1946, pp. 194, 197.

58. Ibid., p. 197. DM, reply to James Farrell, *Politics* 3, March 1946, p. 92. Daniel Bell, *The End of Ideology* (New York: Free Press, 1962), p. 300.

59. DM to Mr. Ludowyk, June 28, 1946.
60. Ibid.
61. Ibid.
62. DM, "Root: Part II," p. 209.
63. See DM, "The Root Is Man," the 1953 Cunningham Press edition, p. 49, footnote 22. This edition, which contains a running commentary by DM in the footnotes, is invaluable for tracing DM's changing opinions and is extremely revealing of his open mind and willingness to engage in analytical self-criticism.
64. Frank Marquart, letter to the editor, *Politics*, August 1946, pp. 251–52. Irving Howe, "The 13th Disciple," ibid., pp. 329–34. Louis Clair (pen name of Lewis Coser), "Digging at the Roots," ibid., pp. 323–24. DM to Anton Pannekoek, Oct. 15, 1946, complaining of the tone of the attacks by Howe and Coser. Irving Howe, *A Margin of Hope* (New York: Harcourt Brace Jovanovich, 1982), pp. 116–17. Interview with Howe, Sept. 25, 1987. Interview with Lewis and Rose Coser, July 29, 1988. See also Lewis Coser's "The WP, *Politics* Magazine and Dwight Macdonald," in *The Legacy of the Workers Party, 1940–1949: Recollections and Reflections*, a Tamiment Library Oral History of the American Left Conference, May 6–7, 1983, p. 83. In the 1988 interview Coser expressed regret at the harsh and accusatory tone of his criticism of DM and credited him with being more sensitive to the need for some kind of ethical basis for radicalism to replace Marxist historicism and relativism. (Transcript in the Tamiment Library, New York University.)
65. Nicola Chiaromonte to DM, n.d. but dated by DM as 1946.
66. Gwynne Nettler to DM, from the University of California at Santa Barbara, July 14, 1946. Later that year DM published a letter from her praising his flexibility, his refusal to give the ideologues the catechism they demanded.
67. Stephen J. Whitfield, *A Critical American: The Politics of Dwight Macdonald* (Hamden, Conn.: Archon Books, 1984), p. 72.

CHAPTER 9: A POLITICAL DESERT WITHOUT HOPE?

1. Nancy Macdonald to Laurette Sejourne, Victor Serge's companion, Sept. 17, 1946. DM to George Woodcock, the British historian and anarchist, Aug. 27, 1946.
2. The following articles appeared in *Politics*: Don Calhoun, "The Political Relevance of Conscientious Objection," July 1944, p. 177. "Non Violence and Revolution," January 1946, p. 17. James Peck, "A Note on Direct Action," January 1946, p. 21. European, "Is a Revolutionary War a Contradiction?" April 1946, p. 128. "Violence and Sociability," January 1947, p. 23.

3. DM, "Why Destroy Draft Cards," *Politics*, March–April 1947, pp. 54–55.

4. Ibid.

5. DM to Art Wiser, Oct. 27, 1947.

6. DM, "Too Big"; "Three on Our Side," review of books about the threat to human and community values, by Victor Gollancz, Grete Hermann, and Aldous Huxley, pp. 395–96; "Periodicals," p. 397 (this review of some magazines was signed with the pseudonym Theodore Dryden; I believe the author was DM, but it might have been Irving Howe), *Politics*, December 1946.

7. DM to George Barbarow, June 8, 1947.

8. Nancy Macdonald to Harriet Landen Smith, Jan. 18, 1946. Nancy to Sasha Muller, March 30, 1947.

9. Interview with Nancy Macdonald, Nov. 11, 1988.

10. Paul Goodman, "The Political Meaning of Some Recent Revisions of Freud," *Politics*, July 1945, pp. 197–203. C. Wright Mills and Patricia Salter, "The Barricade and the Bedroom," *Politics*, October 1945, pp. 313–15. There is a brief rejoinder by Goodman.

11. DM, review of Paul Goodman's *Art and Social Nature*, *Politics*, November 1946, pp. 361–62. DM to Calder Willingham, Nov. 13, 1946. Nancy Macdonald to reader George Demetrian, July 20, 1946. See also Taylor Stoehr, "Growing Up Absurd Again," *Dissent*, Fall 1990, p. 487, for a good discussion of Goodman's view of "sociolatry."

12. DM to Mickey Scharaf, May 6, 1945.

13. DM to Ted Birkey, a longtime *Politics* reader, June 8, 1949. DM to Mark Shechner, a colleague at SUNY Buffalo in the 1970s, June 30, 1975. James Atlas, "Golden Boy," *New York Review of Books*, June 29, 1989, p. 45; this is a review of books about Isaac Rosenfeld.

14. Mark Shechner, "Reich and the Reichians," *Partisan Review*, Spring 1985, p. 101. Shechner cites Frederick Crews and others to support the notion of Reichianism as a kind of counterpolitics. Interview with Shechner, June 6, 1990.

15. Richard King, *The Party of Eros: Radical Social Thought and the Realm of Freedom* (Chapel Hill: University of North Carolina Press, 1972), pp. 31–43, particularly p. 37 for notion of a "new radicalism based on psychology and biology to replace outworn Marxist notions."

16. DM, "Why Destroy Draft Cards."

17. Kurt Glasser to DM, July 10, 1946.

18. DM, "The Russian Culture Purge," *Politics*, October 1946, pp. 301, 302.

19. DM, introduction to "The German Experience—Three Documents," ibid., pp. 314–19.

20. DM, "The *Partisan Review* and *Politics*," ibid., pp. 400–403. DM, "Henry Wallace (Part 2)," *Politics*, May–June 1946, footnote, p. 104. Arthur Schlesinger, Jr., "History of the Week" column in the *New York Post*, Sept.

2, 1951. For a more detailed account of Schlesinger as a cold war activist, see my "Arthur Schlesinger, Jr.: Scholar-Activist in Cold War America, 1946–1956" *Salmagundi*, Spring–Summer 1984, pp. 255–85; "Typhoid Mary" accusation on p. 281.

21. DM, "Notes on the Truman Doctrine," *Politics*, May–June 1946, pp. 85–87.

22. Arthur Schlesinger, Jr., "The Apostle of the Common Man: A Portrait Etched in Acid," *New York Times Book Review*, Feb. 22, 1948, p. 3.

23. DM, *Henry Wallace: The Man and the Myth* (New York: Vanguard Press, 1948), pp. 11–24.

24. Ibid., pp. 109, 173. At the height of the McCarthy period, Kristol wrote: "There is one thing that the American People know about Senator McCarthy; he, like them, is unequivocally anti-Communist. About the spokesmen of American liberalism they feel they know no such thing" ("Civil Liberties, 1952: A Study in Confusion," *Commentary*, March 1952, p. 229).

25. DM to James Henle of the Vanguard Press, n.d. Henle to DM, Sept. 3, 1947, May 14, 1948.

26. Nicola Chiaromonte to DM, Sept. 3, 1948.

27. DM to Chiaromonte, Sept. 14, 1948.

28. Ibid. In "Pacifists and Communism," notes dated July 28, 1949, DM has the grey and black argument with Margaret Marsh, a pacifist who taught at Smith, Pacifist file, box 126, folder 687.

29. DM to Chiaromonte, July 14, 1948. The Pacifist file contains endless articles in which DM consistently confronts and even baits pacifists for being unrealistic and refusing to face up to the Soviet Stalinist reality.

30. DM, special insert on the Waldorf Conference, *Politics*, Winter 1949, p. 32. DM listed Stone as one of the "Stalinoids" supporting this "Communist-sponsored" propaganda extravaganza. I. F. Stone, "Confessions of a Dupe," *The Truman Era* (New York: Vintage, 1988), pp. 66–68. See also Steven M. Gillon, *Politics and Vision: The ADA and American Liberalism* (New York: Oxford University Press, 1987), pp. 55–56. It is of interest that Richard Pells in *The Liberal Mind in a Conservative Age* (New York: Harper and Row, 1985) lists Stone and DM among a small band who fought McCarthyism and constriction of unpopular political views. Opinions here on Stone do seem generational; after his work in exposing the failings of U.S. Vietnam policy he became a hero in the eyes of the liberals and left of the 1960s and thus the attacks on him in the forties and fifties have been dismissed, only to be revived by the most zealous of conservatives. William O'Neill reveals a particularly patronizing attitude toward Stone as Soviet apologist in his *A Better World: The Great Schism: Stalinism and the American Intellectuals* (New York: Simon and Schuster, 1982), pp. 155–56.

31. Interviews with Nancy Macdonald, Nov. 5, 1988; Virginia Chamberlain, Aug. 13, 1987; Danny Rosenblatt, Dec. 13, 1985.

32. Interviews with Nancy Macdonald and Michael Macdonald on several occasions. DM to Cliff Bennett, a *Politics* staffer, July 24, 1947, the date of the fire.

33. DM's notes on this affair are in box 159, folder 14; other related materials in box 170. The relevant journal entries are at the back of the book in which Berryman wrote his 1947 sonnets, pp. 112–15, John Berryman Papers, Manuscript Division, University of Minnesota Library. See also Paul Mariani, *Dream Song: The Life of John Berryman* (New York: Morrow, 1990), p. 196. I am indebted to Mariani for obtaining this material from the Minnesota library.

34. Translator of *Politics*'s French edition Ralph Manheim to DM, July 22, 1947, dealing with the details of problems with Lionel Abel and the editing of the French edition. DM to Mary and Ralph Manheim, n.d., summer 1947. DM to Ralph Manheim, Aug. 17, 1948. Manheim to DM, n.d., summer 1948. Manheim, "The Perspectors," manuscript of the Manheim satirical story on Lionel Abel and May and Harold Rosenberg, in box 33, folder 812, Macdonald Papers. DM to Merleau-Ponty, July 17, 1947, in which DM accuses him of being deluded about Stalinism's "revolutionary nature." Gelo and Andrea Delacourt, the journal's French correspondents, had attacked Merleau-Ponty's Stalinism in the previous issue of *Politics* (May–June 1947, p. 126). DM to Don Calhoun, Sept. 14, 1947, in which DM opposes Sartre's insistence on literary engagement, saying it reminds him of the Brooks-MacLeish thesis. DM, "Small Talk," *Politics*, Winter 1948, p. 56. For extensive treatment of the French intellectuals' tolerance for Stalinism, see Tony Judt's intemperate but thought-provoking polemic, *Past Imperfect: French Intellectuals 1944–1956* (Berkeley: University of California Press, 1992).

35. DM to Cliff Bennett, June 1948. DM quoted in the *Exeter Oxonian*, Feb. 22, 1947.

36. DM, "Too Big," p. 39.

37. DM to Melvin Lasky, Oct. 2, 1947.

38. DM to William Rickel, Dec. 15, 1947.

39. DM to Readers, Nov. 17, 1947; this letter was reprinted in *Politics*, Winter 1948, p. 58.

40. C. Stafford Brown to DM, Dec. 4, 1947. Perhaps one of the more astonishingly perceptive letters was one from a Lamar Rankin, who for some unexplained reason assumed that Nancy had left DM and that he was showing suicidal tendencies because he could no longer endure living without her. Rankin wrote to advise him to forget *Politics* and politics and concentrate on straightening out his personal life. He felt DM was suffering from "unconscious infantile drives" and that he should consult the noted psychiatrist A. A. Brill, who could recommend a good therapist.

41. *Time*, Dec. 8, 1947.

42. Henry Regnery to DM, Dec. 8, 1947.
43. DM, "The Fascinated Readers," *Politics*, Winter 1948, pp. 59–63. This was a summary of the results of the questionnaire; DM's initial response to preliminary results was published in the May–June 1947 issue, pp. 122–24; to one reader he claimed that his circular letter, the *Time* magazine report, and the readers' response had made the letter a "highly successful promotion letter—probably because it de-promoted." He felt he might have stumbled on a new advertising formula: truth.
44. Interviews with Nancy Macdonald.
45. DM to Nelson Lazarevitch, March 27, 1948. See DM, "After Seven Years," a critique of James Agee's *Let Us Now Praise Famous Men*, *Politics*, Spring 1948, p. 124, for similar statement about being trapped in a "social mechanism."
46. DM to Robert Louzon, March 27, 1948.
47. DM to Don Calhoun, March 28, 1947.

CHAPTER 10: GOODBYE TO UTOPIA

1. DM, "On the Elections," *Politics*, Summer 1948, pp. 203–204. This issue did not appear until after the election.
2. DM to Nicola Chiaromonte, Dec. 10, 1948. In 1947 Eliot had visited with the Macdonalds at their apartment on East 10th Street. "Did I tell you that our humble chaise longue . . . recently bore the weight of no less a personage than T. S. himself. Seems he's a great admirer of *Politics* (less, I gathered, of *PR* these days)" (DM to George L. K. Morris, July 13, 1947, Archives of American Art, Washington, D.C., courtesy of Garnet McCoy of the archives).
3. DM to Chiaromonte, Dec. 10, 1948.
4. DM, notes of a letter to a Mrs. Reece, March 1949, a classic example of a begin-again, typeover, retype manuscript of a letter which may or may not ever have been sent. His papers are full of such abortive epistles, suggesting that writing them was a form of therapy rather than communication. In an interview, Michael Macdonald recalled that when he was a child, neither parent read to him. However, DM maintained that he did indeed read to his children (DM, "Mary Poppins Was Not a Junkie," *Discriminations: Essays and Afterthoughts* [New York: Grossman Publishers, 1974], p. 251).
5. DM to Chiaromonte, Dec. 10, 1948.
6. Ibid.
7. Carol Brightman, *Writing Dangerously: Mary McCarthy and Her World* (New York: Clarkson Potter, 1992), p. 308. Rhodri Jeffreys-Jones, *The CIA and American Democracy* (New Haven: Yale University Press, 1989), pp. 68–69. Peter Steinfels, *The Neoconservatives: The Men Who Are Changing America's Politics* (New York: Simon and Schuster, 1979), pp. 29–30. I have

no evidence that Hook was on the payroll of the CIA, but certainly he had the same ends as the agency and obviously cooperated with it.

8. Nicola Chiaromonte to Mary McCarthy, Sept. 3, 1948. These letters are in the Chiaromonte file in the Macdonald Papers because McCarthy was passing on Chiaromonte's letters if they dealt with the EAG.

9. Meeting of Friends of Russian Freedom, March 4, 1949, box 17, folder 428, Macdonald Papers. DM, special insert, "The Waldorf Conference," *Politics*, Winter 1949, pp. 32A–32D. Sidney Hook, *Out of Step: An Unquiet Life in the 20th Century* (New York: Harper & Row, 1987), pp. 383–96. Neil Jumonville, *Critical Crossings: The New York Intellectuals in Postwar America* (Berkeley: University of California Press, 1991), Chapter 1, "The View from the Waldorf," pp. 1–48. Jumonville's account is a detailed, and perceptive interpretation. The *New York Herald Tribune* carried detailed accounts of the conference beginning March 26. It describes Hook's confrontation with Shapley.

10. Interviews with Mary McCarthy, Nov. 13, 1987, Aug. 9, 1988.

11. DM, "The Waldorf Conference."

12. Czeslaw Milosz, "Dwight Macdonald," in *Beginning with My Streets* (New York: Farrar, Straus and Giroux, 1991), pp. 178–88. This essay was originally published in a collection of Milosz's essays in Polish titled *Zaczynajac od moich ulic* (Paris: Instytut Literacki, 1985). An essay similar to this on "The Root Is Man" appeared in the Polish émigré journal *Kultura* (fourth issue, 1954), published in Paris. Milosz put me in touch with his translator, Professor Madeline G. Levine of the University of North Carolina, who initially sent me the unpublished manuscript and gave me the bibliography. Milosz was an "assiduous reader of *Politics*" (Milosz to author, Feb. 5, 1991).

13. DM, "The Waldorf Conference." p. 32C. Neil Jumonville, *Critical Crossings*, p. 33. Interviews with Michael Macdonald.

14. DM, "The Waldorf Conference," p. 32D.

15. DM to David Herron, May 3, 1949.

16. DM to Harry Frankfurt, May 3, 1949.

17. DM, "The Waldorf Conference," p. 32D. See also John P. Rossi, "Farewell to Fellow Traveling: The Waldorf Peace Conference of March 1949," *Continuity*, Spring 1985, pp. 1–31.

18. Eleanor Clark, form letter to DM, March 21, 1949. Elizabeth Hardwick to author, July 21, 1990, in which she states she does not remember contacting DM. DM to Elizabeth Ames, March 3, 1949, in which he refers directly to a conversation with Elizabeth Hardwick and in a postscript says he is sending a copy of his letter to Hardwick. See also Janice R. MacKinnon and Stephen R. MacKinnon, *Agnes Smedley: The Life and Times of an American Radical* (Berkeley: University of California Press, 1988), pp. 320–21.

19. Statement to the board of directors at Yaddo. Clark group statement and letter to DM, March 21, 1949, signed by Breit, Cheever, Clark, Kazin, and Phelan.

20. DM to Elizabeth Ames, Sept. 8, 1953 (a copy of this letter was given to me by the curator of the Yaddo archives). Ames to DM, Sept. 12, 1953. See also Richard H. Pells, *The Liberal Mind in a Conservative Age: American Intellectuals in the 1940s and 1950s* (New York: Harper & Row, 1985), pp. 263–65.

21. DM, "Homage to Twelve Judges: An Editorial," *Politics*, Winter 1949, pp. 1–2. This editorial statement began on the cover of the journal.

22. Allen Tate to DM, spring 1949. DM to Tate, May 3, 1949. William Barrett, "Comment," *Partisan Review*, April 1949, pp. 344–47. Selection of responses to the prize and Barrett's editorial comment appear in *Partisan Review*, May 1949, pp. 512–22. DM to Erik Lee, Jan. 5, 1948, for DM's suspicions of Barrett prior to this incident.

23. Daniel Bell, letter to the editor, *New York Times*, July 17, 1972. Bell was supporting the decision of the American Academy of Arts and Sciences in 1972 to deny their Emerson Thoreau medal to Pound on the grounds of his fascism and anti-Semitism. DM was a member of the academy and was as outraged by this decision as he was supportive of the Bollingen award. He maintained the same principle, that literature is literature and politics is politics and the twain don't meet. DM to Jerome Lettvin, May 30, 1972.

24. The responses of Clement Greenberg, Irving Howe, and others to the Barrett editorial on the Bollingen award appear in "The Question of the Pound Award," *Partisan Review*, May 1949, pp. 512–22. Nicola Chiaromonte to DM, June 7, 1949. See Alexander Bloom, *Prodigal Sons: The New York Intellectuals and Their World* (New York: Oxford University Press, 1986), pp. 146–47; Irving Howe to author, July 27, 1989. Bloom uses Jewish identity to define the New York intellectuals and presents an extended discussion of how that identity affected their perceptions of modernism and the anti-Semitic strain that could be found in so many of its luminaries.

25. DM, "Homage to Twelve Judges."

26. Robert Hillyer, "Treason's Strange Fruit: The Case of Ezra Pound and the Bollingen Award," *Saturday Review of Literature*, June 11, 1949, pp. 9–10ff. Hillyer, "Poetry's New Priesthood," ibid., June 18, 1949, pp. 7–10ff.

27. Felix Giovanelli to DM, April 17, 1949. Allen Tate to DM, April 17, 1949. DM to Tate, May 3, 1949. Hayden Carruth to DM, May 17, 1949. Ezra Pound to DM, April 26, 1949. DM to Pound, April 29, 1949. Mary McCarthy in an interview Nov. 14, 1987, told me the story of DM bringing the Pound letter around to the *PR* office as an illustration of his "naïveté." I argued that it was almost a form of preening, because DM took much pleasure in such exchanges, not to mention notice from eminent intellectual figures.

28. DM, "The Waldorf Conference," p. 32C. Irving Howe to DM, Feb. 14,

1949, declaring his contempt for the recent upsurge in Jewish nationalism. Howe to DM, May 8, 1949. DM to Nicola Chiaromonte, April 14, 1949. Chiaromonte to DM, June 7, 1949. Chiaromonte was very critical of Jewish nationalism in this letter; DM vowed to write an article on the unfortunate upsurge in Semitism and did collect extensive notes in preparation for such a piece.

29. DM, "The Name of the State Is Israel," unpublished notes for an article on Jewish nationalism, box 114, folder 565.

30. DM to a Mr. Myerfield, Feb. 1, 1971, on being a "non-Jewish Jew." DM, Introduction to *Memoirs of a Revolutionist* (New York: Meridian, 1958, p. 6). S. A. Longstaff, interview with DM, quoted at length in "Ivy League Gentiles and Inner-City Jews: Class and Ethnicity Around *Partisan Review* in the Thirties and Forties," *American Jewish History*, Spring 1991, pp. 328–29 and passim.

31. William Phillips to DM, Jan. 20, 1949. DM, notes on the argument at the Rahvs' apartment in December 1949 (they are misdated 1948); he went and discussed the whole issue with Will and Edna Phillips on the evening of Jan. 29, 1949, box 40, folder 989.

32. Robert Bone to DM, April 23, 1949. Bone became an English professor at Columbia's Teachers College and a specialist in black literature. See DM, "The Uncommon People," *Politics*, Winter 1949, p. 63, for DM's praise of Wiener.

33. DM to Robert Bone, June 15, 1949.

34. DM, "The Germans—Three Years Later," in *Memoirs of a Revolutionist*, pp. 75–76. This originally appeared in the *Student Partisan*, an undergraduate publication at the University of Chicago, Winter 1949.

35. DM, "Past Politics," in *Memoirs of a Revolutionist*, p. 4.

36. DM manuscript, "Goodbye to Utopia," delivered as a talk May 19, 1949, in Lecture files, Macdonald Papers.

37. DM to Vincent Ikeotuonye, March 25, 1948.

38. DM, introduction to "Ancestor #5 Alexander Herzen," *Politics*, Winter 1948, pp. 40–42. "The End of Ideology" practically became a school of social science in the fifties, taking its name from Daniel Bell's collection of essays with that title. Schlesinger's *The Vital Center*, Seymour Martin Lipset's *Political Man*, Niebuhr's *Nature and Destiny of Man*, and several other seminal works of the fifties and sixties were representative of the viewpoint. They were really works by liberals and leftists undergoing deradicalization.

39. DM, "Goodbye to Utopia."

CHAPTER 11: THE BREAKUP OF OUR CAMP

1. DM to Nicola Chiaromonte, July 27, 1949; to Gordon Clough, a *Politics* contributor on pacifism, July 27, 1949; to Gelo and Andrea Delacourt, July

18, 1949. He reported in these letters that he was reading Georges Sorel, Hume, and Whitehead.

2. DM to Joan Colebrook, March 23, 1951, Sept. 11, 1951. These observations are a bit later, but DM was constantly retracing his steps, delving into his past, trying to recapture his feelings and experiences.

3. DM, letter to *Politics* subscribers July 1, 1949.

4. Interviews with Virginia Chamberlain, July 18, 1987; Anna Matson and Philip Hamburger, a *New Yorker* writer, Aug. 16, 1989; Paul Magriel, Feb. 22, 1990.

5. DM to Joan Colebrook, April 15, 1951.

6. Interview with Eileen Simpson, April 1, 1986. DM to Joan Colebrook, Dec. 3, 1950. DM's journal entries, October 1951, in box 170.

7. Interview with Nancy Macdonald, Nov. 5, 1988. She gave me encouragement but often found it difficult to discuss these painful personal memories. I would write my interpretations gleaned from DM's letters, journals, jottings, diaries, and she would read them and tell me when she thought I had it right. Interview with Eunice Clark Rodman Jessup, Wilton, Connecticut, March 19, 1987. Ms. Jessup, while hooked up to an oxygen machine and sipping white wine said that when it came to lovemaking the "performance oriented Wasps" thought only of "getting it up, getting it out, getting it off." Although she denied ever having any sexual interest in Dwight, she expressed real admiration and affection for him. Dwight, at one point while staying with the Jessups, did have a brief interest in her but nothing came of it.

8. DM, "Narrative, Journal 1928–43," box 170.

9. DM to Joan Colebrook, Dec. 3, 1950.

10. DM, journal entry, Oct. 31, 1951.

11. DM, "Notes on Love." DM to Joan Colebrook, Feb. 15, 1950. Colebrook to DM, April 10, 1950. There is a wealth of personal material concerning this love affair in box 170 folders labeled Colebrook.

12. DM, letter to subscribers, Oct. 10, 1949. "Talk of the Town, Notes and Comment," *New Yorker*, Jan. 7, 1950. Geoffrey Hellman wrote this facetious squib on DM's announcement, stating that the Cape Cod clamming, swimming, beach activities, DM's refusal to miss the softball games with his two sons, and the fact that his house was in a neck of the woods "so fair as to reveal desk work by contrast for what it is—a flouting of God's will" had something to do with the decision. Hellman also knew that it was more than the "fair woods" of Wellfleet that had captured DM's attention in the summer of 1949.

13. DM, "The Times—One Man's Poison: What an Aroused Critic Would Do if He Was the Editor," *Reporter*, Feb. 14, 1950, pp. 10–13. DM, "Back to Metternich," *New Republic*, Nov. 14, 1949. DM, "Profiles: White Sales and Aristotle," *New Yorker*, Feb. 2, 9, 1952. DM to Nicola Chiaromonte, Dec. 10, 1948, Jan. 6, 1950. DM to Joan Colebrook, Feb. 16, 1950.

14. DM to Chiaromonte, Jan. 6, 1950. These general comments are based on a mass of material in box 170 of the Macdonald Papers that cover the years 1948–54.

15. DM to Joan Colebrook, Dec. 7, 1949.

16. DM, "Religion and the Intellectuals," *Partisan Review*, May–June, 1950, pp. 476–80.

17. DM to Joan Colebrook, Nov. 15, 1949. DM to Chiaromonte, Dec. 21, 1950. At a party one evening DM met the noted sinologist Karl Wittfogel, whose scholarly enthusiasm for his specialty captivated him and made him envious.

18. DM to William Shawn, Feb. 26, April 4, 1950.

19. DM to Chiaromonte, Dec. 21, 1950. DM to Joan Colebrook, Jan. 23, 1951.

20. DM to Dinsmore Wheeler, March 17, 1950.

21. Nancy Macdonald to Danny Rosenblatt, July 30, 1950. DM to Danny Rosenblatt, Aug. 8, 1950.

22. The Clark Foreman incident was related to me by Gloria Macdonald in several interviews. In a letter to Chiaromonte dated Dec. 21, 1950, DM repudiates Mannheim's charge that he was a 100 percent Trumanite.

23. DM, "A New Study of Gandhi," *New York Times Book Review*, Sept. 24, 1950, p. 37.

24. DM, "Love Is Not Enough," unpublished manuscript, dated "Wellfleet 1950," box 7, folder 151. Ironically, there have been recent charges that Bettelheim abused, beat, and terrorized students in his school for autistic children; see "Bettelheim Became the Very Evil He Loathed," letters column, *New York Times*, Nov. 20, 1990.

25. DM to Joan Colebrook, June 19, 1951. DM to Ralph Manheim, Oct. 13, 1950. DM to Colebrook, Nov. 11, 1950.

26. DM to Joan Colebrook, Nov. 2, 1950. Colebrook to DM, November 1950.

27. DM to Joan Colebrook, Nov. 15, 1950.

28. Interview with Elizabeth Pollet, wife of Delmore Schwartz, Jan. 2, 1991. DM to Gertrude Norman, March 23, 1951.

29. DM to Joan Colebrook, March 23, April 24, 1951.

30. DM to Joan Colebrook, ibid., and April 14, 1951.

31. DM to Joan Colebrook, March 23, April 14, April 15, 1951.

32. DM, "A New Theory of Totalitarianism" (a review of Hannah Arendt, *The Origins of Totalitarianism*), *New Leader*, May 14, 1951. DM, "What Is Totalitarianism," Parts 1 and 2, *New Leader*, July 9, 16, 1951, are articles on the general subject but concentrate on Arendt's ideas. I have used quotations from all three pieces.

33. DM to Chiaromonte, Dec. 21, 1950.

34. Evelyn Baldwin to DM, June 19, 1951; DM, "The Defense of Everybody," pt. 2, *New Yorker*, July 18, 1953, p. 45. Roger Baldwin to DM, June 20, 1951.

35. Samuel Walker, *In Defense of American Liberties: A History of the ACLU* (New York: Oxford University Press, 1990), p. 119.

36. Roger Baldwin to DM, June 23, 1951. DM to Baldwin, July 18, 1951. Baldwin to DM, July 23, 1951. Baldwin to DM, July 26, 1951, has DM's marginal note on being objective. DM to William Shawn, July 31, 1951. It took two years before the profile was finally published in July 1953, when Shawn perhaps felt McCarthyism was abating.

37. Baldwin to DM, July 23, 1951.

38. DM to Nancy Macdonald, June 6, 18, 1951.

39. DM to Joan Colebrook, Aug. 28, 1950. Journal entry, Aug. 5, 1952.

40. DM to Joan Colebrook, Sept. 11, Oct. 12, 1951.

41. Journal entries, December 24 and 26, 1951.

42. On his approval of red-baiting, see DM, "The Wallace Campaign: An Autopsy," *Politics*, Summer 1948, p. 183.

43. Robert Warshow, "The 'Idealism' of Julius and Ethel Rosenberg," *Commentary*, November 1953; a copy of this article was in DM's files heavily marked up. Leslie Fielder, *An End to Innocence: Essays on Culture and Politics* (Boston: Beacon, 1955).

44. DM to T. C. Kirkpatrick, Jan. 1, 1948.

45. DM to Chiaromonte, Nov. 7, 1951.

46. DM reported Arendt's admonition in the letter to Chiaromonte of Nov. 7, 1951. So widely known was DM's affair that Delmore Schwartz recorded in his journal that Philip Rahv had said DM was "suffering with a writer's block. He's had an affair and sex has upset him." The Schwartz journal entry is dated prior to his affair with Gloria; during that spring in New York Dwight had pursued several women and Phillips let it be known that he had also been with one of the women Dwight had seen. Elizabeth Pollet, ed., *Portrait of Delmore: Journals and Notes of Delmore Schwartz 1939–1959* (New York: Farrar, Straus and Giroux, 1986), p. 383. Regarding Dwight's other dalliances see DM to Joan Colebrook, March 23, 1951, and his journal entries during these months, which discuss his terrible sexual frustrations. See esp. dream journal entry for August 9, 1951.

47. Ibid.

48. DM, "Abstract 10 Ad Absurdum" (review of *White Collar: The American Middle Classes* by C. Wright Mills), *Partisan Review*, January–February 1952, p. 114.

49. Interview with Daniel Bell, March 9, 1990.

50. See Richard Gillam's two articles, "Richard Hofstadter, C. Wright Mills, and the 'Critical Ideal,'" *American Scholar*, Winter 1977–1978, pp. 69–85, and more particularly, "White Collar from Start to Finish," *Theory and Society*, 1981, pp. 24–25. Gillam sees DM as a jaded radical, suffering from a guilty conscience because he had lost the faith. See also Irving Howe's review in the *Nation*, Oct. 13, 1951, pp. 309–10, for a very positive

response to Mills's book. Harvey Swados wrote an irritated letter to the editors of the *Partisan Review* regarding DM's review, to which DM gave a
petulant response (*Partisan Review*, May–June 1952, pp. 383–84).

51. DM to C. Wright Mills, January 1952.
52. Mills to DM, Jan. 17, 1952.
53. DM to Mills, Jan. 30, 1952.
54. C. Wright Mills to Norman Thomas, March 18, 1952. DM actually repub-
    lished the review with an explanatory note in a collection of his writings
    nearly twenty years later ("The Mills Method," *Discriminations* [New York:
    Grossman, 1974]). He knew that it had ended a friendship and he realized,
    to some degree, why Mills had taken it "personally." But he still saw the
    failing as being in Mills, who was moving toward a more radical politics
    while DM was abandoning it.
55. DM to Joan Colebrook, Sept. 11, 1951. DM, journal entries, Oct. 31, Nov.
    3, Dec. 24, 1951.
56. DM, journal entries, Nov. 3, Nov. 14, 1951.
57. DM to Gloria Lanier, Nov. 25, 1951; journal entry, Dec. 24, 1951.
58. DM, journal entry, Dec. 24, 1951.
59. DM, journal entry, Feb. 4, 1952.

CHAPTER 12: MAKING A NEW CAREER: FROM POLITICS TO CULTURE

1. For a good description of the factional struggles within the Congress for
   Cultural Freedom, see Peter Coleman, *The Liberal Conspiracy: The Congress for Cultural Freedom and the Struggle for the Mind of Postwar Europe*
   (New York: Free Press, 1989), chapter 9, "The Obnoxious Americans,"
   which deals with the strain between the CCF and the American Committee.
2. Minutes of planning conference, American Committee for Cultural Freedom, March 1, 1952, pp. 1–3. Macdonald Papers, box 5, folder 74.
3. DM, ACCF notes, April 11, 1952. His considerable notes, transcripts, and
   papers on the ACCF are in box 5, folder 74; see also the correspondence of
   major participants, such as Daniel Bell and Irving Kristol. I have also consulted the ACCF papers in the Tamiment Library at New York University.
   Interview with Daniel Bell, March 9, 1990.
4. Notes on meeting of the ACCF at the Columbia Club, April 23, 1952.
   Arthur Schlesinger, Jr., to DM, April 29, 1952, expressed worry about a
   deep sickness among some of the New York intellectuals. Interview with
   Daniel Bell, March 9, 1990. Bell to author, May 10, 1989, said DM "gives
   a skewed picture of the debates in the Executive Committee" of the ACCF.
   Bell felt that DM "oversimplified" the issues and pushed them into an "off-
   centered direction." On Schlesinger's changing position during these years,
   based on the fortunes of the Democratic Party, see Wreszin, "Arthur

Schlesinger Jr.: Scholar-Activist in Cold War America: 1946–1956," *Salmagundi*, Spring–Summer 1984, pp. 255–285.

5. DM, "Weil: The Intellectual as Saint," *New Leader*, May 8, 1952.

6. DM, "Profile: The Foolish Things of the World," *New Yorker*, Oct. 4, 1952, p. 40ff. DM, "Weil," pp. 23–24.

7. DM, "God and Buckley at Yale," *Reporter*, May 27, 1952. DM to Gloria Lanier, July 5, 1952.

8. DM to William Shawn, Sept. 4, 1951. Interview with William Shawn, Oct. 27, 1987, appropriately over a drink at the Algonquin Hotel. Shawn had a special table there and all the waiters knew who he was, despite his quiet, diffident nature.

9. Interview with William Shawn. W. H. "Ping" Ferry to author, June 2, 1987. Ferry was an old friend of DM who worked for the Fund for the Republic.

10. DM, "The Book-of-the-Millennium Club," *New Yorker*, Nov. 29, 1952. For this discussion I have used the unedited version that appeared in DM, *Against the American Grain* (New York: Random House, 1962). This version does have an updated appendix.

11. DM, *Politics*, March 1944, p. 61. Dwight had also mentioned Adler, with tongue in cheek, as one of Macy's president Richard Weil's advisors in his program to teach the Bamberger store retailers how to think.

12. DM, "The Book-of-the-Millennium Club," in *Against the American Grain*, pp. 252–53.

13. Ibid., pp. 257–58.

14. William Shawn to DM, Nov. 10, 1952. Alfred Kazin to DM, Dec. 3, 1952. Interview with Saul Bellow, Feb. 12, 1993.

15. "*Critic*: A Prospectus for a New Magazine," 1953. Macdonald Papers, box 12, folder 299.

16. Interview with Irving Howe, Sept. 25, 1987. It was Howe who recalled Goodman's poignant title, "The Break-up of Our Camp." A relevant analysis of the community and its decline may be found in Howe, "The New York Intellectuals," *Commentary*, October 1968, pp. 29–51. Another polemical and provocative account may be found in Russell Jacoby's more recent study, *The Last Intellectuals: American Culture in the Age of Academe* (New York: Basic Books, 1987).

17. DM, "The Lady Doth Protest," *Reporter*, April 14, 1953, pp. 36–40.

18. DM, journal entry, Dec. 17, 1951; DM to Chiaromonte, Nov. 7, 1951.

19. DM to Joan Colebrook, April 21, 1951, about his anxious feeling on the Cape. Nancy Macdonald, *Homage to the Spanish Exiles* (New York, Insight, 1987).

20. DM to Gloria Lanier, n.d., dated by DM later as summer 1953. "I've seen Joan several times, once only alone. Nothing, in fact so much nothing it was embarrassing and depressing."

21. DM, journal entry, n.d. Interviews with Virginia Chamberlain, Michael

Macdonald. Michael Macdonald to author, Nov. 25, 1988. Interview with Anna Matson, Aug. 16, 1989. DM to Gloria Lanier, August 1952. DM to Nicola Chiaromonte, Nov. 7, 1951, Oct. 30, 1952. Chiaromonte to DM, Feb. 2, 1953. These statements about DM's difficulty at expressing emotional connections are drawn from his own journals and letters and the observations of both his sons, Nancy Macdonald, and such close friends as Virginia Chamberlain and Danny Rosenblatt.

22. Michael Macdonald to author, July 13, 1990. DM to Gloria Lanier, July 1953.

23. Interviews with Danny Rosenblatt, Dec. 13, 1985; Paul Magriel, Feb. 22, 1990. Michael Macdonald to author, Sept. 16, 1989. Interviews with Michael Macdonald.

24. Marjorie Levi Bristol, "Rorschach Analysis—Michael Macdonald," Oct. 13, 1948, Macdonald Papers, box 170. See also DM to Mr. Finch, Exeter administrator, Jan. 13, 1954. Nearly everyone I interviewed who knew DM and Nancy well and attended their parties thought they were careless parents who did not supervise their sons' eating habits, their table manners, their bedtime hours.

25. Interview with Mary Day Lanier, April 6, 1988. Interview with Paul Magriel, Dec. 22, 1990.

26. DM to Gloria Lanier, July 1953.

27. DM to William Shawn, Feb. 17, 1953, emphasis added.

28. DM, "Updating the Bible," *New Yorker*, Nov. 14, 1953. I have used the completely unedited version printed in *Against the American Grain*, pp. 262–88.

29. Steven Marcus, "The Politics of Taste," review of *Against the American Grain, Commentary* 35, April 1963, pp. 351–353. Mr. Sparks to DM, June 14, 1954; DM to Mr. Sparks n.d., 1954.

30. DM, "A Theory of Popular Culture," *Politics*, February 1944, pp. 20–22. On reading Nock, see DM to Gloria Lanier, Aug. 18, 1952.

31. DM, "A Theory of Mass Culture," *Diogenes*, Summer 1953, pp. 1–17.

32. DM, journal entry, undated but the internal evidence confirms that it was written in September 1953. Originally in box 54, it has been put in box 170. In it there are further intense recriminations: DM accused himself of blaming everything on Gloria. He also expressed his hate and contempt for Shawn and the *New Yorker* "and the whole set up in this country and i want to resign from the usa and the hell with it . . . just don't have the stomach for it any more. But why why am i so self-defeating . . . christ i am not even interested in myself anymore, it's all a bloody. . . . im too fatigued to go to old gruenthal. . . . and i resent him too. . . . why can't i communicate with anybody???? I feel cut off boxed in self boxed in the worse sort. . . . i'm drowning because i wont cry for help. And then there was the writing. What's wrong? Why don't I want to write? Why does the very thought of

writing fill me with ennui and, worse, with aversion. It's as if I didn't want to put on paper my ideas, didn't want to tell people things, and from an odd combination of pride (why should I bother it's all old stuff to me, why should I bother to explain it all to a lot of dumbbells who ought to have figured it out themselves. I don't WANT to impress them or to affect them or to have anything to do with them) and of humility (what have I got to say after all, nothing new in years, can I really put it all together so as to make something).

33. Ibid.

34. Ibid. John Lukacs in several conversations with me in 1992 and 1993. DM in this period saw himself as close to a mental collapse, "I am about to have what is known as a nervous breakdown, c. f. my article in *Fortune*. . . . the rope, the end, the end of the rope is almost here . . . what shall I do. . . . I just cannot go on any more, everything seems to be a repetition of past pains. . . . i simply cannot go on any more. i cannot go back to N[ancy] and by now i don't feel i have anything to offer G even if she could overcome her indecision . . . , i want g[loria] but have the feeling too much else goes along with that."

35. DM, "Books: In the Land of Diamat," *New Yorker*, Oct. 17, 1953.

36. Milosz to DM, Dec. 12, 1953; Milosz, *Beginning with My Streets*, trans. Madeline G. Levine (New York: Farrar, Straus and Giroux, 1993).

37. DM, "Action on West Fifty-Third Street" Parts 1 and 2, *New Yorker*, Dec. 12, 19, 1953.

38. DM's poetic manuscript of this review is in box 84, folder 250. DM, "Two Acorns, One Oak," *New Yorker*, Jan. 23, 1954, p. 100. It is true that the conventional version did make the same points, but the wording was tamed—"the grandiose diarrhea of language" did not appear, for example. DM was appalled by the overwritten academic jargon interlarded with undigested quotations. He ridiculed an essay on Herman Hesse's *Steppenwolf*; Hesse, he declared, was a "pretentious minor writer" whose work was inflated to "increasingly bigger dimensions until he shares the fate of the frog in the fable." But what drove Dwight to his most malicious parody was the pseudopsychologizing:

Old Freud with his oralerotic cigar O Promethean fisherman
livertorn by vulturebeak of old fashioned conscience
hauling the monster the unconscious the unconscious up into
    the sunlight   landing for a moment a great fish and
now your grandchildren—liverless vultureless conscienceless
    now they let it slide back deep deeper it sinks into
insentient sea flirts and ambiguous tail far down far

down below the   here

> *greeny surface now*   *now*   *now*
>   *there*
>   *gone*
>
> *and the loss proclaimed again.*

> The joy's gone from writing, pleasure's fled poor Tom's cold, but, gentlemen, signorie, messieurs, I have the honor to announce the New Alexandrianism, the systematization of the eccentric, the synthesization of the spontaneous, and a half dozen new literary formulae for escaping from literary formulae, all produced in the antiseptic laboratories of Experimental Writing untouched by human hands.

Translated into *New Yorker*ese, that passage read: "Here the eccentricity, which is spontaneous and hence charming and stimulating in 'The Little Review Anthology' . . . has become contrived, as depressingly systematic as any other mannerism. The escape from literary convention has hardened into a new convention, an Alexandrian formula for abolishing literary formulas."

39. Interview with Eileen Simpson, April 1, 1986. Simpson to author, Feb. 23, April 19, 1991. Interviews with Nancy Macdonald, Michael Macdonald, Eileen Simpson, Norman Mailer, Danny Rosenblatt. Nancy Macdonald to author, February 1991. DM to Gloria Lanier, June 1954. Simpson's account of being in the Macdonald apartment when DM broke the news that he was leaving Nancy is very vivid and clear. However, Michael doesn't remember learning of the news in that manner and can't recall the evening she describes. There is no doubt that the breakup of the marriage took a long time and there were many emotional evenings during the process.

40. DM to Dr. Max Gruenthal, May 5, 1954.

41. Gloria Lanier to DM, June 20, 1954.

42. DM to Gloria Lanier, July 12, 1954. This probably apocryphal story of Gloria hiding her mother in the apartment was frequently repeated by Dwight's friends. The most vivid account was told to me with much amusement by Mary McCarthy, who always loved a story at Gloria's expense. However, at the time of the divorce Dwight agreed to Gloria's idea of having her mother live with them after her father's recent death. And Gloria's mother often spent considerable time with them out in East Hampton. There is no evidence that Dwight did not get along quite well with his mother-in-law.

CHAPTER 13: INNOCENCE ABROAD

1. DM, "The Lie Detector Era: I Know You Done It. The Machine Says So," *Reporter*, June 8, 1954, pp. 10–18. Part 2, "It's a Lot Easier and It Don't

Leave Marks," *Reporter*, June 22, 1954, pp. 22–30. DM won the Benjamin
Franklin Magazine Award, $500.

2. DM, "McCarthy and His Apologists," *Partisan Review*, July–August 1954,
pp. 418–24. Buckley claimed that DM plagiarized the pickpocket line from
some other source. When questioned, however, he could not identify the
alleged originator of the phrase. Norman Podhoretz also suggested to me
that DM was not above plagiarizing, but he too offered no specifics. I have
never encountered any plagiarism in DM's work. Interviews with William
Buckley, May 18, 1991; Norman Podhoretz, April 29, 1991.

3. DM, undated notes in the Divorce file, box 166, folder 104.

4. Interview with Irving Kristol, Dec. 7, 1989. DM to Nicola Chiaromonte,
May 20, 1955.

5. DM to Stephen Spender, April 25, 1955.

6. Spender to DM, May 27, 1955.

7. DM to Spender, May 27, 1955.

8. Ibid.

9. DM to Spender, dated by archivists as between June 6–20, 1955.

10. DM to Spender, June 30, 1955.

11. DM to Nicola Chiaromonte, Jan. 18, 1955.

12. References cited are from the book edition; DM, *The Ford Foundation: The
Men and the Millions* (New York: Transaction, 1988 [1956]), p. 99. The
work originally appeared in the *New Yorker* in a four-part series, Nov. 26,
Dec. 3, 10, 17, 1955.

13. *The Ford Foundation*, pp. 20, 25–28.

14. Ibid., pp. 103, 124. DM to Nancy Macdonald, Aug. 15, 1955. DM to Nicola
Chiaromonte, April 20, 1955.

15. Nancy Macdonald to DM, July 11, 1955. DM to Nancy, July 15, 1955.
Interview with Michael Macdonald, Nov. 9, 1988.

16. For an extended discussion of these events see Peter Coleman, *The Liberal
Conspiracy: The Congress for Cultural Freedom and the Struggle for the
Mind of Postwar Europe* (New York: Free Press, 1989), pp. 74–78. Coleman
insists that after meeting DM in New York, Josselson did not want to hire
him because he was too provincial (a charge repeatedly made by DM's New
York intellectual opponents) and not interested in or knowledgeable about
things outside the American scene, even though he too wanted to replace
Kristol. Coleman offers little evidence for this view, however. DM's correspondence places the blame on Hook and presents Josselson and Nabokov
as making a strong case for him. DM based his understanding on
Schlesinger's inside observations. Schlesinger's recollections are not terribly clear, but he believes that Josselson and Nabokov were a lot more
agreeable to Dwight's editorship than Hook. Schlesinger to author, April
10, 1991. DM to Gloria Lanier, Sept. 15, 1955. Interview with Daniel Bell,
March 9, 1990.

17. Edward Shils, "Letter from Milan: The End of Ideology," *Encounter*, November 1955, pp. 52–58. DM, "No Miracle in Milan," *Encounter*, December 1955, pp. 68–74. Irving Kristol to DM, Oct. 7, 1955. DM to Kristol, Oct. 13, 1955. Stephen Spender to DM, Oct. 5, 1955. DM To Spender, Oct. 18, 1955.

18. Interview with Gloria Macdonald, August 1990. Nancy Macdonald to DM, June 27, 1952. Nancy Macdonald to author, Jan. 16, 1991.

19. DM to Gloria Macdonald, Feb. 6, 1956.

20. DM to Gloria Macdonald, Feb. 6, 9, 1956. Various interviews with Nancy Macdonald, who never trusted Melvin Lasky.

21. DM, notes for a radio talk in London, March 1956.

22. DM, "Scrambled Eggs on the Right," *Commentary*, April 1956, pp. 369, 373.

23. DM to Dexter Perkins, president of the Salzburg Seminar, April 22, 1956, contains a sustained criticism of the seminar; DM found its administrators philistines and its students more prepared for a vocational course in mass journalism than in criticism. The seminar was from February 12 to March 10. See DM's syllabi in Northwestern file, box 184, folder 282. DM to Gloria Macdonald, Feb. 12, 16, 1956. Dr. H. M. Lehfeldt, Gloria's physician, to DM, May 5, 1956.

24. *Time*, June 6, 1956, pp. 65–67ff. DM to Julia Gowing, Nov. 23, 1959. Dwight felt that Barzun and Trilling put on airs. When asked by a British friend what he thought of them during the period when both were being anointed as yea-saying intellectuals, he wrote:

> My opinion of Barzun [is] not very high. I agree with the thesis of *House of Intellect* and most of what he says, but he's so damned civilized, urbane and genteel. He and Trilling are a pair. I mean I'm all for civilization (as against beatniks and angries, nice coinage that last) and that's why I like England, but when an American tries to be urbane-cultivated it's different from an Englishman; something hollow, inauthentic; the clothes don't fit; he tries too hard, denies himself spontaneity, peculiarity—well you catch my drift. We Americans have to be intellectual first, civilized afterwards—like Poe or Edmund Wilson or, at a lower level, Mencken & Nathan. Lawrence wrote in his *Studies in Classic Am Lit* in a real American style, what a relief compared to the watery urbanities of Barzun or Trilling.

25. Phone conversation with Albert Glotzer, spring 1990. For the details of the case, see the material in the Joseph Rauh Papers in the Library of Congress. I have also talked to Jack Widick, a friend of DM's from his Trotskyist days, about this incident several times. See also Alan Wald, *The New York Intellectuals: The Rise and Decline of the Anti-Stalinist Left from the*

1930s to the 1980s (Chapel Hill: University of North Carolina Press, 1987), pp. 276–77.

26. On housing and living in London, Mary Day Lanier Wollheim (Gloria Macdonald's daughter) to author, Oct. 21 1991. DM to Mary McCarthy, May 16, probably 1959. DM to an old friend, Alice El-Tawil, April 16, 1956.

27. DM, "A New Yorker in London," script of a talk that he gave on the radio in London. Some of the observations later appeared in "Reflections," *Encounter*, December 1956.

28. DM to Mary McCarthy, May 16, 1956.

29. Interview with Margot Walmsley, April 7, 1988. Interview with Anastasia (Baba) Anrep, April 4, 1988.

30. Interview with Irving Kristol, Dec. 7, 1989. Interview with Daniel Bell, March 9, 1990. For the Bell skit, see Peter Coleman, *The Liberal Conspiracy*, p. 71, footnote. In an interview, Alistair Reid, a Scotsman who wrote for the *New Yorker*, presented a completely different picture of DM as always ready for an experience of any kind.

31. DM to Gloria Macdonald, n.d., spring 1957; Gloria was traveling in Spain while Dwight remained in London. Malcolm Muggeridge, "I Like Dwight," in *The Most of Malcolm Muggeridge* (New York: Simon and Schuster, 1966), p. 66. All of these critics who were in London at the time told the same story. On a chance encounter with Gertrude Himmelfarb in the hallways of the Woodrow Wilson Center in Washington during the spring of 1991, for instance, I mentioned Dwight and she immediately launched into his provincialism and mispronunciation of Grosvenor Square.

32. DM, "Reader's Indigestion" (review of Colin Wilson's *The Outsider*), *New Yorker*, Oct. 13, 1956, reprinted in DM, *Against the American Grain* (New York: Random House, 1962), as "Inside *The Outsider*."

33. DM, "Reflections," *Encounter*, December 1956, pp. 55–58. Kenneth Alsop, *The Angry Decade* (New York: British Book Center, 1958).

34. DM, "Reflections," pp. 55–56. DM, London journal, Macdonald Papers.

35. DM, "Reader's Indigestion," pp. 73ff. See also Kenneth Alsop, *The Angry Decade*. DM's apology for *The Root Is Man* may be found in a letter to Nicola Chiaromonte, Jan. 15, 1955. Colin Wilson, in "To the Editor of *The Twentieth Century*," December 1958, pp. 587–88, dismissed DM's criticism as the work of a superficial political commentator and journalist, "an eminently nice man . . . without two ideas to rub together." Leslie Fiedler, in "The Un-Angry Young Men," *Encounter*, January 1958, p. 9, in addition to arguing that *Encounter* and its readers and writers were people of the thirties and targets of an angry new generation, described the thirties generation as a blend of homosexual sensibility, upper-class aloofness, liberal politics, and avant-garde literary devices. When one of the newcomers is "boorish rather than well-behaved, rudely angry rather than ironically amused, when he is philistine rather than arty—even when he merely

writes badly, he could feel he was performing a service for literature, liberating it from the tyranny of a taste based on a world of wealth and leisure which has become quite unreal."

36. Interview with Irving Kristol, Dec. 7, 1989. Discussion with Jack Thompson, a literary critic conversant with *Encounter* personnel.

37. DM, "Ten Days in Cairo," *Encounter*, January 1957, p. 5.

38. DM, letter to the editor, *Times* of London, Nov. 29, 1956, not published; carbon in London *Times* file, box 52, folder 1272.

39. DM to Gloria Macdonald, Nov. 13, 15, 16, 1956. "Ten Days in Cairo," pp. 3–13.

40. "Ten Days in Cairo," p. 4.

CHAPTER 14: "OUR BEST JOURNALIST"

1. Nancy Macdonald to DM, Feb. 23, 1957. DM to Gloria Macdonald, March 1957.

2. DM, "Politics Past," *Encounter*, March and April 1957; I have used the identical version printed in *Memoirs of a Revolutionist* (New York: Farrar, Straus and Cudahy, 1957), pp. 3–31. Conservative references on p. 6.

3. T. S. Eliot to DM, Oct. 30, 1956. DM responded to this letter with a long single-spaced page of notations pointing out where they did agree and where they didn't.

4. DM, *Memoirs*, pp. 22, 25.

5. Ibid., pp. 25, 27.

6. Diana Trilling, "Here and Now" column, "Dwight Macdonald's Reminiscences of Radical Politics Before the War," *New Leader*, April 16, 1957, pp. 16–17.

7. Ibid., p. 17. DM, letter to the editor, and Trilling's response, *New Leader*, July 15, 1957.

8. DM, "Books: A Corrupt Brightness," *Encounter*, June 1957, pp. 75–82.

9. Richard Hoggart, *The Uses of Literacy: Changing Patterns in English Mass Culture* (Fair Lawn, N.J.: Essential Books, 1957), pp. 253–55.

10. Edward Shils, "Daydreams and Nightmares: Reflections on the Criticism of Mass Culture," *Sewanee Review*, Fall 1957, pp. 587–608; these quotations pp. 589–92.

11. Kazin cited in Neil Jumonville, *Critical Crossings: The New York Intellectuals in Postwar America* (Berkeley: University of California Press, 1991) p. 156. Jumonville has a very thorough and insightful chapter on the mass culture debate to which I am indebted, pp. 151–86.

12. DM, "Books: A Corrupt Brightness," *Encounter*, June 1957, p. 78.

13. DM, journal entry, Feb. 15, 1952. Interview with Mary McCarthy, Nov. 14, 1987. The McCarthy–Chiaromonte correspondence is in the McCarthy Papers in the Vassar library.

14. Carol Gelderman, *Mary McCarthy: A Life* (New York: St. Martin's, 1988), pp. 230–32. Carol Brightman, *Writing Dangerously: Mary McCarthy and Her World* (New York: Clarkson Potter, 1992), pp. 445, 450, 463–65. For a vivid description of Bocca di Magra and its imminent demise, see Mary McCarthy, *The Hounds of Summer* (New York: Avon, 1981), the title story, pp. 193–237.

15. Mary McCarthy, "The Hounds of Summer," pp. 196–97. This account, which draws an amusing caricature of DM, recalls a condescending observation made by DM in the *Phillips Exeter Monthly* about the locals' built-in resentment toward the summer people on whom they were dependent economically: "To me nothing is more pathetic than the open scorn and secret envy of the natives of any summer resort for the comparatively well-to-do and cultured throng of 'summer people.'"

16. DM to Baba Anrep, Nov. 8, 1957. Interviews with Baba Anrep, April 4, 1988, April 1992.

17. DM "Books: Death of a Poet," *New Yorker*, Nov. 16, 1957.

18. DM to Irving Kristol, Oct. 4, 1957. Kristol to DM, Oct. 10, 1957.

19. DM, "America! America!" in DM, *Discriminations: Essays and Afterthoughts, 1938–1974* (New York: Grossman, 1974). I have used this edition of the essay, but it appeared first in the British publication the *Twentieth Century*, October 1958, and more or less simultaneously in *Dissent*, Autumn 1958.

20. *Time*, Sept. 2, 1957. Ronald Lora, "*By Love Possessed*: The Cozzens–Macdonald Affair," in Louis Filler, ed., *A Question of Quality: Popularity and Value in Modern Creative Writing* (Bowling Green University Popular Press, 1976), pp. 57ff. See also Matthew J. Bruccoli, *James Gould Cozzens: A Life Apart* (New York: Harcourt Brace Jovanovich, 1983), pp. 193–228.

21. John Lukacs has given me a copy of his manuscript review of the Cozzens volume with a note explaining his negotiations with *Commentary*. He was apparently unaware that DM had sent it to *Encounter*. One interesting difference in their appraisals is that Lukacs found the notorious lovemaking scenes the best-written passages in the novel, indicating that Cozzens was "by sex obsessed"; DM thought they were the very worst passages.

22. DM, "By Cozzens Possessed," *Commentary*, January 1958, reprinted in DM, *Against the American Grain* (New York: Random House, 1962).

23. James Gould Cozzens to DM, March 5, 1958.

24. DM to John Hayward, Feb. 7, 1958. Isaiah Berlin to DM, Feb. 6, 1958. Norman Mailer to DM, Feb. 13, 1958. Bruccoli, *James Gould Cozzens*, pp. 218–19. John Hutchins, *New York Herald Tribune*, March 13, 1958.

25. DM, *Memoirs of a Revolutionist* (New York: Farrar, Straus and Cudahy, 1957). The collected pieces were primarily from *Politics*. The Introduction consisted of the two biographical essays he had published in *Encounter*. In

England the book was published by Victor Gollancz under the title *The Responsibility of Peoples and Other Essays in Political Criticism*. At Gloria's urging, DM changed the title for the later paperback edition to *Politics Past*, agreeing with those who said the original title was cheap and that it burlesqued the book's contents.

26. DM to his editor Richard Straus, May 15, 1957. DM to Jeffrey Hart of the Columbia University English department, June 4, 1950, refers to the original title as "wretched." DM to Dinsmore Wheeler, Nov. 18, 1957. On the book's poor sales (1,200 copies), see Richard Straus to DM, Jan. 6, 1958. DM to Straus, Jan. 16, 1958, urges more promotion on the part of the publisher. The paperback edition did better: the first printing sold 5,000 copies; see Arthur Cohen of Meriden Paperbacks to DM, Nov. 14, 1958, and Oct. 12, 1960.

27. Daniel Bell, "Yale Man as Revolutionist," *New Leader*, Dec. 9, 1957, pp. 22–24. Richard Wollheim, "Innocence and Politics," *Encounter*, January 1958. DM, "Patting Hand Gets Bitten," *Encounter*, May 1958, pp. 71–72. Irving Kristol to DM, April 17, 1958, thanking him for his response to Wollheim: "It's an excellent demonstration on how to bite the hand that pats you." For DM's criticism of Bell's review, see DM to Bell, January 1958; Bell to DM, November 1957, January 1958.

28. Hans Meyerhoff, "Offbeat Political Writing," *Commentary*, November 1957, pp. 464–65. DM to Meyerhoff, Jan. 10, 1958 (letter misdated 1959 by Yale archivists).

29. Ian Watt, "Macdonald's Guide to Impractical Politics," *New Republic*, Feb. 24, 1958, pp. 19–21. Watt to DM, March 21, 1958, for unpublished paragraph.

30. Orville Williams (pseudonym for John Lukacs), "Dwight Macdonald: Another Orwell," *America*, May 17, 1958. DM, "Correspondence," *America*, June 21, 1958, p. 341.

31. Harold Rosenberg, "Pop Culture and Kitsch Criticism," *Dissent*, Winter 1958, pp. 14–18.

32. DM, "Letters: The Question of Kitsch," *Dissent*, Autumn 1958, pp. 398–401, contains his response to both Rosenberg's and Paul Goodman's critiques. DM, "Why I Am No Longer a Socialist," *Liberation*, May 1958, pp. 4–7.

33. Paul Goodman, "Our Best Journalist," *Dissent*, Winter 1958, pp. 82–86.

34. DM, "Letters: The Question of Kitsch," and Rosenberg's reply, *Dissent*, Autumn 1958, pp. 398–401.

35. Ibid.

36. Irving Kristol to DM, Feb. 18, 1958. DM to Kristol, March 4, 1958. Stephen Spender to DM, April 11, 1958. DM to "Stephenirvingnicholasmike," April 16, 1958.

37. Kristol to DM, April 17, 1958. Michael Josselson to DM, April 28, 1958.

38. Kristol to DM, April 17, 1958. DM to Kristol, April 21, 1958. Josselson to

DM, April 28, 1958. Kristol to DM, April 28, 1958. DM to Kristol, May 21, 1958.

39. Kristol to DM, June 12, 1958. Sidney Hook in an exchange with Robert Westbrook, who had reviewed his autobiography for the *Nation*, insisted that the Paris office exerted no editorial control over the CCF's publications. However, when Westbrook sent him a copy of Kristol's letter to DM, Hook said he had never known about the incident but Kristol's letter was evidence that in this matter he had yielded to someone's pressure. Hook said he would have expected him to resign.

40. DM to Kristol, June 24, 1958; a "not sent" notation is at the head of the copy of the letter in the Macdonald Papers. See editorial introduction to "America! America!" in *Twentieth Century*, October 1958; DM's letter to the editor in the December 1958 issue. It is also reprinted in DM's *Discriminations*, p. 387. It is the *Dissent* version of the essay that is reprinted with a preface in *Discriminations*, p. 57.

41. Norman Birnbaum, "An Open Letter to the Congress for Cultural Freedom," November 1958, mimeographed copy in Macdonald Papers. The letter was reprinted in *Universities & Left Review*, December 1958.

42. *Universities & Left Review*, December 1958, p. 5. Irving Kristol to DM, Dec. 18, 1958. DM to Kristol, Dec. 19, 1958.

CHAPTER 15: REACTIONARY ESSAYS

1. Interview with Norman Mailer, March 20, 1989.
2. DM, "No Art and No Box Office: Dore Schary's 'Lonelyhearts': The Apotheosis of the Adult Soap Opera," *Esquire*, March 1959, pp. 63–66, reprinted in DM, *Discriminations*, pp. 252–61. This article also found its way into *Encounter*, July 1959. This discussion cites the *Esquire* article because the other versions are slightly edited; quotations in this paragraph from pp. 63–64.
3. Ibid., p. 64.
4. Ibid., p. 66.
5. Ibid.
6. Arnold Gingrich, "A Double Standard for the New Hollywood," *Esquire*, March 1959, p. 6. DM, "Movieland," *Esquire*, June 1959, p. 13.
7. DM to Russell A. Fraser, Dec. 7, 1977, for this account of his "on the whole unfortunate" experience. See Russell A. Fraser, *A Mingled Yarn: The Life of R. P. Blackmur* (New York: Harcourt Brace Jovanovich, 1981), pp. 258–63, for a good account of the establishment and atmosphere of the seminars. See also James Atlas, *Delmore Schwartz: The Life of an American Poet* (New York: Farrar Straus and Giroux, 1977), p. 287; John Haffenden, *The Life of John Berryman* (Boston: Routledge and Kegan Paul, 1982), p. 218. Interview with John Lukacs, July 8, 1991. Interview with Eileen Simpson, April 1, 1986.

8. DM, "Profile: A Caste, a Culture, a Market," *New Yorker*, Nov. 22, 1958, p. 74; Nov. 29, 1958, p. 80.

9. Interviews with Nicholas and Elspeth Macdonald, March 22, July 19, 1989. Michael Macdonald to author, Nov. 15, 1988.

10. These comments are based on several interviews with both sons and with other close friends.

11. DM to Nicola Chiaromonte, Oct. 19, 1959.

12. DM, ed., *Parodies: An Anthology from Chaucer to Beerbohm—And After* (New York: Random House, 1960). References here are from the 1985 Da Capo paperback reprint. Initial quotations are from DM's preface, pp. xi–xvi, and his appendix, "Some Notes on Parody," pp. 557–68. Jason Epstein to DM, May 4, 1960.

13. Walter Allen, "New Wine That Tastes Like Old," *New York Times Book Review*, Dec. 26, 1960. Charles Poor, "Books of the Times," *New York Times*, Dec. 13, 1960. D. J. Enright, *New Statesman*, Nov. 24, 1961. Joseph Carroll, "The Art of Aping," *Nation*, April 29, 1961.

14. DM to reader Wells Kerr about his reactionary essays, n.d., 1963. T. S. Eliot, *Notes Towards the Definition of Culture* (London: Faber and Faber, 1948), p. 16, quoted in DM, "Masscult and Midcult," *Against the American Grain* (New York: Random House, 1962), p. 56. This is the volume in which Eliot acknowledged his indebtedness to DM for his piece on popular culture in the first issue of *Politics*, February 1944.

15. Alvin Toffler, "Don't Knock U.S. Culture," *Life*, Oct. 30, 1964. DM, remarks at "Modern Art and Mass Culture: A Symposium, Brandeis University," June 2, 1962. These remarks were reprinted in the *Wilson Library Bulletin*, June 2, 1962. Interview with John Lukacs, July 8, 1991. A copy of Lukacs's unpublished portrait of DM is in my possession.

16. DM, "Masscult and Midcult," Parts 1 and 2, *Partisan Review*, Spring, Summer 1960, pp. 203–33, 589–631. References to these articles are to the combined version in *Against the American Grain*, pp. 3–75; see p. 37 footnote for list of midcult philistines.

17. There was more than a little truth in Richard Gilman's mean-spirited review of the anthology as written by a man who felt alarm about the situation below and was forever beating back the encroaching enemy. These disparaging references to contemporary art were quoted in Gilman's review of *Against the American Grain* in the *New Republic*, January 5, 1963. See also DM's own comments in his response to Harold Rosenberg's critique of his views on art in *Dissent*, Autumn 1958, p. 398 and DM's "Ernest Hemingway," in *Against the American Grain*, p. 174 in reference to Jackson Pollock.

18. DM, "Mark Twain," *Against the American Grain*, pp. 98ff. DM, "Ernest Hemingway," *Against the American Grain*, pp. 167ff.

19. DM, "Hemingway," p. 174. A friend told DM that Hemingway's sparse

prose was similar to the jazz of Thelonious Monk in that it was what was left out that was important—one listens to the spaces between the notes. DM's parody of Hemingway did indeed have a monkish quality: "He was a big man with a bushy beard and everybody knew him . . . He enjoyed being recognized by the tourists and he liked the bartenders but he never liked the critics very much, he thought they had his number. Some of them did."

20. Maria MacDonald Jolas to DM, Dec. 15, 1959, responding to the essay when it appeared first in the *New Yorker*, Dec. 12, 1959, complained as a friend of Joyce's of DM's failure to discuss the man's artistic achievement in substance while spending far too much time on his personal eccentricities. DM included the Plimpton and Gill criticisms in the collection, pp. 179–84. Janet Flanner (Genêt was her pseudonym for the "Letter from Paris" in the *New Yorker*) in Natalie D. Murray, ed., *Darlinghissima: Letters to a Friend* (New York: Random House, 1985), p. 306. Nelson Algren described DM as a "domesticated peacock" in his *Notes from a Sea Diary* (New York: Putnam, 1965), pp. 87–92. The collection was highly praised in *Time* and *Newsweek* and other mass-circulation magazines, which does suggest that he was not as much "against the American grain" as he liked to think.

21. DM, "Sweet Are the Uses of Usage," *New Yorker*, May 17, 1958, p. 136. DM, "The String Untuned," ibid., March 10, 1962, p. 38 reprinted in *Against the American Grain*. During this period he wrote a series of articles dealing with the decline of language and reviews of books on English usage. See "The Decline and Fall of Good English," *Life International*, April 1962; "The Crisis of American Culture," *Wilson Library Bulletin*, May 1963.

22. "The String Untuned."

23. DM, "Three Questions for Structural Linguists," in James Sledd and Wilma Ebbitt, eds., *Dictionaries and That Dictionary* (Fair Lawn, N.J.: Essential Books, 1962), pp. 260ff. Robert Westbrook, "The Responsibility of Peoples: Dwight Macdonald and the Holocaust," in Sanford Pinsker and Jack Fischel, eds., *America and the Holocaust* (Greenwood, Fla.: Penkevill, 1984), p. 55. For an interestingly similar commentary on the cultural permissiveness of the Kinsey report, see Lionel Trilling, "The Kinsey Report," in his *The Liberal Imagination* (New York: Viking, 1950). Trilling's essay first appeared in the *Partisan Review* in 1948.

24. DM, "Three Questions," p. 262.

25. Wilson Follett, "Sabotage in Springfield," *Atlantic*, January 1962, pp. 73–77, reprinted in *Dictionaries and That Dictionary*, pp. 112–13.

26. Jacques Barzun, "The Scholar Cornered," *American Scholar*, Winter 1963, pp. 176–81.

27. James Sledd, "Reply to Mr. Macdonald," in *Dictionaries and That Dictionary*, pp. 267–74. Sledd to DM, May 1, May 18, 1962. DM to Sledd, May 29, 1962.

28. DM, "Three Questions," pp. 256–64.

29. Unsigned review of *Parodies, Time*, Jan. 13, 1961, pp. 80–81.

30. DM's class poem, *Phillips Exeter Monthly*, June 1924, p. 193. For an exhaustive study of his youth and young adulthood, see Robert Cummings's 1988 Stanford Ph.D. dissertation, "The Education of Dwight Macdonald, 1906–1928: A Biographical Study." This summary of DM's six years as a film critic for *Esquire* as well as his other work on classic directors and Russian cinematography does not do the subject justice. Demands of focus and my own lack of expertise have caused me to make this decision to offer only a cursory treatment of his film criticism. There is considerable material, both published and unpublished, covering DM's long devotion to film. His first and last writings were concerned with film. See: DM, "Vote for Keaton," *New York Review of Books*, Oct. 9, 1980, p. 38. DM, film lecture course, University of Texas, Feb. 2, 1966, pp. 4–5 of transcript in box 144, folders 65–66. Tapes of his lectures are in the Historical Sound Recordings Collection in Yale's Sterling Library. DM, *On Movies* (New York: Da Capo, 1981), pp. 450–51.

31. See DM, *On Movies*, pp. 15–45 and passim.

32. Ibid., pp. 46, 154–55, 178, 318, 424, 443. *Esquire*, May 1960, pp. 149, 284; June 1960, pp. 294, 302; December 1960. *On Movies*, pp. 153–54.

33. DM, *Experiment in Terror* review, *Esquire*, November 1962, reprinted in *On Movies*, p. 58.

34. DM, "Films," *Esquire*, September 1960.

35. *On Movies*, pp. 426–28.

36. Letters to the editor, *Esquire*, August 1960. Undated unaddressed form letter to irritated readers by Gingrich in *Ben Hur* file, box 177, folder 3. DM's response to this criticism appears in *On Movies*, pp. 428–29.

37. *On Movies*, pp. 430–31.

38. Ibid., pp. 429–30. Martin Panzer to DM, Feb. 26, 1962. DM to Panzer, March 9, 1962.

39. Michael Sragow, 1982 review of DM's *On Movies*, unidentified clipping in file of reviews of *On Movies*.

40. DM, "The English Language: The Decline and Fall of Good English," *Life International*, April 9, 1962.

41. DM to Hannah Arendt, Oct. 17, 1962, Hannah Arendt Papers, Library of Congress, describing how elegant the St. Johns Wood house was.

42. DM to Gloria Macdonald, Aug. 31, 1960, on his problems with Mike and enjoyment of Nick.

43. On the Argentine film festival, interview with Felix della Paolera, June 20, 1989. DM to Grillo Greejaw (Felix della Paolera), March 30, 1960. Della Paolera showed DM around at the festival, introduced him to Argentine intellectuals, and remained a good friend, stopping to see him whenever he was in New York. On the Mailer trial, see "Massachusetts vs. Mailer,"

in DM, *Discriminations*, pp. 194–206, especially footnote, p. 197. DM
to Myron Williams, Exeter teacher, April 1, 1961, on London party. DM to
Mary McCarthy, April 6, 1961, describes the Aldermaster March. DM to
William Shawn, Oct. 5, 1960, on Russell profile. DM, "London Letter,"
Partisan Review, March–April 1961, p. 248, on peace march.

44. DM, *On Movies*, pp. 468–69; *Esquire*, September 1961.
45. DM to Mary McCarthy, July 24, 1961. DM to Dinsmore Wheeler, May 25, 1962.

CHAPTER 16: WARMING UP

1. Hannah Arendt, "He's All Dwight," *New York Review of Books*, Aug. 1, 1961, p. 31. This essay may also be found as the Introduction to the Greenwood reprint edition of *Politics*.
2. C. Wright Mills, "Letter to the New Left," *New Left Review*, September–October 1950, reprinted in Mills, *Power, Politics, and People: The Collected Essays of C. Wright Mills* (New York: Oxford University Press, 1963), p. 259. See also James Miller, *Democracy Is in the Streets: From Port Huron to the Siege of Chicago* (New York: Simon and Schuster, 1987), pp. 79–91, for a vivid portrait of Mills. Jack Newfield, *A Prophetic Minority* (New York: New American Library, 1966) makes much of the eclectic, nonsectarian beginnings of the movement.
3. "On the Exhaustion of Political Ideas" was the subtitle of Daniel Bell's book *The End of Ideology* (Cambridge, Mass.: Harvard University Press, 1987).
4. Mrs. Sobell to DM, May 31, 1960; accompanying her letter was a summary of the proceedings of the public meeting in Sobell's behalf at which DM spoke, with a transcript of the comments. See also DM et al., "Free Gold and Sobell" (letter to the editor), *New York Times*, Feb. 16, 1960. DM organized this letter and secured the signatures of strong anti-Communists such as Irving Kristol and Nathan Glazer, who, however, toned down his initial statement, which had censured Judge Kaufman harshly. This material is in the Morton Sobell research file, Macdonald Papers, box 129, folder 704. DM wrote to Arthur Schlesinger on behalf of Sobell and Scales, Oct. 17, 1961, and to President Kennedy on behalf of Scales, June 15, 1962. Gladys Scales wrote to thank DM for his efforts, n.d., 1962. Junius Scales research file, Macdonald Papers, box 129, folder 702.
5. Aryeh Neier, executive secretary of Students for a Democratic Society (SDS), to DM, May 24, 1960. DM to Neier, May 31, 1960. This correspondence is in the SDS papers, Tamiment Library, New York University. The *SDS Voice*, July 1960, made a brief note that DM spoke at the convention only by mentioning the title of his talk, "The Relevance of Anarchism." Bob Ross, then a freshman from the University of Michigan, remembers

DM referring to himself as a "conservative anarchist." Ross recalls being impressed by DM's elegant erudition and, in particular, by his "distant amusement over his factional past." However, the seventeen-year-old Ross was not persuaded by the relevance of DM's message. DM appeared to him "a war weary intellectual with a message of skepticism," while he and his "youthful comrades were impatient activists filled with piss and vinegar." Telephone converation with Bob Ross, April 20, 1992.

6. DM, "What Is Anarchism," *Oxonian*, March 3, 1954, pp. 30–33.

7. DM, "The Candidates and I," *Commentary*, April 1960, pp. 287–94.

8. Ibid. In a rather bizarre article in *Dissent*, Hannah Arendt also criticized state-enforced integration in Little Rock and offered a complex argument that maintained the state should not interfere with forms of discrimination, followed by a defense of "states' rights" as one of the most authentic sources of power. ("Reflections on Little Rock," *Dissent*, Winter 1959, pp. 45–56; critiques of the article appeared in the same issue and Arendt replied in the Spring 1959 issue, pp. 179–81). For an interesting analysis of Arendt's position, see Elizabeth Young-Bruehl, *Hannah Arendt: For Love of the World* (New Haven, Conn.: Yale University Press, 1982), pp. 308–18.

9. Theodore Roszak, "The Counter Culture: Part I, Youth and the Great Refusal," *Nation*, Feb. 25, 1968, p. 405.

10. Staughton Lynd, "Marxism-Leninism and the Language of *Politics* Magazine: The First New Left . . . and the Third," in George Abbott White, ed., *Simone Weil: Interpretations of a Life* (Amherst, Mass.: University of Massachusetts Press, 1981), pp. 127–28. Telephone interview with Todd Gitlin, July 15, 1992.

11. DM to Robert Taber, April 8, 1960. Taber was a CBS newsman and an eyewitness to the Cuban revolution who had written DM soliciting his name for a newspaper ad in support of Castro; see Robert Taber, "Castro's Cuba," *Nation*, Jan. 23, 1960, pp. 63–71. In this letter DM even repudiated Castro's charge that all American intervention going back to the Spanish-American War and the Philippines was imperialism.

12. On Bourne as one of his "culture heroes," see DM, "Portrait of a Man Reading: Dwight Macdonald," an unpublished manuscript initially prepared for the *Herald Tribune Book World*, May 31, 1958, in box 38, folder 177. Randolph Bourne, "Twilight of Idols," in Carl Resek, ed., *War and the Intellectuals*, pp. 58–59. Noam Chomsky, *American Power and the New Mandarins: Historical and Political Essays* (New York: Vintage Books, 1969), pp. 6, 323. Chomsky also says he reread DM's *Politics* essays and found that "they had lost none of their power or persuasiveness."

13. DM to Barbara Deming, May 1, 1961. Barbara Deming, "The Peacemakers," *Nation*, Dec. 16, 1960, pp. 471–74.

14. DM to Barbara Deming, May 1961.

15. Conrad Lynn, summary of the meeting held at Donald Harrington's Com-

munity Church May 24, 1960, transcript in the Sobell file, Macdonald
Papers. DM to William Shawn, March 19, June 12, 1962. DM to John
Lukacs, Feb. 20, 1962, recommending Harrington as a potential professor
at Chestnut Hill College in Philadelphia, where Lukacs taught.

16. DM, "Our Invisible Poor," *New Yorker*, Jan. 19, 1963, pp. 82–132. Michael
Harrington, *The Long-Distance Runner: An Autobiography* (New York: Holt,
1988), p. 224.

17. DM, "Our Invisible Poor," p. 84.

18. Ibid., pp. 128–32. The term "the culture of poverty," introduced by the
anthropologist Oscar Lewis in several books, became controversial. Critics
charged that it was a way of blaming the victim, for in effect it claimed that
the poor over time developed a completely different value system and attitude toward life. A good survey of the controversy and the leading participants may be found in Michael B. Katz, *The Undeserving Poor: From the
War on Poverty to the War on Welfare* (New York: Pantheon, 1989), chapters
1 and 2.

19. DM, "Our Invisible Poor," p. 132. DM to Ned Chase, Jan. 30, 1963. DM
insisted that government spending was the only way to deal with the problem, which does appear inconsistent with his growing anti-state position.

20. DM to Hans Kohn, Jan. 31, 1963. See Arthur Schlesinger, Jr., one of the
most articulate of our historical myth-makers, *A Thousand Days: John F.
Kennedy in the White House* (Boston: Houghton Mifflin, 1965), p. 922.
Nicholas Lemann, *The Promised Land: The Great Black Migration and How
It Changed America* (New York: Knopf, 1991), pp. 130–31, 376, bases his
view that Kennedy read Macdonald and not Harrington on an undated
memo by Walter Heller and Robert P. Lampman in the Kennedy Library.
Garry Wills, "A Tale of Three Cities," *New York Review of Books*, March 2,
1991, p. 11, credits DM's review. TRB, *New Republic*, Feb. 2, April 4,
1963. Schlesinger to author, March 26, 1992, said that Kennedy "read and
reread Dwight's review" and "read in" Harrington's book only after reading
the review. See DM, "Afterword" to the edited version reprinted in *Discriminations*, pp. 97–98.

21. DM to Mary McCarthy, June 27, 1962. DM to Ping Ferry, June 27, 1962.
On FDR's toleration of and collaboration with J. Edgar Hoover, see William
W. Keller, *The Liberals and J. Edgar Hoover*, (Princeton, N.J.: Princeton
University Press, 1989).

22. DM, letter to the editor, *New Republic*, Oct. 24, 1962, in response to an editorial comment defending American policy toward Franco. DM to John
Lukacs, Nov. 3, 1962.

23. DM, "Mr. Schlesinger's *Realpolitik*," *New York Review of Books*, Special
Issue no. 1, (February 1963), republished in DM, *Discriminations: Essays
and Afterthoughts* (New York: Grossman, 1974), pp. 301–306.

24. Edmund Morgan, *The Puritan Dilemma: The Story of John Winthrop*

(Boston: Little, Brown, 1958), pp. 8–9, 31–32. I can think of no text that better illustrates how deeply history is embedded in the times that it is written, and certainly Morgan is one of our most distinguished and intelligent historians. Arthur Schlesinger, Jr., *The Age of Jackson* (Boston: Little, Brown, 1945), pp. 387–88; *A Thousand Days*, pp. 679–80.

25. Arthur Schlesinger, Jr., "One Vote for Anarchy," *Show*, November 1961. DM to Cynthia Grenier, Nov. 29, 1961.

26. DM to Nicola Chiaromonte, Nov. 9, 1962. DM to Dinsmore Wheeler, May 25, 1962.

27. DM to Chiaromonte, July 5, 1963.

28. DM to Chiaromonte, Oct. 9, 1963. DM to Mary McCarthy, Sept. 24, 1963.

29. Hannah Arendt, "A Reporter at Large: Eichmann in Jerusalem," *New Yorker*, Feb. 16, 23, March 2, 9, 16, 1963; the articles were published as a book, *Eichmann in Jerusalem: A Report on the Banality of Evil* (New York: Viking, 1963). This subtitle became one of the most controversial aspects of the book. The suggestion that Eichmann was the man next door was too much for many to take. But the conduct of the average citizen and his responsibility was the crux of the issue. Gordon Zahn dealt with this in his "The Private Conscience and Legitimate Authority," *Commonweal*, March 30, 1962.

30. Gordon Zahn, "The Private Conscience."

31. DM to Gordon Zahn, May 22, 1962. DM, letter to the editor, *Commonweal*, July 6, 1962, p. 378.

32. Mary McCarthy, "The Hue and Cry," *Partisan Review*, Winter 1964, pp. 82–94. Norman Podhoretz, "Hannah Arendt on Eichmann: A Study in the Perversity of Brilliance," *Commentary*, September 1963, pp. 201–208. Bruno Bettelheim, "Individual and Mass Behavior in Extreme Situations," *Journal of Abnormal and Social Psychology*, October 1943, pp. 417–52, reprinted in an edited form in *Politics*, August 1944, pp. 199–209. David Rousset, *Les Jours de notre mort* (The Days of Our Death) (Paris: Editions du Pavois, 1947); DM printed a selection from it in *Politics*, July–August 1947, pp. 151–57. Eugene Kogon, *Der SS Staat* (Frankfort: Verlag der Frankfurter Hefte, 1946).

33. DM to William Phillips, July 16, 29, 1963. Lionel Abel's initial critique of Arendt was "Pseudo Profundity," a review of her collection of essays *Between Past and Future: Six Exercises in Political Thought*, in *New Politics*, Fall 1961, pp. 124–31. William Phillips to DM, June 29, July 16, 1963.

34. "Eichmann in Jerusalem: An Exchange of Letters Between Gershom Scholem and Hannah Arendt," *Encounter*, January 1964. DM, "Hannah Arendt and the Jewish Establishment," *Partisan Review*, Spring 1964, pp. 262–69; reprinted in *Discriminations*, pp. 308–17.

CHAPTER 17: BACK TO THE BARRICADES

1. DM to Mary McCarthy, Oct. 28, 1964. In this letter he reminds her of Schlesinger's remark made to her and repeated to him.

2. DM in "Kennedy and After," a symposium in the *New York Review of Books*, Dec. 26, 1963.

3. DM, "A Critique of the Warren Report," *Esquire*, March 1965. When it was reprinted in *Discriminations: Essays and Afterthoughts* (New York: Grossman, 1974), he conceded that more recent scholarship had led him "away from the solipsistic desert habitat of Lee and Jack into some more populous and fertile conspiratorial swamp." But he never made that concession public and continued to go on record as agreeing with the determination of the Warren Commission. DM cited Edward Jay Epstein's *Inquest: The Warren Commission and the Establishment of Truth* and Josiah Thompson's *Six Seconds in Dallas: A Micro-Study of the Kennedy Assassination* as two books that had shaken his earlier certainty.

4. DM, "A Critique of the Warren Report," in *Discriminations*, p. 137, footnote.

5. Ibid., p. 131.

6. Ibid., p. 135, footnote. Copy of notes on Lane's book for Viking in the Warren Commission Report file, box 138, folder 780.

7. Ibid., p. 136, footnote. DM, "Kennedy and After."

8. Interview with Norman Mailer, March 20, 1989. I. F. Stone to DM, Nov. 15, 1964. Sargent Shriver to DM, Dec. 23, 1964. DM to Nicola Chiaromonte, Nov. 18, 1965, expresses surprise at Stone's agreement with his own judgment on Oswald and noting Murray Kempton's good analyses. Robert C. Cottrell, *Izzy: A Biography of I. F. Stone* (New Brunswick, N.J.: Rutgers University Press, 1992), p. 253.

9. Note in the Macdonald Papers of a phone conversation with Nat Hentoff, Nov. 3, 1963.

10. Robert Scheer, "Genesis of United States Support for the Regime of Ngo Dinh Diem," in *How the United States Got Involved in Vietnam: Report to the Center for the Study of Democratic Institutions*, Santa Barbara, California, 1965. A selection from this information appears in Marvin E. Gettleman, Jane Franklin, Marilyn Young, and H. Bruce Franklin, eds., *Vietnam and America: A Documented History* (New York: Grove, 1985), pp. 118–23. Morton Stavis, "Kennedy's Private War," *New York Review of Books*, July 22, 1971, p. 20.

11. Thong had spent his adult life in the United States studying at Ohio University and Cornell University. He was a member of the Tan Dai Viet, the reformed branch of the Dai Viet Party. He opposed Diem because of Diem's discrimination against the Tan Dai Viet. The Democratic League of Vietnam was an ad hoc front composed of members of the Tan Dai Viet and distant remnants of the VNQDD (Vietnamese Nationalist Party) since most of

the surviving members of the VNQDD joined the Viet Minh, Ho Chi Minh's revolutionary party. Thong had become increasingly anti-Communist in America and refused to denounce U.S. aggression. I am indebted to Marvin Gettleman, one of the chief editors of *Vietnam and America: A Documentary History* for this information.

12. Julius Lester to author, July 30, 1967. *New York Times*, Oct. 10, 1963, p. 1, on Madame Nhu's visit and protesters.

13. DM to W. H. "Ping" Ferry, June 25, 1964. Norman Podhoretz in his *Breaking Ranks* (New York: Harper, 1979), pp. 181–82, describes both DM and Robert Lowell as Johnny-come-latelys to the anti-war position only after it became fashionable on the left. This memory has stuck with him, as he mentioned it in an interview as well, April 29, 1991. The evidence I have doesn't support that notion, however.

14. DM to David McReynolds, Nov. 27, 1964. Ad Hoc Committee's open letter, *New York Times*, Oct. 18, 1963, section 5.

15. David McReynolds's solicitation letter, Nov. 25, 1964, with copy of "Call for a December 19 Public Demonstration," in Vietnam file, Macdonald Papers.

16. A. J. Muste and David Dellinger to DM, Jan. 13, 1965. DM to Muste, Jan. 18, 1965 (not sent). DM to Muste, Jan. 20, 1965.

17. Vietnam file, Macdonald Papers.

18. DM to Robert Kennedy, March 27, 1965. Kennedy to DM, May 6, 1965. DM to Kennedy, May 12, 1965.

19. Fred Halstead, *Out Now!: A Participant's Account of the American Movement Against the Vietnam War* (New York: Monad, 1979), p. 53. See also Nancy Zaroulis and Gerald Sullivan, *Who Spoke Up?: American Protest Against the War in Vietnam, 1963–1975*, (Garden City, N.Y.: Doubleday, 1984), pp. 37–39. McGeorge Bundy, the team captain, didn't show up, on the pretext that he was taking care of business in the Dominican Republic, which brought boos from the crowd.

20. Editorial board of the *Partisan Review*, "On Vietnam and the Dominican Republic," Summer 1965, pp. 397–98. DM's response, "On Vietnam," *PR*, Fall 1965, pp. 635–38.

21. Ibid., p. 638.

22. *Daily Texan*, May 6, 1965. DM to David McReynolds, May 14, 1965, promises to send him potential recruits to the War Resisters League from the Texas campus.

23. Lewis Mumford, "Lewis Mumford Writes to the President," *Minority of One*, May 1965, p. 24. Eric Goldman, *The Tragedy of Lyndon Johnson* (New York: Dell, 1974), p. 431. Lewis Mumford, "Presidential Address," May 19, 1965, in *Proceedings of the American Academy of Arts and Letters and the National Institute of Arts and Letters*, 2nd series, New York, 1966. Donald L. Miller, *Lewis Mumford: A Life* (New York: Weidenfeld & Nicolson,

1989), pp. 515–16, quotes Mumford's letter to LBJ, but the author is mistaken when he says it was printed in only two newspapers, as it appeared in *A Minority of One* cited above.

24. Eric Goldman, "The White House and the Intellectuals," *Harper's*, January 1969, pp. 31–45, 32. This article also appears, substantially expanded, in Goldman, *The Tragedy of Lyndon Johnson*, pp. 459–563. I have used the *Harper's* version except where the book's elaboration and comment are important to the narrative and interpretation.

25. Robert Lowell's letter of May 30, 1965, to President Johnson appeared in the *New York Times* June 3, 1965. Ian Hamilton, *Robert Lowell: A Biography* (New York: Random House, 1982), pp. 321ff. Goldman, "White House," *Harper's*, pp. 33–34.

26. Goldman, "White House," *Harper's*, p. 35. Goldman's edited version cut out the impressive array of signatories (*The Tragedy of Lyndon Johnson*, pp. 527–28). Telegram in *New York Times*, June 4, 1965.

27. Goldman, "White House," *Harper's*, p. 37. DM, "A Day at the White House," *New York Review of Books*, July 15, 1965; I am using the version in his *Discriminations*, p. 142.

28. Lewis Mumford, "Lewis Mumford Writes to the President," *A Minority of One*. In an open letter to President Johnson, Mumford charged that Johnson's aims in the war were emptied of meaning by his "totalitarian tactics." See Donald L. Miller, *Lewis Mumford: A Life* (New York: Weidenfeld & Nicolson, 1989), pp. 513–15, for an account of Mumford's protest against the war. Mumford, "Presidential Address," *Proceedings of the American Academy of Arts and Letters and the National Institute of Arts and Letters*, p. 11, he likens U.S. conduct to the Soviet Union. DM, "A Day at the White House," *Discriminations*, p. 144 footnote, for Schlesinger's response to Mumford's address likening the audience to that of a Nuremberg Rally. DM quoted in *Time*, June 25, 1965.

29. Goldman, "White House," *Harper's*, p. 36.

30. DM papers dated June 14, 1965. Goldman, "White House," *Harper's*, p. 44. DM to Nicholas Macdonald, July 6, 1965, said that seven had signed his statement; in a letter to the *New York Times* June 24, 1965, he said nine; by that I presume he meant seven plus his and Hess's signatures.

31. DM to Nicholas Macdonald, who was in Paris on his honeymoon, July 6, 1965. Goldman, "White House," *Harper's*, pp. 43–44. Goodman, letter to the editor, *Harper's*, March 1965, p. 8.

32. Goldman, "White House," *Harper's*, pp. 43–44.

33. DM, "A Day at the White House," p. 148.

34. Ibid., p. 143.

35. Ibid., pp. 147 and 152. On his argument with Bellow, see DM to Mark Shechner, Nov. 3, 1975. Robert Hatch to DM July 14, 1965, reported that Bellow was still fuming about the affair up on Martha's Vineyard. Bellow

said that the intellectuals had no right to protest since they had been so "anti-political all these years" and that there were "more important issues to be debated . . . such as the unfairness of the income tax (I swear to God) and that if you live in the country you have to support its institutions, even if you don't like some of its policies. . . . He can't keep off the subject of Lowell or the New York intellectuals whom he sees as enemies." Interview with Saul Bellow, February 11, 1993. Saul Bellow to author, Oct. 21 and Nov. 19, 1992. Bellow, in his talk with me, conceded that his position had been stupid. He wrote with reference to the occasion that he had "a weakness for stupid loftiness." See Saul Bellow, "Writers, Intellectuals, Politics," *The National Interest*, Special Issue, Spring 1993, p. 131. In any case Bellow insisted that Dwight did not buttonhole him, they did not have a violent argument, and that he had not been irritated by Dwight's antics.

36. Goldman, *The Tragedy of Lyndon Johnson*, pp. 510–21, is an extended analysis of American intellectuals that did not appear in his *Harper's* article; it critiques the snobbish elitism found among a group of American intellectuals of which he clearly believes DM was a preeminent member. See S. A. Longstaff, "Ivy League Gentiles and Inner-City Jews: Class and Ethnicity Around *Partisan Review* in the Thirties and Forties," *American Jewish History*, Spring 1991, p. 333. Irving Howe, "Strangers," *Yale Review*, June 1977, pp. 486–87, an address delivered at the American Historical Association Annual Meeting, 1989. Howe's essay is reprinted in his *Celebrations and Attacks: Thirty Years of Literary and Cultural Commentary* (New York: Horizon Press, 1979), pp. 15–16.

37. Goldman, *Tragedy*, pp. 562–63. Lady Bird Johnson, *A White House Diary* (New York: Holt, Rinehart, 1970), p. 287. *Izvestia*, June 17, 1965, as reported in *Current Digest of the Soviet Press* 17, no. 24.

38. Goldman, *Tragedy*, pp. 562–63.

CHAPTER 18: THE CRITIC AT WAR

1. "End Your Silence," *New York Times*, June 27, 1965.

2. Interview with Danny Rosenblatt, Dec. 13, 1985. Interview with Nicholas and Elspeth Macdonald, March 22, 1989. When Nick was reminded of the call and Rosenblatt's account of it, he expressed irritation at Dwight's showing more interest in talking to Mary about events in the States than inquiring about their trip: "He uses a transatlantic call to make political points; you would think he would have talked about the wedding."

3. "Statement of Dwight Macdonald," Aug. 12, 1965, before congressional hearings held by Congressman William Fitz Ryan held at the Carnegie Endowment International Center at Rockefeller Plaza. Manuscript in the Vietnam Research file, box 135, folder 753. Phone conversation with Tom

Hughes, who works for the Carnegie Endowment and was present at the hearings.

4. DM to Dorothy Day, Nov. 4, 1964.

5. David McReynolds to DM, Oct. 5, 1965; see marginal notes. See also Nancy Zaroulis and Gerald Sullivan, *Who Spoke Up?* (Garden City, N.Y.: Doubleday, 1984), pp. 67–82, for an account of the constant factional infighting.

6. Tom Wolfe, "Tiny Mummies! The True Story of the Ruler of 43d Street's Land of the Walking Dead!", *New York* magazine of the *New York Herald Tribune*, April 11, 1965; "Lost in the Whichy Thicket: *The New Yorker* II," ibid., April 18, 1965. "Aftermath: Tom Wolfe & the *New Yorker*," ibid., April 25, 1965, was a selection of letters to the editor, mostly by *New Yorker* writers, suggesting that their complaints far outnumbered those of people not connected to the magazine.

7. DM, "Parajournalism, or Tom Wolfe and His Magic Writing Machine," *New York Review of Books*, Aug. 26, 1965; "Parajournalism II: Wolfe and the *New Yorker*," ibid., Feb. 3, 1966. DM published a much-edited version of the first piece in his *Discriminations* (New York: Grossman, 1974), pp. 228–40. Part II has a lengthy analysis of the Leopold story; there is a note in the Wolfe file of the Macdonald Papers stating that Robert Silver phoned Leopold in Puerto Rico, Jan. 10, 1966, but it does not tell what response he got.

8. Norman Mailer in his *Armies of the Night* (New York: New American Library, 1968), claimed that DM had a long-standing "love affair with the *New Yorker*." DM, "Parajournalism I," p. 4. Wolfe, "Whichy Thicket," p. 18.

9. Transcript of Tom Wolfe in conversation with Dennis Wholey, "The Age of Involvement," WBAI-FM, Aug. 16, 1965, pp. 22–23, in the "Parajournalism" file, box 94, folder 404.

10. Colin Wilson to the editors of *Twentieth Century*, December 1958, p. 586. Tom Wolfe, letter to the editor, and DM's response, *New York Review of Books*, Feb. 3, 1966. Although DM got support from the network of writers for and supporters of Shawn and the *New Yorker*, he knew he had carried the affair on too long, had become "rather too much involved," and had weighted his rebuttal down by a far too defensive and detailed refutation of errors. What amounted to a trivial issue had been elevated into "the Dreyfus trial," with DM writing the "J'*accuse*" and "getting his finger caught in the works." DM to Mark Harris, April 11, 1966. Irving Kristol, *Observer*, April 15, 1966. *Time*, Jan. 31, 1966.

11. "A Band of New York Intellectuals Meets with Prof. Schlesinger for a Talk on Vietnam," *New York Times Magazine*, Feb. 6, 1966, pp. 12 passim. Why Schlesinger got top billing is not quite clear. The meeting had taken place Jan. 15, 1966, and it was reported in the daily *Times* Jan. 16, p. 5.

12. DM comment and questioning of Schlesinger, "A Talk on Vietnam," p. 74.
13. Ibid.
14. Ibid., pp. 74–75.
15. Michael Wreszin, "Arthur Schlesinger, Jr.: Scholar-Activist in Cold War America, 1946–1956," *Salmagundi*, Spring–Summer 1984, p. 79. This dismissal of the protests irked many in the audience and some of the panel. Norman Mailer wrote a letter to the *New York Times* saying that the image of an American immolating himself was a nightmarish vision that must have an impact on senators and congressmen. He insisted that they "have the feeling that something is going on that's never gone on before in America, and that's one of the reasons that Senators Frank Church of Idaho and Robert Kennedy are beginning to think there is something perhaps very seriously wrong, or even tragic with the Vietnamese war." *New York Times Magazine*, February 20, section VI, p. 16. This material reminds me of an interview I had with Schlesinger at the Century Club when I was working on an immoderate critique of his Cold War politics. We were discussing a recent harsh review he had written of Carey McWilliams's autobiography. He justified the criticism on the grounds that since McWilliams had never been abroad, what could he know about foreign policy? McWilliams had spent years as a political activist, much of the time in defense of migrant workers among others, while Schlesinger during the same years was traveling abroad at his parents', and later Averell Harriman's expense, making me wonder what he could possibly know about the demonstrators or their effects. Schlesinger is a fine writer, and when he is not attached to the party in power, comes up with penetrating insights. During the cold war, however, he often gave the liberalism he claimed to represent a bad name.
16. Nicola Chiaromonte to DM, Feb. 28, 1966.
17. *Daily Texan*, Feb. 25, 1966. See Vietnam file, box 136, folder 766, for a collection of clippings on DM's political and cultural commentary while at the University of Texas.
18. DM's response is from an undated clipping from the *Daily Texan*, box 98, folder 408.
19. DM to Mary McCarthy, April 11, 1966.
20. Ibid. Interviews with Gloria Macdonald.
21. DM, "Delmore Schwartz (1913–1966)," *New York Review of Books*, Sept. 8, 1966, pp. 14–15.
22. DM to Stephen Spender (not sent), Sept. 7, 1966. DM to Robert Hivnor, a poet and friend of DM and Schwartz, Aug. 17, 1966.
23. DM to D. Stanley Watson, July 18, 1966. DM had been introduced to Watson, a diplomat stationed in Mexico City, on a trip after his teaching stint at Texas. They had gotten into a very heated argument, but, as was often the case, had become friends.
24. DM to Nicola Chiaromonte, Sept. 12, 1966. John Lukacs to DM, Sept. 24,

1966. Arnold Gingrich, "Look Out, Politics, Here Comes Macdonald," an introduction to his new column, *Esquire*, January 1967.

25. Nicola Chiaromonte to Mary McCarthy, Nov. 6, 1966, McCarthy Papers, Vassar College library.

26. Ibid. Chiaromonte almost invariably blamed DM's failures on "that woman," Gloria.

27. DM, "Politics," *Esquire*, January 1967. See discussion in chapter 3 of DM's 1936 vote as confirmed by Fred Dupee.

28. DM, "Birds of America," *New York Review of Books*, Dec. 1, 1966. DM, "Politics," *Esquire*, June 1967, compared *MacBird*'s anti-establishment tone with that of *Ramparts* magazine.

29. Lionel Abel, "Arguments: Much Ado," *Partisan Review*, Winter 1967, pp. 110–14. *Times Literary Supplement* (London), Feb. 9, 1967. Max Ways, "Intellectuals and the Presidency," *Fortune*, April 1967, p. 216. The intensity of the outraged response to *MacBird*'s suggestion of conspiracy, and that an American president or security institutions could possibly be involved, was later echoed by the anger expressed by many about Oliver Stone's film *JFK*, which also implicated Johnson.

30. DM to Nicola Chiaromonte, July 5, 1967.

31. Henry Fairlie is cited in Eric Goldman, *The Tragedy of Lyndon Johnson* (New York: Dell, 1974), p. 520.

CHAPTER 19: BRINGING THE WAR HOME

1. DM to Stephen Spender, Jan. 12, 1962. Spender to DM, Jan. 17, 1962.

2. Arthur Schlesinger, Jr., J. K. Galbraith et al., letter to *New York Times*, May 9, 1966. DM to Michael Josselson, March 30, 1967, mentions the Bell request. I am indebted for aid in hacking a path through this political thicket to Wilfrid Sheed and Peter Steinfels for their unpublished manuscript "The *Encounter* Affair," submitted to *Esquire* in 1967 but rejected. A copy is in the Macdonald Papers. See also Peter Coleman, *The Liberal Conspiracy* (New York: Free Press, 1989), Chapter 14, "A Black Operation," pp. 219–34. Coleman was an active member of the Australian branch of the Congress for Cultural Freedom and his book is largely a defense of its operations and relationship with the CIA. It is, in effect, an official account and highly praised by Kristol, Lasky, and other defenders of *Encounter* and the CCF.

3. Thomas W. Braden, "I'm Glad the CIA Is Immoral," *Saturday Evening Post*, May 20, 1967, p. 12.

4. DM to Stephen Spender, Sept. 20, 1966. DM to Malcolm Muggeridge, March 31, 1967. DM to Michael Josselson, March 30, 1967.

5. For Buckley's knowledge, see Sheed and Steinfels, "The *Encounter* Affair," p. 8. Steinfels said in a phone conversation April 8, 1993, that the excuse

given by *Esquire* editors was their lawyers believed the material on Lasky might be libelous in England, but he felt their real reason was that they just didn't think the piece was a hot item when ready for publication. Paul Goodman, "The Devolution of Democracy," *Dissent*, Winter 1962, p. 6. Nicholas Nabokov, letter to the editor, *Dissent*, Summer 1962, p. 308.

6. Irving Kristol to DM, April 11, 1967.

7. Stephen Spender to DM, April 12, 1967. Melvin Lasky to DM, April 12, 1967.

8. DM, "Politics," *Esquire*, June 1967, p. 78.

9. Interview with Daniel Bell, March 9, 1990. Interview with Melvin Lasky, April 4, 1990.

10. DM, "Politics," *Esquire*, September 1967, p. 46.

11. DM to William Phillips, June 14, July 3, 1967. At first DM declined to sign the statement because it was vague and moralistic, but on second thought he decided to lend his signature and suggested making the letter stronger. It appeared in *Partisan Review*, Summer 1967. Nicola Chiaromonte to DM, June 13, 1967. DM to Chiaromonte, June 21, 1967.

12. Chiaromonte to DM, Sept. 4, 1967. DM to Chiaromonte, Oct. 10, 1967. Chiaromonte to DM, Oct. 16, 1967.

13. Chiaromonte to DM, Oct. 16, 1957. Actually these two old friends were not so far apart, it was just that it was the Americans fighting in Vietnam, not the French, and DM did not want to do anything to discourage any form of resistance. He was impatient with the mindless romanticism of ill-informed young radicals, but the movement needed the troops.

14. See "Confrontation: The Old Left and the New," *American Scholar*, April 1967, pp. 567–88; this is an edited transcript of a panel discussion between DM, Richard Rovere representing the old left, and Tom Hayden and Ivanhoe Donaldson representing the New Left at the YMHA, March 6, 1967. In reading this transcript it seemed to me that Dwight did badger and not listen very attentively. It occurred to me he might have been tight. However, Daniel Aaron, the moderator, did not have that feeling at all. He felt the youngsters were obstreperous, impolite, and outrageous and thus Dwight's dismissive tone justified. Unfortunately the YMHA seems to have lost its tapes of that session. DM, comments in symposium, "Liberal Anti-Communism Revisited," *Commentary*, September 1967, pp. 54–56. Norman Podhoretz recalls this as the first public example of DM's having more or less betrayed his anti-Communist principles, interview, April 29, 1991.

15. Sandy Vogelgesang, *The Long Dark Night of the Soul: The American Intellectual Left and the Vietnam War* (New York: Harper & Row, 1974), pp. 108–9. "Misled and misleading men" is Vogelgesang's description of the feeling of many leftist intellectuals, not her own feeling. There was a widespread belief that Harrison Salisbury was right when he charged that Wash-

ington was led by men who "lived only for some growing disaster which would bury all the smaller errors of the past."

16. Mary McCarthy to DM, April 4, 1967. DM to McCarthy, April 15, 1967. While DM maintained amiable relations with Noam Chomsky, both he and McCarthy suspected him of being a hysteric and inordinately self-righteous. DM also worried that Chomsky's hatred of American policy allowed him to whitewash the Soviet Union. On the income tax, McCarthy persuaded DM that it was not a feasible idea; she wondered just how Chomsky did it if deductions were taken from his salary checks at MIT. She assumed he didn't make enough on his royalties and speaking engagements to make a difference. DM to David McReynolds, Oct. 16, 1967.

17. DM was so taken with Mailer's account, and understandably flattered by his prominent role in it, that for the remainder of his life he used the book as a core reading in a course on American political fiction, at SUNY Buffalo, the University of Massachusetts at Amherst, and Hofstra.

18. Norman Mailer, *The Armies of the Night: History as a Novel, the Novel as History* (New York: New American Library, 1968), p. 25.

19. Clark Loomis Herbert, "On Hearing a Notorious Writer (SO-Publicized) Rail at a Washington D.C. Theater Benefit to End the War In Vietnam," box 22, folder 549; DM's reply is with it.

20. Mailer, *Armies*, pp. 67, 93–94.

21. Ibid., pp. 87–89, 105, 185.

22. Robert Lowell, *Notebook 1967–68* (New York: Farrar, Straus and Giroux, 1969), pp. 27ff. Mailer, *Armies*, p. 123.

23. DM, "Politics," *Esquire*, June 1968, pp. 46–50.

24. Ibid., p. 47. This was the second of two columns devoted to Mailer and the Pentagon march; the first, which appeared in the May 1968 issue, pp. 41–45, is also printed in *Discriminations: Essays and Afterthoughts* (New York: Grossman, 1974), pp. 210–16, and begins with a parody of Mailer.

25. Interview with Diana Trilling, May 21, 1992. Ms. Trilling is one of the people who encouraged me to undertake this biography when I began in the mid-1980s. She thought DM was a significant participant. After working on her own memoirs and going over the struggles of the sixties, she still thinks he is significant as a representative figure but feels he had a weak grasp on reality and an ill-informed and even destructive frivolity. DM to Martin Luther King, June 12, 1967. DM, unidentified manuscript for response to a questionnaire on what journals and newspapers he was reading. He made a big pitch for *I. F. Stone's Weekly*, "a multum in parvo job by a man who knows everybody in Washington, as a correspondent there for decades for the *Nation* and other magazines, who reads everything, including the *Congressional Record*, and who has extreme, sensible and human opinions about Vietnam and other issues."

26. DM, introduction to Mark Levine, ed., *The Tales of Hoffman: The Chicago Seven Trial* (New York: Bantam, 1970).

27. Frank Lindenfeld, "A Manifesto for Participants in the Congress on the Dialectics of Liberation, and for Others," written during the summer of 1967. DM to Lindenfeld, Nov. 11, 1967. Lindenfeld to author, May 15, 1992. Lindenfeld reprinted "The Root Is Man" in part in his anthology *Radical Perspectives on Social Problems*, in three editions, the last being 1986.

28. Norman Podhoretz, *Breaking Ranks* (New York: Harper & Row, 1979) is by far the most articulate account of this record of assimilation followed by a strong celebration of Americanism.

29. Interview with Alistair Reid, March 19, 20, 1992. Interview with Gloria Macdonald, Aug. 9, 1990. DM to Delphino, Aug. 14, 1967, commiserates with one of his youthful fans, a hippie type from Cleveland, over the arrest and subsequent jail term for a mutual friend, a Cleveland bookseller, on marijuana charges. DM said he was not attracted to pot, that four joints in four years was about his tempo. He was much more addicted to booze and wished it was the other way around, he said. Gloria felt he was "much nicer and more gentle and perceptive and sensitive on pot than on booze, also much less trouble socially."

30. DM, "Politics," *Esquire*, December 1967, pp. 21ff. DM gathered a great deal of material on the conference, and his son Mike, who accompanied him, took careful and extensive notes which DM used.

31. DM on *Firing Line* May 1, 1967, transcript given to me by William F. Buckley. Interview with Buckley, May 18, 1991. Buckley, "On the Right," *New York Post*, Dec. 16, 1967, Jan. 11, 1968. "On Civil Disobedience, 1967," *New York Times Magazine*, Nov. 24, 1967; DM's contribution was headed with his statement that "Legal Protest Is No Longer Tactically Effective." DM, letter to the editor, "Judge for Yourself," *Village Voice*, Dec. 21, 1967; this was part of an exchange with Buckley, who had challenged DM's sincerity in advocating law-violating civil disobedience.

32. Irving Howe, "An Exchange on the Left," *New York Review of Books*, Nov. 23, 1967. Michael Harrington, "A Question of Philosophy, A Question of Tactics," *Village Voice*, Dec. 7, 1967.

33. DM, "Politics," *Esquire*, November 1967, pp. 40–41.

34. Ibid.

35. Ibid., p. 44.

36. Ibid., p. 47.

37. DM's account of the encounter with Thelwell and Lester is given in a letter to David McReynolds, March 29, 1971. Julius Lester confirmed and commented on this episode in letter to author, July 30, 1987.

CHAPTER 20: *KULTURKAMPF*

1. DM to Richard Barnet, Jan. 15, 1968. *Michigan State News*, Jan. 22, 1968. Doug Hoekstra and George Graeber, letters to the editor, *Michigan State News*, January 30, 1968. In a letter to Barnet after a long discussion of anti-war activity, Dwight apologized for his leaving early as he had to go to Michigan State to give his "Need for an Elite Culture" lecture.

2. DM to Marcus Raskin, Jan. 25, 1968.

3. David F. Ricks, "Living Up Against the Wall," unpublished manuscript, June 1968, sent to DM for support by Ricks, professor of psychology, Teachers College, Columbia University. Fred Dupee, "The Uprising at Columbia," *New York Review of Books*, Nov. 26, 1968.

4. Jacques Barzun to DM, April 4, 1964. DM to Barzun (not sent), April 30, 1964. Phone interview with Janet Groth, a young *New Yorker* secretary who applied for graduate work at Columbia, May 27, 1992.

5. This account of the Columbia uprising is based primarily on Jerry L. Avorn, *Up Against the Ivy Wall* (1969); George Keller, "Six Weeks That Shook Morningside Heights," *Columbia College Today*, Spring 1968; Fred Dupee, "The Uprising at Columbia"; Diana Trilling, "On the Steps of Low Library," originally published in *Commentary* and reprinted in her *We Must March My Darlings* (New York: Harcourt Brace Jovanovich, 1973), and the articles and interviews below.

6. Alfred Kazin, *New York Jew* (New York: Random House, 1979), pp. 191–3. A distinguished group of scholars responded to Mordecai Richler's review of Kazin's book, rejecting the negative comments concerning Lionel Trilling's account as a "grotesque misrepresentation." Mordecai Richler, "Literary Ids and Egos," *New York Times Book Review*, May 7, 1978, section 7, p. 1. Quentin Anderson, Eric Bentley, Leslie Fiedler, Charles Frankel, Howard Mumford Jones, Stanley Kaufmann, Frank Kermode, Steven Marcus, Robert K. Merton, Edward Said, Arthur Schlesinger, Jr., Barbara Probst Solomon, Fritz Stern, Aileen Ward, Michael Wood, Paul Zweig, letter to the editor, *New York Times Book Review*, June 25, 1978, section 7, p. 56. Robert Penn Warren wrote a separate letter scoring Kazin's treatment of Trilling; Dwight's name is almost conspicuous by its absence. The notion of Trilling as having sold out his Jewish identity in order to make it in the academic world is rejected by William M. Chace in his perceptive and thoughtful *Lionel Trilling: Criticism and Politics* (Palo Alto, Calif.: Stanford University Press, 1980), p. 18, footnote. On Trilling as deradicalizer see Kazin, *New York Jew*, p. 192; and Cornel West, "Lionel Trilling: Godfather of Neo-Conservatism," *New Politics*, Summer 1986, pp. 233–42; and Mark Krupnick, *Lionel Trilling and the Fate of Cultural Criticism* (Chicago: Northwestern University Press, 1986).

7. DM, "An Exchange on the Columbia Student Strike of 1968," *New York Review of Books*, July 11, 1968, reprinted in *Discriminations*, p. 456. Diana

Trilling, "Interview: Dwight Macdonald," unpublished version in Macdonald Papers. A much shortened version of this interview was published in *Partisan Review*, anniversary edition, 51, 1984.

8. DM, "An Exchange," p. 456.
9. Ibid. In "The Future of Democratic Values," *Partisan Review*, July–August 1943, p. 322, DM had cited his favorite passage from *Charterhouse of Parma* in which Stendhal describes the liberation of Milan from Austrian rule.
10. Diana Trilling, "On the Steps of Low Library," in *We Must March My Darlings*, p. 90. Fred Hechinger, "The Harsh Legacy of the Class of '68," *New York Times*, June 9, 1968, Education column on DM and others. Daniel Bell and Irving Kristol, eds., *Confrontation: The Student Rebellion and the Universities* (New York: Basic Books, 1969) p. viii.
11. Interview with Diana Trilling, May 21, 1992. DM, "Exchange on Columbia," Parts 1 and 2, *New York Review of Books*, July 11, Aug. 22, 1968. Diana Trilling, "Interview: Dwight Macdonald," *Partisan Review*, anniversary edition, 1984, pp. 816–17.
12. DM, "An Exchange," pp. 456–57.
13. Ibid., pp. 457–58.
14. DM to Wells Kerr, Nov. 4, 1968. DM to Richard Barnet, Jan. 15, 1968.
15. Fred Dupee, "The Uprising at Columbia," *New York Review of Books*, Nov. 26, 1968, pp. 21, 30, 34. Interview with James Shenton, June 4, 1992. Lionel Trilling, "Columbia Crisis Project," May 22, 28, 1968, pp. 1–44 and passim; see pp. 5ff. Columbia Oral History Collection, Columbia University Library.
16. Marvin Harris, "Big Bust on Morningside Heights," *Nation*, June 10, 1968, p. 760.
17. Diana Trilling, "On the Steps of Low Library," pp. 99ff, did not see sufficient reasons for the student revolt. She conceded that when one marries a university professor one marries into his university, and there developed, perhaps, a "distorting intimacy."
18. Lionel Trilling in Stephen Donadio, "Columbia: Seven Interviews," *Partisan Review*, Summer 1968, p. 386. Dan Wakefield, in *New York in the Fifties* (New York: 1992), pp. 24–33, expresses Mark Van Doren's and Trilling's interest in their students' work and of the professional help and advice they offered.
19. See *Who Rules Columbia*, published by the North American Congress on Latin America, June 1968, pp. 26–27, passim.
20. This solicitation letter was printed under DM's 87th Street address and opened "Dear Friend." It is reprinted in *Discriminations*, pp. 450–52.
21. Diana Trilling to DM, May 22, 1968.
22. DM, "An Exchange," pp. 456, 461. See John E. Englund, faculty member at Teachers College, letter to the editor, *New York Review of Books*, July 5,

1968. See *New York Times*, May 23, 1968, pp. 8, 50. Orest Ranum to author, Oct. 6, 1992, says that at the time, he deliberately refrained from describing exactly what had been burned because he didn't know. He did not think the university would "use" his case to defuse the SDS occupations, "but the right-wing faculty would do so if it could." There was a charge that the university paid him off for keeping quiet about the fact that what was actually destroyed were a year's graduate notes. Ranum says they gave him a sum "roughly equivalent to a semester's income, since they had no insurance that covered such an incident."

23. Irving Howe, "The New York Intellectuals: A Chronicle and a Commentary," *Commentary*, October 1968, p. 39; this essay is reprinted in his *Decline of the New* (New York: Harcourt, Brace and World, 1970) the criticism of DM appears in the footnote, p. 236. DM, "An Exchange," pp. 456. DM, letter to the editor, *New York Review of Books*, June 20, 1968, supports the students.

24. See Jack Newfield, "Mailer as History: On the Steps of the Zeitgeist," *Village Voice*, May 30, 1968, pp. 26–27.

25. A partial text of Richard Hofstadter's address from which these passages were taken is housed in the Office of Public Information, Low Library, Columbia University. The text was also printed at some length in the *New York Times*, June 5, 1968, under a photograph showing DM at the podium at the countercommencement in front of Low Library.

26. DM, sporadic notes in the Macdonald Papers. George Keller, "Six Weeks." DM refers to his speech in his rejoinder to Ivan Morris, "An Exchange," p. 462. See also the *New York Times*, June 5, 1968, which carries a fine photo of DM at the podium in front of Alma Mater.

27. Fred Hechinger, "The Harsh Legacy of the Class of '68." DM, Taylor, Fromm, letter to the editor, *New York Times*, June 17, 1968.

CHAPTER 21: THE CULTURAL COLD WARS

1. James M. Mathias, secretary of the Guggenheim Foundation, to DM, May 14, 1968. DM to James M. Mathias, May 28, 1968.

2. Interviews with Gloria Macdonald, Jan. 29, 1986, Aug. 9, 1990; Joel Carmichael, May 14, 1987.

3. Phone conversation with Helen Weinberg, East Hampton neighbor, June 10, 1992. Interviews with Sabina Lanier and Helen Weinberg, June 15, 1992. Undated clipping from a spring issue of *Living Now* in the East Hampton file, Macdonald Papers.

4. DM, "Politics," *Esquire*, March and July 1968. DM to Rust Hills, April 19, 1968.

5. DM, "Politics," *Esquire*, July 1968.

6. DM to Michael List, a Berkeley radical, Dec. 4, 1970. DM, "Politics," *Esquire*, March 1969, pp. 18ff.

7. Michael Harrington, letter soliciting support for the Ad Hoc Committee to Defend the Right to Teach, Sept. 12, 1968. "Freedom To Teach," flyer signed by Michael Harrington and Tom Kahn as co-chairmen, appeared in the *New York Times*, Sept. 20, 1968. A good if partisan anthology of significant documents and writings relevant to the struggle is Maurice R. Berube and Marilyn Gittell, eds., *Confrontation at Ocean Hill–Brownsville: The New York School Strikes of 1968* (New York: Frederick A. Praeger, 1969).

8. DM to Bertha Gruner, Oct. 16, 1968.

9. Nat Hentoff, "Review of the Press: Ad Hoc Committee on Confusion," *Village Voice*, Sept. 26, 1968, pp. 15ff. DM, "An Open Letter to Michael Harrington," *New York Review of Books*, Dec. 5, 1968, pp. 48–51.

10. Ibid.

11. Ibid. Sol Stern, "'Scab' Teachers," *Ramparts*, Nov. 17, 1968, pp. 17–25, reprinted in Berube and Gittell, *Confrontation*, pp. 176–92. Michael Harrington, "A Reply: An Open Letter to Men of Good Will with an Aside to Dwight Macdonald," *New York Review of Books*, Jan. 2, 1968. DM, "Reply to a Non-Reply," ibid., Jan. 16, 1968, p. 11. Brendan Sexton to DM, Dec. 11, 1968.

12. Bertha Gruner to DM, Sept. 29, 1968. DM to Bertha Gruner, Oct. 16, 1968. Nathan Glazer, note to DM, spring 1969. Stern, "'Scab Teachers'" in Berube and Gittell, *Confrontation*, pp. 182–83 on "A Socialist Lobby."

13. DM, "Ocean Hill Visited," letter to the editor, *New York Times*, Nov. 18, 1968.

14. Ibid. He mentions Kazin's previous tour in his letter.

15. Ibid. DM to Mayor John Lindsay, Oct. 16, 1968. The Ocean Hill–Brownsville file in the Macdonald Papers is full of denunciatory notes and letters to DM. DM to Nicola Chiaromonte, n.d.; there was a growing strain in their relationship due to DM's close association with the New Left and apparent acceptance of its willingness to collaborate with the North Vietnamese. Phone conversation with Murray Kempton, June 17, 1992.

16. Given the shambles the New York public school system is in today and the widespread corruption exposed among members of its local boards, many have charged that the decentralization people were wrong. Of course, there is no evidence that the centralized system would prove any better. In any event, Dwight's response was consistent with his individualistic philosophy, which stressed smaller units and local control. He could not and would not have been likely to take any other position, even if he had had more knowledge of the intricacies of the system.

17. Nathan Glazer, "Blacks, Jews, and the Intellectuals," *Commentary*, April 1969, pp. 33–39. Glazer note to DM, spring 1969. The main thesis of Glazer's article was that it wasn't simply blacks and Wasps who had promoted anti-Semitism but many Jewish intellectuals on the left, such as Norman Mailer and Noam Chomsky. I recall at this time sitting with

Richard Hofstadter in the Columbia Faculty House when he was approached to give money for the Eldridge Cleaver defense. He reached in his wallet and wrote out a personal check for $50. When I asked him why he supported Cleaver, he responded that the young black man was too bright to be made a convict or an expatriate. DM gave $50 to the Panther Defense Fund; see Marie Runyon to DM, May 21, 1970. The executive committee of the fund included I. F. Stone, Martin Duberman, Gloria Steinem, and Mailer, among others.

18. Dennis Wrong, "The Case of the 'New York Review,' *Commentary*, November 1970, pp. 49ff; on Stone, p. 56. Maurice Isserman, *If I Had a Hammer: The Death of the Old Left and the Birth of the New Left* (New York: Basic Books, 1987), pp. 192–93, for the evolution of Shachtman and his associates in the Socialist Party in the late 1960s. Alan Wald, *The New York Intellectuals: The Rise and Decline of the Anti-Stalinist Left from the 1930s to the 1980s* (Chapel Hill: University of North Carolina Press, 1987), pp. 327–29. Julius Jacobson, "Neo-Stalinism: The Achilles Heel of the Peace Movement and the American Left," *New Politics*, Summer 1976.

19. Curt Gentry, *J. Edgar Hoover: The Man and the Secrets* (New York: Norton, 1991), pp. 618–19.

20. This account is based primarily on Nora Sayre, "Strikes and Lulls: Yale, May 1970," in her *Sixties Going on Seventies* (New York: Arbor House, 1973), p. 164 and passim; "May Day at Yale: A Narrative of Events," *Yale Alumni Magazine*, May 1970, pp. 16–35; Michael Macdonald's notes taken during the weekend; sporadic notes of DM and other memorabilia in the Yale May Day file, box 139, folders 789–92. Yale, unlike Harvard, was not situated in a consumer wonderland like Harvard Square—a kiddie culture for the rich with every conceivable toy at easy reach. Students at Yale were wary upon leaving the campus, gates were locked, campus security was a major preoccupation of the university community.

21. The Brewster and Coffin statements and the *New York Times*'s coverage are quoted in "May Day at Yale," pp. 17, 20, 23, 26. The *Times* denounced Brewster for sowing "moral confusion" among the students. The *Times* was consistently hostile to Brewster in its editorials, often contradicting its own news coverage.

22. Undated handwritten note written by DM, presumably for his memoirs, in the Additional Materials file, Macdonald Papers.

23. See "Rumors," *Yale Alumni Magazine*, May 1970, p. 31. Tom Hayden, *Reunion: A Memoir* (New York: Random House, 1988), pp. 416–17.

24. Hayden, *Reunion*. "May Day at Yale," pp. 27–28.

25. "May Day at Yale," pp. 32–33.

26. Michael Macdonald notes. Phone interviews with Nora Sayres, Oct. 6, 1986, June 22, 1992.

27. DM to Nicola Chiaromonte, May 27, 1970. Tom Hayden, *Reunion*, p. 417.

CHAPTER 22: DWIGHT'S LAST TAPES

1. Interview with former student Louis Lombardi July 21, 1988. DM to Peter Kenez, a former colleague at the University of Wisconsin at Milwaukee, Dec. 13, 1970. The material on DM's teaching career comes largely from interviews, particularly with former students Jean-Noel Mahoney, Michael Silverblatt, Myrna Greenfield, and Louis Lombardi, and colleagues Alan Spiegel, Mark Shechner, Leslie Fiedler (Buffalo), Theodore Gill (John Jay), and Adolph and Susan Rosenblatt (University of Wisconsin, Milwaukee). In addition, the Macdonald Papers have extensive files of course descriptions, grades, syllabi, writing assignment instructions, comments on papers.

2. Interviews with former students Myrna Greenfield (John Jay), March 11, 1988; Jean-Noel Mahoney (Buffalo), June 8, 1990.

3. Phone conversation with Rita Lipsitz, assistant to the chairman of the English department, SUNY Buffalo, July 7, 1992.

4. Interview with Daniel Okrent, July 8, 1992. Interview with Gloria Macdonald, Aug. 9, 1990. DM to Norman Birnbaum, writer and professor of sociology, Nov. 30, 1970, refers to his "manic" periods. Many of the people I have talked to mentioned DM's complaints about the writing block. But opinion varies as to whether he was subject to serious bouts of depression. Some agreed that was the case during the last decade of his life, while others deny ever seeing it. Gloria did agree that he was often very depressed.

5. DM to Miriam Chiaromonte, Jan. 19, 1972.

6. John Lukacs to DM, Jan. 2, 1969.

7. DM to John Lukacs, Aug. 25, Sept. 12, Oct. 7, 1972. John Lukacs gave me his unpublished portrait of DM, April 1985.

8. John Lukacs to DM, Jan. 2, 1969. Alexander Herzen, *My Past and Thoughts*, edited by DM (New York: 1973); DM wrote the preface, which was an update of an earlier *Politics* piece. DM, "Luce and His Empire," *New York Times Book Review*, Oct. 1, 1972, pp. 1–2. DM, "Norman Cousins' Midcult 'World': A Discursive Review," *Columbia Forum*, Fall 1972; letters on the Cousins article and DM's replies in the Winter 1973 issue; reprinted in *Discriminations*, pp. 174–93.

9. Roger Sale, *New York Times Book Review*, Nov. 3, 1974, was a peculiarly sneering dismissal of DM which reflected a generation gap. Louis Berg, "Momentary Opinions," *Commentary*, May 1975, pp. 79–80, felt that DM had no claims to credibility on political questions but he liked the prose. Total sales were in the neighborhood of 2,000 copies. DM gathered up a great many remaindered copies and attempted to sell them at lectures and other literary gatherings, to the amusement and sometimes contempt of those solicited.

10. Ruth Gay to DM, June 21, 1973. DM to Ruth Gay, July 8, 1975. DM to Herman Kahn, chief of the Yale Archives, June 19, 1974.

11. Interviews with Janet Groth, who worked as a secretary at the *New Yorker*, and staff writer Alistair Reid, March 20, 1992.

12. DM to Elizabeth Phillips, Nov. 24, 1973. DM did publish a selection of Poe's poetry in *The Poems of Edgar Allan Poe* (New York: Crowell, 1965), for which he wrote a lengthy introduction.

13. DM, undated note but most certainly in 1976.

14. DM, notes for memoirs, "A Summary of My Lazarus Return to Life, Nov. 25th to this day Dec 21, 1978," box 9, Additional Materials file.

15. Interview with Dr. Stephen Smiles, Aug. 22, 1992.

16. Discussion with Gloria Watts, a Wellfleet neighbor of DM's, Aug. 26, 1987.

17. Interview with Theodore Gill, March 9, 1988. *New York Times*, April 24, 1975.

18. Interview with Gloria Macdonald, Jan. 19, 1986. Phone conversation with John Lukacs, 1990.

19. Interview with John Lukacs, July 1, 1991. Lukacs, unpublished profile of DM, originally written for a book of remembrances by well-known friends, copy given to me by Lukacs.

20. This discussion is based on interviews with William Arrowsmith, Michael Silverblatt, Gloria Macdonald, Tom Dardis, and John Lukacs. Interview with Theodore Gill, Louis Lombardi, Myrna Greenfield at John Jay College.

21. For accounts of the London incident, see Colin Welch, review of Malcolm Bradbury, *No, Not Bloomsbury, Spectator*, May 16, 1987; a defensive response by Geoffrey Wheatcroft, *Spectator*, May 23, 1987; Wheatcroft, *Absent Friends* (London: Hamish Hamilton, 1989) , chapter 12, which is devoted to a sympathetic portrait of DM's insensitive manner. I confirmed this incident with Welch, Wheatcroft, and Peregrine Worsthorne. Phone conversation with Helen Rattray. Interview with Helen Weinberg.

22. John Lukacs, *Outgrowing Democracy: A History of the United States in the 20th Century* (Lanham, Md: University Press of America, 1984), pp. 316–17.

23. Interview with Theodore Gill, March 9, 1988. Interview with Myrna Greenfield (student), March 11, 1988. Interview with Louis Lombardi (student), July 21, 1988. Interviews with Mary McCarthy, November 14, 1987 and August 9, 1988. Interview with Gloria Macdonald, January 29, 1986. Interview with Leslie Fiedler, June 6, 1990. It is clear from my discussions that the term "alcoholism" is a political or perhaps ideological term, and that its mention makes people immediately opinionated or defensive. There is no question that in the last years of DM's life he suffered from the effects of severe alcohol abuse, which seriously affected his health. At the same time he was able to cope with his life and be an interesting and pleasant companion to his friends.

24. This account was given to me by Annabel Anrep in a letter of May 30, 1989.

25. Interview with Helen Weinberg, June 12, 1992.

26. DM, "East Hadleyburg?" *East Hampton Star*, Dec. 23, 30, 1971. These two lengthy letters to the editor were reprinted as "The Great East Hampton

Library Mess" in *American Libraries*, a professional journal, September 1972, pp. 869–74.

27. Most of this material comes from the National Institute of Arts and Letters file in the Macdonald Papers.

28. DM to James Atlas, Oct. 8, 1975, Feb. 17, 1976. For a splendid account of this editing job, see James Atlas, "Unsentimental Education," *Atlantic*, June 1983.

29. DM, "Vote for Keaton," *New York Review of Books*, Oct. 9, 1980, pp. 33–38.

30. Details based on DM's "Admission Record—General," Metropolitan Hospital, Dec. 17, 1982, and obtained from the hospital; mention of his hospitalization and diagnosis in East Hampton is based on phone conversations with his physician Dr. Kenneth Cairns of the Hampton Medical Center. Dwight had a very friendly relationship with Dr. Cairns. His general practitioner in New York, Dr. Stephen Smiles, also confirmed much of the information. All three physicians agreed that Dwight suffered from alcohol abuse, but were reluctant to say that he was an alcoholic. Dwight referred to himself as such whenever Gloria asked him why he drank so much. I am indebted to Dr. Martin Kelly of Newton, Massachusetts, and Boston for interpreting the medical nomenclature in the admission record and discussing with me its implications, August 1992.

31. Phone conversation with Helen Rattray, summer 1992.

32. Dr. Cairns was aware of the injections DM administered to himself and there is a reference to his use of adrenaline in the Metropolitan admission records. The public injection was at a book party at Joel Agee's apartment; Agee dates the party in May 1981; phone conversation with Agee, March 3, 1993.

33. Robert Cummings generously made a copy of the interviews that he and Robert Westbrook had with DM at Yale University Jan. 5 and 18, 1982, just short of a year before DM died. I refer to the humanistic radical tradition analyzed in Casey Nelson Blake's remarkable book, *Beloved Community: The Cultural Criticism of Randolph Bourne, Van Wyck Brooks, Waldo Frank, and Lewis Mumford*; and also in the work of Jeffrey C. Isaac, in his *Arendt, Camus, and Modern Rebellion*.

Index